Pulse Methods
in 1D and 2D
Liquid-Phase NMR

Pulse Methods in 1D and 2D Liquid-Phase NMR

Edited by Wallace S. Brey

DEPARTMENT OF CHEMISTRY
UNIVERSITY OF FLORIDA
GAINESVILLE, FLORIDA

ACADEMIC PRESS, INC.
Harcourt Brace Jovanovich, Publishers

SAN DIEGO   NEW YORK   BERKELEY
BOSTON   LONDON   SYDNEY   TOKYO   TORONTO

COPYRIGHT © 1988 BY ACADEMIC PRESS, INC.
ALL RIGHTS RESERVED.
NO PART OF THIS PUBLICATION MAY BE REPRODUCED OR
TRANSMITTED IN ANY FORM OR BY ANY MEANS, ELECTRONIC
OR MECHANICAL, INCLUDING PHOTOCOPY, RECORDING, OR
ANY INFORMATION STORAGE AND RETRIEVAL SYSTEM, WITHOUT
PERMISSION IN WRITING FROM THE PUBLISHER.

ACADEMIC PRESS, INC.
1250 Sixth Avenue
San Diego, California 92101

United Kingdom Edition published by
ACADEMIC PRESS INC. (LONDON) LTD.
24-28 Oval Road, London NW1 7DX

BOOKNET
(List 93/3 Rat1)

Library of Congress Cataloging-in-Publication Data

Pulse methods in 1D and 2D liquid-phase NMR.

  Includes index.
  1. Nuclear magnetic resonance spectroscopy.   2. Nuclear
magnetic resonance, Pulsed.   I. Brey, Wallace S.
QD96.N8P85   1987   543'.0877   87-1224
ISBN 0-12-133155-5   (alk. paper)

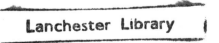

PRINTED IN THE UNITED STATES OF AMERICA
88 89 90 91   9 8 7 6 5 4 3 2 1

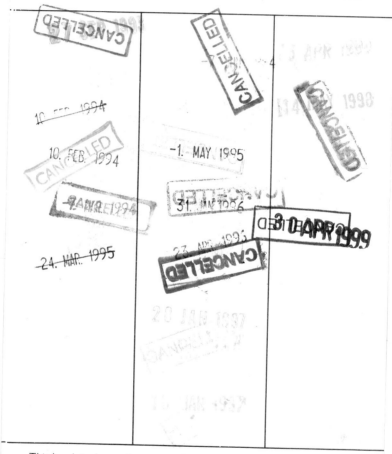

Contents

PREFACE ix

1 Basic Methods and Simple Pulsed Experiments
WALLACE S. BREY

I.	Introduction	2
II.	Spin-Echo Experiments	6
III.	Polarization-Transfer Experiments	31
IV.	Experimental Aspects of Pulsed NMR	83
	References	104

2 Density-Operator Theory of Pulses and Precession
MALCOLM H. LEVITT

I.	Introduction	111
II.	Overview	113
III.	Theory and Use of Product Operators	125
IV.	A Typical Two-Dimensional Experiment—Double-Quantum Spectroscopy of Two Coupled Spins-$\frac{1}{2}$	140
V.	Summary	145
	References	145

3 Polarization Transfer and Editing Techniques
OLE W. SØRENSEN AND HANS J. JAKOBSEN

I.	Introduction	150
II.	Theory	152
III.	Polarization Transfer and Purging	169
IV.	Heteronuclear Spectral Editing Techniques	207
	Appendixes	248
	References	257

4 Principles of Multiple-Quantum Spectroscopy
THOMAS H. MARECI

I.	Introduction	259
II.	Operator Formalism	266
III.	Multiple-Quantum Coherence for a System of Two-Spin-½ Nuclei	267
IV.	Order-Selective Detection	292
V.	Multiple-Quantum Coherences for Systems Containing Three or More Spin-½ Nuclei	310
IV.	Summary	337
	References	338

5 Applications of Two-Dimensional NMR Experiments in Liquids
GEORGE A. GRAY

I.	Introduction	343
II.	Instrumental Requirements	345
III.	J-Resolved 2D NMR	351
IV.	Chemical Shift Correlation 2D NMR	364
V.	Shift Correlation by Relayed Coherence-Transfer 2D NMR	402
VI.	Magnetization-Transfer 2D NMR	411
VII.	Zero- and Multiple-Quantum 2D Methods	420
	References	429

6 Applications of 2D NMR to Biological Systems
J. H. PRESTEGARD

I.	Introduction	435
II.	The Basic 2D NMR Experiments for Spectral Assignment and Structure Elucidation	437
III.	Advanced Methods for Biological Samples	449
IV.	Structural Analysis of Biomolecules	470
	References	485

7 Multiple-Resonance and Two-Dimensional NMR Techniques in Analysis of Fluorocarbon Compounds and Polymers
DERICK W. OVENALL AND RAYMOND C. FERGUSON

I. Introduction	489
II. Resolution Enhancement and Proton Decoupling	491
III. Two-Dimensional NMR	494
IV. Experimental Details	504
V. Summary	505
References	505

8 Recent Developments in Pulsed NMR Methods
WALLACE S. BREY

I. RELAY and Observation by Hydrogen Detection	508
II. Homonuclear Correlation	520
III. Isotropic Mixing Experiments	528
IV. Applications of Zero-Quantum Coherence	531
V. Nuclear Overhauser Enhancement	535
References	542

APPENDIX: READING LIST	547
INDEX	549

Preface

Pulse Methods in 1D and 2D Liquid-Phase NMR is written to enable the practicing NMR spectroscopist to understand and apply the varied and powerful new techniques developed in the past few years for obtaining spectra with greatly increased information content and from smaller and smaller samples. The intent is to describe both theory and practice in simple and detailed fashion so that the methods may be critically evaluated and effectively used in any potential application. As methods become more complex they require more instrument time, and it is important to be able to judge whether the investment of this time is justified. It is also essential for the spectroscopist to be in a position to evaluate the capabilities of the instrumentation available, as well as the additional requirements for utilization of particular new methods.

The material in this book assumes a knowledge of continuous-wave NMR methods as well as an elementary understanding of the normal pulsed Fourier-transform spectroscopic procedures, together with a knowledge of such related phenomena as the nuclear Overhauser effect. Although much of the treatment is necessarily mathematical, this aspect of the presentation has been simplified as much as possible.

To the extent feasible the editor has aimed for consistency in terminology and symbolism throughout the volume, but in some instances where conventions in the literature diverge, the benefit of presenting several versions has been recognized. Further, the editor has striven to minimize duplication between chapters except in areas where repetition or treatment from differing viewpoints seemed desirable.

Chapter 1 includes an elementary presentation of a number of aspects of pulsed NMR and of the "building blocks" which are combined to form more complex two-dimensional experiments. In Chapter 2, the density-operator approach to the description of the response of nuclear systems to pulses and of their behavior under conditions of free precession is developed and justified. In Chapter 3, polarized transfer experiments and their applications are discussed in detail, and the applications of density-operator theory are further illustrated. Principles of multiple-quantum spectroscopy and applications of multiple-quantum techniques are considered from a theoretical point of view in Chapter 4. Chapters 5, 6, and 7 include a variety of specific applications of 2D methods, along with descriptions of the experimental considerations involved. In Chapter 8, we have taken the opportunity to review some developments in 2D methods which have materialized very recently, and are, as well, related to the contents of several of the earlier chapters. Finally, the reader may note that some exercises have been incorporated in the earlier chapters; these are designed to afford practice in dealing with some of the basic principles of the subject.

It is a pleasure to acknowledge the active cooperation of all the contributors to this volume and the help that each has provided in assembling it, as well as the special assistance of the editor's wife, Mary Louise, in reading and correcting the entire manuscript, and of Dr. Ole W. Sørensen in reading most of the chapters and providing helpful suggestions. The editor also appreciates the comments, support, and assistance of students in his laboratory, particularly the help of Paul J. Kanyha and H. Daniel Plant.

1

Basic Methods and Simple Pulsed Experiments

WALLACE S. BREY
DEPARTMENT OF CHEMISTRY
UNIVERSITY OF FLORIDA
GAINESVILLE, FLORIDA 32611

I. Introduction	2
A. The Rotating Frame	2
B. Effects of Pulses	3
C. Evolution of Magnetization	4
II. Spin-Echo Experiments	6
A. Echoes and Refocusing	6
B. J Spectroscopy	12
C. Modulated Spin Echoes for Multiplicity Determination and Spectral Simplification	17
D. The Significance of Refocusing Periods	30
III. Polarization-Transfer Experiments	31
A. Advantages of Polarization Transfer	31
B. Selective Polarization Transfer (SPT)	32
C. Spectral Analysis by Selective Polarization Transfer	44
D. Decoupling in Selective Polarization-Transfer Experiments	51
E. Nonselective Polarization Transfer—The INEPT Experiment	54
F. DEPT and Related Sequences	66
G. Inverse Polarization Transfer	73
H. Forward and Reverse Polarization Transfer Involving Quadrupolar Nuclei	76
I. Some Experimental Considerations in Polarization-Transfer Experiments	82
IV. Experimental Aspects of Pulsed NMR	83
A. Nature and Properties of Pulses	83
B. Bilinear Pulse Clusters	100
C. Calibration of Decoupler Pulses	102
References	104

I. Introduction

In this section we shall describe several terms and concepts necessary for the understanding of chapters which follow and shall define some conventions which will be employed later in the book. These subjects have been collected here for the benefit of those readers who may wish to skip some of the earlier chapters or to read the chapters in an order other than that in which they appear. Some readers may already be familiar with most of the basic ideas presented here; for others who require additional details, a list of introductory books on Fourier transform nuclear magnetic resonance is provided at the end of the volume.

In subsequent parts of the first chapter, we will discuss several of the relatively simple pulse experiments which serve as building blocks to construct the many different pulse sequences that have been devised in order to gain the greatest possible amount of information from Fourier transform spectroscopy. We will show how these experiments can be represented in terms of the motions of magnetization vectors and the changes in populations of energy levels. Some experimental aspects of pulsed spectroscopy will be described and the idea of multiple quantum coherence will be introduced. Practice in analyzing one-dimensional experiments should make easier the task of the reader in understanding the more elaborate and complex two-dimensional experiments to be described in later chapters.

A. THE ROTATING FRAME

The equations which describe the behavior of nuclear spins in relation to spatial coordinates referred to the laboratory—that is, the world outside the probe—are very cumbersome, and substantial simplification is achieved if nuclear spin motions are referred to a coordinate system which rotates about the direction of the fixed magnetic field, the z direction, with a frequency equal to the frequency of the applied rf field. The transverse coordinates in the rotating frame are often denoted by x' and y', but we shall not use the primes; coordinates x and y should always be understood to be in the rotating frame unless otherwise stated.

If several nuclear species are involved, with various resonance frequencies corresponding to their magnetogyric ratios, then one may think of several coordinate frames all rotating about the same axis, one at each of several frequencies, but having no phase coherence with one another. For the case of two species, such as carbon and hydrogen, the term "doubly rotating frame" has been coined. However, this refers not to a single frame, but to two frames which have a common z axis. Furthermore, in a system having both carbon and hydrogen nuclei, the precession phase of the hydrogen nuclear magnetization cannot be related to the precession phase of the carbon magnetization. If the nuclei of the two species are spin–spin coupled to one another, the spin state of one species does affect the precession frequency of the other species, so that there is a correlation between differences in precession frequency if not a correlation in phase.

B. Effects of Pulses

A pulse of radio frequency applied to an NMR sample generates a magnetic field perpendicular to the direction of the constant field and fixed in direction in the rotating frame. The pulse thus produces on the nuclear moments a torque which tends to rotate the sample magnetization about the direction of the field produced by the pulse. The angle of rotation depends on the duration of the pulse t, according to the relation $\theta = \gamma B_1 t$, where γ is the magnetogyric ratio of the nuclei and B_1 is the strength of the magnetic field of the pulse. If it is desired to turn all the nuclear spins and their resultant magnetization by a particular angle such as 90° ($\pi/2$ radians) or 180° (π radians), the duration of the pulse is adjusted to meet this condition. In a normal single-pulse FT experiment, a pulse corresponding to a flip angle of something up to 90° turns the magnetization from the z direction toward the xy plane, and the subsequent precession of the magnetization induces a voltage in the receiver coil, which is measured to generate the free-induction decay.

It is evident that the larger the value of B_1, the shorter need be the pulse which turns the magnetization through a given angle. At the same time, the shorter the pulse, the wider the distribution of the frequency components of the rf about the center frequency of the pulse. For example, a pulse of about ten microseconds or less is required to provide a sufficiently wide frequency range to cover the chemical shift range of carbon in a high-field spectrometer; such a pulse, which affects all nuclei of a particular species more or less uniformly, is termed a *nonselective* or *hard* pulse. This is the type of pulse usually used for the observed nucleus in a single-pulse experiment. If one wishes to disturb only those nuclei corresponding to one resonance in the spectrum, then a much weaker rf field corresponding to a 90° pulse length of tens of milliseconds to seconds is usually appropriate. This is a *soft* or *selective* pulse. In an intermediate region, it may be possible to design a pulse which has an appreciable effect only on all the components of a particular spin multiplet but does tip them all to the same extent—this is a *semiselective* pulse.

For larger samples such as are often employed for less receptive nuclei, the value of B_1 is not uniform over the sample. The problem of nonuniformity is even more acute for biomedical imaging experiments, for which the sample may be quite heterogeneous. Hence some attention will be directed to a consideration of means of compensating for the fact that what is a θ pulse at one point in a sample may produce a tip angle for nuclei at other points which is much larger or smaller than θ.

By custom, the direction defined by the magnetic field of the first pulse in a sequence of pulses is usually labeled the x axis, although this rule is not always followed. A subsequent pulse may be along the same axis, that is, in the x direction in the rotating frame, or may be shifted in phase so that its magnetic field lies along some other transverse direction in the rotating frame. Thus a pulse of phase y could, for example, be obtained in principle by interrupting the rf for a quarter of a cycle so that the maxima of the magnetic field generated by rf during the pulse are 90° behind the maxima that would have occurred had there been no interruption. It is important to distinguish a "90° phase shift" from a "90° pulse length," and it is also important to remember that "rf phase" only has meaning relative to some other phase of the same signal. Thus the phase of a pulse applied to hydrogen nuclei in a sample cannot be related to the phase of a pulse applied to the carbons.

A symbol such as $(\pi/2)_\phi(I)$ or $90°_\phi(I)$ will be used to represent a pulse of phase ϕ in the rotating frame applied to the nuclei of species I and of duration long enough to rotate the magnetization through a right angle. The phase of a pulse may be specified as along one of the x or y coordinate axes or as an angle by which the pulse axis is displaced from the x axis. In many current instruments, the phases are limited to x, y, $-x$, and $-y$ by the nature of the phase shifter installed in the hardware, but intermediate phase capabilities are becoming available and are particularly needed in experiments in which multiple-quantum coherence is to be generated.

It will be convenient to establish a convention for the direction in which a pulse rotates magnetization, and we will assume that the sense of the rotation is clockwise to an observer looking outward from the origin along the particular axis of the coordinate system, or counterclockwise to an observer looking along the specified axis toward the origin. Thus equilibrium magnetization in the $+z$ direction will be considered to be rotated by a $90°_x$ pulse to the $-y$ axis or by a $90°_y$ pulse to the $+x$ direction. Some examples of the application of this convention are shown in Fig. 1-1. The convention is opposite to that frequently employed, but the difference is really quite arbitrary, for the actual direction of rotation depends on, among other things, the sign of the magnetogyric ratio of the nuclear species.

C. Evolution of Magnetization

Consider the magnetization of a sample at equilibrium in a fixed magnetic field. The individual nuclei are precessing about the z axis—the direction

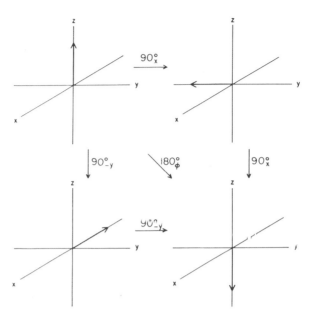

Fig. 1-1. Examples of the effects of pulses on the resultant magnetization vector of the nuclei in a sample, according to the convention used in this book. The effect of an ideal 180° pulse on equilibrium magnetization is independent of its phase.

of the field—with random phase, that is, with a random distribution of x and y or transverse components of magnetic moment. The resultant magnetization of the sample is directed along the z axis. If this resultant magnetization is tipped away from the z direction by an rf pulse, the magnetization vector precesses about the z axis with the Larmor frequency.

For this system, the rotating frame is defined by one circularly rotating component of the alternating signal from the rf transmitter. Rf is applied to the sample during the pulse but not after completion of the pulse. However, the transmitter does continue to oscillate without interruption after the end of the pulse, and the rf signal is, in practice, usually fed continuously to the reference channel of the phase detector. If the rf is exactly on resonance, the nuclear precession is at exactly that frequency, and if the rf is then applied to the sample again during a later pulse, the rf will be in phase with the magnetization. However, when nuclei in the sample have precession frequencies somewhat different from the rf frequency, the phase angle between the precession of these nuclei and the rf increases with lapse of time. This "evolution" of the system may be the result of magnetic field inhomogeneity, chemical shift differences, or spin–spin splitting. The last of these is removed if the nuclei responsible for the splitting are irradiated at their resonance frequency or *decoupled*, whereupon the precession frequencies of all the components of a multiplet are collapsed to that frequency corresponding to the chemical shift alone.

II. Spin-Echo Experiments

A. Echoes and Refocusing

1. *The "Hahn" Experiment*

The first "building-block" experiment we shall describe involves the application of *spin echoes*, discovered by Hahn in the early days of NMR. Echoes represent nuclear coherence or magnetization appearing apparently spontaneously at some time interval after the end of a series of two or more pulses applied in succession to a sample.

In Hahn's original experiment, an echo was generated following a sequence of two successive 90° pulses (1). It is possible to visualize somewhat more easily a slightly different version of the spin-echo experiment in terms of the motions of vectors representing nuclear magnetization, and that is the form of the experiment which will be described here. The reader should remember that, throughout this account, each magnetization vector represents the resultant of the magnetic moments of all the nuclei of a given type within the sample, that is, all those spins having the same precession frequency, often referred to as a *spin isochromat*.

In the version of the experiment we will first describe, a "hard," nonselective 90° pulse is applied to the nuclei to be observed, such as those of ^1H. After a time interval τ, a second pulse is applied, this time long enough to rotate the magnetization by 180°. An echo signal is observed to grow out of the baseline at some time following the second pulse and to reach a maximum intensity at time τ after that pulse. This sequence is represented in Fig. 1-2.

To explain what has occurred in this experiment, we begin with the situation at equilibrium in a fixed magnetic field of strength B_0. In this condition, the magnetization vectors of all groups of nuclei—various nuclei may be located in differing local magnetic fields—are parallel to the field direction, the z direction, as shown in Fig. 1-3a. The effect of the initial pulse is to turn the magnetization by 90°; according to the convention used here and described in Section I,B above, the magnetization is then in the $-y$ direction, as in Fig. 1-3b. The effect of the pulse is assumed to be the same on each nucleus, regardless of its chemical shift or of where it may be in a spatially varying magnetic field. Following the initial pulse, the spin vectors precess about the z axis, but both the magnitude of the constant field affecting each nucleus and, consequently, the precession frequency of

1. BASIC METHODS AND SIMPLE PULSED EXPERIMENTS

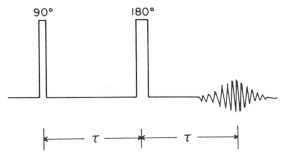

Fig. 1-2. Pulse sequence for a spin-echo experiment. The phase of the 180° pulse may be either x or y; the effect of the distinction on the results is described in the text and shown in Fig. 1-3.

that nucleus depend on the chemical shift and on local inhomogeneities. In Fig. 1-3, we focus attention on the precession of two spin isochromats, labeled **s** and **f**.

During the first period of length τ, in which the spin vectors precess without external interference, the "faster" spins **f**, which have a higher resonance frequency than the spins **s** and thus precess more rapidly, get somewhat ahead of the "slower" **s** spins, as shown in Fig. 1-3c. The 180°

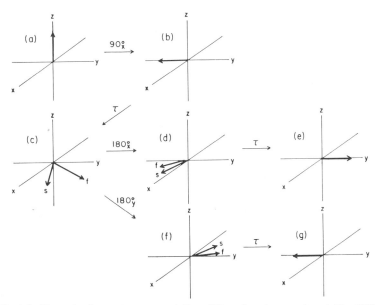

Fig. 1-3. Magnetization vector representation of the spin-echo experiment. The 180° pulse may have the same phase as the 90° pulse, leading to the result in (e), or be shifted in phase by 90°, leading to the result in (g).

pulse, with the same phase as the initial pulse and thus directed along the x axis, reverses the signs of the y components of the vectors as shown in Fig. 1-3d. The nuclei then precess at their usual rate for the second period of length τ; in this interval, the **f** spins, which were placed behind the **s** spins by the 180° pulse, catch up with the **s** spins so that at the end of this interval both groups are aligned along the $+y$ axis, as in Fig. 1-3e.

The overall course of this series of events may be described as fanning out in the xy plane of the magnetizations of the nuclei precessing at different speeds, followed by refocusing after the 180° pulse. The echo is generated as the various magnetizations combine to produce a resultant in a particular direction in the rotating frame, and the precession of this magnetization induces an rf signal in the spectrometer receiver coil. In many of the pulse sequences to be described later, a 180° pulse is similarly employed to refocus the magnetizations of groups of nuclei having different precession frequencies; very often the reason for doing this is the need to eliminate the effects of differing chemical shifts.

In Fig. 1-3e, the refocused magnetization is seen to be in the direction opposite to that of the magnetization after the initial 90° pulse. Thus, if the spectrometer is arranged so that the signal that is obtained by collecting the FID immediately following the initial 90° pulse and processing it is positive, the signal obtained from the echo is negative. If the spin-echo sequence is modified by shifting the phase of the 180° refocusing pulse so that it is applied along the y axis, 90° away from the phase of the first pulse, the magnetization vectors follow the path shown in Figs. 1-3f and 1-3g and refocus in the direction in which they started, so that the signal observed is positive rather than negative.

Consider now how the results differ within a series of such experiments carried out with varying lengths of the interval τ. During the periods of free precession, the nuclei tend to dephase irreversibly by *spin-spin relaxation processes*, and this causes the maximum amplitude of the echo to decrease with increasing τ, and the time constant of the decrease is the spin-spin relaxation time, T_2. The spin-echo method can accordingly be used to measure T_2 for lines for which T_2 cannot be estimated from the shape of the line because the width is determined by factors such as magnetic field inhomogeneity. Diffusion of the nuclei between regions of differing magnetic field also causes the echo amplitude to be reduced, because the change in field alters the nuclear precession frequencies, interfering with refocusing. The equation expressing the dependence of the amplitude of the echo upon both spin-spin relaxation and diffusion is (2)

$$A_{2\tau} = A_0 \exp\left(-\frac{2\tau}{T_2} - \frac{2}{3}\gamma^2 G^2 D \tau^3\right) \qquad [1]$$

1. BASIC METHODS AND SIMPLE PULSED EXPERIMENTS

where D is the diffusion coefficient, G is the magnetic field gradient through which diffusion is occurring, and A_0 is the amplitude of an FID with τ equal to zero.

2. Carr-Purcell-Meiboom-Gill Procedure

Although one way of evaluating T_2 or D from spin-echo amplitudes is to repeat the entire "Hahn" experiment a number of times, varying the value of τ, Carr and Purcell (2) showed how this determination could be carried out in a single experiment by applying a series of 180° pulses at intervals $\tau, 3\tau, 5\tau, 7\tau, \ldots$ following the first 90° pulse; echoes are then observed at times $2\tau, 4\tau, 6\tau, 8\tau \ldots$. If all pulses have the same phase as the 90° pulse, the echoes alternate in sign. From the decrease of echo amplitude with time, T_2 and D can be evaluated, using the equation

$$|A_t| = A_0 \exp\left(-\frac{t}{T_2} - \frac{1}{3}\gamma'G'D_\iota'_\iota\right) \qquad [2]$$

where t is the elapsed time since the 90° pulse. By making the interval 2τ between pulses small, the contribution of the second term can be reduced and diffusional effects on the decrease of the echo amplitude can be minimized.

A drawback to this method arises from the accumulation of errors as successive 180° pulses continue to turn the magnetization in the same sense. For example, after 15 pulses, an error of 3° in the length of each one would result in a cumulative error of 45° in the position of the magnetization, causing the echo intensity to be much less than it should be, since the signal arises only from the component of magnetization in the xy plane. Meiboom and Gill (3) suggested applying the 180° pulses with their phase shifted by 90° from the phase of the initial 90° pulse, so that their magnetic field is in the y direction in the rotating frame, tending to turn z magnetization into x magnetization, $-x$ magnetization into z magnetization, and so on. Not only are the resulting echoes all positive, but, as is shown in Fig. 1-4, if the nominal 180° pulses are either a little too long or too short, an error in the length of one pulse is compensated by the effect of the next pulse, which is of the same length but turns the magnetization in the opposite direction. Even if the pulse length is adjusted to precisely 180° at one point in a sample, the effective angle may vary over the volume of the sample, so that this technique may be necessary to compensate for rf inhomogeneity, whether it results from the design of the transmitter coil in the probe or from inhomogeneity of the sample.

The Carr-Purcell-Meiboom-Gill method was used fairly extensively to measure T_2 in systems which have only a single resonance or where the

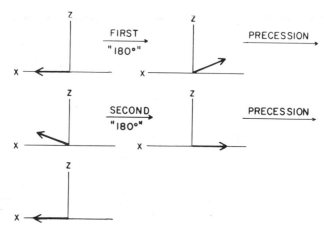

Fig. 1-4. Effect of the Meiboom-Gill modification of the Carr-Purcell spin-echo sequence. Because successive nominally 180° pulses turn the magnetization in opposite directions, the pulse error is compensated in every even-numbered echo.

signal represents an average over all the chemical shifts of a particular nucleus in the sample. The application of Fourier transform methods to conventional high-resolution NMR spectroscopy led to the realization that spin echoes could be Fourier-analyzed by a procedure parallel to that which is applied to a free-induction decay in order to obtain, for example, relaxation rates of individual resonances in a multiline spectrum (4). In high-resolution spectroscopy, however, the effort to reduce the effects of diffusion may cause the interval between successive Carr-Purcell refocusing pulses to be as short as several milliseconds, too short to allow acquisition after each echo maximum of a sufficient number of points to provide the desired resolution. It may then be necessary to acquire only the last half-echo from a long train of echoes and to repeat the experiment with pulse trains of varying lengths to establish the rate of signal decay by T_2 processes.

Although evaluation of genuine spin–spin relaxation times of individual resonances in a spectrum is, in principle, possible by spin-echo experiments, many complications are involved in attempts to measure T_2 or rate constants for other processes in *coupled* spin systems. These problems are discussed in detail by Freeman and Hill (5) and by Vold and Vold (6). The effects of deuterium exchange on transverse relaxation in the complex spectra of proteins have been analyzed (7), and applications of spin echoes to kinetic measurements have been described (8).

3. Other Uses and Consequences of Echoes

Spin-echo methods have the special advantage of minimizing the effects of the distortion which results when acquisition of data is begun before the receiver has fully recovered from a pulse. Rance and Byrd (9) have utilized this advantage of echo acquisition to overcome receiver recovery problems in obtaining undistorted broad resonances of phosphorus in heterogeneous systems such as phospholipid bilayers. The problem of pulse breakthrough is particularly difficult for broad resonances, since most of the information about lineshape is found in the first part of the FID. A critical aspect of the Rance-Byrd procedure is a phase-cycling scheme which compensates for pulse imperfections. Pulse sequences involving acquisition of echoes have also been employed in an effort to suppress effects of acoustic ringing in spectroscopy of nuclei which resonate at low frequencies by allowing the ringing to die out during the delay before the echo (10).

Several methods involving spin echoes have been proposed to improve the efficiency of data collection in ordinary FT NMR. In multiple acquisition, the cycle rate is limited by the need for spin-lattice relaxation to be nearly complete before the start of the subsequent pulse cycle. If the signal decays rapidly because of dephasing of the nuclei from factors such as magnetic field inhomogeneity, so that $T_2^* \ll T_2 \approx T_1$, useful data lasts only a short time compared to T_1, and time is wasted waiting for T_1 recovery. In the DEFT (for *d*riven *e*quilibrium *F*ourier *t*ransform) procedure (11), an attempt was made to minimize this inefficiency by performing a spin-echo experiment in which data are collected both during the FID following the first pulse and during the rising half of the first echo—an echo can for many purposes be regarded as two FIDs back-to-back—and then, just as the peak of the echo is reached, applying a 90° pulse to turn the magnetization back to the z direction in preparation for the next cycle. For carbon-13, experiments have shown, however, that the genuine T_2 is often so short as to defeat the aim of DEFT, because the magnetization will have decayed by the time the echo appears (12). Subsequent analysis has also demonstrated that sensitivity enhancement similar to that of DEFT is obtained from the refocusing effect associated with the steady-state response to a series of pulses at constant short intervals (13).

Opella and Cross (14) selected nonprotonated carbon resonances by applying low-level off-resonance noise-modulated decoupling during the cycle delay, broadening the resonances of protonated carbons and thus decreasing their T_2 values. Acquisition of an echo, accompanied by full decoupling to narrow the nonprotonated resonances, then discriminated against the carbons with short T_2.

Recently, spin-echo measurements have seen extensive applications in proton biomedical imaging, where the contrast that is observed between tissues is much greater in terms of relaxation time difference than in terms of mere variation in proton density. Any series of pulses of the same phase, repeated at uniform short intervals, causes spin echoes to be created. In data processing in the "sensitive point" method of biomedical localization of NMR resonances, it has been found essential to use only the echo signal generated by rapid pulsing and to discard the normal FID (15).

Effects of spin echoes may also appear where not expected, as, for example, when acquisition cycles in ordinary pulsed NMR are repeated in times short compared to the T_2^* of the sample, leading to baseline roll and phase anomalies in the transformed spectrum. Turner (16) discusses methods for elimination of this problem, including cycling of the phase of the transmitter.

B. J Spectroscopy

1. Effects of Homonuclear Spin–Spin Coupling in Spin-Echo Experiments

The 180° nonselective or "hard" pulse used as described above to generate echoes has the effect of refocusing nuclear spins which differ in precession rate because of chemical shift differences, magnetic field inhomogeneity, or heteronuclear spin–spin coupling. When homonuclear spin–spin coupling is present, however, an additional effect is produced: the 180° pulse affects not only the nuclei being observed but also the nuclei causing the splitting, reversing the directions of their spins. Since the spin states of the nuclei responsible for splitting the resonance of the observed nuclei are now interchanged, the nuclei being observed interchange precession rates.

As an example, consider an AX system following the 180° pulse as represented in Fig. 1-3d. Not only have the magnetizations of the f and s nuclei been flipped around the x axis, but those nuclei which were the faster precessors before the pulse because they were coupled to, say, an α spin, now precess more slowly because they are coupled to a β spin and vice versa. Thus the labels on the two magnetization vectors must be interchanged. After some period of free precession, these two vectors are again in phase, but this does not necessarily happen along the $-y$ or the $+y$ axis or in any other generally definable direction, but rather in a direction which depends on the magnitude of the coupling constant and thus does not match the direction in which the effects of chemical shifts are refocused to an echo.

If the pulse interval τ in a spin-echo sequence is varied in a series of spectra of a homonuclear coupled system, the response of each line is found to have a phase which varies cyclically at a frequency which has a contribution of $J/2$ from each coupling constant affecting the nucleus. Freeman and Hill (17) showed that useful information can be obtained by analysis of the pattern of echo modulation. They employed a narrow-band filter to pick out the resonance at a particular frequency and sampled each echo at

1. BASIC METHODS AND SIMPLE PULSED EXPERIMENTS

its peak. From the τ dependence of the amplitude of the responses, the magnitudes of the couplings involving the resonant nucleus could thus be determined, in simple cases by inspection, or in more complex cases by Fourier transformation. The resulting spectrum was termed a partial J spectrum; an example is shown in Fig. 1-5. In this illustration, the observed nucleus has two resolvable couplings which give responses at $[J(BX) - J(AX)]/2$ and $[J(BX) + J(AX)]/2$, an example of the general rule that multiple couplings yield signals at frequencies proportional to their sums and differences.

Freeman and Hill described some important considerations in obtaining J spectra. If a spectrum is not first order or if the refocusing pulses are not exactly 180°, additional resonances appear at frequencies which are functions of the chemical shifts as well as of the coupling constants. To obtain a valid spectrum, the pulse repetition frequency must be low compared to the chemical shift difference δ between the coupled nuclei; at very high repetition rates, J modulation disappears entirely. On the other hand, the repetition rate must be high enough to satisfy the Nyquist criterion for adequate sampling of the J modulation pattern. Combining these conditions leads to

$$\pi\delta \gg \frac{1}{2\tau} \geq \sum_k |J_{jk}| \qquad [3]$$

where $\Sigma_k |J_{jk}|$ is the sum of the absolute values of all couplings of the observed nucleus to other nuclei.

A very significant feature of J spectra is the narrow linewidth, which leads to good resolution of adjacent peaks. The linewidth is limited either

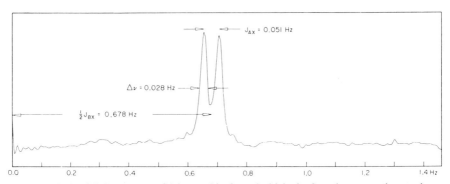

Fig. 1-5. Partial J spectrum of 3-bromothiophene-2-aldehyde. Samples were taken at the peak of each of 150 Carr–Purcell spin echoes at 0.4-sec intervals and then Fourier transformed. Responses appear at $[J(BX) - J(AX)]/2$ and $[J(BX) + J(AX)]/2$. Because of the narrow lines, very high resolution is obtained. [Reprinted with permission from R. Freeman and H. D. W. Hill, *J. Chem. Phys.* **54**, 301 (1971). Copyright 1971 American Institute of Physics.]

by the intrinsic lineshape or by spin–lattice relaxation, since the effects of magnetic field inhomogeneities are refocused. Thus multiplets may be resolved for samples of larger volume and poor field homogeneity, and, for smaller samples, very small splittings arising from spin–spin coupling can be measured. Long-range coupling constants in a series of esters have been measured with a precision of the order of ±0.003 Hz (18). Modulation of spin echoes by coupling has been employed for accurate evaluation of $J(H-H)$ in large molecules such as lysozyme, where overlap of resonances limits the usual types of spectral analysis (19). The resonance examined was normally a doublet; to better estimate peak intensity, Campbell *et al.* collapsed the doublet by selectively irradiating, in the receiver-off intervals during data acquisition, the resonance responsible for the splitting. The echo interval, $\tau = 1/4J$, to give zero intensity was then determined.

2. *Heteronuclear J Spectroscopy*

The key to homonuclear J spectroscopy is the modulation produced by the nonselective or "hard" pulse which affects all the nuclei of a particular species, as, for example, all the hydrogen nuclei in a sample. Because the 180° pulse does not affect nuclei of other species and thus does not interchange the spin states of nuclei responsible for heteronuclear splitting, heteronuclear couplings are not, in general, effective in modulating spin echoes in the experiment described in the previous section. Exceptions to this rule occur when the heteronuclei are strongly coupled to one another, but, when the heteronuclei are hydrogens, this effect may be eliminated by decoupling, provided the decoupling is coherent (20).

When it is desired to introduce into spin echoes modulation associated with heteronuclear coupling, either of two methods may be employed. These methods were initially proposed in connection with two-dimensional J spectroscopy, but are also applicable to one-dimensional spectra. The first, a rather straightforward procedure, is to apply a 180° pulse to the heteronucleus at the same time as, or immediately following, the 180° pulse to the observed nucleus, as shown in the seqeuence in Fig. 1-6 (21). This has the same effect on the spin states of the heteronuclei as does the nonselective pulse on the nucleus responsible for spin–spin splitting in the homonuclear case. Since the usual application of this procedure is to the observation of a nuclear species for which multiplets are produced by coupling to hydrogen, it is often referred to as the *proton-flip* method. If a J spectrum of a less-abundant nucleus such as carbon-13 or nitrogen-15 is obtained in natural abundance, neighboring nuclei of the same species are so rare as to produce no significant homonuclear modulation effect. However, several recent applications have reversed the roles of the two nuclear species, with

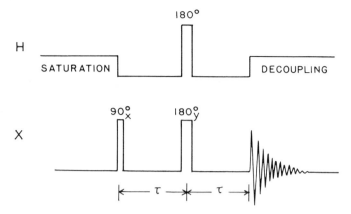

Fig. 1-6. Pulse sequence for the spin-flip or proton-flip method for providing heteronuclear modulation of spin echoes.

the modulation of proton echoes by a nonhydrogen species to be observed, so that a more general name is the *spin-flip* method. For experiments in which protons are observed, there are also likely to be complications from homonuclear modulation by other coupled protons.

The echo modulation caused by the spin-flip method is *phase modulation*, for, at the end of the precession period—the time when acquisition of the free induction decay begins—the angle between the direction of a magnetization vector in the xy plane and the direction it would have in the absence of spin–spin interaction depends on the value of τ. To see this in more detail, notice that the precession frequency of a multiplet of the observed nucleus is $\nu_0 + \Sigma_i m_i J_i$, where ν_0 is the precession frequency in the absence of spin–spin interaction, m_i is the quantum number describing the spin state of a heteronucleus coupled to the observed nucleus by the interaction constant J_i, and the summation is over all such coupled nuclei (22). The phase angle accumulated in the interval τ before the 180° spin-echo pulse is therefore $2\pi\tau(\nu_0 + \Sigma_i m_i J_i)$. After the 180° pulse to the heteronuclei, such as hydrogen nuclei, the precession frequency expression is modified because of the change in sign of each of the m_i values. As a consequence of the refocusing effect of the 180° pulse on the chemical shift, the precession resulting from the term $2\pi\tau\nu_0$ in each of the two intervals of length τ automatically amounts to an odd multiple of 180° if the refocusing pulse is of phase x or to a multiple of 360° if that pulse is of phase y, the two alternatives which are shown in Fig. 1-3. Consequently, the total phase angle accumulated in the two time intervals, each of length τ, includes only the contribution from spin–spin interaction, which is $2\pi(2\tau) \Sigma_i m_i J_i$, and this value differs for each peak in a multiplet. If the decoupler is turned on, the various multiplet components existing at just the instant of turn-on are collapsed onto the direction of the chemical shift, converting the phase modulation into *amplitude modulation*, since only the amplitude of the detected magnetization and not its direction is affected by the spin–spin interactions operative during previous stages of the experimental sequence.

The second method, an alternative to the spin-flip experiment, is the *gated-decoupler* method, in which the components of the spin multiplet are allowed to dephase during one half of the period between the 90° pulse and the echo, whereas further dephasing during the other half of the interval is eliminated by applying broad-band decoupling to the hydrogen nuclei (23). One version of this procedure is represented in Fig. 1-7. Application of decoupling at the time of the 180° pulse to the observed nuclei causes the components of the multiplet for that nucleus to collapse to the direction of the chemical shift, and they then continue to precess at the chemical shift frequency ν_0, since the decoupler removes the differential effects of the hydrogen spin states. Each signal component present just before the decoupler is turned on contributes an amount to the resultant magnetization equal to the projection, onto the direction of the vector representing the chemical shift, of the magnetization which corresponds to that component. The projection is proportional to the cosine of the accumulated phase angle, which for an AX system is $\cos 2\pi(J/2)\tau$. When the echo signal is observed, the magnetization has the same direction it would have in the absence of multiplet splitting, that is, the direction of the vector representing the magnetization if only chemical shift precession were effective. However, its amplitude depends on how much the nuclei had dephased before the decoupler was turned on, and therefore, for any value of τ, the amplitude varies with the value of the coupling constant. Thus the signal has constant

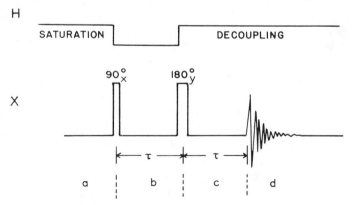

Fig. 1-7. Pulse sequence for the gated-decoupler method for providing heteronuclear modulation of spin echoes.

phase but is amplitude-modulated as a function of τ at a frequency of $J/2$. A similar result is obtained if the sequence in Fig. 1-7 is modified by having the decoupler on during the first τ interval and during acquisition but off during the second τ period. If the interval τ is represented instead by $t_1/2$, as is common in 2D spectroscopy, so that the entire refocusing period for chemical shift has a length of t_1, the terminology commonly used in 2D spectroscopy, then the modulation frequency as a function of t_1 is $J/4$.

Results of the gated-decoupler and "proton-flip" methods have been compared (24, 25). Since the evolution period in the proton-flip sequence is twice as long as in the gated-decoupler sequence, the resolution afforded by the proton-flip method is twice as good as that by the gated-decoupler method. However, the proton-flip method seems more prone to give spectra which deviate from simple patterns because of non-first-order effects.

3. Two-Dimensional J Spectra

At this point, let us consider an experiment in which a spectrum for each of a series of values of τ is obtained by Fourier transforming the whole of an echo—in contrast to the Freeman–Hill method of sampling only at the top of each echo after a particular frequency has been chosen by a filter. The resulting spectra for a particular multiplet might appear as in Fig. 1-8. From each of the spectra, the intensity value at the maximum of one of the resonance peaks is read, and these values are then arranged in a series as an ordered function of the time interval from the 90° pulse to the particular echo from which the point is obtained. A second Fourier transformation with respect to time elapsed after the 90° pulse is then carried out. The frequency components in the resulting transform are the modulation frequencies arising from the coupling constants of the nucleus responsible for the resonance. If this process is repeated for every resonance in the spectrum, or indeed for every point in the spectrum, the final transforms can be plotted in a two-dimensional array, with chemical shift in one dimension and the multiplet pattern of the nucleus resonating at a particular chemical shift spread out in the other dimension, as shown in the example in Fig. 1-9. This is a simple example of *two-dimensional* or double-Fourier-transform spectroscopy. We shall return later to the details of this and other experiments which are related to it, and also to the use of the gated-decoupler and proton-flip procedures in 2D spectroscopy.

C. Modulated Spin Echoes for Multiplicity Determination and Spectral Simplification

If the magnitude of the spin–spin coupling between heteronuclei can be estimated fairly well, we may reverse the approach of J spectroscopy from

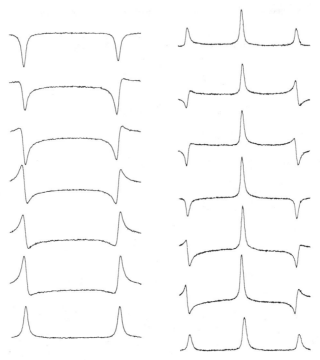

Fig. 1-8. Modulated spin-echo spectra of the protons of $CH_2ClCHCl_2$ in acetone d-6 at 300 MHz. The values of τ in the pulse sequence, as in Fig. 1-2, increase from bottom to top and correspond to 0, 1/8J, 1/6J, 1/4J, 1/3J, 3/8J, and 1/2J. The frequency scale for the doublet differs from that for the singlet.

that of a method of improving resolution or determining J more precisely and employ the modulation of spin echoes to establish, for example, how many hydrogen atoms are attached to a particular carbon atom observed in a decoupled spectrum (26). This technique is a very advantageous alternative to off-resonance decoupling for ascertaining the number of components or *multiplicity* that a resonance would have in a coupled spectrum. Since the ratio of largest to smallest one-bond coupling constant between hydrogen and carbon-13 generally is not much greater than 1.5 to 1, it is possible to determine whether a carbon resonance arises from a methyl carbon, a methylene carbon, a methine carbon, or, by default, a quaternary carbon.

1. *Echo Modulation and Multiplicity Determination*

Consider first the gated-decoupler pulse sequence, which is shown in Fig. 1-7. In this diagram, four intervals are labeled as follows: *a* is the

1. BASIC METHODS AND SIMPLE PULSED EXPERIMENTS   19

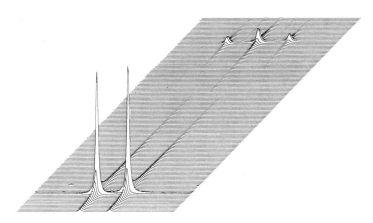

Fig. 1-9. Two-dimensional ^{19}F J spectrum of $CF_2ClCFCl_2$, 15% solution in $CDCl_3$, at 282 MHz; 512 spectra were obtained in the "F_1 dimension" with the time between pulses in the sequence incremented by 10 msec between successive spectra. Spectral width in the "F_2 dimension"—the frequency range of each of the individual spectra—was ±1000 Hz. The spectral information is displayed in a stacked plot, and the fluorine chemical shift scale runs from lower left to upper right.

relaxation interval to allow the system to come to equilibrium before the first pulse is applied; b is the interval between the first pulse and the refocusing pulse; c is the interval between the refocusing pulse and the start of acquisition; and d is the period of acquisition of the FID. Later we will refer to these interval designations to describe other patterns for the on-time of the decoupler, but for the present we will discuss what happens if the decoupler is on during the periods a, c, and d as shown in the figure. Whatever pattern is used, having the decoupler on during a establishes a nuclear Overhauser enhancement; whether this is helpful or harmful to sensitivity depends on the characteristics of the observed nucleus.

In Fig. 1-10 we follow, for the Fig. 1-7 version of the gated-decoupler experiment, the precession of the multiplet spin vectors in AX, A_2X, and A_3X spin systems, assuming that any effects of magnetic field inhomogeneity or chemical shift are refocused and that the spectra are first order. When the decoupler is turned on at the time of the 180° carbon pulse, the magnetization vectors are projected onto the position of the chemical shift, labeled ν_0, which corresponds to the midpoint of the magnetization vectors of the multiplet. The total magnetization giving rise to the signal is the sum of the projections of the several components on this direction. Decoupling during interval c does not interfere with refocusing the effects of influences other than spin–spin coupling, such as field inhomogeneity and chemical shift, but it does prevent the spin–spin coupling from further modifying the

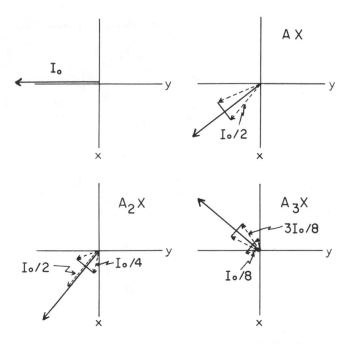

Fig. 1-10. Scaled representations of the collapse of multiplet components by decoupling. The resultant in each case is the sum of the projections of the individual components on the direction of the center of the multiplet, that is, the chemical shift position.

resonance signal by influencing the speed of precession. Diagrams related to Fig. 1-10 have been given by Brown, Nakashima, and Rabenstein (27), but in these the irradiation is assumed to be exactly on resonance for the observed nucleus, and for this circumstance the chemical shift vector does not precess in the rotating frame, so that the multiplet magnetization vector midpoint remains along one of the coordinate axes.

Let us first examine an AX system. If the total multiplet intensity is I_0, the length of each doublet component vector is $I_0/2$. The phase angle acquired by each component of the doublet during the dephasing time interval of length Δ is $2\pi(J/2)\Delta$, and the component along ν_0 is the cosine of this angle multiplied by $I_0/2$, and in turn multiplied by an exponential decay factor describing spin–spin relaxation with effective time constant T_2^* during the period 2τ, leading to the expression

$$I(\text{AX}) = 2(I_0/2)e^{-2\tau/T_2^*}\cos 2\pi(J/2)\Delta = I_0 e^{-2\tau/T_2^*}\cos \pi J\Delta \quad [4]$$

The reader will recall from the discussion above that the direction of the magnetization vector at the start of acquisition, and therefore the sign of the spectral peak intensity, depends on whether the 180° pulse is along the x axis or is phase-shifted by 90° to be along the y axis. The reader should also note that the time between the initial 90° pulse and the refocusing 180°

1. BASIC METHODS AND SIMPLE PULSED EXPERIMENTS 21

pulse is defined as τ; thus the time between the 90° pulse and the first echo is 2τ. *We shall use the symbol Δ to represent the period of spin-spin multiplet dephasing.* In the gated-decoupler experiment, this is usually equal to τ, whereas for the proton-flip experiment, it is equal to 2τ. As mentioned above, t_1 is equivalent to 2τ in either method and is the customary notation when the pulse interval is systematically varied in a series of spectra in a 2D experiment.

Exercise: Show that for A_2X and A_3X systems, the corresponding expressions, neglecting the relaxation factor, are

$$I(A_2X) = \tfrac{1}{2}I_0(1 + \cos 2\pi J\Delta) = I_0 \cos^2 \pi J\Delta$$

$$I(A_3X) = \tfrac{1}{4}I_0(3 \cos \pi J\Delta + \cos 3\pi J\Delta) = I_0 \cos^3 \pi J\Delta$$

The generalized intensity function turns out to be

$$I(A_nX) = I_0 \cos^n \pi J\Delta \qquad [5]$$

The intensity functions for $n = 0, 1, 2$ are plotted, neglecting the effects of relaxation, in Fig. 1-11, along with the intensity for a noncoupled nucleus, which is not modulated and thus is constant. Of course, if a nucleus has only a small coupling constant, as for a typical quaternary carbon, the intensity does vary, but the rate of variation produced by long-range coupling is negligible on the time scale of the variation for typical one-bond coupling constants. Coupling to additional nuclei with intermediate J values of course complicates the intensity-time relation.

The functions shown in Fig. 1-11 may be applied in various ways to distinguish X-nucleus resonances of various multiplicities. In discussing these techniques, we shall focus on cases involving carbon as the X nucleus,

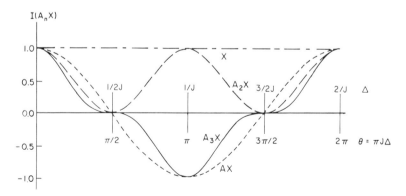

Fig. 1-11. Intensities of decoupled multiplets as a function of the product $J\Delta$, where Δ is the time of spin–spin multiplet dephasing, equal to 2τ for the proton-flip experiment. The total time scale of the plot corresponds to 16 msec for a C–H coupling constant of 125 Hz. [After Jakobsen *et al.* (40).]

TABLE 1-1

POSSIBLE TIMING SEQUENCES FOR HYDROGEN BROADBAND DECOUPLING
DURING EXPERIMENTS TO OBSERVE AN X NUCLEUS IN A SPIN-ECHO SEQUENCE

Sequence number	Decoupler status during intervals in Fig. 1-7				Resulting spectrum
	a	b	c	d	
(1)	On	On	Off	On	Decoupled, amplitude-modulated, with NOE
(2)	On	Off	On	On	Decoupled, amplitude-modulated, with NOE
(3)	On	On	On	On	Normal decoupled, with NOE
(4)	On	On	Off	Off	Coupled, each component phase-modulated, with NOE
(5)	Off	On	Off	On	Decoupled, amplitude-modulated, with minimum NOE
(6)	Off	Off	On	On	Decoupled, amplitude-modulated, with minimum NOE

The modulated spin-echo method has been extended to include the carbon-13 spectra of deuterated molecules (32). The time dependence of the intensities for $J(C-D)$ differs from that in the $J(C-H)$ case discussed above because the deuterium spin quantum number is 1. At Δ of $1/2J$ (or $\theta = 180°$), signals of CD groups have intensities of $-(1/3)I_0$; those of CD_2 groups, of $+(1/9)I_0$; and those of CD_3 groups, of $-(1/27)I_0$, permitting a distinction to be made among the three types. In another approach, the

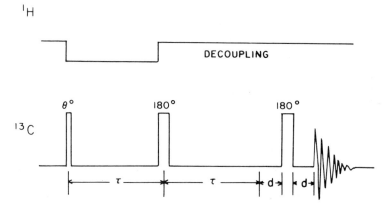

Fig. 1-13. Pulse sequence of the Patt-Shoolery "APT" experiment. The delay d is very short. The angle θ may be set to the value of the "Ernst angle" to save time in the acquisition of multiple transients.

resonances of all nondeuterated carbons can be eliminated by designing the experiment so that phases of transitions of deuterated carbons differ by 180° on successive scans; an add–subtract sequence then cumulates these signals and cancels all others (33).

It may be noted that multiplicity determination can be done in another quite similar way but without the 180° refocusing pulse, simply by delaying the start of acquisition and of decoupling for some period after an initial 90° pulse (34). However, because of chemical shift precession, there result quite large, frequency-dependent phase variations, for which correction may be difficult. Recently, Bigler (35) has proposed starting data acquisition immediately after the initial 90° excitation pulse but delaying decoupling for an appropriate period to provide multiplicity sorting of the resonances. This leads to some baseline distortion, but the Ernst angle for maximum sensitivity can be used without special consideration of the variations from 90° and there is no need to be concerned with eliminating imperfections from the refocusing pulse.

To this point, we have implicitly assumed that all one-bond X–H couplings in a molecule are about the same in magnitude, and this is not true, even for carbon. Pei and Freeman (36) showed how the difference between the values of $^1J(C-H)$ for sp^2 carbons (about 125 Hz) and sp^3 carbons (about 165 Hz) can be utilized to distinguish between the two types of methylene groups. When Δ is 3.5 msec, corresponding to $1/2J$ for the mean of these values of J, the peaks for both types of carbon atom have about 21 percent of normal intensity, and can be differentiated because they have opposite signs, while CH and CH$_3$ resonances have only a few percent of normal intensity. Assignments of carbon-13 resonances of chloroalkanes have been made on the basis of the effects of chlorine substitution in aliphatic hydrocarbons on $^1J(C-H)$, using spin-echo spectra with suitably chosen echo delay times (37).

Spin-echo multiplicity determination in the spectrum of a nucleus such as carbon-13 by utilizing the modulation produced by proton coupling has the advantage that the instrumental requirements are minimal. In the proton decoupler channel, it is only necessary to have the capability of gating the output. Phase shifts and level changes are not essential, and it is not necessary to know the pulse flip angle if the gated-decoupler method is used. Phase shifting of the observe channel is also not essential, but it may be employed to advantage if available. Patt and Shoolery (30) recommend phase cycling the first 180° pulse in their version of the experiment according to the sequence $+y$, $+y$, $-y$, $-y$, with the second 180° pulse concurrently going through the sequence $+y$, $-y$, $+y$, $-y$. More extensive phase cycling can be utilized, with the 90° pulse, the 180° pulse, and receiver each cycled through all four phases, giving a total of 64 combinations.

2. Generation of Subspectra

Another approach to multiplicity determination is by separation of the resonances into *subspectra* (38, 39), a procedure also referred to as *spectral editing*. For example, if the results for $\pi J\Delta = \pi$ (that is, $\Delta = 1/J$) are added to those for the experiment in which $\Delta = 0$, the CH and CH_3 contributions are canceled, leaving, ideally at least, only C and CH_2 resonances. If, however, one of these results is subtracted from the other, rather than added to it, only CH and CH_3 resonances appear. To avoid effects of spectrometer instabilities, it is best to alternate cycles of the two types, storing the two FIDs in different sections of memory. These FIDs may be subsequently combined to obtain the desired subspectra. Experience indicates that combining the FIDs gives better results than combining the spectra after processing.

To make a further separation of multiplicities, a value of θ of $\pi/2$ ($\Delta = 1/4J$) can be used to null all resonances but those of quaternary carbons, permitting these to be distinguished from methylenes, while CH resonances may be picked out by utilizing the fact that they are three times as intense as the CH_3 resonance for a value of θ equal to 54°44' (40). If the difference is taken between the FIDs for spectra obtained at 54°44' and at 180° − 54°44', the same methyl-methine distinction appears, and methylene and quarternary carbon resonances are suppressed.

A different version of spin-echo editing is that of Cookson and Smith (41), who wished to obtain quantitative results for carbon types in complex hydrocarbon mixtures. They measured a series of five spectra, including a "conventional" spin-echo spectrum in which the decoupler is on throughout the sequence except for a relatively long relaxation delay, as well as gated spin-echo spectra in which values of Δ were $1/4J$, $1/2J$, $3/4J$, and $1/J$. By taking particular linear combinations of these spectra, they were able, in turn, to suppress the resonances of all save one of the carbon types. For example, to obtain resonances of only methylene carbons, they used the combination

$$0.5[I(\text{conventional}) + I(1/J)] - I(1/2J)$$

where, for instance, $I(1/2J)$ represents the intensity obtained with $\Delta = 1/2J$. For various carbon multiplicities, the terms in this expression appear as follows, with I_0 denoting the intensity of the conventional spin echo:

$$C \qquad 0.5[I_0 + I_0] - I_0 = 0$$
$$CH \qquad 0.5[I_0 - I_0] - 0 = 0$$
$$CH_2 \qquad 0.5[I_0 + I_0] - 0 = I_0$$
$$CH_3 \qquad 0.5[I_0 - I_0] - 0 = 0$$

1. BASIC METHODS AND SIMPLE PULSED EXPERIMENTS

Exercise: Determine what peaks appear for the combination

$$[I(\text{conventional}) - I(1/J)] - \sqrt{2}\,[I(1/4J) - I(3/4J)]$$

From the analysis of the process of complete editing of spectra obtained by other pulse sequences, which will be discussed further in Chapter 3, it has been found best to carry out four experiments with different values of the time, such as are given in Table 1-2a, where the ideal relative intensities of X resonances of different multiplicities are also listed for each experiment (42). To obtain a fully separated spectrum, the various experimental results may be combined with the weighting coefficients in an appropriate column of Table 1-2b. The results using these coefficients are not normalized; further details of the necessary arithmetic may be found in reference (42).

In the method just described, so-called "cross talk"—weak signals from resonances of the multiplicity appearing in a spectrum intended to include only signals of another multiplicity—may appear and may confuse the interpretation. One source of cross talk is a distribution of J values within the sample. Madsen, Bildsøe, Jakobsen, and Sørensen (42) showed how these errors can be reduced so that a moderate range of J values such as

TABLE 1-2

INTENSITY FACTORS FOR EDITED SPIN-ECHO SPECTRA

(a) Ideal Relative Intensities of Peaks of Various Multiplicities as a Function of the Interval Δ

Experiment label	Δ	Multiplicity type			
		X	HX	H_2X	H_3X
A	0	1	1	1	1
B	$1/J$	1	-1	1	-1
C	$1/3J$	1	$\frac{1}{2}$	$\frac{1}{4}$	$\frac{1}{8}$
D	$2/3J$	1	$-\frac{1}{2}$	$\frac{1}{4}$	$-\frac{1}{8}$

(b) Weighting Coefficients Used to Generate Subspectra from Spin-Echo Modulated Experiments in (a)

Experiment label	Multiplicity type desired			
	X	HX	H_2X	H_3X
A	-1	-1	1	1
B	-1	1	1	-1
C	4	8	-1	-2
D	4	-8	-1	2

120–160 Hz for C–H coupling can be covered. In their method, called ESCORT, for "*error self-compensation reached by tau-scrambling*," experiments with the precession angle $\theta = \pi J \Delta$ are replaced in part by those with $2\pi - \theta$, so that experiment C becomes five scans using $\Delta = 1/3J_{av}$ plus one scan at $5/3J_{av}$ and experiment D becomes two scans at $2/3J_{av}$ plus one scan at $4/3J_{av}$. A special quaternary-only sequence consists of three scans at $1/2J_{av}$ plus one scan at $3/2J_{av}$.

In "edited" spectra of this sort, obtained by combination of the results of several experiments, other sources of extra peaks are modulation by long-range coupling, which will be considered later, and the effects, when the length of the interval Δ is varied, of differences in relaxation times of different nuclei concerned. The latter problem may be handled by keeping the time of precession of the observed nuclei fixed while varying the J-modulation period, using a sequence such as those shown in Fig. 1-14.

3. *Other Types of Spin-Echo Difference Spectroscopy*

The generation of subspectra, as described in the previous section, is a form of spin-echo *difference spectroscopy*. Spectral simplification, including elimination of some resonances and enhancement of others, can often be achieved by other versions of spin-echo difference spectroscopy which have been devised. One application is to the analysis of complex, many-line spectra of overlapping multiplets, such as observed for the proton spectra of large molecules typified by proteins (43). Bolton (44) described a procedure in which spectra of a series obtained with differing values of the delay time are added together. The modulated peaks cancel, leaving only singlets and the center lines of triplets.

In another homonuclear spin-echo experiment, Pei and Freeman (45) employed a selective 180° pulse with a length of several hundred milli-

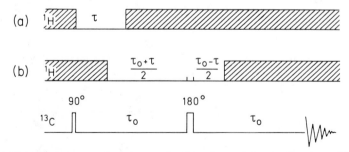

Fig. 1-14. Pulse sequences for subspectral editing by modulated spin-echo or ESCORT experiments to eliminate differential ^{13}C relaxation in the period before detection. (a), (b) Alternative patterns for gating the proton decoupler. [From Madsen, Bildsøe, Jakobsen, and Sørensen (42).]

seconds, and repeated the experiment with the radio frequency of the pulse regularly incremented through a series of steps. The difference between each FID and one obtained with the 180° pulse far off resonance was then calculated. A stacked plot of the processed spectra resembles a 2D correlation spectrum, with the irradiation frequency as one coordinate.

Yet another illustration is the simplification of proton spectra of biological tissues with accompanying improvements of sensitivity and resolution, described by Rothman and co-workers (46). Along with a spin-echo procedure, selective decoupling is applied in every other scan to the protons which are responsible for spin–spin splitting of a given multiplet, and alternate scans are subtracted. If Δ is $1/J$, modulation of a doublet or quartet leads to a negative peak (see Fig. 1-11); with decoupling to remove the modulation, the peak is positive. As a result, resonances for which the spin–spin interaction has been removed are reinforced upon subtraction, but all other multiplets see the same effect on every cycle and are eliminated by subtraction. For triplets, the two outer components can be selected in similar fashion, provided a value of Δ of $1/2J$ is used. Lohman and Bulthuis (47) have pointed out that, when lines are broad and T_2 short, as is characteristic of *in vivo* studies, sensitivity may be enhanced by using smaller values of Δ.

Homonuclear carbon-13 spin-echo modulation can serve to sort out resonances in labeled compounds (48). Singlets from carbon-13 adjacent to carbon-12, which confuse the interpretation of the normal spectrum by their presence near the middle of carbon-13–carbon-13 multiplets, can be eliminated by difference spectroscopy. Similarly, peak separation arising from chemical shift effects which occur on substitution of oxygen-18 is distinguishable from spin–spin multiplicity.

Heteronuclear spin-echo difference spectroscopy, in which the 180° refocusing pulse to the heteronucleus in the spin-flip method is applied only on every other cycle and alternative FIDs are subtracted, is also useful in picking out particular resonances, as well as in the indirect observation of other less sensitive nuclei with enhanced sensitivity by way of observing the proton spectrum. Carbon-13 satellites, which frequently are obscured by other peaks in the proton spectrum, may be uncovered by carrying out a spin-echo experiment on the protons with a delay time of $1/2J(C-H)$ and with a broad-band carbon-13 180° pulse on every other cycle—or the equivalent, alternate pulse pairs, $90°_x 90°_x$ and $90°_x 90°_{-x}$ (49). In the difference spectrum, the peaks for protons attached to carbon-12 as well as those for hydrogens attached to other nuclei such as oxygen or nitrogen are canceled. If carbon-13 decoupling is applied during acquisition, the satellites can be collapsed to a single peak (50). The observation of carbon-13-labeled metabolites has been facilitated by using a proton spin-echo sequence arranged so that those hydrogen resonances not modulated by spin–spin coupling to a directly attached carbon-13 are canceled (51, 52). The spin-echo approach has also been used to observe indirectly, via the hydrogen spectrum, phosphorus in phosphate esters (53) and nitrogen-15 labels in enzyme-catalyzed reactions (54), both experiments utilizing modulation by spin–spin coupling through two bonds, a more difficult procedure because of the small and variable values of J.

It is necessary to be concerned about the effects of homonuclear coupling within the proton system of a molecule when spin-echo difference spectroscopy is used for the indirect observation of other nuclei. There are several approaches to overcoming the effects of echo modulation from coupling between protons. One is the choice of the echo delay time so that the homonuclear modulation is exactly or very nearly exactly refocused, and another is selective decoupling of the nuclei responsible for the spin–spin interaction with the observed nucleus. A third method involves designing the 180° proton pulse to flip selectively the spins of only the protons to be observed, allowing the magnetization components from the homonuclear interaction to refocus in the normal spin-echo manner.

Those versions of spin-echo difference spectroscopy in which protons are observed as indicators of the presence of other nuclei to which they are coupled do require special instrumental arrangements. In order to apply suitable pulses or decoupling signals to the other nuclei, there must be an rf source of suitable frequency and power level, and the probe must have a coil connected to this source and tuned—or double-tuned—to the appropriate frequency.

D. THE SIGNIFICANCE OF REFOCUSING PERIODS

Because periods of the type $-\tau-180°-\tau-$ appear in a great many pulse sequences, it is important to understand clearly their effects. We summarize here the generalizations that apply to periods of this form:

1. All chemical shift precession in the rotating frame which occurs in the first τ interval is reversed in the second τ interval, and the chemical shifts of nuclei which were in phase at the beginning of the first interval are refocused at the end of the second interval.

2. Spin–spin coupling evolution is also reversed and multiplets come back into phase *if the 180° pulse does not affect the nuclei responsible* for the spin–spin splitting.

3. Spin–spin coupling evolution continues throughout the entire period (and modulates the echo) *if the 180° pulse reverses the direction of the moments of the spins responsible* for that splitting because either (a) the 180° pulse is nonselective and the spin system is homonuclear or (b) 180° pulses are simultaneously applied to both members of a heteronuclear spin system.

III. Polarization-Transfer Experiments

A. ADVANTAGES OF POLARIZATION TRANSFER

A process of polarization transfer—sometimes called population transfer—is one in which an interaction such as spin–spin coupling, dipolar coupling, or the nuclear Overhauser effect serves as the means by which magnetic order or coherence is transferred from one nuclear species to another. The order that is to be transferred may be initially present at equilibrium or it may be created by appropriate manipulations by the experimenter. In this section we shall deal primarily with transfers by means of spin–spin interactions.

One purpose of polarization transfer via spin–spin coupling is enhancement of the intensity of the resonance of a nuclear species of low magnetogyric ratio. Typically the polarizing nucleus is hydrogen, fluorine, or phosphorus, which has a relatively high magnetogyric ratio, and the nucleus for which the intensity is enhanced is one of the "less susceptible" variety, such as carbon-13 or nitrogen-15 or silicon-29. For example, the Boltzmann distribution leads to a population difference between the two spin states of a hydrogen nucleus which is four times as large as the population difference between the two states of a carbon-13 nucleus. If the two kinds of nucleus interact in such a way that the energy of the carbon-13 nucleus depends on the spin states of hydrogen nuclei, the four-fold larger population difference of the hydrogen nuclei can be transferred to the carbon nuclear system. Assuming that the transfer is fully effective and is not offset by rapid relaxation of the carbon-13 nuclei, there results an intensity enhancement of carbon transitions by a factor of the order of four, and consequently a sixteen-fold reduction in the number of repetitions required in a time-averaging experiment. For carbon-13, this benefit is not tremendously greater than that obtained from the NOE, but the corresponding advantage is much greater for nitrogen-15 or silicon-29 or silver-109, for which the negative magnetogyric ratio may lead to disastrous consequences if circumstances are such that the NOE enhancement does not reach its maximum possible value because the relaxation of the nucleus is only partially dipolar.

An auxiliary benefit associated with a typical polarization-transfer experiment, such as that from hydrogen to nitrogen-15, is that successive cycles in a time-averaged sequential pulse experiment may be repeated with relaxation delays based on the relaxation time of the polarizing nucleus, rather than on that of the nucleus being observed. Relaxation times of carbon, nitrogen, or silicon are frequently much longer than those for hydrogen, often reaching several minutes when there is no directly attached proton. This advantage does fall off, however, for large molecules, where longer correlation times may make T_1 values for carbon-13 rather short.

A third aspect of the utilization of polarization transfer is its frequent function as an indicator that the nuclear species involved are related in such a way that coherence can be transferred. Thus, experiments may be designed to determine that the polarizing and recipient species are spin–spin coupled, or that the two nuclear species are in close enough proximity to one another to permit a nuclear Overhauser enhancement of one by the other. In a solid or in an oriented liquid, dipolar interactions may also be investigated.

Polarization transfer has been of particular importance in the study of less abundant or low-gamma nuclei in solids where these factors combine with very long T_1 values to make observation difficult. In fact, some of the methods used in liquid-phase NMR derive from ideas first developed for solids (55).

Certain versions of polarization transfer may be used to determine the signs and the precise magnitudes of coupling constants. The technique also affords alternatives to the spin-echo procedures for separating resonances according to their multiplicity, which were described in the preceding section. In 2D spectroscopy, as we shall see later, polarization transfer is often the basis for experiments such as those that determine which resonances of one nucleus, such as carbon-13, are coupled to particular resonances of another nucleus, such as hydrogen. Homonuclear as well as heteronuclear polarization transfer experiments can be designed to emphasize resonances correlated with a particular transition that has been irradiated.

There is an important limitation on the application of polarization transfer, and that is that the spin states of the nuclei involved must be preserved for a reasonable length of time. For instance, difficulties are encountered in polarizing nitrogen-15 nuclei by hydrogen atoms which are attached to the nitrogen if these atoms are undergoing rapid exchange.

B. Selective Polarization Transfer (SPT)

1. *The AX Spin System*

In the first polarization transfer technique we shall describe, a specific transition in the spectrum of the polarizing nuclear species is chosen for pulsed irradiation and the polarization of the receiving nucleus is limited to those transitions having spin-state energy levels in common with the irradiated transition (56). We illustrate this by considering an AX spin system, with the energy-level diagram as shown in Fig. 1-15, drawn to the scale appropriate to the CH moiety. The spin state designated $\alpha\beta$, for example, is that in which the carbon is in state α and the hydrogen is in

1. BASIC METHODS AND SIMPLE PULSED EXPERIMENTS   33

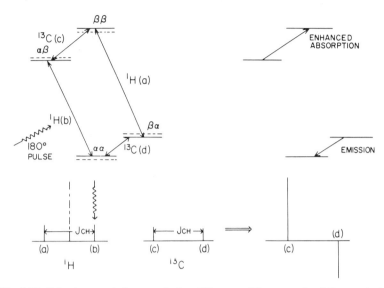

Fig. 1-15. Selective population transfer in a CH system. The energy-level diagram is shown at the upper left, with dashed lines representing levels in the absence of spin–spin coupling and solid lines displaced from the dashed lines, on an exaggerated scale, to represent the effects of coupling. In the lower left is the normal spectrum. The right side of the figure shows the spectral effects resulting from the transfer of population when hydrogen transition b is subjected to a 180° pulse.

state β. The dotted lines indicate the energy levels in the absence of spin–spin coupling; the solid-line energy levels are displaced from the dotted-line levels in a direction corresponding to a positive value of $J(CH)$.

We now suppose that a 180° *selective* pulse, indicated in the figure by a wavy arrow, is applied in the proton spectrum at the position of the carbon-13 satellite corresponding to the transition marked (b). A selective pulse is one that is sufficiently long and weak that it perturbs one line of the proton spectrum without appreciably affecting the other transitions. In a typical spectrometer, this pulse can be obtained from the proton decoupler with the power level 50 to 70 dB below the full output of 10 to 20 W; the pulse length is usually somewhere between a few tenths of a second and several seconds. In a spectrometer equipped for gated decoupling, the decoupler can simply be turned on at the appropriate level during the relaxation interval between cycles, with the delay time appropriately adjusted to give a 180° pulse at the power level being used. Some instrumental software includes provision for a "presaturation" period, and this may be used as the interval for the polarization pulse if one wishes the relaxation recovery delay to be longer than the time for the 180° pulse. Two cycles of a ^1H-^{13}C polarization transfer experiment are represented in Fig. 1-16.

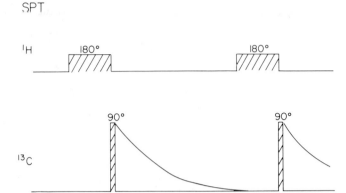

Fig. 1-16. Time sequence of two transients in the selective polarization transfer (or selective population transfer) experiment.

The effect of the 180° (or π) pulse described in the last paragraph is to reverse the magnetization direction and interchange the spin states of all the hydrogen nuclei which have the desired chemical shift and which are coupled to a carbon atom that is in the α spin state. The systems that were formerly in the lower-energy $\alpha\alpha$ state are now in the higher-energy $\alpha\beta$ state, and those that were in the $\alpha\beta$ state are in the $\alpha\alpha$ state. Since the intensity of a transition depends on the excess of population of the initial state over that of the final state, as long as the observe pulse does not mix the spin states of coupled homonuclei, increasing the $\alpha\beta$ population enhances the intensity of transition (c), as shown in the spectrum in the lower right of Fig. 1-15. In contrast, the population of the $\alpha\alpha$ state is decreased below that of the $\beta\alpha$ state and an emission line is observed for transition (d). Notice that a nonselective pulse, one applied equally to satellites (a) and (b), would not alter the relative intensities in the carbon-13 spectrum. However, we should note that relative intensities in more complex spin systems may indeed be affected by a nonselective pulse.

Exercise: Describe qualitatively the nature of the carbon-13 spectrum that would be observed after a 180° pulse is applied to satellite (a) in the hydrogen spectrum of Fig. 1-15.

In order to predict quantitatively the relative intensities in the SPT spectrum of a CH system, let us consider in more detail the populations of the levels involved, referring to Fig. 1-17. We take the zero of energy to be at the vertical center of the diagram, and express the populations of the various levels in terms of the energy differences of those levels from the arbitrary zero. The ratio of population of a level to that of a hypothetical reference level at zero energy is, at equilibrium, equal to $\exp(-\Delta\varepsilon/kT)$, where $\Delta\varepsilon$ is the energy of the given level above the reference level. However,

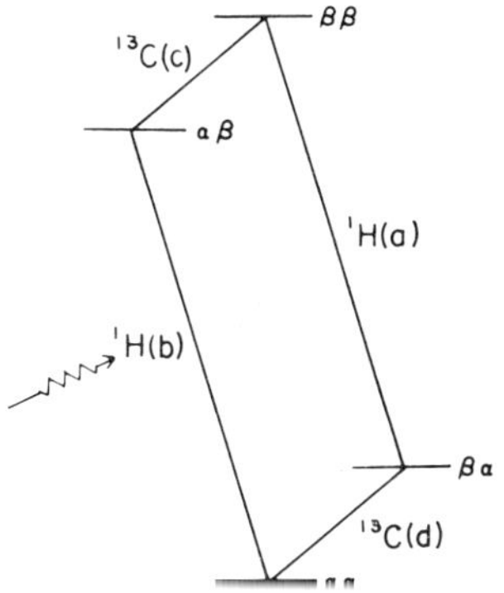

1. BASIC METHODS AND SIMPLE PULSED EXPERIMENTS   35

		State population		Relative carbon intensity	
		Normal	After 180° pulse	Normal	After 180° pulse
^{13}C(c) ββ, αβ		$-\frac{5}{2}\delta$ $-\frac{3}{2}\delta$	$-\frac{5}{2}\delta$ $+\frac{5}{2}\delta$	δ	$+5\delta$
^{13}C(d) βα, αα		$+\frac{3}{2}\delta$ $+\frac{5}{2}\delta$	$+\frac{3}{2}\delta$ $-\frac{3}{2}\delta$	δ	-3δ

Fig. 1-17. Populations of the several states of a CH system at equilibrium and after a 180° SPT pulse to hydrogen transition b. The intensity of a transition is proportional to the difference in population between the states it connects. The value of δ is $h\nu_C/2\pi$.

since the value of $\Delta\varepsilon = h\nu$ is much less than kT, we may expand the exponential as a series $1 - (h\nu/kT) + \frac{1}{2}(h\nu/kt)^2 + \cdots$ and neglect all but the first two terms. The populations may then be expressed as deviations from equilibrium in multiples of $h\nu_C/kT$, using $h\nu_H/kT$ as four times $h\nu_C/kT$, the latter represented by δ.

Each carbon transition represents, in these units, an energy interval of 1δ with a corresponding population difference, while each hydrogen transition represents an energy difference of 4δ and a proportionately large population difference. The total energy difference between the lowest level and the highest level is 5δ, and thus each extreme level is 2.5 units from our arbitrary zero level. Small energy differences associated with spin-spin splitting are neglected in estimating populations, but they are, of course, responsible for separation of the lines in the spectral multiplets.

Following the application of the 180° pulse to transition (b), the populations of levels $\alpha\alpha$ and $\alpha\beta$ are interchanged. The intensity of transition (c) is proportional to the population excess in the lower level involved in that transition over that in the upper level, which is $+(5/2)\delta - (-5/2)\delta = 5\delta$, and the intensity of transition (d) is proportional to the population excess $(-3/2)\delta - (3/2)\delta = -3\delta$. These populations and intensities are indicated in Fig. 1-17.

An alternative way of representing an SPT experiment is in terms of a vector diagram depicting the manner in which the magnetization vectors corresponding to the various spin states of the AX system undergo changes, as shown in Fig. 1-18. At equilibrium, state (a) in the figure, the hydrogen

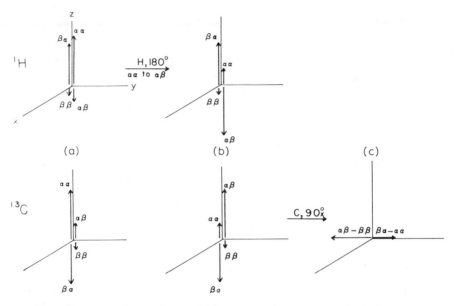

Fig. 1-18. Vector diagram for the SPT experiment, showing the relative direction of magnetization corresponding to each spin state of the nuclei involved (a) at equilibrium, (b) after the 180° pulse, and (c) after the 90° read pulse. The important characteristic of polarization transfer experiments is that antiphase magnetization of the polarizing nucleus is first generated and is then transferred to antiphase magnetization of the polarized nucleus.

magnetizations of states $\alpha\alpha$ and $\beta\alpha$ are in the positive z direction and those of the systems in the $\alpha\beta$ and $\beta\beta$ states are oriented in the negative z direction. At the same time, the carbon magnetizations of the two states with α spin for the carbon, the $\alpha\alpha$ and the $\alpha\beta$ states, are in the positive direction, whereas those for the $\beta\alpha$ and $\beta\beta$ states are in the negative z direction. The intensity of a transition between two states is proportional to the difference in populations between those states, represented in the figure by the difference in length of the vectors which are directly above and below one another.

The effect of the 180° pulse on the hydrogen transitions from $\alpha\alpha$ to $\alpha\beta$ is to reverse the directions of hydrogen magnetization for the atoms which are in the two states involved in this transition, so that the $\alpha\alpha$ level and the $\alpha\beta$ level interchange populations. The effect on the carbon-13 magnetization is indirect, representing an interchange of the labels on the carbons in states $\alpha\alpha$ and $\alpha\beta$, since the hydrogens coupled to these carbons have switched states, together with a corresponding interchange of spin-state population. In Fig. 1-18c, a 90°(C) observe or "read" pulse along the x axis has turned the magnetization vectors into the xy plane. The spectrometer now senses the $\beta\alpha - \alpha\alpha$ and the $\alpha\beta - \beta\beta$ resultants with magnitudes proportional to the differences in length of the corresponding vectors in the diagram and with phases differing by 180°, so that if one of these resonances is positive in the spectrum, the other is negative.

In any type of polarization transfer experiment which involves a spin–spin coupled AX system, *such as the INEPT sequence* to be described later, the process involved is similar: the magnetizations corresponding to the two transitions of the polarizing nucleus are brought into an antiphase relationship to one another, and this relationship is then transferred to the nucleus to be polarized. However, we must realize that there is *no net enhancement*, no increase of the total resultant magnetization of the polarized nuclear species.

2. Degenerate Systems

For degenerate systems, that is, systems in which the observed nucleus is equally coupled to two or more isochronous nuclei, selective population transfer experiments may lead to substantial increases in resonance intensity, as demonstrated by Jakobsen and co-workers (57). An example is the carbon in a methyl group, as in methyl iodide. Figure 1-19 is the energy-level diagram of a CH_3 system, with populations represented on a scale similar to that used in Fig. 1-17. Because irradiation at the carbon-13 resonance frequency does not change the spin state of the hydrogens, the diagram can be divided into two portions, one corresponding to a total spin of $\frac{3}{2}$ and the other to a total spin of $\frac{1}{2}$ for the group of three hydrogen nuclei. Since the spin-$\frac{1}{2}$ levels have a statistical weight of 2, the overall diagram turns out to be consistent with the 1:3:3:1 intensity ratio of the lines produced by coupling to any X_3 unit comprised of spin-$\frac{1}{2}$ nuclei. As in Fig. 1-17, this diagram again does not show the small spin–spin energy effects that differentiate the component transitions in the carbon-13 or proton multiplets.

The equilibrium populations expressed in terms of $\delta = h\nu_C/kT$ have again been calculated on the basis that the reference level is located at the vertical center of the diagram. Consider first the relative intensities in the normal carbon spectrum, corresponding to the six transitions labeled C1 to C4, of which C2 and C3 are degenerate. For line C1, the intensity is proportional to the population excess in the lower level, $-(11/2)\delta - [-(13/2)\delta]$ or simply δ. Likewise, the intensity of C4 is proportional to δ. For C2 there is a contribution from the $I(H_3) = \frac{3}{2}$ system equal to $-(3/2)\delta - [-(5/2)\delta]$ or δ, plus a double-weight contribution from the $I(H_3) = \frac{1}{2}$ system equal to 2δ, for a total of 3δ.

A 180° pulse is now applied to the carbon-13 satellite which corresponds to the transitions H1. This has the effect of reversing the spin directions of

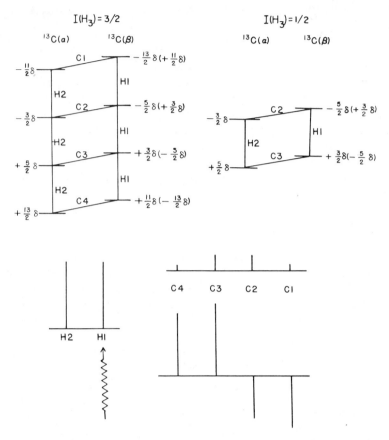

Fig. 1-19. Population diagram for SPT in a methyl group. For simplicity, the effects of spin–spin coupling have not been shown in the energy levels. Relative populations at equilibrium are shown for the several energy levels in terms of $\delta = h\nu_C/2\pi$ as differences from the average value. Numbers in parentheses show the relative populations after a 180° pulse is applied to the low-frequency carbon-13 satellite in the proton spectrum as indicated at the lower left. The spin states corresponding to a total spin of $\frac{1}{2}$ for the three hydrogens, shown at the upper right, have twice the weight of the states with total spin of $\frac{3}{2}$, shown at the upper left. At the lower right, the appearance of the SPT spectrum is compared with the appearance of the normal 1:3:3:1 quartet.

all hydrogen nuclei attached to those carbons which have β spin. The new populations of the levels are shown in parentheses. The relative intensities of the carbon transitions observed following the selective proton pulse are again proportional to the excess in population in the lower level over that in the upper level, assuming no relaxation has occurred. For C1, the lower level is deficient in population by 11δ compared to the upper level, so that an emission line of relative intensity 11 is observed. For C4, the population in the lower state is 13δ greater than in the upper state, leading to an absorption line intensity of 13 units. C2 has three components, one for hydrogen spin $\frac{3}{2}$ and two for hydrogen spin $\frac{1}{2}$, each with a population deficiency 3δ, yielding an emission line proportional to 9δ. Finally for C3, there is an absorption line of 15 units of intensity. Therefore, the intensity ratio of the carbon-13 quartet is $13:15:-9:-11$. If the numbers in the usual intensity ratio, $1:3:3:1$, are subtracted from the corresponding numbers in the SPT ratio, the result is $12:12:-12:-12$. Thus, if the experiment is started with the populations of the four levels equal, that is with the carbons saturated instead of at equilibrium, there is obtained a group of four lines of equal amplitude.

A general calculation of intensities within multiplets following heteronuclear polarization transfer, assuming magnetization initially at equilibrium, leads to the triangle shown in Table 1-3a, analogous to Pascal's triangle, which is normally applied to multiplet intensities. In this diagram, the relative intensities are expressed in terms of Γ, the ratio of the magnetogyric ratio of the polarizing nucleus to that of the polarized nucleus. Versions specifically applicable to carbon-13 (57), nitrogen-15, and silicon-29 (58), each polarized by hydrogen, are also included in Table 1-3, employing Γ values of 4, 10, and 5, respectively. In Fig. 1-20, a coupled silicon-29 spectrum obtained by the SPT method is compared with the corresponding spectrum obtained by normal methods without NOE.

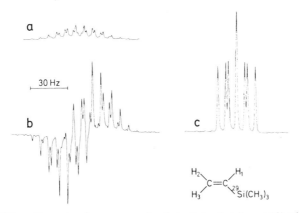

Fig. 1-20. Silicon-29 spectra, four scans each, of vinyltrimethylsilane, 75% v/v in acetone-d6. (a) Normal proton-coupled spectrum without NOE. (b) SPT-enhanced coupled spectrum with polarization transfer from the trimethylsilyl protons, using $\gamma B_2/2\pi = 2.5$ Hz, $\tau = 0.2$ sec, and applying the SPT 180° pulse to the high-frequency ^{29}Si satellite of the methyl resonance in the proton spectrum. (c) SPT-enhanced selectively decoupled spectrum obtained using the pulse sequence shown in Fig. 1-24b, with $\tau_1 = 16.3$ msec, $\tau_2 = 8.7$ msec, and a selective decoupling amplitude of 62 Hz. [From Jakobsen et al. (66).]

TABLE 1-3

THEORETICAL INTENSITIES OF MULTIPLET COMPONENTS RESULTING FROM POLARIZATION TRANSFER FROM SEVERAL EQUIVALENT NUCLEI OF SPIN $\frac{1}{2}$ TO A SINGLE NUCLEUS OF SPIN $\frac{1}{2}$[a]

(a) General intensity ratios, expressed in terms of Γ, the ratio of magnetogyric ratio of the polarizing nucleus to that of the polarized nucleus

$$
\begin{array}{ccccccccc}
 & & & & 1 & & & & \\
 & & & 1-\Gamma & & 1+\Gamma & & & \\
 & & 1-2\Gamma & & 2 & & 1+2\Gamma & & \\
 & 1-3\Gamma & & 3(1-\Gamma) & & 3(1+\Gamma) & & 1+3\Gamma & \\
1-4\Gamma & & 4(1-2\Gamma) & & 6 & & 4(1+2\Gamma) & & 1+4\Gamma \\
1-5\Gamma & 5(1-3\Gamma) & & 10(1-\Gamma) & & 10(1+\Gamma) & & 5(1+3\Gamma) & 1+5\Gamma \\
1-6\Gamma & 6(1-4\Gamma) & 15(1-2\Gamma) & & 20 & & 15(1+2\Gamma) & 6(1+4\Gamma) & 1+6\Gamma \\
\end{array}
$$

(b) Carbon-13; $\Gamma = 4$

$$
\begin{array}{ccccccccc}
 & & & & 1 & & & & \\
 & & & -3 & & 5 & & & \\
 & & -7 & & 2 & & 9 & & \\
 & -11 & & -9 & & 15 & & 13 & \\
-15 & & -28 & & 6 & & 36 & & 17 \\
-19 & -55 & & -30 & & 50 & & 65 & 21 \\
-23 & -90 & -105 & & 20 & & 135 & 102 & 25 \\
\end{array}
$$

(c) Silicon-29; $\Gamma = 5$

$$
\begin{array}{ccccccccc}
 & & & & 1 & & & & \\
 & & & -4 & & 6 & & & \\
 & & -9 & & 2 & & 11 & & \\
 & -14 & & -12 & & 18 & & 16 & \\
-19 & & -36 & & 6 & & 44 & & 21 \\
-24 & -70 & & -40 & & 60 & & 80 & 26 \\
-29 & -114 & -135 & & 20 & & 165 & 126 & 31 \\
\end{array}
$$

(d) Nitrogen-15; $\Gamma = 10$

$$
\begin{array}{ccccccccc}
 & & & & 1 & & & & \\
 & & & -9 & & 11 & & & \\
 & & -19 & & 2 & & 21 & & \\
 & -29 & & -27 & & 33 & & 31 & \\
-39 & & -76 & & 6 & & 84 & & 41 \\
-49 & -145 & & -90 & & 110 & & 155 & 51 \\
-59 & -234 & -285 & & 20 & & 315 & 246 & 61 \\
\end{array}
$$

1. BASIC METHODS AND SIMPLE PULSED EXPERIMENTS   41

TABLE 1-3—continued

(e) General intensity ratios, expressed in terms of Γ, if the natural polarization is eliminated

			0			
		$-\Gamma$		Γ		
	-2Γ		0		2Γ	
	-3Γ	-3Γ		3Γ	3Γ	
-4Γ	-8Γ		0		8Γ	4Γ
-5Γ	-15Γ	-10Γ		10Γ	15Γ	5Γ
-6Γ	-24Γ	-30Γ	0	30Γ	24Γ	6Γ

[a] Each diagram is equivalent to the familiar Pascal triangle for spin–spin multiplet relative intensities.

The question may now be asked: what is the quantitative advantage to be expected from polarization transfer in a degenerate system? For nuclei such as nitrogen-15 and silicon-29 with a negative magnetogyric ratio and for which the NOE may very well be unfavorable, it seems justified to make a comparison with nonenhanced spectra that are obtained without NOE. Qualitatively, it is clear that the greatest advantage appears in the outer lines: for the lowest rows in the diagrams of Table 1-3, corresponding to the case of polarization by six hydrogens as in two methyl groups, the multiplicative factor for the outermost lines is $1 \pm 6\Gamma$; for a trimethylsilyl or *tert*-butyl unit, the factor is $1 \pm 9\Gamma$, again the largest for any component of the multiplet. The outer lines are likely to be lost in noise in a normal spectrum, so that the chance of recognizing the true multiplicity is much better in a polarization transfer experiment.

After polarization by nine hydrogens, as in the trimethylsilyl group, the most intense lines, the fourth and seventh in the ten-line pattern, have multiplicative factors of $1 \pm 3\Gamma$, which for silicon-29 amounts to $+16$ and -14. However, a more significant comparison is of these lines with the most intense lines in the normal coupled spectrum; this ratio is approximately 10. If one sums the line intensities as if the spectrum were being decoupled, with negative peaks made positive and all peaks collapsed to a single line—we will see later how this is done—the ratio of intensity with polarization transfer to that without is 12.3. Full NOE for silicon would correspond to an intensity of -1.5 times the normal peak intensity. Some experimental results for silicon-29 experiments are shown in Table 1-4.

It is possible to transfer polarization from a nucleus of smaller magnetogyric ratio to one of larger ratio. Polarization of nitrogen-14 in methylisocyanide, a molecule in which the nitrogen-14 resonance is relatively narrow, has been transferred to the isocyanide carbon (59). Despite the fact

TABLE 1-4

OBSERVED SIGNAL/NOISE VALUES FOR SILICON-29 SPECTRA

Compound	Conditions	Individual multiplet peaks, coupled spectrum										Decoupled spectrum	
		1	2	3	4	5	6	7	8	9	10	With NOE	Minimum NOE
$(CH_3)_3SiOSi(CH_3)_3$	No irradiation	—	9	36	83	127	129	83	35	8	—		548
	With NOE	—	—	15	35	54	55	36	16	—	—	251	
	With SPT	−8	−85	−256	−386	−183	187	374	260	85	8		
$(CH_3)_3SiOCH_2CH_3$	No irradiation	—	—	39	81	124	128	81	39	—	—		2180
	With NOE	—	—	31	67	108	106	67	31	—	—	2040	
	With SPT	36	261	774	1100	546	−571	−1140	−798	−274	−36		

that $\gamma(^{13}C)/\gamma(^{14}N)$ is 3.5, appreciable intensity enhancement in one of the carbon-13 multiplet components was observed, in part because of the degeneracy associated with the nitrogen quantum number of 1. Rapid relaxation might be expected to defeat the effects of population transfer during the relatively long period needed for selective excitation, but calculations by Bildsøe (60) have shown that the 180° pulse experiment gives generally better inversion than a continuous irradiation experiment, regardless of relaxation rates or magnitude of J.

3. *Some Extensions and Applications of SPT*

In the SPT experiment, the cycle rate for accumulation of successive spectra is governed by the relaxation time of the polarizing nuclear species, rather than by that of the observed species. There may thus be considerable time advantage gained if the relaxation time of the polarizing species is relatively short. Indeed, it is not necessary for the polarizing species to return completely to equilibrium between cycles. Pachler and Wessels (61) have suggested what they term a "progressive saturation" SPT experiment, using a relaxation delay of 1.25 times T_1 of the irradiated nucleus, chosen to maximize the signal obtained in a fixed time. Since oxygen affects substantially the T_1 of protons, degassing the sample may lengthen the relaxation time. This reduces the sensitivity enhancement of the progressive saturation method but permits the polarization pulse to be longer and therefore more selective, without as much loss of proton polarization by relaxation during the pulse. Combination of this method with the conventional SPT method enabled the evaluation of small long-range proton couplings in carbon-13 spectra (62).

Bolton and James (63) have pointed out another aspect of SPT experiments which may be useful. If the resonance of the nucleus observed is broad because of an inhomogeneous magnetic field, the irradiation bandwidth can be kept narrow and selective enough to invert only those nuclear spins in locations at a certain value of the field. Consequently, only nuclei located at points having very nearly that particular value of the field will be polarized, and the observed linewidth will be governed by the exciting linewidth, rather than by the field inhomogeneity.

Difference methods can be utilized in heteronuclear SPT spectroscopy, somewhat as described above for spin-echo experiments, with the inversion pulses applied only on every other scan and with the FIDs from alternate scans subtracted in the computer memory. This eliminates resonances which have not been polarized, including those from impurities, solvent, and foldbacks, and thus aids, for example, in picking out protonated species from a deuterated-solvent background (64). The difference procedure (65) also eliminates "native" magnetization present if the experiment is started with the spin system at equilibrium. A method which is more effective, although a bit more complex to set up, involves applying the 180° pulse alternately to low-frequency and to high-frequency satellites, again with subtraction of every other FID (66). In this approach, every cycle contributes to the signal, as compared to the simpler procedure in which those cycles without an inversion pulse cancel unwanted signals but add nothing to the desired signal.

The precision with which satellite positions can be located and the use of long-range coupling interactions for polarization are illustrated by an experiment carried out in the author's laboratories. The molecule involved is $CF_3CCl=CCl_2$, for which carbon spectra, obtained on an XL-100 spectrometer at 25 MHz in a 12-mm sample tube, are shown in Fig. 1-21. The sample consists predominantly of four isotopomers, one of these having three carbon-12 nuclei and each of the others having a carbon-13 nucleus in one of the three possible positions. Thus the fluorine resonance has three pairs of carbon-13 satellites, corresponding, in order of increasing magnitude of the splitting, to three-bond, two-bond, and one-bond spin–spin coupling. After manual adjustment of the fluorine-19 irradiation frequency to locate both satellites in each of the three pairs, as judged by the frequency which yields maximum carbon enhancement, it is possible to calculate the exact fluorine chemical shift in each isotopomer. Comparison of the results, presented in Table 1-5, for the several species shows that there are substantial one-bond and two-bond isotope effects, but no three-bond shift beyond the experimental error.

C. Spectral Analysis by Selective Polarization Transfer

1. Qualitative Spectral Analysis

As examples of the use of SPT experiments, we will cite first several qualitative applications, both heteronuclear and homonuclear, in spectral peak assignment. Kakinuma and co-workers (67) identified carbonyl carbon-13 resonances in a complex antibiotic molecule by polarization from methyl-group protons. Li, Johnson, Gladysz, and Servis (68) confirmed the presence of some trimethylsilylated metal complexes and obtained silicon-29 shifts with improved accuracy by polarization transfer to silicon from the methyl hydrogens in $(CH_3)_3Si-$ units of the complexes. Hallenga and Hull (69) used SPT following the Redfield pulse sequence for water suppression to determine connectivities between amide and alpha protons in peptides in aqueous solution, irradiating the alpha protons and observing the NH

1. BASIC METHODS AND SIMPLE PULSED EXPERIMENTS   45

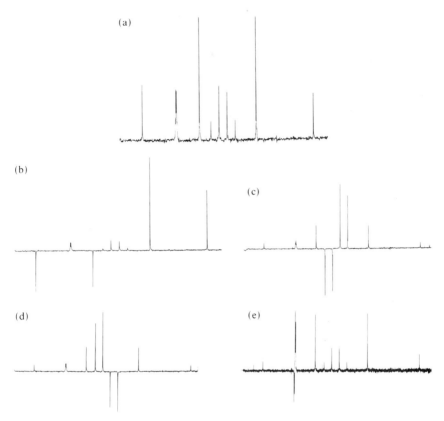

Fig. 1-21. ^{13}C spectrum, at 25 MHz, of $CF_3CCl=CCl_2$. (a) The normal spectrum, consisting of three overlapping quartets. (b) SPT spectrum with the 180° pulse applied to the high-frequency one-bond ^{13}C satellite in the ^{19}F spectrum. (c) SPT spectrum with the 180° pulse applied to the high-frequency two-bond satellite in the ^{19}F spectrum. (d) SPT spectrum with the 180° pulse applied to the low-frequency two-bond satellite. (e) SPT spectrum with the 180° pulse applied to the high-frequency three-bond satellite.

region. SPT has the advantage, over homonuclear proton decoupling methods, that the water resonance is only very weakly excited by the relatively lower irradiation power level, and, over presaturation methods, that the delay between cycles can be much shorter.

2. Determination of Signs of Coupling Constants

Selective population transfer can be used to establish connectivities of transitions in first-order spin systems and thus to determine relative signs of spin–spin coupling constants in AMX or in more complex spin systems. One of the first compounds to be investigated was 2,3-dibromothiophene

(65, 70). In the normal proton-coupled carbon-13 spectrum of this molecule, C3 appears as a doublet of doublets because of coupling to H4 and H5, as shown in Fig. 1-22a. In the proton spectrum itself, each of the carbon-13 satellites of H5 is a doublet because of coupling to H4, as shown in Fig. 1-22e. When a 180° pulse is applied to the H5 satellite line of the highest frequency, A in Fig. 1-22e, the C3 multiplet is perturbed as shown in Fig. 1-22b. Because relaxation is relatively rapid, none of the lines becomes negative, but one is reduced in intensity because of the negative-going

TABLE 1-5

DETERMINATION OF CARBON-13 ISOTOPE EFFECTS ON FLUORINE-19 CHEMICAL SHIFTS IN 1,2,2-TRICHLORO-3,3,3-TRIFLUOROPROPENE-1 BY FLUORINE–CARBON POLARIZATION-TRANSFER EXPERIMENTS[a]

Isomer	J_{CF} (Hz)	Carbon shift (ppm from TMS)	ν_F (± 0.03 Hz)
$^{13}CF_3$, Cl, C=C, Cl, Cl	274.04	120.87	70,720.64
CF_3, Cl, ^{13}C=C, Cl, Cl	39.38	122.12	70,732.00
CF_3, Cl, C=^{13}C, Cl, Cl	2.32	130.53	70,732.95
CF_3, Cl, C=C, Cl, Cl	—	—	70,732.95

[a] The frequencies in the last column represent, on an arbitrary scale, the relative resonance frequencies of the four isomers, determined for the first three as the average of the two satellite irradiation frequencies corresponding to maximum polarization transfer.

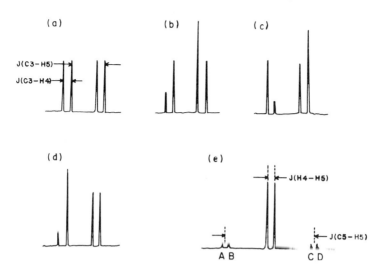

Fig. 1-22. Schematic representation of carbon and hydrogen spectra of 2,3-dibromothiophene. (a) Normal ^{13}C spectrum of carbon 3. (b) Resonance of C3 after SPT pulse is applied to highest-frequency satellite of H5. (c) Resonance of C3 after the SPT pulse is applied to the second-highest-frequency satellite of H5. (d) Resonance of C3 after the SPT pulse is applied to the highest-frequency satellite of H4. (e) Proton spectrum of H5. Intensities of the resonances are not to scale. [Adapted from Sørensen, Hansen, and Jakobsen (70).]

polarization effect. The result that a high-frequency line becomes *less* positive and a low-frequency line separated from it by the coupling to the nucleus which is irradiated becomes *more* positive is parallel with the situation for an AX system, which was described above. The behavior, both of this system and of the AX system, illustrates the general rule: negative enhancement of a peak occurs in the same half, to lower or higher frequency, of the splitting pattern caused by the irradiated nucleus as the frequency half—*with respect to the corresponding splitting*—on which the polarizing nucleus is irradiated. Conversely, positive enhancement occurs for peaks in a relative frequency direction opposite to that of the irradiation.

However, we have in the present example additional information: of each H4 doublet in the C3 pattern, it is the high-frequency line which is modified in intensity. We can now compare the frequency sense of the intensity perturbation observed in the pattern of nucleus C3 with respect to the splitting produced by the indifferent nucleus H4 with the frequency sense, again with respect to the H4 splitting, of the selective irradiation applied to nucleus H5. If the intensity effect appears on the same side of the H4 splitting as that of the irradiation, as found in this sample, the coupling constant of the observed nucleus to the indifferent nucleus has the same sign as that of the irradiated nucleus to the indifferent nucleus. We may express this for 2,3-dibromothiophene as $J(H4-H5) \times J(H4-C3) > 0$. This experiment is equivalent to the determination of which transitions share a common spin state, just as would be done in a tickling experiment.

If the second-highest-frequency satellite line (B) of H5 is irradiated, the result for C3 is as shown in Fig. 1-22c. Since the spin state of H4 which is involved has now been changed, the effect appears on the *low-frequency* side of each H4 doublet. However, the irradiation of H5 is at the high-frequency side of the C3 doublet, so that the positive-negative sense of the *intensity* change is the same as in the previous experiment. It is important to notice that the two coupling constants compared are always those of an unaffected nucleus—a "third party" which is coupled to both the irradiated and observed nuclei—to the irradiated nucleus and to the observed nucleus.

Exercise: Predict the intensity perturbation of the C3 multiplet of 2,3-dibromothiophene in each of two SPT experiments in which (a) peak C or (b) peak D, respectively, of the H5 satellites is irradiated.

Exercise: What conclusion can be drawn about relative signs of coupling constants from the result shown in Fig. 1-22d for the irradiation of the highest frequency satellite of the H4 resonance?

Exercise: Consider an AMX spin system with $J(AM) = -12$ Hz, $J(AX) = +4$ Hz, and $J(MX) = +20$ Hz. Construct a cube-shaped energy-level diagram for this system, labeling the energy levels $\alpha\alpha\alpha$, $\alpha\alpha\beta$, and so on, and calculate the effect of spin-spin coupling on the energy of each level. Draw spectra for A, M, and X nuclei and assign a spectral line to each transition on the diagram. Apply the rules above to the effects of inverting the populations (a) of the highest-frequency A transition and (b) of the lowest-frequency M transition and show that the predictions are consistent with the population changes in the various energy levels.

Examples of the application of heteronuclear SPT in the analysis of spectra and in the determination of relative signs of coupling constants include studies of tellurium compounds (71), of various proton–carbon couplings including long-range interactions (72), of silicon-29–proton coupling (58, 66), of fluorine-19–nitrogen-15 coupling (73), and of proton–nitrogen-15 coupling (74–76).

J. P. Jacobsen has used SPT methods in the investigation of nuclear coupling in liquid crystals, polarizing either carbon-13 (77) or deuterium (78) by proton irradiation. For deuterium, both quadrupole splitting and dipolar coupling influence the results; by the use of short observe pulses to avoid mixing effects, the deuterium quadrupole splitting and the hydrogen-deuterium dipolar coupling were deduced to be of opposite sign in

CHDCl$_2$. In oriented ^{13}CH$_3$COOH, the relation between the sign of the C–H dipolar interaction and that of the H–H dipolar interaction was obtained.

3. The Flip-Angle Effect

There are several circumstances under which polarization transfer experiments for analysis may be complicated by results more involved than those which have been described. Of course, if the polarizing nucleus is part of a strongly coupled spin system, then the irradiation cannot be truly selective (79). In addition, there are complexities when the observe pulse acts upon other nuclei which are coupled to the nucleus being polarized. These complications illustrate the flip-angle effect, following the rule that, *when a nonselective read pulse is applied to a homonuclear spin system in a nonequilibrium state, the intensities of the lines obtained depend on the flip angle of that read pulse.*

In the present context, the flip-angle effect is particularly significant for polarization transfer between nuclei of the same species, as where both polarizing and polarized species are hydrogen (80, 81), but it is also encountered when molecules containing pairs of coupled nuclei such as carbon-13 atoms (82, 83) or nitrogen-15 atoms (84), either isotopically enriched or in natural abundance, are polarized by hydrogen. The basic theory for the flip-angle effect, applicable more widely than to SPT experiments, was developed by several groups (85, 86) and extended and applied by a number of others (87–89).

Both theory and experiment show that, if a perfect and nonselective 90° *read* pulse is used, the intensity changes produced by homonuclear polarization transfer are spread throughout the observed multiplet so that its components have their normal intensity ratios, and relative sign information is not available. At the other extreme are the results for small flip angles of less than about 30° in which relative intensities reflect the state of the spin system and thus do provide information about the results of selective population transfer. Use of the small flip angle amounts to limiting the perturbation to transitions directly connected to the polarized transition, whereas larger flip angles cause the polarization to spread to other transitions. For flip angles greater than 90°, polarization transfer effects are again evident, although they are of opposite sense to those for read pulses shorter than 90°; the specific read pulse flip angle for maximum transfer depends on the nature of the spin system as well as on the relaxation properties of the observed nucleus (83). The example shown in Fig. 1-23 demonstrates that relative sign information can be obtained only if the read-pulse flip angle is chosen to be other than 90°.

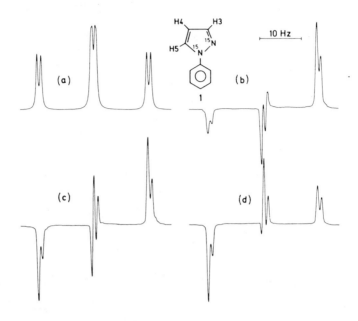

Fig. 1-23. (a) Simulated proton-coupled ^{15}N spectrum for the N2 atom in 1-phenyl[^{15}N$_2$]pyrazole. (b–d) Experimental SPT ^{15}N spectra for N2, obtained with a 180° pulse applied at the center frequency of the two H3 transitions at highest frequency. Flip angles for the ^{15}N read pulse were varied as follows: (b) 60°; (c) 90°; (d) 120°. The perfectly symmetrical pattern in (c) yields no information about relative signs. Parts (b) and (d) show that $^1K(^{15}$N1-^{15}N2$) \times {}^3K(^{15}$N1-^{1}H3$) < 0$ and $^3K(^{15}$N2-^{1}H4$) \times {}^3K(^1$H3-^{1}H4$) > 0$, where K is the reduced coupling constant. [From Berkhoudt and Jakobsen (84).]

Bauer and Freeman (90) showed how proton multiplets can be picked out of a complex spectral pattern by utilizing the flip-angle effect. A single proton resonance is inverted by application of a 180° pulse which is given a Gaussian shape in order to be selective (see Section IV,A,2 for an explanation). A nonselective read pulse of 90° is then applied, distributing the enhancement throughout the multiplet, and the difference between the resulting spectrum and the normal spectrum consists of the lines belonging to the same multiplet as the polarizing line. A composite 90° pulse (see Section IV,A,1) employed for observation compensates for off-resonance effects. The consequences of strong coupling were also analyzed and were found to include population changes in addition to the polarization transfer changes, distinguishable because the population changes always have the same sign, regardless of the irradiation frequency, while polarization effects depend in sign on the frequency of a selective pulse.

D. Decoupling in Selective Polarization-Transfer Experiments

If sensitivity enhancement rather than investigation of spin–spin coupling is the primary aim of polarization transfer, then it is desirable to have the further gain in signal strength afforded by decoupling. However, since the components of a multiplet appear in antiphase and there is no net magnetization transfer to the multiplet as a whole, acquisition with decoupling immediately after the 90° pulse, which amounts in effect to integrating the entire multiplet, removes the polarization transfer advantage. The result is complete cancellation of the signal if the original magnetization has been suppressed, or a peak of normal, unenhanced, intensity if it has not been suppressed.

A procedure devised by Harris for an AX system, in which a second selective 180° pulse is applied, but now to the observed nucleus rather than the polarizing nucleus, is designed to invert the population difference that is responsible for the emission line, so that it becomes an absorption line and can be combined by decoupling with the other absorption transition (91). Bolton and James (92) developed a sequence comprising three successive selective pulses, 180°(A)—180°(X)—180°(A), which achieves a similar result. These are, needless to say, inconvenient procedures and not applicable to larger spin systems.

A much more generally useful technique is the delay of acquisition and decoupling for some interval following the 90° pulse in order to allow the various magnetization components of a multiplet to precess until they come into phase with one another. The optimum delay time between the completion of the polarization transfer process and the beginning of decoupling depends on the value of J and the number of coupled nuclei, and the values shown in Table 1-6 have been published (93). These values were originally

TABLE 1-6

Values of the Refocusing Delay to Give the Maximum Decoupled Signal as a Function of the Number of Coupled Protons, n^a

	n							
	1	2	3	4	6	9	12	18
$\Delta_{opt}{}^b$	0.50	0.25	0.196	0.167	0.134	0.108	0.093	0.076
Enhancement^c	1.00	1.00	1.16	1.30	1.55	1.87	2.15	2.61

^{a} Adapted from Doddrell et al. (93). Copyright 1981, American Chemical Society.
^{b} Expressed as a multiple of $1/J$.
^{c} Expressed as a multiple of γ_H/γ_X.

compiled primarily for use with the INEPT sequence, which is described in the next section, and in which similar antiphase magnetization is produced. If various multiplets are observed simultaneously, as in that sequence or sometimes in SPT experiments, the chemical shift precession during the waiting interval, the extent of which differs from multiplet to multiplet, causes a large frequency dependence of the phase. This can be eliminated by a 180° refocusing pulse, applied to the observed nucleus and accompanied by a 180° hydrogen pulse designed to relabel the spin states of the observed nucleus in order to prevent that part of the precession which results from spin-spin interaction from being refocused along with the chemical shifts. This sequence is shown in Fig. 1-24a. Recently Sarkar and Bax (94) have proposed that, in the same way as described above for modulated spin

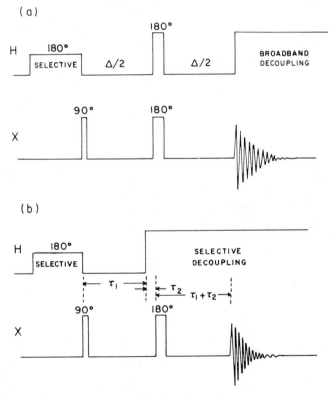

Fig. 1-24. (a) SPT pulse sequence with refocusing pulse and broadband decoupling. The refocusing period Δ is optimized on the basis of the number of coupled spin-$\frac{1}{2}$ nuclei and the magnitude of the coupling constant according to Table 1-6. (b) SPT pulse sequence with selective decoupling, as employed in Fig. 1-20c; τ_1 is optimized according to the values in Table 1-6. Introduction of the delay τ_2 is found to improve the performance.

echoes, acquisition in an SPT experiment begin immediately after the 90° pulse, with decoupling delayed for the appropriate interval to allow multiplet components to come into phase. Of course, if only one multiplet is being observed, as is often the case for nitrogen-15 or silicon-29, there is no phase problem resulting from differences in chemical shift precession.

Because of the antiphase nature of spin multiplets from polarization transfer, complex overlapping spectral patterns may be difficult to analyze without some form of selective decoupling. A case in point is that illustrated in Fig. 1-20b, which shows the intensity enhancement achieved in the coupled silicon-29 spectrum of vinyltrimethylsilane by application of a selective 180° pulse to the silicon satellites of the methyl resonances in the proton spectrum (66). An alternative method of simplification of the ^{29}Si spectrum of the molecule, continuous selective decoupling of the olefinic hydrogens, which would not cause phase complications because these hydrogens are not being used for polarization, might be attempted. However, these resonances occur about 70 Hz apart at 300 MHz, too close for truly selective decoupling and too far apart to permit both to be irradiated without perturbing the methyl hydrogens. If the methyl protons in vinyltrimethylsilane, utilized as a polarization source because they are both numerous and equivalent, are decoupled during acquisition, the silicon nuclei dephase under the influence of coupling to the vinyl protons, and the resulting spectrum cannot be phased. This difficulty can be overcome by use of a 180° refocusing pulse on silcon halfway between the 90° read pulse and the start of acquisition. The resulting spectrum is shown in Fig. 1-20c, and the pulse sequence used for selective decoupling with SPT is shown in Fig. 1-24b. The delay τ_1 allows refocusing of the antiphase silicon multiplet before decoupling is begun and is optimized according to the values in Table 1-6. The dephasing of silicon nuclei because of coupling to the vinyl hydrogens during the first τ_2 interval is reversed during the second τ_2 interval. Better results are obtained if the refocusing pulse is delayed one or two milliseconds past the start of proton decoupling. An alternative procedure would be to apply the silicon 180° pulse at the middle of τ_1, along with a 180° hydrogen pulse to prevent refocusing of the spin–spin effect, followed later by the simultaneous start of decoupling and acquisition. However, this would usually require programming three different decoupler output levels, which is beyond the capability of many currently used spectrometers.

An experiment combining SPT and acquisition of modulated, decoupled spin echoes was devised by Cookson and Smith (95). Following a selective 180° pulse to hydrogen, carbon-13 spin echoes were created and observed, with broadband decoupling operative during the refocusing portion of the echo interval. The selective irradiation frequency was stepped at regular intervals through the hydrogen region. Normal signals were removed by subtracting alternate scans with and without SPT or by inclusion, just before data acquisition, of a 90° carbon-13 pulse with phase alternating between $+x$ and $-x$ in successive scans and adding all scans, so that the response obtained in any spectrum was from the carbon nucleus responsible for the particular satellite on which the irradiation for that spectrum happened to fall. In this way a correlation pattern between hydrogen and carbon-13 could be obtained; at the same time, the phase of the echoes could be used for multiplicity assignment.

E. Nonselective Polarization Transfer—The INEPT Experiment

1. *The Pulse Sequence for INEPT*

Disadvantages of selective polarization transfer as a general means of sensitivity enhancement are the need to locate the position of an appropriate satellite in the spectrum of the polarizing nucleus, usually hydrogen, and the fact that only one multiplet at some one chemical shift can be polarized in one experiment. A pulse sequence for nonselective polarization transfer, shown in Fig. 1-25a, was devised by Morris and Freeman (96) and dubbed by them INEPT (*i*nsensitive *n*uclei *e*nhancement by *p*olarization *t*ransfer). This experiment requires the ability to alter the phase of the proton pulse, a pulse normally generated by the decoupler channel, but many spectrometers now in use do have this capability. In most versions of INEPT, all pulses in the sequence are nonselective.

We shall here employ a vector model to analyze the INEPT experiment for a heteronuclear AX system, following the diagrams in Fig. 1-26. To simplify the discussion, we will refer to the polarizing or A spins as hydrogen and to the observed or X spins as carbon.

The first 90° proton pulse turns both kinds of hydrogen nuclei, those coupled to α carbons and those coupled to β carbons, into the xy plane, where they then precess at different rates. The key to success in the experiment is the choice of the first precession interval, τ, to be $1/[4J(AX)]$, so that the protons are 90° out of phase with one another at the end of this period, when the 180° proton pulse is applied.

Exercise: If one variety of hydrogen nucleus precesses 20 Hz faster than a second variety, how long will it require for them to get (a) 20 revolutions out of phase? (b) 40 revolutions out of phase? (c) 5 revolutions out of phase? (d) 0.1 revolution out of phase?

The function of the concurrent 180° pulse on the nucleus to be polarized, carbon for example, is to interchange labels on the hydrogen nuclear spin

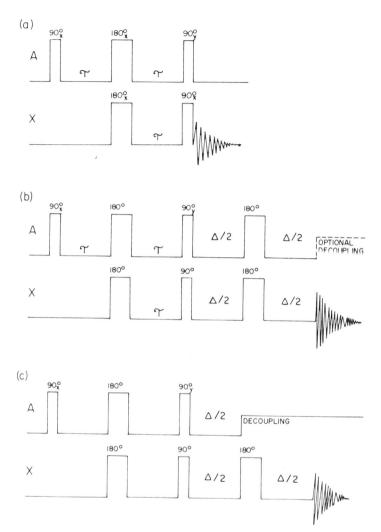

Fig. 1-25. Versions of the INEPT pulse sequence. (a) Sequence yielding a coupled spectrum, with antiphase multiplets. (b) With dual refocusing pulses, yielding either a decoupled spectrum or a coupled, in-phase spectrum. (c) With broadband decoupling gated on at the time of the 180° refocusing pulse on the observe nucleus.

states, as in the SPT and J-spectroscopy experiments; in the absence of this pulse, the effects of spin–spin splitting in the proton spectrum would be refocused, along with chemical shift effects, by the 180° proton pulse. The hydrogen nuclei now continue to dephase during the second τ interval, reaching a full 180° phase difference at time 2τ, with one component oriented along the $+x$ axis and the other along the $-x$ axis. Thus the desired antiphase magnetization has been achieved, and the $90°_y(H)$ pulse now turns the magnetizations into the $-z$ and $+z$ directions, respectively, leading to a situation like that shown in Fig. 1-18b for an SPT experiment. Polarization is transferred to the carbon nucleus in this step, just as it is in the SPT sequence, and the final 90° carbon pulse serves as the "read" pulse. Because the hydrogen pulse sequence in the INEPT experiment is essentially a spin-echo sequence, hydrogen chemical shift differences are refocused, and antiphase character results independently of the proton resonance frequency.

For an A_nX system, the relative intensities of the multiplet components in an INEPT spectrum are the same as those in an SPT experiment, with the same antiphase or up–down character for each multiplet. Natural polarization may be removed by presaturation of the X nuclei, either by having an interval between successive sequences which is short compared to the T_1 of the X nuclei or by a burst of 90° pulses to randomize the magnetization of these nuclei.

Another method of canceling natural polarization, as well as eliminating undesired solvent signals or signals from unlabeled species in a labeling experiment, is to alternate between x and y the phase of the first 180° pulse on the polarizing nucleus. This has the effect of interchanging between $+x$ and $-x$ the orientations of the vectors shown in Fig. 1-26f. In turn, the orientations of the two vectors along the $+z$ and $-z$ axes shown in Fig. 1-26g are interchanged, and the transferred polarization has the reverse sense. Alternatively, the interchange of vectors in Fig. 1-26g may be obtained by cycling between $+y$ and $-y$ the phase of the second 90° pulse on the polarizing nucleus. Whichever method of phase alternation is employed, addition and subtraction of the FIDs from successive scans—this is often referred to as "alternating the receiver phase"—causes cancellation of normal signals and leaves only the signals resulting from polarization transfer.

Exercise: (a) Work out the vector diagram corresponding to Fig. 1-26 for the case in which the 180° pulse on H has phase y. (b) Work out a diagram corresponding to Fig. 1-26 for an A_2X system.

If natural polarization is eliminated, the intensity ratios within a multiplet are the same as those for selective polarization transfer, and from Table 1-3e these are $-1:+1$, $-1:0:+1$, and $-1:-1:+1:+1$, for a doublet, triplet, or quartet, respectively. The absolute intensities depend, as for selective polarization transfer, on the ratio γ_A/γ_X.

The antiphase character of each coupled multiplet—one half consisting of positive peaks and the other half consisting of negative peaks—which

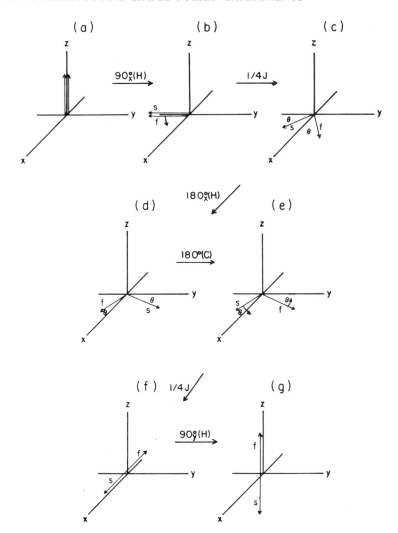

Fig. 1-26. Vector model of the INEPT pulse experiment. In (g), antiphase magnetization of the protons has been reached as in Fig. 1-18b, but here it has been achieved independently of the chemical shift of the protons; polarization is then transferred to the nuclei coupled to the protons, such as carbon-13.

results from this version of INEPT is evident in the example shown in Fig. 1-27. The antiphase relation may sometimes be helpful in picking out the components which together form a particular multiplet, but negative-going peaks in one multiplet may cancel positive-going peaks in another. If acquisition is delayed for an appropriate interval Δ after the read pulse,

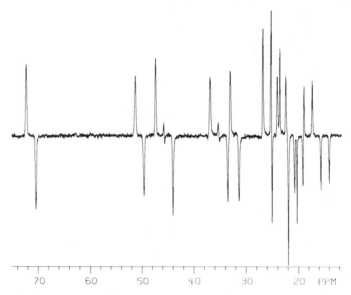

Fig. 1-27. INEPT carbon-13 spectrum of menthol at 75 MHz. The spectrum has been phased so that the high-frequency halves of the multiplets are positive. The central peaks of the CH_2 triplets show nearly zero intensity. Overlap in the high-field region of the spectrum makes it difficult to pick out the multiplets; the decoupled spectrum would, of course, assist in sorting out the overlapping patterns.

components of a multiplet come back into phase so that the peaks are all positive and mutual cancellation does not occur (97, 98). Note that this use of Δ, although consistent with the earlier use in Section III,C,1 in that it designates a period of free precession during which the components of a multiplet change relative phase, refers to a period of coming back into phase, rather than a period of dephasing. Some authors assign the symbol Δ to half the rephasing period, rather than to the length of the full period, and the reader must be careful not to be confused by the variation in convention.

The dependence of rephasing time on J causes problems when values of J vary throughout a spectrum. Fortunately, in carbon-13 spectroscopy of hydrocarbons, J varies only over a fairly limited range, especially in the aliphatic region, so that the problem is not serious. In addition, however, there is the difficulty that no value of rephasing time brings doublets, triplets, and quartets into phase simultaneously. The optimum value of Δ for doublets is $1/2J$ or about 4 msec for hydrocarbons, that for triplets is $1/4J$ or about 2 msec, and quartet components never do all get quite into phase. To see all multiplets in reasonably good form, a suitable compromise for Δ is $1/3.3J$ or 2.5 msec for hydrocarbons.

The presence of a delay before acquisition results in a large phase variation across the spectrum, because centers of multiplets at various chemical shifts will have precessed through different angles by the time acquisition is started (98).

2. Refocused INEPT

The phase problem following a delay after the last INEPT pulse is considerably alleviated if a 180° refocusing pulse is inserted midway between the read pulse and the start of acquisition (99), as described for SPT in Section III,D. In order to avoid refocusing the coupling-caused evolution at the same time, a concurrent 180° pulse on the A nuclei is needed to relabel the X spin states, just as for the SPT experiment or for the first portion of the INEPT sequence. Use of the two 180° pulses eliminates off-resonance effects for both nuclei. The resulting sequence, called *refocused* INEPT or INEPTR, is diagrammed in Fig. 1-25b.

If it is desired to obtain a decoupled spectrum from an INEPT sequence, a delay before acquisition is mandatory, preferably with a refocusing pulse, since, just as in the SPT experiment, turning on the decoupler immediately following the read pulse destroys all the effect which the transferred polarization should have on the intensity. As an alternative to simply having the decoupler on during the acquisition period of INEPTR to produce a decoupled spectrum, broad-band decoupling of nucleus A may be started at the time of the 180° refocusing pulse on the X nucleus and acquisition begun after a time interval Δ following the start of decoupling, as diagrammed in Fig. 1-25c and as described for selective decoupling with SPT. Optimum-intensity values of Δ for use with decoupling were given in Table 1-6.

The dependence of the sign and intensity of a decoupled signal on the number of polarizing nuclei responsible for splitting the resonance into a multiplet may be utilized as a way of determining the number of peaks that would be present in a multiplet had it not been decoupled, an alternative to the method of *multiplicity determination* described in Section III,D. The equations for decoupled signal intensity, neglecting relaxation effects, are (99)

$$I(AX) \propto \sin \pi J_{AX}\Delta \qquad [6]$$

$$I(A_2X) \propto \sin 2\pi J_{AX}\Delta \qquad [7]$$

$$I(A_3X) \propto 0.75 \, (\sin \pi J_{AX}\Delta + \sin 3\pi J_{AX}\Delta) \qquad [8]$$

These functions are plotted in Fig. 1-28. As an illustration of their use, if Δ is chosen to be $1/2J_{AX}$ so that $\pi J_{AX}\Delta$ is equal to 90°, only AX resonances

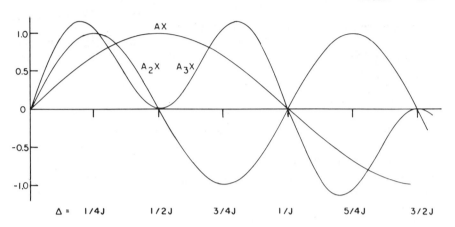

Fig. 1-28. Dependence of the intensity of the decoupled signals for various multiplets on the duration of the total refocusing interval Δ after a transfer of polarization, based on Eqs. [6]-[8].

are observed. Or, since the intensities of both AX and A_3X are symmetrical about 90° and the intensity of A_2X is antisymmetric, combination, with opposite sign, of spectra for $\pi J_{AX}\Delta$ of 45° and 135° gives only A_2X peaks. In Table 1-7 are shown expressions for combining spectra in this fashion, both in terms of angles and in terms of the magnitudes of the coupling constant for the interaction through which polarization is transferred, for the ideal situation, as well as expressions for real cases in which the coefficients should be modified to take account of pulse imperfections if optimum suppression of nondesired peaks is to be achieved. It should be

TABLE 1-7

EDITING COMBINATIONS TO OBTAIN SUBSPECTRA FROM REFOCUSED INEPT EXPERIMENTS, BASED ON FIG. 1-28

Subspectrum	Flip-angle function[a]	Coupling-constant function
Ideal pulses		
AX	$I(90°)$	$I(\Delta = 1/2J)$
A_2X	$\frac{1}{2}[I(45°) - I(135°)]$	$\frac{1}{2}[I(\Delta = 1/4J) - I(\Delta = 3/4J)]$
A_3X	$\frac{1}{2}[I(45°) + I(135°)] - \sqrt{2}I(90°)$	$\frac{1}{2}[I(\Delta = 1/4J) + I(\Delta = 3/4J)] - \sqrt{2}I(\Delta = 1/2J)$
Nonideal pulses		
AX	$I(90°) - z[I(45°) + xI(135°)]$	
A_2X	$\frac{1}{2}[I(45°) - xI(135°)]$	
A_3X	$\frac{1}{2}[I(45°) + xI(135°)] - yI(90°)$	

[a] For the ideal case, $x = 1$, $y = 1/\sqrt{2}$, and $z = 0$.

remembered that, while Δ can be varied depending upon the multiplicity of the peaks to be selected, the appropriate interval τ in the first part of the INEPT sequence, before polarization transfer, is always $1/4J$, whatever the multiplicity may be.

Exercise: Work out the vector diagram for the INEPT sequence applied to an A_2X system.

Rutar (100) has suggested an editing procedure for carbon-13 spectra in which a $180^\circ_x(H)$ pulse is inserted just before acquisition, using $\Delta = 1/2J$. This pulse reverses the phase of CH_2 resonances but does not affect CH or CH_3 carbons. Thus combinations of acquisitions with and without this rephasing pulse give only CH_2 peaks if subtracted and give CH plus CH_3 peaks if added.

Spectra combined in the way described may be either coupled or decoupled; when applied to coupled spectra, the aim of *spectral editing* of this type is primarily to sort out overlapping multiplets in crowded spectra. Methods of spectral editing will be discussed in much more detail in Chapter 3.

3. *Some Applications and Extensions of INEPT*

Particular advantages of polarization transfer for carbon-13 observation in paramagnetic transition-metal complexes have been pointed out by Doddrell *et al.* (101). In these species, there is very little NOE, so that normal observation of carbon-13 with broadband decoupling is of relatively low sensitivity; the quality of the spectra obtained in reasonable time is therefore greatly enhanced by the use of polarization transfer. Examples representing typical specific applications of INEPT to carbon-13 spectroscopy of organic molecules may be found in the literature (102, 103).

In addition to carbon-13, other spin-$\frac{1}{2}$ nuclei have been polarized by hydrogen via the INEPT sequence. The nucleus most frequently involved has been nitrogen-15, for which the one-bond coupling of 70–100 Hz can easily be utilized when the nitrogen atom is protonated. Materials that have been examined by INEPT include blocked glycylglycine (97), formamide (104), amine–borane adducts and aminoboranes (105), cyclic peptides (106), substituted pyridines and pyrimidines (107), enaminones (108), and gramicidin-S (109).

Silicon-29 in $(CH_3)_4Si$ was polarized through INEPT by Doddrell *et al.* (93), as was the tin nucleus in $(CH_3)_4Sn$. When observed by INEPT, using as the avenue for polarization relatively small three-bond couplings in amine complexes, the metals silver-109 and rhodium-103 showed very large sensitivity gains and time savings as compared to spectra obtained by previous methods (110). Schraml (111) polarized silicon nuclei in trimethylsilated compounds using single-frequency proton decoupling throughout the course of the INEPT sequence at a frequency corresponding to 0.2 ppm from TMS with a constant power level corresponding to 3.5 ppm. This resulted in enhanced silicon intensities by polarization transfer, at the same time producing an off-resonance decoupled pattern.

Provided suitably resolved coupling is present, spin-$\frac{1}{2}$ nuclei other than hydrogen can be used as the source for polarization transfer. The most likely candidates as polarizing species for sensitivity enhancement, since they have relatively large magnetogyric ratios, are fluorine-19 and phosphorus-31. However, the use of nuclei other than hydrogen is limited in practice by the need for a source of rf of appropriate power at the resonance frequency of the polarizing nucleus, plus a probe tuned to accept this frequency on one of its coils. Brevard and Schimpf (112) employed a special low-frequency probe fitted with a phosphorus decoupling channel to demonstrate polarization transfer from phosphorus-31 to isotopes of Fe, Rh, Ag, W, and Os in phosphine complexes of ions of these metals. Since the natural T_1 values of phosphorus-31 in these species are relatively long, 7 to 12 sec, it was advantageous to add a small amount of Cr(acac)$_3$ in order to reduce the recycle times.

A homonuclear version of refocused INEPT has been applied to suppress the signal of solvent water as compared to signals from those spins in the sample which are involved in spin–spin coupling (113). The second 90° pulse is applied along the y axis but only on alternate acquisitions, serving to invert only the coupled spins. On those acquisitions in which this pulse is absent, a $90°_{-y}$ purging pulse is applied just before acquisition. Alternate addition and subtraction ("alternating the receiver phase") cancels peaks from uncoupled spins.

Several selective versions of INEPT have been devised. Bax and co-workers (114, 115) showed how two- and three-bond connectivities can be established by polarization through long-range couplings by means of a sequence in which all the proton pulses are selective, while at the same time X-nucleus sensitivity is enhanced. This approach is especially applicable to nonprotonated carbons or nitrogens and overcomes some of the difficulties which result because of the long interval required when a small value of J is used for polarization. If all proton spins are initially turned into the xy plane as in normal INEPT, they dephase during the waiting period, and the second 90° pulse creates homonuclear multiple-quantum coherence which is not transformed into X-nucleus magnetization. In the selective experiment, in which the proton pulses are applied to one X satellite in the proton spectrum, the dephasing because of homonuclear coupling is refocused just before the second 90° pulse; those nuclei not irradiated by the selective 180° pulse do not "relabel" the X-nucleus spin states, and consequently the effect of these protons on the X nuclei is refocused. All magnetization is thus transferred to the X nucleus. However, this method does not eliminate the loss of magnetization as a result of relaxation during the delay periods, including the rephasing interval before the decoupler is turned on.

In another variety of selective experiment, Davis *et al.* (109, 116) preceded the INEPT sequence by a modified SPT sequence: $90_x^\circ(X)$; $180^\circ(H, \text{selective})$; $90_x^\circ(X)$. The X-nuclei pulses were nonselective and the proton pulse was applied, as usual in selective sequences, to an X-nucleus satellite in the proton spectrum. At the end of the SPT sequence, the nonselected proton magnetization is at equilibrium, and the nonselected X magnetization is in the $-z$ direction, so that they are affected in the usual way by a subsequent INEPT cycle, providing sensitivity enhancement. For the selected H and X spins, polarization transfer by INEPT is ineffective, and the X signal can be distinguished by the fact that it is only of normal intensity.

INEPT has been applied as a prelude to several other experiments, simply as a means of enhancing sensitivity by increasing the initial magnetization. One type of investigation where the sequence has been used in this way is in the measurements of carbon-13 or nitrogen-15 relaxation times. Marion and co-workers (117) used a modified version of the sequence in which the refocusing pulses were omitted and the last 90° pulse on hydrogen was replaced by a 180° pulse, as a spin-preparation method for inversion-recovery measurements on nitrogen-15 in enriched peptides. The mechanisms contributing to relaxation rates determined in this way, however, were found to differ from those active in relaxation of nitrogen-15 measured in the conventional way. Spin–lattice relaxation rates of carbon-13, silicon-29, and nitrogen-15 in tetramethyldisilazane were measured, using INEPT followed by a $90_y^\circ(X)$ or $90_{-y}^\circ(X)$ pulse to convert x-axis magnetization into z magnetization (118). The z magnetization was then allowed to decay toward equilibrium for a variable time period, in the usual way for an inversion recovery sequence, after which it was observed by a final 90° (X) pulse.

Polarization transfer from hydrogen to carbon following a proton saturation–recovery or inversion–recovery experiment has been used by the Oxford group to measure relaxation rates of individual proton resonances in a crowded spectrum, taking advantage of the greater chemical shift dispersion of carbon-13 to remove the overlap present in the proton spectrum (119, 120). Phase alternation of the final 90° pulse on hydrogen accompanied by alternate addition and subtraction of FIDs canceled signals not arising from polarization transfer. This method, of course, measures only the relaxation rates of hydrogens attached to carbon-13, and offers low sensitivity.

Two-dimensional experiments involving nuclei other than hydrogen are often especially time-consuming because of the large number of acquisitions required. Applied ahead of some of these experiments, INEPT may serve to enhance sensitivity. In combination with the multiple quantum experiment, INADEQUATE, in either one- or two-dimensional form, INEPT can be used to reduce the time required to obtain adequate signal-to-noise (121). In this combination, although a rephasing period may be allowed following the end of the polarization-transfer portion and before decoupling is begun, it is important to realize that the INADEQUATE experiment time must be counted from the instant when transverse carbon-13 magnetization is first created by the pair of 90° polarization-transfer pulses.

Decoupled INEPT has also been employed to enhance carbon-13 magnetization in 2D exchange studies on decalin, bullvalene, and N,N-dimethylformamide complexed to aluminum ion. The necessarily very long performance time of these experiments was reduced, primarily because of a reduction in the relaxation delay, an advantage over NOE even if the full Overhauser enhancement could be achieved (122).

The INEPT sequence may itself be modified to convert it into a *two-dimensional technique*. Thomas *et al.* (39) suggested that a 2D J spectrum could be obtained by carrying out a series of experiments in which the interval Δ is varied regularly through the series, and a second FT is performed with respect to Δ, using corresponding points from the several spectra. Values of Δ in the experiments were integral multiples of its initial value, 1/2SW, where SW is the desired sweep width in the J dimension. Since chemical shift effects are eliminated in this dimension, this width need be no greater than the order of the largest coupling constant to be measured. There is an evident parallelism between this procedure and the procedure described in Section II,B,3 for extracting coupling constants from the modulation frequencies of simple spin echoes arranged in a 2D array. In addition to sensitivity enhancement of the resulting spectrum because of the recycle-time advantage when polarization transfer is effective, there is also an advantage in resolution because of the elimination of the effects of field inhomogeneity by the refocusing step in INEPT.

Davis, Agosta, and Cowburn (123) investigated long-range ^{15}N-^{1}H coupling constants in natural abundance samples by appending to a regular INEPT sequence (without the usual refocusing pulse because only one nitrogen chemical shift was involved) a further time interval t_1 in the middle of which were simultaneous 180° pulses, with the proton pulse selective (25 msec, $\gamma_H B_2 = 20$ Hz). This selective pulse prevents the spin–spin splitting effects of the irradiated protons on the nitrogen from being refocused by the nitrogen 180° pulse, whereas the spin–spin splitting effects of all other protons are refocused. The resulting nitrogen spectrum shows only coupling to the protons which have been selectively flipped. The nitrogens are polarized by one- or two-bond couplings, the values of which determine the delays to be used in the first part of the INEPT sequence. The value of t_1 was varied in 32 increments of 40 msec, giving a 2D J experiment with a spectral width of 12.5 Hz.

Two methods of obtaining chemical shift correlations between carbon nuclei and the protons attached to them—not quite full-fledged 2D experiments—by modification of INEPT have been described, one by Reynolds et al. (124) and the other by Pearson (125). They differ from normal 2D experiments in the use of limited data sets. Both use INEPT with a fixed overall evolution period from the initial 90° pulse to the pair of polarization-transfer 90° pulses and with a fixed value of the interval before acquisition, and both displace the initial pair of 180° pulses from the center point of the 90°-90° interval. The two sequences are shown in Fig. 1-29; they differ

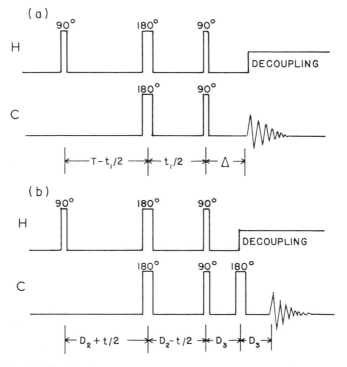

Fig. 1-29. Modified INEPT pulse sequences for obtaining heteronuclear shift-correlated spectra using a 180° pulse which is incrementally stepped through a fixed evolution period to give full ^{1}H-^{1}H decoupling. (a) Version of Reynolds et al. (124). (b) Version of Pearson (125). Appropriate phase-cycling schemes are given in the original papers.

primarily from one another in the arrangement of the rephasing period before decoupling and in the phase cycling which is employed. In both experiments relatively few values of the variable time—the displacement of the 180° pulses from their normal location—are employed. Reynolds varies this interval regularly, acquiring 16 or 24 spectra, with the resulting low resolution in the hydrogen dimension improved by zero-filling before Fourier transformation. Pearson acquires two to eight spectra using a range of time values, not necessarily regularly incremented, and processes the results by nonlinear least-squares fitting rather than by a second Fourier transformation, claiming considerable improvement in precision by this method.

Moving the pair of 180° pulses while keeping the evolution period fixed achieves two results: it introduces a net proton evolution time, the difference between the interval before the pulses and the interval after the pulses, so that the amount by which proton chemical shift evolution is effective depends on the length of this net time and the distance of the particular absorption from the carrier, and it decouples hydrogens, unless they are strongly coupled, for there is no J modulation of the signal (126).

Extensions of polarization transfer to systems involving nuclei of higher spins as well as "reverse INEPT" will be discussed following the presentation of DEPT sequences in the next section. Further, several extensions of INEPT directed at improving the quality of edited spectra will be described in Chapter 3.

F. DEPT and Related Sequences

A group of pulse sequences involving heteronuclear polarization transfer with optional application to spectral editing has been developed by Pegg, Bendall, Doddrell, and co-workers in Australia. The best known member of this family of sequences is termed DEPT (127, 128), an acronym for *d*istortionless *e*nhancement by *p*olarization *t*ransfer, although, as we shall see, this description extends only to certain applications of the sequence.

In order to analyze DEPT and related experiments, we must first understand something about a phenomenon referred to as *multiple-quantum coherence*. This subject will be encountered in every chapter in this volume, and it will be dealt with in detail in Chapter 4. Here, the effect will be introduced only briefly and in a very general way.

Consider first an ordinary single-quantum transition, particularly as we might encounter it in a cw spectrum. One approach to describing what happens as the transition occurs is based on a quantum-mechanical model, in which we emphasize the absorption from the rf radiation of photons having the energy needed to induce transitions in the nuclear spin state from α to β, or from parallel to the magnetic field to antiparallel. The transitions may accordingly be detected by measuring the absorption of energy by the sample from the surroundings. In an alternative model, however, we think of the nuclear behavior as analogous to that we expect of macroscopic magnets, with the nuclei precessing about the B_0 direction, initially at random angles, and subject to a torque if B_1 alternates at the same frequency as the frequency of precession. Since the direction of the torque moves with B_1, nuclei that are flipped have a special relation to B_1, end up in phase with one another, and subsequently precess coherently about B_0, so that there is a rotating dipole which induces a voltage in the receiver coil. Thus a "coherence" has been generated by the spin flip process.

A multiple-quantum coherence is associated with a particular transition, just as is a single-quantum coherence. For example, in a weakly coupled two-spin system, as represented in Fig. 1-30, one may consider transitions in which both spins flip simultaneously, either in the same sense ($\alpha\alpha \to \beta\beta$ or $\beta\beta \to \alpha\alpha$), corresponding to a "double-quantum" change with the total spin quantum number changing by two units, or in the opposite sense ($\alpha\beta \to \beta\alpha$ or $\beta\alpha \to \alpha\beta$), corresponding to a "zero-quantum" change in which the total spin quantum number remains constant. However, since these transitions are forbidden, the corresponding coherence cannot be generated, at least by pulse methods, by inducing the transition, but it must be produced by first creating a related single-quantum coherence and then transferring this coherence in an appropriate way. Indeed, almost any nonselective rf

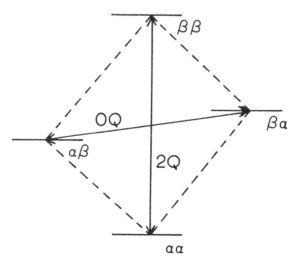

Fig. 1-30. Energy-level diagram of AX system showing double-quantum (2Q) and zero-quantum (0Q) transitions. Single-quantum transitions are indicated by dashed lines.

pulse applied to a sample in which a single-quantum coherence exists, except perhaps a 180° pulse, is likely to convert some or all of the single-quantum coherence to multiple-quantum coherence. Since a multiple-quantum coherence cannot be directly created by a single-pulse process, it follows that it cannot be directly detected: there is no precessing magnetic dipole moment which affords a handle to the outside world. To find out what has happened to a multiple-quantum coherence during a particular time interval, it must be converted back into observable single-quantum coherence by a suitable pulse. An important feature of double-quantum and zero-quantum coherences in an AX system is that the transition frequencies are independent of the spin–spin coupling magnitude, and thus their precession frequencies depend only on the chemical shifts of the two nuclei involved.

The Australian group has preferred not to use the term "multiple-quantum coherence" but has instead described the effect as "correlated motion in the doubly rotating frame in the transverse plane" (128). Whichever name is applied, the significant aspects of this phenomenon are its experimental characteristics, including the methods of creation and detection and the consequences of an interval of free precession.

Returning now to the DEPT family of sequences shown in Fig. 1-31, we find them to be alike in several respects, and they can be viewed as special cases of the most general version, the UPT or *u*niversal *p*olarization *t*ransfer experiment shown in Fig. 1-31c (129, 130). Each of the sequences consists of five pulses, of which the first is a $90°_x$ pulse on the polarizing nucleus, just as in INEPT. For simplicity, we shall refer to the polarizing nucleus as hydrogen and to the nucleus to be polarized as "X." After an interval $1/2J$, where J represents the coupling constant $J(\text{HX})$ by which polarization is to be transferred, there is a 180° pulse on hydrogen accompanied by a pulse of angle ϕ on the X nucleus. After a second interval $1/2J$, a pulse of angle θ with phase y is applied to hydrogen, along with a 180° pulse to X. Acquisition of the X spectrum begins after a third $1/2J$ interval, accompanied by decoupling of hydrogen if desired. Alternation of the phase of the angle θ between $+y$ and $-y$ along with subtraction of alternate scans may be applied to cancel magnetization not associated with the desired polarization transfer process.

During the first interval, the hydrogen nuclei precess in the xy plane until neighboring components of a multiplet are 180° out of phase with one another. The following pair of pulses creates multiple-quantum coherence involving the coupled H and X spins. This coherence precesses during the second $1/2J$ interval, after which some or all of it is converted back into observable single-quantum coherence by the last pair of pulses. The third $1/2J$ period serves to rephase the components of each multiplet; decoupling

1. BASIC METHODS AND SIMPLE PULSED EXPERIMENTS 69

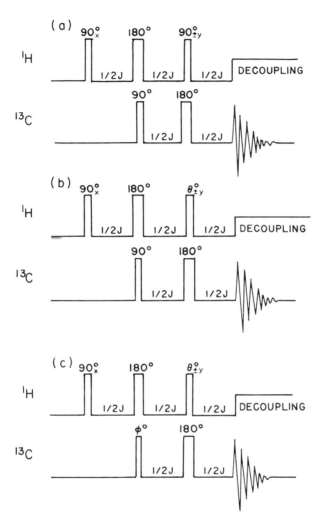

Fig. 1-31. Polarization transfer pulse sequences in the DEPT family. (a) EPT, to obtain only methine carbon resonances. (b) DEPT, for multiplet selection in spin-$\frac{1}{2}$ systems by varying the tip angle θ. (c) UPT, for multiplet selection in systems involving spins greater than $\frac{1}{2}$, by varying the angles ϕ and θ.

may then be started or the coupled spectrum can be recorded with all multiplets simultaneously in phase. In a sense, the first and third periods, along with the two 180° pulses, form a symmetric pattern about the middle period of multiple quantum evolution. The particular type of polarization transfer achieved depends on the values of the angles ϕ and θ, which determine what type of multiple quantum coherence is formed and what part of the coherence is returned to observable single-quantum magnetization.

If the angles ϕ and θ are both 90°, as in Fig. 1-31a, transfer from a single spin-$\frac{1}{2}$ nucleus to a second spin-$\frac{1}{2}$ nucleus is selected, as from ^{1}H to ^{13}C in a methine entity. This version of the sequence is termed EPT, for *e*xclusive *p*olarization *t*ransfer (131–133). Phase alternation of the last 90°(H) pulse with subtraction of alternate scans cancels methylene and methyl carbon resonances, whether the spectrum is coupled or decoupled.

In the DEPT sequence proper, Fig. 1-31b (127, 128), the angle ϕ is kept at 90° but the angle θ is varied. Polarization transfer may be optimized for a particular value of n in A_nX, where n is the number of equally coupled spin-$\frac{1}{2}$ nuclei which are to polarize a single spin-$\frac{1}{2}$ X nucleus, by choosing an appropriate value of θ. The effective angle of the pulse θ is related in a simple way to the interval Δ in the INEPTR sequence:

$$\theta = \pi J \Delta \qquad [9]$$

However, in contrast to INEPT, the DEPT sequence applies only to transfer from one or more spin-$\frac{1}{2}$ nuclei to one spin-$\frac{1}{2}$ nucleus.

To "edit" spectra by the use of DEPT, that is, to obtain spectra having only resonances from one multiplicity of hydrogen, it is possible to run a series of experiments with various values of θ and then to combine the results with suitable weighting factors. One choice of angles and combinations was given in Table 1-7. Another set of angles that can be used includes 30°, 90°, and 150°, for which, based on Eqs. [6] to [8], all the weighting coefficients are ±1 (134–136):

$$I(\text{CH}) = I(90°)$$
$$I(\text{CH}_2) = I(30°) - I(150°) \qquad [10]$$
$$I(\text{CH}_3) = I(30°) + I(150°) - I(90°)$$

Editing based on DEPT is discussed extensively in Chapter 3 and in a number of papers in the literature (127, 128, 137–142).

Although the pulse angles to be used in DEPT-related sequences are independent of J, the intervals between pulses depend on the value of J, and some error will result in selecting resonances of a particular multiplicity if there is a range of J values or if the J value is not properly estimated. However, the errors introduced by this cause into EPT and DEPT results are smaller than those for INEPT. DEPT also has the advantage, for purposes of editing, that it includes fewer pulses than INEPTR, so that the presence of pulse deviations is less significant. DEPT has a constant overall time interval for the different spectra to be combined, although there are ways of modifying INEPT to provide constant total duration. For coupled spectra, there is the substantial difference that DEPT furnishes multiplets with intensity ratios approximating those in a normal spectrum, a particularly significant point for A_2X groups in which the middle peak of the triplet is missing in INEPT spectra. However, for multiplets with many components, the INEPT advantage of greater relative enhancement of outer lines is lost. Finally, the overall time for an INEPT sequence is shorter, important if relaxation times are sufficiently short so that magnetization transfer advantages may be lost during the course of the sequence.

Various modifications and extensions of DEPT have been proposed, and it has been used as a prelude to other experiments in order to create increased amounts of magnetization and to decrease recycle time by taking advantage of the speedier relaxation of hydrogen nuclei. For example, the combination of DEPT with INADEQUATE has been described by Sparks and Ellis (143). They found that it was necessary to presaturate the carbon-13 nuclei, using 64 90° pulses 0.1 sec apart, so that artifacts would not be produced by the presence of initial transverse magnetization. At the end of the DEPT portion of the sequence, since transverse magnetization has already been created, the usual first 90° pulse in the INADEQUATE part of the combination is not required. By selecting the value of θ, it is possible, just as in the normal DEPT sequence, to choose among CH, CH_2, and CH_3 resonances.

Coxon (136) used DEPT as preparation for spin-flip heteronuclear J-resolved two-dimension spectroscopy, both for sensitivity enhancement and for the benefit of the editing feature in sorting out complex overlapping spectra. In the pulse sequence he employed, shown in Fig. 1-32, τ_1 is a very short delay to compensate for the variable time required by pulse θ in obtaining the several data sets, and τ_2 is a second short delay, equal to τ_1 plus the length of the pulse θ, used in order to make the refocusing time equal to the dephasing time. Some results for methyl 2,3-anhydro-4,6-O-benzylidene-α-D-mannopyranoside are shown in Fig. 1-33.

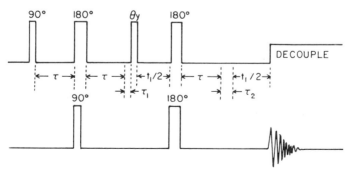

Fig. 1-32. Pulse sequence for 2D DEPT J-resolved ^{13}C spectral editing. [After Coxon (136).]

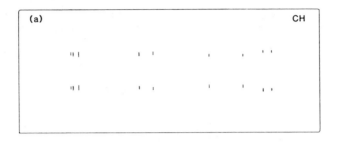

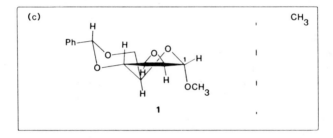
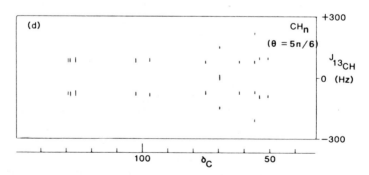

Fig. 1-33. Contour plots of the magnitude spectra obtained by the J-resolved pulse sequence shown in Fig. 1-32, for a solution of 0.4 M methyl 2,3-anhydro-4,6-O-benzylidene-α-D-mannopyranoside in $CDCl_3$. [From Coxon (136).]

Extension of DEPT itself to two dimensions is possible by inserting at an appropriate place in the sequence a variable time period, with 180° pulses on either one or both of the nuclei at the middle of the period. The added 180° pulse, if suitably located, may serve in place of one of the 180° refocusing pulses normally present (144, 145). Nakashima, John, and McClung (134, 135) utilized a sequence of this type with spectral editing procedures to construct selective CH, CH_2, and CH_3 maps.

A selective version of DEPT, in which an initial 180° pulse is applied to a carbon-13 satellite in the proton spectrum, much as in the SPT experiment, may be used if relatively few C-H correlations are of interest and a full 2D data set is not required (146, 147). Thus the first segment of DEPT, normally a hard 90°(H) pulse followed by $1/2J$ to create antiphase magnetization, is replaced by a selective 180°(H) pulse which yields antiphase magnetization for one multiplet, followed immediately by a pair of 90° hard pulses to the two species involved to create multiple-quantum coherence.

G. Inverse Polarization Transfer

Although polarization-transfer experiments were originally designed to enhance sensitivity in spectra of less receptive spin-$\frac{1}{2}$ nuclei by utilizing the larger magnetogyric ratio and more rapid relaxation of hydrogen nuclei, there are circumstances under which transfer of polarization to a nucleus of higher gamma value is advantageous. We shall first consider some experiments in which this method is used to achieve simplification of complex overlapping spectra, observation of X-nucleus satellites in the proton spectrum, or determination of connectivity patterns of other nuclei to hydrogen. A related advantage of "inverse" polarization transfer in solving problems of these types is the greater signal obtained per hydrogen nucleus than per X nucleus.

In a procedure called "inverse EPT," the usual role of the two nuclei can simply be reversed, without modifying the sequence, so that carbon-13, for example, is the source of polarization and hydrogen the recipient of polarization transfer (133, 148). A proton spectrum containing only resonances from ^{13}CH groups, with ^{12}CH as well as all CH_2 and CH_3 hydrogen resonances quite well suppressed, can then be obtained by presaturating the hydrogens, phase-alternating the last 90° (C) pulse, and subtracting alternate scans.

"Inverse" or "reverse" INEPT was used by Freeman, Mareci, and Morris (49) to observe the carbon-13 satellites in a hydrogen spectrum obtained through the decoupling coil in a 10-mm carbon-13 probe. In their pulse sequence, which is shown in Fig. 1-34a, the interval τ now depends on the number of hydrogens in the multiplet, in much the same way as does the

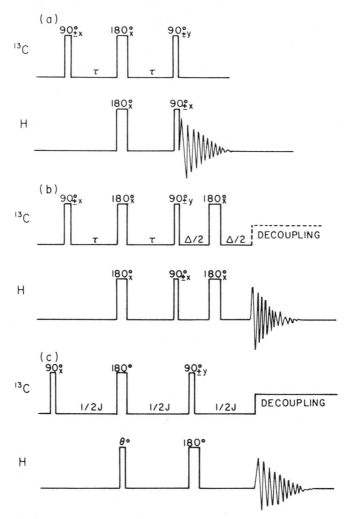

Fig. 1-34. Pulse sequences for inverse polarization transfer. (a) Inverse INEPT for observation of carbon-13 satellites in a coupled proton spectrum. (b) Inverse INEPT with refocusing, for observation of a decoupled proton spectrum. For this to be useful, the initial carbon and 90° proton pulses are phase-alternated by 180° in parallel in order to cancel the signal from protons attached to ^{12}C. (c) Inverse DEPT.

refocusing delay in refocused INEPT, as shown in Fig. 1-28, as well as on the value of J for the interaction by which polarization is transferred. The interval used in these experiments was $1/4J_{CH}$ and would be somewhat shorter if CH_2 and CH_3 groups were also to be observed. Furthermore, if long-range couplings are to be observed, the interval should be proportionately longer than for transfer by one-bond couplings. In the original experiments, the parent signals for the hydrogens attached to carbon-12 were suppressed in part by presaturation of the hydrogen nuclei by a train of 90° pulses at 3-msec intervals, and the remaining signal was removed by phase alternating in parallel the initial 90° carbon pulse and the final 90° proton pulse. The signals obtained by polarization transfer were thus reversed in sign twice between successive scans and reinforced when added, but the signals not affected by polarization transfer, those from hydrogens attached to carbon-12, were reversed in sign only once and were canceled by addition. Appending to the sequence a waiting period with a pair of refocusing pulses before acquisition, as shown in Fig. 1-34b, would permit the proton spectrum to be carbon-decoupled, provided appropriate hardware is available.

Shaka and Freeman (149) utilized ^{1}H-{^{13}C} polarization transfer, with selective excitation of a particular carbon resonance by a DANTE sequence, to simplify complex hydrogen spectra by emphasizing a particular proton multiplet. Because of the greater carbon chemical shift dispersion, selective excitation of that nucleus is much easier than it is for hydrogen. The protons, which had been presaturated, were decoupled during excitation. The length of the succeeding interval required to develop antiphase magnetization or maximum polarization transfer again depends on the number of coupled hydrogens; the overall length is $1/2J$ for CH and $1/4J$ for CH_2. Since the carbon transmitter is set at exact resonance, there is no chemical shift evolution during the INEPT waiting period and the 180° pulses normally at the midpoint of that period, as well as those at the midpoint of the $1/2J$ acquisition delay, can be dispensed with, avoiding in this way complications from strong coupling. Decoupling of the carbon during acquisition, using square-wave modulation, is easily accomplished since the carbon transmitter is on resonance. Again the proton signals from carbon-12 sites are suppressed by presaturation pulses on the carbon, along with phase alternation of the carbon excitation pulse in parallel with the receiver reference phase.

Bendall *et al* (148) developed an inverse DEPT sequence which is suitable for transferring magnetization from one spin-$\frac{1}{2}$ nucleus to one or more spin-$\frac{1}{2}$ heteronuclei, and which can be used for editing proton spectra. The change from regular DEPT is that the pulse of variable length in this sequence is now the *first* pulse on the hydrogen nuclei, the nuclei to *receive* the polarization, rather than the *last* pulse on the hydrogen nuclei, the nuclei to act as the *source* of polarization. The sequence was applied to transfer polarization from carbon-13 in an enriched sample; limitations in this method for natural abundance samples arise because of sensitivity problems associated with the need to suppress adequately the "parent" peak and to apply a decoupling signal to carbon during acquisition.

H. Forward and Reverse Polarization Transfer Involving Quadrupolar Nuclei

Polarization transfer to or from nuclei with spin greater than $\frac{1}{2}$ requires that there be resolved spin–spin interaction between the polarization source and the recipient nucleus. Thus the application is restricted in scope to a relatively few quadrupolar nuclei, such as deuterium and boron. Another important consideration in carrying out such experiments is that the pulse sequence may require adaptation to the requirements of the higher-spin nuclei. INEPT needs adjustment of the interpulse intervals, but otherwise operates pretty much without limit as to the number of nuclei or the magnitude of their quantum number. Although it can be extended somewhat, DEPT proper is limited to transfer between a single spin-$\frac{1}{2}$ nucleus and one or more spin-$\frac{1}{2}$ nuclei; UPT is much more versatile.

INEPT has been used to transfer polarization from hydrogen to quadrupolar nuclei, such as the nitrogen-14 atom in the ammonium ion, in which relatively slow relaxation is associated with tetrahedral symmetry (150). Since the rate at which the polarizing hydrogen nucleus precesses depends on the spin of the nucleus to which it is coupled, the initial period, between the 90° and the 180° pulses and intended to produce a 90° angle between the vectors representing the components of the multiplet, is $1/8J$, rather than the $1/4J$ appropriate to the spin-$\frac{1}{2}$–spin-$\frac{1}{2}$ case. The five components of the nitrogen-14 pentet, which arises from coupling to four hydrogens, were observed in ammonium nitrate solution to have intensity ratios of $-1:-2:0:2:1$, just as expected when initial saturation is assumed (Table 1-3e). For the outer lines in the pentet, an intensity enhancement of 24.5 was measured, compared to a theoretical maximum of 27.9. For the decoupled spectrum, a delay of $1/6J$ after the last pulse was used, but no refocusing proton pulse was required, since only one multiplet was being observed; an intensity enhancement of 7.1 was measured, compared to 8.99 calculated.

Another application of INEPT has been to the polarization by hydrogen of boron isotopes in borohydride ion in aqueous solution (151). In the same paper in which this was reported, Pegg *et al.* worked out many of the details of the theory of INEPT-mediated transfer from n hydrogen atoms to a nucleus X of arbitrary spin quantum number. They found the enhancement of the components of the X multiplet varies with the interval τ—one-half of the precession period—as $\Sigma_{m_I>0}\, m_I \sin 4\pi J m_I \tau$, where m_I represents each possible value of the projection of the spin quantum number I of nucleus X. For the two isotopes of boron, the intensities are accordingly given by the proportionalities

^{10}B $(I=3)$:   Int. $\propto 3 \sin 12\pi J\tau + 2 \sin 8\pi J\tau + \sin 4\pi J\tau$ [11]

^{11}B $(I=\frac{3}{2})$:   Int. $\propto \frac{3}{2} \sin 6\pi J\tau + \frac{1}{2} \sin 2\pi J\tau$ [12]

The time intervals to obtain maximum enhancement are given by the roots of

$$\sum_{m_I} m_I^2 \cos 4\pi J m_I \tau = 0 \quad [13]$$

with optimum enhancement at the smallest root for τ. The value of Δ, the total refocusing delay, for optimum enhancement of the decoupled signal is $(1/\pi J)$ arcsin $n^{-1/2}$, where n is the number of spin-$\frac{1}{2}$ nuclei involved in the transfer.

The UPT sequence was applied to the same type of transfer of polarization from hydrogen to boron in borohydride (129, 130). For ^{10}B, the signal intensities are given in terms of the two variable angles in UPT as

$$\text{Int.} \propto 2 \sin 2\theta + \sin 4\theta \quad [14]$$

$$\text{Int.} \propto 3 \sin 6\phi + 2 \sin 4\phi + \sin 2\phi \quad [15]$$

For ^{11}B, the θ dependence is the same as for ^{10}B, but the ϕ dependence is given by

$$\text{Int.} \propto \tfrac{3}{2} \sin 3\phi + \tfrac{1}{2} \sin \phi \quad [16]$$

For this nucleus, the quintuplet in the coupled spectrum was found to have intensity ratios of 0.95:4.05:6.0:3.85:1.0, compared to the theoretical 1:4:6:4:1.

Pegg and Bendall (152) and Chandrakumar (153) have provided analyses of transfers involving higher-spin nuclei. Chandrakumar used the product operator method and limited the analysis to spin-$\frac{1}{2}$-spin-1 combinations with transfer in either direction. He showed that transfer from spin-$\frac{1}{2}$ to spin-1 is possible with DEPT if the time interval is set to $1/4J$ rather than $1/2J$, although phase anomalies exist, even with perfect pulse timing. These anomalies are removed if the versions DEPT$^+$ or DEPT^{++}, described in Chapter 3, are employed, provided the angle θ is 90°.

Polarization transfer from quadrupolar nuclei to spin-$\frac{1}{2}$ nuclei is more generally useful than the "forward" transfer, since it can be arranged to take advantage of the rapid relaxation of the quadrupolar nuclei as a means of accelerating the rate at which transients of a spin-$\frac{1}{2}$ nucleus may be obtained. Experiments of this sort have been reported with ^{2}H or ^{11}B as the source of polarization of carbon or hydrogen. Other possible applications are to selective observation of hydrogens near metal atoms in organometallics or hydrogens near labels in isotopically enriched compounds.

As an example, 3- to 30-fold sensitivity enhancement of the resonance of protons attached to boron in the molecule $(CH_3)_2SBH_3$ was obtained by Rinaldi and Baldwin (154) by INEPT transfer from the spin-$\frac{3}{2}$ nucleus ^{11}B. Although the signal is expected to lose a fraction 0.36 of the usual intensity because of transfer from the lower-gamma nucleus, the much faster relaxation permits more rapid acquisition. The effect is similar to that of $Cr(acac)_3$, but there is no line broadening or influence on the chemical shift. By application of phase cycling and saturation procedures, the broad resonances of the protons attached to boron were accentuated in relation to the sharp lines for other protons, including those of the solvent. Relative intensities of the polarization transfer quartet of hydrogen corresponded to the theoretical $-3:-1:+1:+3$ ratio.

Transfer of polarization from deuterium to carbon-13 is particularly useful for locating deuterium atoms which have been introduced into large organic molecules, where the greater chemical shift dispersion of carbon-13 is advantageous. The relaxation of deuterium is usually rapid enough to be helpful in speeding acquisition, but not so rapid as to wash out the spin–spin interaction. The rapid recycling rate tends to saturate the resonances of protonated carbons within the solute as well as those in a protonated solvent. Both INEPT and DEPT procedures may be employed, although INEPT has the advantage of smaller loss of magnetization by relaxation during the pulse sequence because it occupies less time.

Rinaldi and Baldwin (155, 156) have demonstrated deuterium–carbon transfer in molecules such as d_6 benzene, and phenylethanol deuterated in the aromatic ring. In the application of the standard INEPT sequence, refocused for decoupling, the initial delay period between the 90° and 180° pulses on deuterium is set equal to $1/4J(CD)$, since the deuterium nuclei, regardless of how many are attached to a given carbon, precess at a rate determined by the spin of the coupled spin-$\frac{1}{2}$ ^{13}C nucleus. However, in the refocusing interval, Δ, the behavior of the system depends on the number of deuterium nuclei, since now it is the carbon-13 which is precessing under the effect of the deuterium spins. Thus for a CD system, the multiplet components get into or out of phase twice as fast as for the CH system, since their frequencies lie at $\pm J$ with respect to the rotating frame, rather than at $\pm\frac{1}{2}J$ as for the CH system. The decoupled intensities for various spin systems are modulated by the interval Δ according to the expressions shown in Table 1-8. No single value of Δ gives optimum sensitivity for all three multiplet types, but $3/20J$ is a reasonable compromise, yielding all resonances positive and each type of about the same intensity. Of course, a choice of Δ can be used to some extent to provide spectral editing, but about the only effective separation is that obtained for Δ of $1/3J$, where only CD carbons appear in the spectrum.

In order to predict relative intensities within a multiplet and to evaluate the intensity of an INEPT spectrum relative to a normal spectrum, we shall first consider a CD system, for which an energy level diagram is shown in Fig. 1-35. In preparing this diagram, we have assumed that the magnetogyric

1. BASIC METHODS AND SIMPLE PULSED EXPERIMENTS

TABLE 1-8

RELATIVE INTENSITIES IN MULTIPLET COMPONENTS AND DECOUPLED SPECTRA FOR POLARIZATION TRANSFER FROM $I=1$ TO $I=\tfrac{1}{2}$ (^{2}H TO ^{13}C)

Spin system	Spectrum	$m(^2\text{H})$	Intensity^a Refocused INEPT	DEPT
CD	Coupled	0	0	$(2/3)\sin 2\theta$
		±1	$(2/3)\sin 2\pi J\tau_2$	$(1/3)\sin 2\theta$
	Decoupled		$(4/3)\sin 2\pi J\Delta$	$(4/3)\sin 2\theta$
CD$_2$	Coupled	0	0	$(2/9)\sin 2\theta + (1/3)\sin 4\theta$
		±1	$(4/9)\sin 2\pi J\tau_2$	$(2/9)\sin 2\theta + (2/9)\sin 4\theta$
		±2	$(4/9)\sin 4\pi J\tau_2$	$(1/9)\sin 2\theta + (1/18)\sin 4\theta$
	Decoupled		$(8/9)\sin 2\pi J\Delta + (8/9)\sin 4\pi J\Delta$	$(8/9)\sin 2\theta + (8/9)\sin 4\theta$
CD$_3$	Coupled	0	0	$(1/4)\sin 2\theta + (2/9)\sin 4\theta + (5/36)\sin 6\theta$
		±1	$(4/9)\sin 2\pi J\tau_2$	$(3/16)\sin 2\theta + (7/36)\sin 4\theta + (5/48)\sin 6\theta$
		±2	$(4/9)\sin 4\pi J\tau_2$	$(7/72)\sin 2\theta + (1/9)\sin 4\theta + (1/24)\sin 6\theta$
		±3	$(2/9)\sin 6\pi J\tau_2$	$(5/144)\sin 2\theta + (1/36)\sin 4\theta + (1/144)\sin 6\theta$
	Decoupled		$(8/9)\sin 2\pi J\Delta + (8/9)\sin 4\pi J\Delta + (4/9)\sin 6\pi J\Delta$	$(8/9)\sin 2\theta + (8/9)\sin 4\theta + (4/9)\sin 6\theta$

^{a} Each intensity factor must also be multiplied by the ratio of γ_D to γ_C.

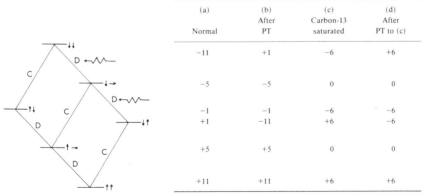

	Relative Populations			
	(a) Normal	(b) After PT	(c) Carbon-13 saturated	(d) After PT to (c)
	−11	+1	−6	+6
	−5	−5	0	0
	−1	−1	−6	−6
	+1	−11	+6	−6
	+5	+5	0	0
	+11	+11	+6	+6

Fig. 1-35. Energy-level diagram for a CD system.

ratios of ^{13}C and ^{2}H are in the ratio of 5/3, and that the carbons are initially at equilibrium in the magnetic field. In order to deal with integers rather than fractions, the energy differences of the carbon and deuterium transitions have been assigned the relative numbers 10 and 6, respectively. The populations "after polarization transfer" refer to the situation after the deuterium polarization for those systems in which the carbon spin which has the orientation represented by the downward arrows (at the upper right in the diagram) has been reversed. Notice that there is no change in the state for which $m_I = 0$, because the effects from the two transitions cancel, and that the populations are interchanged between the states for which $m_I = +1$ and -1.

From the differences in population between the initial and final levels of a transition, the relative intensity of the transition can be calculated: 10 for each of the normal lines, and -2, $+10$, and $+22$ for the perturbed lines. If we calculate the change in intensity for each line—or if we assume that the carbons are initially saturated so that only polarization-transfer-induced signals appear—the ratio of the three lines becomes $-12:0:+12$ or $-1:0:+1$. To obtain the intensity of the decoupled spectrum after refocusing, we can add the two lines of magnitude 12, yielding a sum of 24, compared to the sum of 30 for the three unperturbed lines. Thus we expect a ratio of 4/5 for the INEPT intensity versus the normal intensity; this reduction, of course, is more than compensated by the more rapid acquisition of data in the INEPT experiment.

Exercise: By construction of energy-level diagrams like that in Fig. 1-35, show that the relative intensities in the multiplets for CD_2 and CD_3, assuming the carbons initially saturated, are $-1:-1:0:+1:+1$ and $-1:-2:-2:0:+2:+2:+1$, respectively. It will be necessary to allow for the relative degeneracies of the various transitions (1:2:3:2:1 for CD_2, for example). Reference 156 gives similar diagrams, in a slightly different form. Consider how the intensities of the component lines in the CD_2 multiplet should be added to obtain the decoupled intensity for the carbon resonance.

DEPT or UPT can be used for deuterium-carbon transfer in much the same way as INEPT, and the pulse length θ is related to the refocusing interval Δ for INEPT in the same way as for proton-carbon transfer: $\theta = \pi J \Delta$. An angle of 60°, corresponding to Δ of $1/3J$, gives only CD responses while an angle of 27° corresponds to the Δ interval of $3/20J$ recommended for observing all multiplet types and 45° corresponds to the maximum CD intensity. Nakashima, McClung, and John (157) have analyzed the use of DEPT for spectral editing in considerable detail. In their experiments, they added an initial 90°(C) pulse, coincident with the 90°(^{2}H) pulse, in order to suppress natural carbon magnetization. Their expressions for relative

1. BASIC METHODS AND SIMPLE PULSED EXPERIMENTS   81

intensities of the various components of coupled multiplets are given in Table 1-8. The intensities for decoupled systems, plotted in Fig. 1-36, are simply the sums of the intensities for the three, five, or seven lines of the multiplets for one, two, or three deuterium atoms, respectively. These results are consistent with the theoretical predictions and experimental confirmations described by others, including those of Bendall, Pegg, and co-workers (152, 158).

In practice, partially rather than fully deuterated molecules are likely to be encountered. It is then necessary to make distinctions such as those between CDH_2, CDH, and CD, each of which has one deuterium atom, as well as to differentiate carbons according to the number of attached deuteriums. Two methods have been proposed for making both determinations in one experiment. In one procedure, a DEPT sequence designed to separate the resonances according to the number of attached protons is followed by a spin-echo sequence which is adjusted on the basis of $J(CD)$ to edit according to the number of deuteriums (33, 159). In the other method,

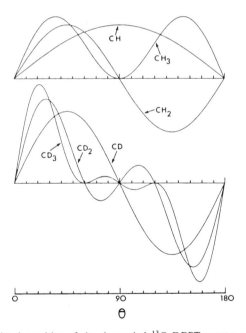

Fig. 1-36. Relative intensities of the decoupled ^{13}C DEPT spectra of CD_n systems as functions of θ (lower plot), together with the relative intensities of decoupled ^{13}C DEPT spectra of CH_n systems (upper plot). The intensities for decoupling of spin 1 have a period one half that for the intensities of decoupling spin $\frac{1}{2}$. [From Nakashima, McClung, and John (157).]

UPT editing for the number of attached deuterium atoms is followed by gated proton decoupling, which modulates the spin echoes according to the number of hydrogens in the group (160).

I. Some Experimental Considerations in Polarization-Transfer Experiments

In small molecules, the T_1 of hydrogen is usually much shorter than the T_1 of carbon-13, nitrogen-15, or silicon-29, and the shorter recycle times made possible by polarization transfer provide a substantial advantage in use of these methods. For larger molecules, the T_1 for carbon-13 is relatively short, and at the same time the proton T_2 may be so short that proton magnetization is rapidly lost; as a result, there may not be much gain in sensitivity achieved by polarization transfer methods (104).

In comparing the sensitivity benefit from polarization transfer with that from NOE, one finds that the increase per scan is nearly the same for both methods when protonated carbons are being observed. For silicon-29 and nitrogen-15, where the NOE is negative and may be unfavorable, the advantage of polarization transfer, provided the nucleus is coupled to hydrogen, is usually very great. However, there is likely to be difficulty in the polarization-transfer experiment when a large range of magnitudes of coupling constants is encountered, as often is true if several different nitrogen-15 nuclei are to be observed, or when chemical exchange of the hydrogens attached to nitrogen smears out the coupling interactions.

For nonprotonated nuclei, longer-range interactions must be utilized for polarization transfer, and, since the J values for these are fairly small in magnitude, long interpulse intervals are required, during which some polarization may be lost by proton relaxation. Here, the shorter INEPT sequence is to be preferred over the DEPT sequence. In any event, even for carbon-13, one is likely to be better off in most instances with polarization transfer than by relying on the NOE, which is usually small for nonprotonated nuclei, since they are usually physically distant from protons. If carbon-13 nuclei are to be polarized by long-range couplings, there is also the problem that the magnitudes of H-H couplings are about the same as those of C-H long-range couplings, and precession as a result of the homonuclear couplings may cause loss of magnetization.

Schenker and von Philipsborn (161) have described calculations which enable prediction of INEPT and DEPT intensities for different combinations of coupling constants in various spin systems, and thus aid in optimizing the experiments for particular molecular types. They also evaluated off-resonance effects for an AX system (162) and recommended phase cycling schemes which are similar to those analyzed in Chapter 3.

Polarization transfer between less commonly encountered nuclei may require instrumental modifications. If the frequencies of the two nuclei are even moderately close, such as those of deuterium and carbon, filters may be required to prevent pulses applied to one nucleus from disturbing the channel for the other nucleus. For inverse transfers, obtaining a decoupled spectrum may require some sort of special arrangement for X-nucleus decoupling, and, in some spectrometers, may require reconfiguration of the hardware or software so that a proton spectrum may be acquired despite the fact that the proton pulses are coming from the decoupler channel.

Pulse imperfections are particularly likely to be troublesome in 180° pulses as compared to those with smaller angles. Composite pulses, described in a later section, may substantially improve the performance of the refocusing pulses, and it has been found desirable to cycle the phase of each 180° pulse through all four 90° steps. In another approach, Bax and Sarker (163) have proposed that the refocusing pulse before decoupling simply be omitted, and have suggested that the resulting baseline distortion that is observed on attempting to phase the spectrum is less troublesome than errors introduced by pulse imperfections. To what extent this is true probably depends on the characteristics of a particular spectrometer.

Finally, a helpful discussion of the considerations involved in setting up INEPT and DEPT experiments has been provided by Morris (164).

IV. Experimental Aspects of Pulsed NMR

A. Nature and Properties of Pulses

Special analysis of the behavior of rf pulses is important for several reasons. The ideal which was aimed at in the earlier days of FT spectroscopy was a pulse of rectangular shape, designed particularly for broadband excitation independent of resonance offset, with constant rf phase and amplitude, and with leading and trailing edges as straight and steep as possible. Even for simple broadband excitation, however, most of the power of a rectangular pulse is outside the spectral region and therefore wasted. As pulse sequences become more complex, any imperfections in the pulses tend to have cumulative results, such as artifacts in 2D spectra. Problems with flip angles of 180° or more are substantially more troublesome than those for angles of 90° or less. When it is desired to excite a particular region of the spectrum selectively or according to some particular profile, then modification of pulse shape turns out to be particularly useful. Other means of improving performance include the substitution of composite pulses for simple pulses and the cycling of pulse phases as a sequence is repeated. In this section, we shall discuss these methods, the use of clusters of pulses such as "bilinear" groupings, and procedures for pulse calibration.

1. Compensation of Errors by the Use of Composite Pulses

Pulse "imperfections" may arise because the pulse length is not exactly that required for the angle through which the magnetization is to be rotated, or because of the variation of the effective B_1 over the sample volume, as a consequence either of probe design or of the heterogeneity of the sample. Closely related are differences in pulse effects because there is a frequency offset, which may vary for different resonances in a sample, between the rf carrier and the nuclei to be excited; here it is not a matter alone of deviation of the effective flip angle from the desired value, but the axis about which the magnetization is rotated depends on the direction of the effective field, which varies for nuclei at different resonance frequencies.

"Composite" is the term applied to a pulse which is divided into segments by the introduction of one or more phase shifts during its course. A great many combinations can be generated which have different relative lengths as well as relative phases of the several pulse segments. Composite pulses may, to varying degrees, compensate for B_1 errors, for resonance offsets, or for both. An extensive review of the theory of composite pulses has recently been written by Levitt (165), and numerous methods of deriving optimum structures of composite pulses for various purposes have been proposed. Here we shall first describe some of the more commonly used forms, which are based on phase shifts in steps of 90°, steps that are available on most spectrometers, and portray the mode of action of a few types by vector models, then mention more complex types, and finally indicate the applicability and limitations of various types of composite pulses.

Levitt and Freeman (166) introduced a composite 180° pulse in the form $90°_x\, 180°_y\, 90°_x$ as an improved means of inverting magnetization which is in the z direction. An example of this use is the initial pulse in the inversion-recovery method of measuring relaxation times. To see how this composite works, suppose the effective pulse length is $90° - \Delta$ for a nominal 90° pulse and therefore $180° - 2\Delta$ for a nominal 180° pulse. The first part of this composite rotates $+z$ magnetization in the zy plane nearly to the direction of the y axis, but short of this axis by an angle Δ. The $180°_y$ pulse now rotates the magnetization about the y axis so that the residual $+z$ component after the first pulse becomes a $-z$ component—at least, nearly so, with a small error in the nominal 180° angle producing at most a small deviation from the zy plane. Neglecting this small "second-order" error, the magnetization vector is now at an angle Δ below the y axis, and the second 90° segment of the composite turns the vector the remaining portion of the way to the $-z$ axis. A $90°_y\, 180°_x\, 90°_y$ pulse is similarly effective for rotation about the y axis, as shown in Fig. 1-37, yielding a final result which is the same, except for possibly a very slight, second-order phase difference caused by the error in the 180° segment of the composite.

1. BASIC METHODS AND SIMPLE PULSED EXPERIMENTS   85

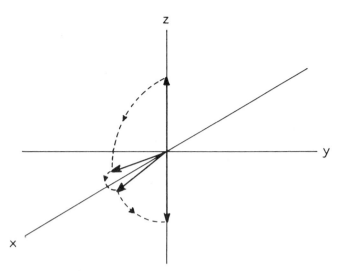

Fig. 1-37. The vectors trace out the path of the magnetization for a nominal $90°_y\ 180°_x\ 90°_y$ composite pulse where the components rotate the magnetization through an angle somewhat less than the nominal. The 180° pulse brings the magnetization not quite back to the xz plane, since it has less than the nominal effect, and thus there remains a small phase error in the magnetization at the end of the composite rotation.

The Levitt-Freeman 180° composites are examples of the general form $90°_\phi\ \theta°_{\phi\pm 90}\ 90°_\phi$, with $\theta = 180°$ and give good compensation for B_1 errors, but leave much to be desired in correcting for off-resonance effects. Better offset compensation is obtained if θ is greater than 180°, and solution of applicable equations shows that for $\theta = 270°$ inversion is independent of offset for small offsets (167, 168). For larger offsets, however, the value of θ should be less than 270°, with 240° often recommended as a good compromise for a moderate range of offsets (169). Values less than 240° may be better for compensation at large offsets, but with these values the compensation is not very uniform over the offset range (168). Compensated inversion may also be achieved by a $270°_x\ 180°_{-x}\ 90°_x$ or a $90°_x\ 180°_{-x}\ 270°_x$ composite (170).

There have been proposed longer spin-inversion composites which give better combinations of corrections for B_1 and offset than do the three-pulse composites. Among these are a five-pulse sequence, $90°_x\ 200°_y\ 80°_{-y}\ 200°_y\ 90°_x$, which inverts all resonances over a width that may be nearly twice as great as the equivalent of the rf field strength (168). Levitt and Ernst (171) developed a recursive expansion method for constructing composite pulses to yield arbitrary flip angles, which led them to evaluate, along with the other 180° composites, the sequence $90°_y\ 90°_x\ 90°_{-y}\ 180°_x\ 90°_{-y}\ 90°_x\ 90°_y$. This sequence was found to be outstandingly effective in correcting for combinations of resonance offsets and rotation angle errors. It can be represented in an alternative manner as a sequence of 90° segments in the symbolic form Y X $\bar{\text{Y}}$ X X $\bar{\text{Y}}$ X Y, where each letter denotes a nominal 90° rotation about the indicated axis, with an overbar indicating rotation about the negative axis. This notation has been adopted in the development of complex pulse composites, especially where these are applied to produce broadband, frequency-independent excitation for decoupling.

The effort to find pulse composites with *dual compensation*—for both offset and B_1 errors—was extended by Shaka and Freeman (172) and Starčuk and Sklenář (173) on the basis of the idea of a "reversed nutation" pulse, proposed by Levitt. A normal, or "right-handed" 90° nutation about the $-x$ axis is indicated by the symbol $\bar{\text{X}}$, while a reverse, or "left-handed," nutation can be represented by $[\bar{\text{X}}]^{-1}$. Although reverse nutation cannot be achieved directly, since the sense of precession of a nucleus is determined by the sign of the magnetogyric ratio, it can be accomplished indirectly by the trick of removing a 90° pulse from the beginning or end of a sequence which is cyclic or nearly cyclic. Thus one might begin with any one of the sequences 4X 4$\bar{\text{X}}$, 4$\bar{\text{X}}$ 4X, 4Y 4$\bar{\text{Y}}$, or 4$\bar{\text{Y}}$ 4Y, each of which is cyclic over a moderate range of offsets, and remove one 90° pulse from either the beginning or end, leading to combinations such as $3\bar{\text{X}}\,4\text{X} = [\bar{\text{X}}]^{-1}$ or $3\bar{\text{Y}}\,4\text{Y} = [\bar{\text{Y}}]^{-1}$.

The effect of a composite such as $\text{X}[\bar{\text{X}}]^{-1}$ or $[\bar{\text{X}}]^{-1}\text{X}$ should be a 180° rotation, pretty much independent of offset. Rotation about the y axis can be inserted to compensate for pulse length error or B_1 variation, yielding $[\bar{\text{X}}]^{-1}\text{Y}[\bar{\text{Y}}]^{-1}\text{X}$. Substitution in this composite of the reverse nutation expression from the preceding paragraph leads to the expression $3\bar{\text{X}}\,4\text{X}\,\text{Y}\,3\bar{\text{Y}}\,4\text{Y}\,\text{X}$, which corresponds to $270°_{-x}\,360°_x\,90°_y\,270°_{-y}\,360°_y\,90°_x$, termed GROPE-16 (172). Alternatively, one might follow Levitt (165), who chose $[\bar{\text{X}}]^{-1}$ to be 4X 3$\bar{\text{X}}$ and $[\bar{\text{Y}}]^{-1}$ to be 4$\bar{\text{Y}}$ 3Y, leading to 4X 3$\bar{\text{X}}$ Y 4$\bar{\text{Y}}$ 3Y X. Experimental tests of inversion by these composites showed that, apparently fortuitously, they perform much better than expected.

That inversion performance could be further improved by changing the lengths of the pulse segments away from integral multiples of nominal 90°, although using only 180° phase shifts, was demonstrated by Starčuk and Sklenář (173). Thus if the starting point is $90°_x\,270°_{-x}\,360°_x$ (that is, X 3$\bar{\text{X}}$ 4X), one may consider the effects of varying the value of a in composites of the form $90°_x(270° - a)_{-x}(360° - a)_x$. For a from about 45° to 70°, good performance is obtained with respect to bandwidth and accuracy of spin inversion. Increase of a above this range increases the bandwidth but decreases the uniformity of spin inversion over the bandwidth. Division of the initial 90° pulse into two segments to give $22.5°_{-x}\,112.5°_x\,202.5°_{-x}\,292.5°_x$ results in performance about equivalent to Levitt's five-pulse composite mentioned above (168).

If the magnetization, rather than being inverted by 180°, is to be turned into the xy plane, a simple 90° pulse has a reasonable compensation for frequency offset. Variation in frequency offset corresponds to rotation about a variable tilted axis, and the magnetization ends up close to the xy plane, but with a phase depending on the offset value. Normally, in single-pulse experiments, this phase variation is linear and can therefore be compensated by the phase correction in the final spectrum. Further compensation can be effected by using a $10°_{-x} \tau 100°_x$ composite, in which the interval τ has a length of several microseconds, where the exact value depends on the magnitude of B_1 and is chosen to minimize z magnetization after the pulse (169).

To compensate to first order for inhomogeneity, a pulse pair such as $90°_y 90°_x$ or $90°_x 90°_y$ is applicable (169). The first half of the pair turns the magnetization approximately to the xy plane, and the second 90° rotation takes any residual $\pm z$ component to the $\mp y$ direction. Again, the error is converted into a phase distortion which is fairly close to linear and thus readily compensated in data processing.

Composite pulses can also be used as a means for rotating magnetization about the z axis, a transformation which cannot be accomplished at all by a single pulse. For example, a combination $90°_x \theta°_y 90°_{-x}$ can be shown to be equivalent to a rotation of $\theta°$ about the z axis (174). This type of rotation can be used to simulate a phase shift of $-\theta°$, where hardware does not permit shifts of an arbitrary angle in the rf phase. As will be discussed in Chapter 4, this may be valuable in generating multiple-quantum coherences of a desired order. Since the central part of the pulse is sensitive to rf inhomogeneity, a long z pulse may be used in place of a static field gradient to randomize magnetization of a sample (165).

We turn now to some more elaborate versions of composite pulses for rotations about the x and y axes. The recursive method of generating pulses of better performance from 90° flip-angle segments, developed by Levitt and Ernst (171), was mentioned above. Other early studies of composites were based on similar geometrical arguments or on an approach which was essentially that of trial and error, with numerical simulations to test the results. Several, more advanced, mathematical tools have been developed to facilitate design of sequences with optimized rotation performance as well as reduced phase error. Elimination of phase error was emphasized by Tycko and co-workers (175), who, using the Magnus expansion, found a composite $385°_x 320°_{-x} 25°_x$ which is closely equivalent to a 90° rotation, as well as a seven-step composite for spin inversion: $336°_x 246°_{-x} 10°_y 74°_{-y} 10°_y 246°_{-x} 336°_x$. Shaka's mathematical analysis (176, 177) using numerical nonlinear optimization has led to development of rather elaborate inversion composites. Thus there was found an eleven-segment sequence using only 180° phase shifts, with the form 86.8° $\overline{163.8°}$ 155.1° $\overline{316.4°}$ 105.9° $\overline{67.5°}$ 254.1° $\overline{133.0°}$ 313.7° $\overline{125.3°}$ 70.4°, where the overbar indicates a phase shifted 180° from that of the first pulse. This composite gives good frequency offset compensation over a bandwidth four times the equivalent of B_1, and a 31-segment composite was devised with a bandwidth of $6.4B_1$. The overall time required by this composite is of the order of a millisecond.

For further flexibility, the limitation of phase shifts to multiples of 90°, a restriction imposed by the four-step phase shifting hardware of many spectrometers currently in use, has been relaxed in several calculations aimed at optimizing composites, either with or without the extension to nonintegral multiples of the 90° flip angle. Thus Tycko and Pines (178) used phase shifts in multiples of 60° and an iterative scheme to develop composites that compensate for both B_1 variation and resonance offset. Tycko and co-workers proposed for spin inversion a sequence of the form $180°_0\ 180°_{105}\ 180°_{210}\ 360°_{59}$ (175) or for 90° rotation from the z axis $180°_{60}\ 90°_{120}\ 180°_{60}$ (179) or $270°_0\ 360°_{169}\ 180°_{33}\ 180°_{178}$ (175). In these representations, the subscripts denote the phase shift in degrees from the x axis in the rotating frame. For a 90° rotation from the z direction, Levitt (165) concludes that the composite $90°_{180}\ 180°_{300}$ compensates to second order for rf inhomogeneity.

Although improvements in high-resolution liquid-state NMR instrumentation are making the need for compensation less critical for experiments performed on these instruments, the demands of spatial imaging experiments where the sample is necessarily quite inhomogeneous have accentuated the requirement for more effective compensation over wider ranges of B_1 and have stimulated the development of mathematical methods for analysis of composite pulses, as illustrated by recent work by Lurie (180, 181). He found, for example, a three-segment pulse for inversion from the z axis, $88°_0\ 352°_{110}\ 88°_0$, which has properties comparable to Tycko's four-segment pulse (175).

In the application of composite pulses, the functions to be performed in a particular experiment must be carefully analyzed. For example, whether the pulse is to be applied to magnetization in equilibrium in the z direction or to magnetization already in the xy plane makes a significant difference in the requirements. One must keep in mind that the effect of a composite pulse is often to convert amplitude distortion into phase distortion, which may or may not be acceptable in subsequent steps. If relaxation times are short, the overall length of the composite may be a consideration in its choice.

Composite 180° pulses of the type $90°_x\ 180°_y\ 90°_x$ are quite suitable for inversion–recovery T_1 measurements where the magnetization is initially along the z direction, and the resulting phase distortion has no adverse consequence. However, it was originally thought that they would be unsuitable as refocusing pulses in spin-echo experiments, until Levitt and Freeman (167) showed that they provided effective compensation of B_1 variation for this purpose if only alternate echoes were observed, since phase distortion appeared only on odd-numbered echoes. In Section II,A,2, we described the use of the Meiboom–Gill modification to compensate for the cumulative effects of errors in the 180° refocusing pulses, but this method of compensation breaks down when there is echo modulation as a result of homonuclear spin–spin coupling. The composite 180° pulse, however, is effective in this error compensation. Furthermore, use of the composite pulse in the experiment in which there is only heteronuclear coupling affords even better compensation for pulse-length errors than the Meiboom–Gill phase shift alone. It has also been suggested (175) that, if composites are used for both 180° and 90° pulses in spin-echo experiments, both should have minimum phase distortion. If offset compensation is required, one may use a simple 90° pulse and a composite 180° pulse, and then correct for phase distortion by the usual linear phase correction.

Schenker and von Philipsborn (162) tested the use of composite 180° pulses in the INEPT and DEPT experiments and found that, applied to carbon-13, they improved polarization transfer at large offsets. For that nucleus, the 180° pulse is applied to magnetization at equilibrium; for hydrogen, on the other hand, where the 180° pulse acts as a refocusing pulse, a composite leads to results poorer than those from a simple pulse.

Levitt and Ernst analyzed the use of composite pulses in the DEPT experiment (182). They pointed out that the final pulse of flip angle θ on the hydrogens could be replaced by a pair of 90° pulses, the first of which becomes a part of the excitation sequence and the second of which is a "read" pulse. Selection of a particular multiplicity is then effected by changing the phase of the excitation sequence, and appropriate composites may be substituted for the several pulses on hydrogen.

Levitt and Ernst also discussed in considerable detail the best methods of compensation in the INADEQUATE experiment, in which connectivities are determined by selecting out double-quantum resonances and suppressing the normal single-quantum responses. There are two coherence-transfer steps in this experiment, one to generate multiple-quantum coherence and the second to transform it back to observable single-quantum coherence; combined with the considerable spectral widths in carbon-13 spectroscopy to which INADEQUATE is usually applied, this situation leads to large sensitivity to off-resonance effects and pulse imperfections. The sequence they recommend is the following:

$$[180_x^\circ\ 180_{-x}^\circ\ 90_x^\circ - \tau/2 - 270_x^\circ\ 180_{-x}^\circ\ 90_x^\circ - \tau/2 - 90_x^\circ]_x$$
$$-[180_y^\circ\ 180_{-y}^\circ\ 90_y^\circ\ 270_{-y}^\circ\ 180_y^\circ\ 90_{-y}^\circ\ 90_x^\circ]_\xi \mathrm{Acq}_\psi$$

Enclosed within the first set of square brackets is the excitation sequence, and enclosed within the second set of brackets is the read sequence. The phases of the pulses within each of the two segments are given in relation to the overall phase of the segment, which is indicated by the subscript χ or ξ following the final bracket of the segment. These phases and the acquisition phase ψ are cycled in the usual way required for INADEQUATE, as described in Chapters 2 and 4. On inspection of this sequence, it may be observed that, omitting the intervals in the excitation sequence, the two sequences are related, with the read sequence components derivable from the excitation sequence by adding 90° to the phase of each of the first three units and subtracting 90° from the phase of each of the next three units.

In two-dimensional J spectroscopy, composite 180° pulses were found to be useful in suppressing artifacts when applied to carbon-13 in the heteronuclear experiment. These spurious signals arise from CH_2 or CH_3 groups in which not all hydrogens within the group have been inverted by the pulse on hydrogen, and the signals cannot be canceled by phase cycling (183). In 2D heteronuclear correlation experiments, to be described in subsequent chapters, a 180° pulse on the X (or nonhydrogen) nucleus in the middle of the initial evolution period may serve to relabel spin states in such a way as to cause the differing extents of precession which result from spin–spin interactions to be refocused, thus effectively decoupling the X nucleus from the hydrogens. A composite can be employed without difficulty here, since the magnetization of the X nucleus, usually carbon-13, is initially at equilibrium, and use of the composite does reduce artifacts. The situation is like that for the first 180° pulse on the nuclei to be polarized in the INEPT experiment, mentioned above.

In some circumstances, phase cycling, to be described somewhat more in Section IV,A,4, can be used to eliminate phase anomalies introduced by the use of composite pulses (184, 185). Phase cycling is sometimes an alternative means of compensating for pulse imperfections, but it is preferable to employ composite pulses and phase cycling as complementary techniques which support and reinforce one another.

2. *Selective Excitation and Tailored Pulses*

We now consider methods of selectively exciting nuclei within a narrow frequency range. Applications in which this is required are varied, including spectral analysis, selective elimination of a solvent resonance, excitation at frequencies on either side of a solvent peak, and excitation of individual resonances for NOE measurements. In spatial imaging, the needs for selectivity are based on the requirement that excitation be limited to nuclei within a narrow range of magnetic field in order to provide adequate spatial resolution; these needs have recently been analyzed in relation to pulse designs by several groups (186, 187).

One approach to selective excitation is to employ a very weak pulse of long duration, a so-called "soft" pulse (188), as described above for selective population transfer. This type of pulse is, of course, limited primarily to excitation in the vicinity of a single frequency, rather than permitting excitation over a band of frequencies or at several selected frequencies; what may not be realized is the circumstance that any rectangular pulse, when Fourier-transformed, gives rise to a sinc function, $(\sin x)/x$, which has side lobes that produce low-level excitation at various frequencies away from the carrier. These side lobes are more troublesome at larger flip angles, especially above 90°. Indeed, a "hard" rectangular pulse, while providing broadband excitation, is inefficient because of the substantial dispersion of energy into side lobes outside the region over which a uniform effect is required.

As an alternative to a soft pulse for selective excitation, Bodenhausen, Freeman, and Morris (189, 190) proposed DANTE, comprising a series of m short hard pulses, each of tip angle α such that $m\alpha$ amounts to the desired tip angle, usually 90°. This has the effect of generating sidebands offset from the carrier by a frequency equal to the pulse frequency, or $1/t_p$, where t_p is the time interval between centers of the pulses. For nuclei precessing in the rotating frame at a frequency equal to this offset, the effect of the pulses in tipping the magnetization toward the xy plane is cumulative, because the nuclei precess by exactly 360° in phase angle between successive pulses, while for nuclei precessing at any other frequency, the successive flips are in various directions and the net turn away from the z axis is very small.

As an example, suppose a series of 50 pulses, 2 msec apart, is applied, for a total time of 100 msec. A bandwidth of 10 Hz at an offset of $1/(2 \times 10^{-3}$ sec) or 500 Hz from the carrier is achieved. The selectivity is better, the longer the total pulse time and the shorter each pulse, but very short pulses are likely to have poor shape. A trick of combining positive-going and negative-going pulses of slightly different lengths to give the desired net result, equivalent to that of a very short pulse, has been proposed. Higher-order sidebands at integral multiples of $1/t_p$ are also produced by DANTE, but phase cycles can be devised to minimize these and concentrate power in the first sideband, as well as to emphasize excitation on one side of the carrier and reduce the power delivered in the other frequency direction from the carrier. Of course, one may use DANTE to irradiate at two frequencies by placing the carrier exactly in the middle. One advantage of excitation with hard pulses in the manner used in DANTE is that it permits the transmitter or decoupler output to be left at the usual level, rather than requiring it to be switched between levels.

A rather different approach to concentration of the excitation from a pulse in a specific region is the use of a shaped or tailored pulse, for which the amplitude and perhaps also the phase is varied according to a preset program. One possibility of generating excitation in a rectangular window would seem to be to apply an impulse in the shape of the Fourier transform of a rectangle, the sinc function. However, the sinc function has a complicated shape including the side lobes described above, and somewhere these lobes must be truncated, leading to some distortion in the excitation pattern (191).

When the exciting pulse is amplitude modulated according to an approximate sinc function, sidebands are produced at equal distances on both sides of the carrier. Temps and Brewer (192) showed how frequency modulation could be used with the sinc function to place the excitation band on only one side of the carrier. However, this requires a synthesized signal generator capable of accepting simultaneously amplitude and frequency modulation. Several groups have carried out calculations showing how small modifications in the shape of the sinc function should lead to better performance for a selective 90° pulse (193, 194).

Another shape function for a pulse is the Gaussian, which was found to be much more selective than a rectangular pulse of the same power and to give an absolute-value-mode pattern that is a slightly distorted Gaussian, somewhat flattened in the middle (195). Although the Gaussian impulse has no side lobes about which to be concerned, it, too, must be truncated somewhere, and Bauer and co-workers chose the point at which it has dropped to 1% of the maximum. An additional advantage of this pulse shape is that it produces an almost linear variation of phase with frequency offset, whereas the phase variation resulting from a rectangular pulse varies discontinuously with offset frequency. The profile of the Gaussian pulse is given by the expression

$$B_1(t) = B_1^0 \exp[-a(t-t_0)^2] \qquad [17]$$

McCoy and Warren (196) proposed using a Gaussian pulse further shaped by multiplication by a Hermite polynomial. For a 90° pulse, the polynomial to give optimum shape was calculated to be $(1 - 0.667t^2)e^{-t^2}$; this function was found to yield a satisfactory 180° pulse as well.

Silver, Joseph, and Hoult (197) solved the Bloch equations in the Riccati form and noted that, for a 90° pulse, introduction into the data collection of a short time delay of about one-tenth the width at half height of the sinc-shaped excitation function produced results quite satisfactory and almost equivalent to those from the Riccati solutions. For a 180° pulse, however, it was necessary to employ a function of the form of the hyperbolic secant:

$$\Omega = \omega_{1x} + i\omega_{1y} = -\gamma(B_{1x} + iB_{1y}) = \Omega_0[\text{sech}(\beta t)]^{(1+i\mu)} \quad [18]$$

where μ is a real constant. This function requires simultaneous amplitude and phase modulation. The value of μ was taken to be 5.0 for a 180° pulse, and the width of the selected region is $\pm\beta\mu$. It was found, rather surprisingly, that, above a critical power threshold, the profile of the 180° inversion is independent of the power level used in the pulse.

Baum, Tycko, and Pines (198) calculated the effects on excitation bandwidth of pulses with continuously varying phase. Although this type of pulse is not simple to generate experimentally, they were able to deduce from their results the forms of some composite pulses consisting of phase-shifted segments which should give uniform broad-band excitation.

In addition to designing the pulse shape which is to be used, it is also necessary to devise a means for generating that shape. The earliest method described for producing an NMR excitation profile of a desired pattern, with particular application to selective excitation, was that of Tomlinson and Hill (199). A discrete Fourier synthesis of the desired frequency spectrum was computed, stored in digital form in a computer memory, and used to modulate, through a function generator, the amplitude, width, or phase of a series of equally spaced pulses. The pulse interval was about 500 μsec, and the pulses were up to 30 μsec in length. The receiver was turned off for an interval of 40 μsec surrounding the pulse. Compared to use of a filter to select a narrow band of frequencies, the Fourier synthesis method has the advantage of producing no phase effects as well the ability, at least in principle, to generate several frequencies simultaneously. The technique was applied to solvent suppression, selective T_1 determination, and selective decoupling.

Subsequent practice in pulse shaping, used in generating most of the tailored waveforms which have been described, has been to store the profile information in RAM or PROM of some type, after dividing the pulse into a number of steps, usually between 64 and 256. The output from successive addresses is then clocked out at desired time intervals and fed through a smoothing filter to eliminate the effects of the steps between pulse increments. In a sense, this procedure can be viewed as using a series of pulses with no interval between, in place of Tomlinson and Hill's series of pulses with appreciable intervals between them.

Brandes and Kearns (200, 201) recently demonstrated modulation of a 10-MHz rf signal with an audio signal which had been shaped by a filter. The modulated rf was then mixed with the carrier from the NMR transmitter. Frequencies near the carrier were obtained with a low-pass audio filter, while a high-pass filter suppressed frequencies in the vicinity of the carrier, and a band-pass filter produced excitation at two frequencies symmetrically placed with respect to the carrier. The resulting phase characteristics were good, but the lengths of the pulses required were longer than those of corresponding soft pulses.

3. *Phase Shifting*

In Section I,B we introduced the idea of shifting the phase of an rf pulse in order to modify the axis, in the rotating frame, about which nuclear magnetization is rotated by the pulse, and we have described the use of phase shifts in several experiments and in the design of composite pulses. The reader should be aware by this point that a phase shift corresponds to an interruption in the regular oscillation of the rf wave such that subsequent cycles begin at times which are regularly displaced from starting times which would be extrapolated from the wave form before the shift: a 90° shift corresponds to a time displacement equal to one-quarter of a period of the wave in the positive direction or three-quarters of a period in the negative direction. More complex experiments to be described later in this book will involve extensive application on phase shifts.

The method of phase shifting in most commercial spectrometers produced in the period of about 1974 to 1984 was based on an analog device limited to 90° phase increments. It is particularly the increase in interest in generation of multiple-quantum coherences of order higher than two, which requires phase shifts in increments of less than 90°, that has led manufacturers of instruments recently to incorporate shifters of higher resolution. However, the need to retrofit older spectrometers with the necessary phase-shift capability has resulted in the design for this purpose of a number of units which have been described in the literature, and we shall list several of these. In addition, some tricks for achieving phase shifts by sequences of pulses will be described.

Phase shifting by hardware is usually done at the intermediate frequency of the spectrometer, because it can then be effective for any nucleus to be observed, since the shift is preserved when the frequency on which it has been effected is mixed with another frequency, as well as when the base frequency is multiplied upward.

A phase shift may be introduced by placing in the circuit a *delay line*, which simply causes a time lag in signal propagation, a lag which is of a fixed magnitude for a given frequency. To achieve capability for shifts of various amounts, delay lines of differing lengths may be arranged in parallel, one for each shift desired, and multiplexed under computer control, or a smaller number of delay units may be added in various combinations in series, again with the switching arrangement under computer control. Shifters of either of these types may be commercially obtained, but are expensive, in part because of the lengthy cut-and-try procedures involved in setting them up and calibrating them.

Several types of analog shifters designed primarily for NMR have been described. Cosgrove *et al.* (202, 203) used a quadrature hybrid with LC circuits including voltage-controlled "varactor" diodes to adjust the phase shift. Boyd and co-workers (204) devised an analog shifter in which the output is a series of rectangular pulses from a monostable vibrator at a frequency determined by the input frequency but with the trigger to the monostable circuit delayed by an amount governed by the length of a voltage-controlled pulse from another monostable.

In general, analog circuits for phase shifting are relatively simple to construct, but they require calibration and may drift with time. Digital shifters are more complex, but, like most digital circuitry, provide an unambiguous, stable output.

The essential principle of a digital shifter is division of a high-frequency clock, operating at some multiple of the desired frequency, down to that desired frequency, but with the initial trigger for the low-frequency pulse supplied at some controllable point in the series of high-frequency pulses. Thus two or more low-frequency signals might be derived from the high frequency by having the trigger occur at different numbers of high-frequency pulses from an "origin." Suppose, for example, a counter is activated by a 140-MHz input and resets after 12 counts. The reset rate is then (140/12) MHz, and feeding the pulsed output through a shaper gives a sine wave of 11.66 MHz frequency. If the counter output is compared with a computer-set word in a comparator, the output of the comparator may be arranged to occur at any integral multiple of one-twelfth of a total cycle, or 15°. Thus the pulsed output is advanced by this phase angle compared to the direct output from the counter. This scheme was described by Bodenhausen and co-workers (205, 206). A somewhat similar arrangement was employed by Frenkiel and Keeler (207) with a series of N flip-flops, clocked at frequency f, with the input to the first stage the complement of the output of the last stage. The ouptut changes at a frequency equal to $f/2N$ and the phase shift is $360°/2N$. Outputs from various stages in the chain have different phases, and all the outputs are continuously available for use through an appropriate switching arrangement.

A phase shift corresponds to a rotation about the z axis, which cannot be achieved by a single pulse, but can be brought about by a $90°_x$ $\theta°_y$ $90°_{-x}$ composite (174). However, success of this approach requires a very uniform rf field, and multiple-quantum spectroscopists have found that suppression of unwanted orders of coherence is usually unsatisfactory when a "z pulse" is used in practice. Decorps et al. (208) employed a series of very small incremental pulses, alternately with phase x and with phase y, to implement a z rotation. Thus a series of $2n$ pulses, from 10 to 40 in number, represented by $\{[(\theta/n)\cos\phi]_0[(\theta/n)\sin\phi]_{90}\}^n$, is equivalent to a single θ°_ϕ pulse. This composite was termed a stacked orthogonal alternate pulse. A third trick to avoid hardware phase-shifter requirements is a *phase pulse* (209). Here the synthesizer frequency is simply shifted for a short period to a slightly different frequency and then returned to the original frequency. During the interval, the signal gains or loses a fraction of a cycle. Thus a frequency shift of 1 Hz for 1 msec corresponds to a phase difference of $10^{-3} \times 360°$ or $0.36°$. This method works only if the synthesizer has sufficient resolution and speed of response to effect the shift during the available time interval between pulses.

4. *Phase Cycling and Alternatives to Phase Cycling*

In both 1D and 2D experiments, *phase cycling* is employed to compensate for experimental imperfections or to emphasize desired magnetizations or coherences and cancel those which are to be eliminated. In a phase cycle, the phases of the pulses in an experiment are changed according to a regular pattern, either individually or in groups, as the experiment is repeated. In describing phase cycles, the most frequent practice has been to assume that the phase of the initial pulse is held constant and that phases of subsequent pulses are varied in relation to this initial phase, generally designated x. However, it is equally feasible, in description or in practice, to hold the phase of any later pulse or of the receiver constant and to cycle the phase of a preceding pulse in the opposite direction, since phases only have significance relative to one another.

Control of the *receiver phase* is achieved by the way in which data from the phase detector outputs are handled. A 90° phase shift is provided simply by interchanging the sections of computer memory to which the real and imaginary outputs of the dual phase detector are channeled. Furthermore, a 180° phase shift is simply obtained by negation of the sign of the output signal by subtraction in the computer memory rather than addition. Should any finer phase-discrimination steps relative to some preceding pulse be required in the receiver, then the receiver phase can be held constant and the pulse phase varied in smaller intervals as needed.

a. *Phase Cycling for Baseline Correction.* The first systematic use of phase cycling was in conventional pulsed FT spectroscopy as a means of compensating for dc voltage offsets generated by the hardware. Some number of pulses, usually one, two, or four, is applied with one phase relation of transmitter rf and phase detector setting. The phase of the

1. BASIC METHODS AND SIMPLE PULSED EXPERIMENTS 97

transmitter signal is then shifted by 180° for another group of pulses of the same number, and the "receiver phase" is shifted by 180° as described above. If the signals from the first group of pulses are positive, those from the second group are negative, and combination with opposite signs produces reinforcement; at the same time, the dc levels, of opposite sign, cancel one another.

b. *Phase Cycling for Quadrature Detection.* The two channels, adjusted to detect signals 90° different in phase from one another, used in quadrature detection, can usually not be exactly balanced in gain; the result is that, when the signals from the two channels are combined, there is a blip at zero frequency and there are images—signals reflected across the transmitter position and appearing with a frequency displacement from zero of incorrect sign. A scheme of phase cycling and data handling which overcomes this problem was developed by Stejskal and Schaefer (210, 211) and by Hoult and Richards (212), who named it CYCLOPS, for *cyc*lically *o*rdered *p*hase *s*equence.

This scheme can be represented diagrammatically as follows:

Transient	Transmitter phase	Output of phase detector A	Output of phase detector B
1	x or $0°$	Add to data table I	Add to data table II
2	y or $90°$	Add to data table II	Subtract from data table I
3	$-x$ or $180°$	Subtract from data table I	Subtract from data table II
4	$-y$ or $270°$	Subtract from data table II	Add to data table I

Data table I may be regarded as the v mode or absorption data; data table II is then the u mode or dispersion data. In the first two pulses, the functions of the two phase detector channels are interchanged. Thus the channel which contributes to the absorption mode on the first transient contributes to the dispersion mode on the second transient and vice versa, and each data set receives equal contributions from the two channels. The second pair of pulses replicates the first pair with all signs changed, and therefore extension to four pulses corresponds to the incorporation of baseline correction in the sequence. This combination of two pairs of pulses to give a four-pulse sequence is a simple example of a procedure commonly encountered, in which one cycle is nested inside another.

It is to be noted that in a commercial spectrometer, the direction—that is, the sign—of the phase shift in the transmitter is not necessarily established and may be either positive or negative. Thus the location of "$-y$" in the pulse sequence—whether it be in the second or fourth transient—may require empirical determination, based on the combination which gives the best result for image suppression.

The spectroscopist who wishes to program new sequences on a spectrometer with commercial software must be concerned with the way in which quadrature phase cycling has been built into the software, that is, whether it continues to be operative in the background or not during more extensive pulse sequences than a simple one-pulse experiment. If it is not disabled, then it must be taken into account in the design of a phase cycling scheme.

c. *Exorcycle.* This phase-cycling sequence, proposed by Bodenhausen, Freeman, and Turner (213), derives its name from the original purpose of exorcising *ghosts* and *phantoms* from 2D *J* spectra. The cycle is given by the following diagram:

Transient	Phase of 90° pulse on observe nucleus	Phase of 180° pulse on observe nucleus	Phase of receiver
1	0°	0°	0°
2	0°	90°	180°
3	0°	180°	0°
4	0°	270°	180°

Ghosts arise from magnetization that does not experience the 180° refocusing pulse, and they are canceled by combination of steps 1 and 2 or of steps 3 and 4. Phantoms come from transverse magnetization resulting from an imperfect refocusing pulse, and this magnetization is canceled by combination of steps 1 and 3 or of steps 2 and 4, by the inversion of the phase of the refocusing pulse.

d. *Phase Cycles in Specific Experiments.* Later chapters include discussions of phase cycles for various 1D and 2D experiments. Here we mention two. For the *J*-modulated echo experiment (or APT), it has been recommended that the phase of the 180° pulses be cycled together through all four 90° steps and that the receiver be cycled independently through all four steps, giving a 16-step cycle (16). In the INADEQUATE experiment embodying composite pulses, Levitt and Ernst (182) employed, in addition to the basic 16-step cycle required to select the double-quantum coherence, CYCLOPS on the incoming data and Exorcycle on the composite pulses in the excitation sequence and on the read cluster, giving a total of 1024 steps. Since this is an experiment in which one is looking for carbon-13 satellites of carbon-13 resonances, use of so extended a cycle does not lengthen the experiment beyond the time that otherwise would be required to obtain satisfactory signal to noise.

e. *Analysis of Coherence Pathways.* Both of the functions of phase cycling, compensation for imperfections and selection of coherences of particular order, may be described in terms of *coherence pathways*. Analysis of the coherence pathway for a complex experiment makes it possible to deduce the phase cycle which is appropriate for that experiment.

The coherence pathway is the sequence of *coherence levels* through which the system passes from the start of an experiment until the signal is detected. A coherence level is equal to the number of quanta in the transition corresponding to that coherence; the coherence level is fixed during a period of free precession and is changed for a system only by the application of a pulse. For a single spin $\frac{1}{2}$, coherence levels of $+1$ and -1 are possible, and for a two-spin-$\frac{1}{2}$ system, levels of $+2$, $+1$, 0, -1, and -2. Since only single-quantum coherence is detectable, only those coherence pathways ending on a level with magnitude 1 are observable, and specifically the -1 level is required for detection.

Several recent papers have provided systematic approaches to the analysis of coherence pathways, for various experiments and in various terms. Bain (214) equated the coherence level to the "z component of the Liouville space angular momentum" or the "z component of the superspin angular momentum," and treated homonuclear and heteronuclear correlation experiments as well as the cycle described above for elimination of quadrature images. Bodenhausen, Kogler, and Ernst (215) simply used the term *coherence order* in place of coherence level in their analysis of homonuclear correlation, 2D exchange and NOE spectroscopy, multiple-quantum filters, and multiple-quantum spectroscopy. Nakashima and McClung (216) showed that the spherical-tensor basis represents a convenient embodiment of the superspin approach, developed a computer program for coherence-pathway analysis using this basis, and discussed a zero-quantum experiment and a 2D DEPT experiment.

f. *Alternatives to Phase Cycling.* Coherences of different orders have differing sensitivity to inhomogeneities of the magnetic field or of the rf signal. Thus the rate of defocusing or refocusing in an inhomogeneous field is proportional to the order of the coherence. Bax *et al.* (217) utilized this effect in selecting coherence-transfer echoes of particular orders in an experiment that can be represented

$$90°\text{---preparation---}90°\text{---evolution---}90°\text{---detection}$$

If a field-gradient pulse is applied only during the evolution, all orders except zero-quantum coherence are dephased, and what was zero-quantum coherence at the end of the evolution period is observed in the detection period. If pulses of the same length and magnitude are applied during the evolution period and during the detection, single-quantum coherence rephases during the detection period at the same rate at which it dephased during the evolution period and is observed. If the pulse during detection is n times as long as the pulse during evolution, then coherence which is of order n during evolution will be refocused as single-quantum coherence to be observed in the echo.

Barker and Freeman (218) showed that field-gradient pulses could be used, not only for selection of multiple quantum order, but also in place of phase cycling to eliminate the "quadrature glitch," to disperse unwanted residual xy magnetization which would otherwise lead to spectral artifacts, and to provide sign discrimination in the second dimension in a 2D experiment so that the carrier frequency can be placed in the middle of the spectral range. Limitations on the pulsed magnetic field gradient method are that the field lock must be disabled during the pulse and that subsequent steps in an experiment may lead to refocusing of the magnetization dispersed by the pulse.

Inhomogeneous z pulses have been used in a somewhat similar way. Counsell, Levitt, and Ernst (219) replaced the field gradient pulse by a composite z pulse of the form $90°_{\psi+\pi} \theta°_{\psi+\pi/2} 90°_{\psi}$. The angle θ is chosen to be quite large so that the effect of rf inhomogeneity is emphasized. Pairs of such pulses on either side of a coherence-transfer step, as described for magnetic field gradient pulses above, produce a type of rotary coherence-transfer spin echo, and adjustment of the relative values of θ in the two components of the pair permits selection of coherences of particular order.

B. Bilinear Pulse Clusters

A cluster of three pulses, termed a bilinear or BIRD pulse, was proposed by Garbow, Weitekamp, and Pines (220) as a means of selectively inverting nuclei with a particular value of the coupling constant. One version of the sequence, as applied to a C–H system, is

$$^{1}\text{H} \quad 90°_x—\tau—180°_x—\tau—90°_x$$

$$^{13}\text{C} \qquad\qquad\quad 180°$$

The name derives from the fact that the interaction effective in the system involves a linear term of the form $J(\text{IS})I_xS_z$ or $J(\text{IS})I_yS_z$.

If τ in the bilinear cluster has the value $1/2\,^1J(\text{CH})$, the effect of the sequence is equivalent to a 180° pulse on the protons directly bonded to carbon. The first 90° pulse turns the ^1H magnetization into the $-y$ direction, and, during the first period of length τ, the nuclei precess until the two components of the proton doublet differ in phase by 180°. The 180°(H) pulse turns the magnetization vectors around the x axis, but the 180°(C) pulse interchanges their labels, so that they refocus along the $-y$ axis after the second precession period—the effect to this point is somewhat like a heteronuclear spin-echo experiment. The second 90° pulse then turns the magnetization of these protons from the $-y$ to the $-z$ direction. Meanwhile the remote proton magnetization has been turned from $-y$ to $+y$ by the 180°(H) pulse and back to $+z$ by the final pulse; the very much smaller coupling constant associated with the remote interaction will have had little effect during the period of length τ.

If the phase of the 180°(H) pulse is changed to y or the phase of the last 90°(H) pulse to $-x$, the sequence inverts remote protons but returns the ^{13}C-bound proton magnetization to its initial direction.

Wimperis and Freeman (221) developed a related pulse cluster, which they termed TANGO:

$$^1\text{H} \qquad 135°_x\text{—}\tau\text{—}180°_x\text{—}\tau\text{—}45°_x$$

$$^{13}\text{C} \qquad\qquad\qquad 180°$$

If τ is equal to $1/2\,^1J(\text{CH})$, this cluster acts as a 360° pulse for remote protons and a 90° pulse for adjacent protons. Another version acts as a 90° pulse for remote protons and a 180° pulse for directly bonded protons:

$$^1\text{H} \qquad 45°_x\text{—}\tau\text{—}180°_y\text{—}\tau\text{—}45°_{-x}$$

$$^{13}\text{C} \qquad\qquad\qquad 180°$$

Enhanced effectiveness of BIRD clusters in discriminating direct and remote couplings for systems where the one-bond coupling constants vary over a range has been achieved by use of composite pulses or by expanding some of the constituent pulses so that they have an intensity ratio corresponding to the binomial coefficients, a technique used in solvent suppression (222). One resulting version is the following:

$$^1\text{H} \qquad \tau\text{—}45°_x\text{—}\tau\text{—}180°_y\text{—}\tau\text{—}90°_{-x}\text{—}\tau\text{—}180°_y\text{—}\tau\text{—}45°_x$$

$$^{13}\text{C} \qquad\qquad\qquad 180° \qquad\qquad\qquad 180°$$

This inverts only remote protons and can be seen to be equivalent to two TANGO sequences in succession. Replacement of $180°_y$ pulses by $180°_x$ pulses produces a sequence selective for inversion of local protons.

Bilinear pulses have been inserted in a variety of experiments to achieve selectivity based on the magnitude of nuclear spin–spin coupling constants, and a number of these applications will be described in subsequent chapters.

C. Calibration of Decoupler Pulses

Experiments in which protons are subjected to pulses from the decoupling channel of a spectrometer almost invariably require knowledge of the decoupler output level in order that pulse lengths can be properly adjusted, and it has been found that the results of these experiments are generally quite sensitive to the tip angle adjustment. The orginal method for calibration of decoupler output in continuous mode of operation involved the use of off-resonance decoupling with measurement of residual splitting (223), but this method has several disadvantages, an important one of which is that continuous-duty output is not necessarily equal to pulsed output.

Various ingenious methods have been devised for decoupler pulse calibration, all of them, of course, indirect in that they are based on the effects observed in the spectrum of a nonproton species as a consequence of a decoupler pulse. In one of these methods, the sequence shown in Fig. 1-38a is applied to an isolated X–H system, where X is usually ^{13}C (39, 224). A null is observed in the signal when the angle θ corresponds to 90°, for

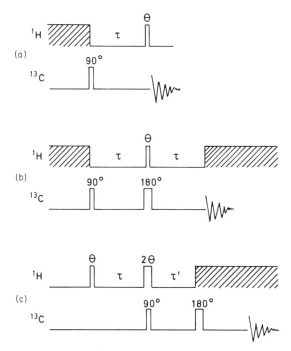

Fig. 1-38. Pulse sequences for calibration of the decoupler field strength. (a) Thomas–Bax (39, 224). (b) SEMUT. (c) Refocused SINEPT. [From Nielsen, Bildsøe, Jakobsen, and Sørensen (226).]

then all magnetization has been converted into a mixture of zero-quantum and double-quantum coherence. For reasonable accuracy, the decoupler frequency should be fairly close to resonance, and decoupling during the pulse delay time enhances the sensitivity by providing NOE enhancement of the carbon signal. The disadvantages of this method are that there is a phase shift across the spectrum if the observe nucleus has a number of resonances and that the FID must be acquired without decoupling lest the components of the antiphase multiplets cancel.

Addition of a 180° pulse on the observe nucleus, with a subsequent refocusing period, eliminates the phase problem and permits acquisition with broadband decoupling (225). This sequence, shown in Fig. 1-38b, is the same as the SEMUT sequence, which is discussed in detail in Chapter 3. Nielsen and co-workers (226) have analyzed in detail the applicability and relative calibration accuracy of SEMUT as well as of the sequence SINEPT, which is also described in Chapter 3. Strong coupling of protons

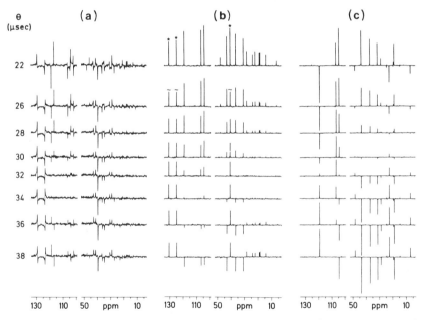

Fig. 1-39. Carbon-13 spectra at 75.43 MHz of estrone 3-methyl ether (150 mg in 3 ml $CDCl_3$) by the three pulse sequences shown in Fig. 1-38. Each spectrum represents 192 scans. (a) Sequence of Fig. 1-38a, with $\tau = 3.23$ msec, optimum for the CH doublet at 117.5 ppm. (b) Sequence of Fig. 1-38b, with $\tau = 3.84$ msec, optimum for $J = 130$ Hz. Signals marked with an asterisk are from quaternary carbons and have been truncated. (c) Sequence of Fig. 1-38c with $\tau = 3.70$ msec and $\tau' = 2.88$ msec. The decoupler frequency was positioned at 3.95 ppm for (a) and (b) and at 1.75 ppm for (c). [From Nielsen, Bildsøe, Jakobsen, and Sørensen (226).]

may interfere with the use of SEMUT, and it should be used on a CH group, rather than on CH_2 or CH_3. Refocused SINEPT, when used in a way similar to the other sequences to find a null in the observe nucleus signal, is suitable for any spin system and has the further advantage of increased sensitivity for nuclei such as ^{15}N and ^{29}Si because it utilizes polarization transfer. In Fig. 1-39, the results of calibrations by the three methods with the same sample are compared.

A method much like SINEPT, but with proton pulses exactly on resonance, has been described and was analyzed for a variety of spin systems (227). Specialized decoupler calibration methods have also been described for operation at low power levels, such as might be used for NOE experiments (228) and for irradiation of spin-1 nuclei such as deuterium (229, 230).

References

1. E. L. Hahn, *Phys. Rev.* **80**, 580 (1950).
2. H. Y. Carr and E. M. Purcell, *Phys. Rev.* **94**, 630 (1954).
3. S. Meiboom and D. Gill, *Rev. Sci. Instrum.* **29**, 688 (1958).
4. R. L. Vold, J. S. Waugh, M. P. Klein, and D. E. Phelps, *J. Chem. Phys.* **48**, 3831 (1968).
5. R. Freeman and H. D. W. Hill, *in* "Dynamic Nuclear Magnetic Resonance Spectroscopy" (L. M. Jackman and F. A. Cotton, eds.), Chap. 5. Academic Press, 1975.
6. R. L. Vold and R. R. Vold, *Prog. NMR Spectrosc.* **12**, 79 (1978).
7. R. Cassels, C. M. Dobson, F. M. Poulsen, R. G. Ratcliffe, and R. J. P. Williams, *J. Magn. Reson.* **37**, 141 (1980).
8. J. Frahm, *J. Magn. Reson.* **47**, 209 (1982).
9. M. Rance and R. A. Byrd, *J. Magn. Reson.* **52**, 221 (1983).
10. S. L. Patt, *J. Magn. Reson.* **49**, 161 (1982).
11. E. D. Becker, J. A. Ferretti, and T. C. Farrar, *J. Am. Chem. Soc.* **91**, 7784 (1969).
12. R. R. Shoup, E. D. Becker, and T. C. Farrar, *J. Magn. Reson.* **8**, 298 (1972).
13. R. Freeman and H. D. W. Hill, *J. Magn. Reson.* **4**, 366 (1971).
14. S. J. Opella and T. A. Cross, *J. Magn. Reson.* **37**, 171 (1980).
15. K. N. Scott, H. R. Brooker, J. R. Fitzsimmons, H. F. Bennett, and R. C. Mick, *J. Magn. Reson.* **50**, 339 (1982).
16. C. J. Turner, *Prog. NMR Spectrosc.* **16**, 311 (1984).
17. R. Freeman and H. D. W. Hill, *J. Chem. Phys.* **54**, 301 (1971).
18. R. L. Vold and R. R. Vold, *J. Magn. Reson.* **13**, 38 (1974).
19. I. D. Campbell, C. M. Dobson, R. G. Ratcliffe, and R. J. P. Williams, *J. Magn. Reson.* **31**, 341 (1978).
20. A. Kumar and R. R. Ernst, *Chem. Phys. Lett.* **37**, 162 (1976).
21. A. Kumar, W. P. Aue, P. Bachmann, J. Karhan, L. Müller, and R. R. Ernst, *Proc. 19th Congr. Ampere, Heidelberg* p. 473 (1976).
22. A. Bax, "Two-Dimensional Nuclear Magnetic Resonance in Liquids." Reidel, Dordrecht, Netherlands, 1982.
23. G. Bodenhausen, R. Freeman, and D. L. Turner, *J. Chem. Phys.* **65**, 839 (1976).
24. G. Bodenhausen, R. Freeman, R. Niedermeyer, and D. L. Turner, *J. Magn. Reson.* **24**, 291 (1976).
25. G. Bodenhausen, R. Freeman, R. Niedermeyer, and D. L. Turner, *J. Magn. Reson.* **26**, 133 (1977).
26. D. L. Rabenstein and T. T. Nakashima, *Anal. Chem.* **51**, 1465A (1979).

27. D. W. Brown, T. T. Nakashima, and D. L. Rabenstein, *J. Magn. Reson.* **45**, 302 (1981).
28. C. Le Cocq and J.-Y. Lallemand, *J.C.S. Chem. Commun.* p. 150 (1981).
29. R. R. Ernst and W. A. Anderson, *Rev. Sci. Instrum.* **37**, 93 (1966).
30. S. L. Patt and J. N. Shoolery, *J. Magn. Reson.* **46**, 535 (1982).
31. P. Crews, S. Naylor, B. L. Myers, J. Loo, and L. V. Manes, *Magn. Reson. Chem.* **23**, 684 (1985).
32. P. Schmitt, J. R. Wesener, and H. Günther, *J. Magn. Reson.* **52**, 511 (1983).
33. D. M. Doddrell, J. Staunton, and E. D. Laue, *J. Magn. Reson.* **52**, 523 (1983).
34. F. A. L. Anet, N. Jaffer, and J. Strouse, *21st Exp. NMR Conf., Tallahassee, Fla., 1980.*
35. P. Bigler, *J. Magn. Reson.* **55**, 468 (1983).
36. F. Pei and R. Freeman, *J. Magn. Reson.* **48**, 318 (1982).
37. K.-F. Elgert, R. Kosfeld, and W. E. Hull, *J. Magn. Reson.* **56**, 1 (1984).
38. M. R. Bendall, D. M. Doddrell, and D. T. Pegg, *J. Am. Chem. Soc.* **103**, 4603 (1981).
39. D. M. Thomas, M. R. Bendall, D. T. Pegg, D. M. Doddrell, and J. Field, *J. Magn. Reson.* **42**, 298 (1981).
40. H. J. Jakobsen, O. W. Sørensen, W. S. Brey, and P. Kanyha, *J. Magn. Reson.* **48**, 328 (1982).
41. D. J. Cookson and B. E. Smith, *Org. Magn. Reson.* **16**, 111 (1981).
42. J. C. Madsen, H. Bildsøe, H. J. Jakobsen, and O. W. Sørensen, *J. Magn. Reson.* **67**, 243 (1986).
43. I. D. Campbell, C. M. Dobson, R. J. P. Williams, and P. E. Wright, *FEBS Lett.* **57**, 96 (1975).
44. P. H. Bolton, *J. Magn. Reson.* **45**, 418 (1981).
45. F. Pei and R. Freeman, *J. Magn. Reson.* **48**, 519 (1982).
46. D. L. Rothman, F. Arias-Mendoza, G. I. Shulman, and R. G. Shulman, *J. Magn. Reson.* **60**, 430 (1984).
47. J. A. B. Lohman and J. Bulthuis, *J. Magn. Reson.* **63**, 593 (1985).
48. T. T. Nakashima and D. L. Rabenstein, *J. Magn. Reson.* **47**, 339 (1982).
49. R. Freeman, T. H. Mareci, and G. A. Morris, *J. Magn. Reson.* **42**, 341 (1981).
50. M. R. Bendall, D. T. Pegg, D. H. Doddrell, and J. Field, *J. Am. Chem. Soc.* **103**, 934 (1981).
51. K. M. Brindle, J. Boyd, I. D. Campbell, R. Porteous, and N. Soffe, *Biochem. Biophys. Res. Commun.* **109**, 864 (1982).
52. D. L. Foxall, J. S. Cohen, and R. G. Tschudin, *J. Magn. Reson.* **51**, 330 (1983).
53. J. S. Cohen, C. Chen, and A. Bax, *J. Magn. Reson.* **59**, 181 (1984).
54. K. M. Brindle, R. Porteous, and I. D. Campbell, *J. Magn. Reson.* **56**, 543 (1984).
55. A. Pines, M. G. Gibby, and J. S. Waugh, *J. Chem. Phys.* **59**, 569 (1973).
56. K. G. R. Pachler and P. L. Wessels, *J. Magn. Reson.* **12**, 337 (1973).
57. H. J. Jakobsen, S. Aa. Linde, and S. Sørensen, *J. Magn. Reson.* **15**, 385 (1974).
58. S. Aa. Linde, H. J. Jakobsen, and B. J. Kimber, *J. Am. Chem. Soc.* **97**, 3219 (1975).
59. H. J. Jakobsen and H. Bildsøe, *J. Magn. Reson.* **26**, 183 (1977).
60. H. Bildsøe, *J. Magn. Reson.* **27**, 393 (1977).
61. K. G. R. Pachler and P. L. Wessels, *J. Magn. Reson.* **28**, 53 (1977).
62. R. Pachter and P. L. Wessels, *Org. Magn. Reson.* **14**, 374 (1980).
63. P. H. Bolton and T. L. James, *J. Magn. Reson.* **36**, 443 (1979).
64. T. Bundgaard and H. J. Jakobsen, *J. Magn. Reson.* **18**, 209 (1975).
65. A. A. Chalmers, K. G. R. Pachler, and P. L. Wessels, *J. Magn. Reson.* **15**, 415 (1974).
66. H. J. Jakobsen, P. J. Kanyha, and W. S. Brey, *J. Magn. Reson.* **54**, 134 (1983).
67. K. Kakinuma, N. Imamura, N. Ikekawa, H. Tanaka, S. Minami, and S. Omura, *J. Am. Chem. Soc.* **102**, 7493 (1980).
68. S. Li, D. L. Johnson, J. A. Gladysz, and K. L. Servis, *J. Organomet. Chem.* **166**, 317 (1979).
69. K. Hallenga and W. E. Hull, *J. Magn. Reson.* **47**, 174 (1982).
70. S. Sørensen, R. S. Hansen, and H. J. Jakobsen, *J. Magn. Reson.* **14**, 243 (1974).
71. M. L. Martin, M. Trierweiler, V. Galasso, F. Fringuelli, and A. Taticchi, *J. Magn. Reson.* **42**, 155 (1981).

72. T. G. Dekker, K. G. R. Pachler, and P. L. Wessels, *Org. Magn. Reson.* **8**, 530 (1976).
73. H. J. Jakobsen and W. S. Brey, *J.C.S. Chem. Commun.* p. 478 (1979).
74. H. J. Jakobsen and W. S. Brey, *J. Am. Chem. Soc.* **101**, 774 (1979).
75. O. W. Sørensen, S. Scheibye, S.-O. Lawesson, and H. J. Jakobsen, *Org. Magn. Reson.* **16**, 322 (1981).
76. H. J. Jakobsen, P.-I. Yang, and W. S. Brey, *Org. Magn. Reson.* **17**, 290 (1981).
77. J. P. Jacobsen, *J. Magn. Reson.* **37**, 365 (1980).
78. J. Voigt and J. P. Jacobsen, *J. Magn. Reson.* **45**, 510 (1981).
79. P. L. Wessels and R. Pachter, *J. Magn. Reson.* **38**, 365 (1980).
80. K. G. R. Pachler and P. L. Wessels, *J.C.S. Chem. Commun.* p. 1038 (1974).
81. K. Bock, R. Burton, and L. D. Hall, *Can. J. Chem.* **54**, 3526 (1976).
82. A. A. Chalmers, *J. Magn. Reson.* **38**, 565 (1980).
83. O. W. Sørensen, H. Bildsøe, and H. J. Jakobsen, *J. Magn. Reson.* **45**, 325 (1981).
84. T. Berkhoudt and H. J. Jakobsen, *J. Magn. Reson.* **50**, 323 (1982).
85. P. Meakin and J. P. Jesson, *J. Magn. Reson.* **11**, 182 (1973); **13**, 354 (1974).
86. S. Schäublin, A. Höhener, and R. R. Ernst, *J. Magn. Reson.* **13**, 196 (1974).
87. R. E. D. McClung and N. R. Krishna, *J. Magn. Reson.* **29**, 573 (1978).
88. D. Canet, J. Brondeau, J.-P. Marchal, and H. Nery, *J. Magn. Reson.* **36**, 35 (1979).
89. G. Bodenhausen and R. Freeman, *J. Magn. Reson.* **36**, 221 (1979).
90. C. Bauer and R. Freeman, *J. Magn. Reson.* **61**, 376 (1985).
91. R. A. Craig, R. K. Harris, and R. J. Morrow, *Org. Magn. Reson.* **13**, 229 (1980).
92. P. H. Bolton and T. L. James, *J. Am. Chem. Soc.* **102**, 1449 (1980).
93. D. M. Doddrell, D. T. Pegg, W. Brooks, and M. R. Bendall, *J. Am. Chem. Soc.* **103**, 727 (1981).
94. S. K. Sarkar and A. Bax, *J. Magn. Reson.* **62**, 109 (1985).
95. D. J. Cookson and B. E. Smith, *J. Magn. Reson.* **54**, 354 (1983); **60**, 125 (1984).
96. G. A. Morris and R. Freeman, *J. Am. Chem. Soc.* **101**, 760 (1979).
97. G. A. Morris, *J. Am. Chem. Soc.* **102**, 428 (1980).
98. D. M. Doddrell and D. T. Pegg, *J. Am. Chem. Soc.* **102**, 6388 (1980).
99. D. P. Burum and R. R. Ernst, *J. Magn. Reson.* **39**, 163 (1980).
100. V. Rutar, *J. Magn. Reson.* **53**, 135 (1983).
101. D. M. Doddrell, H. Bergen, D. Thomas, D. T. Pegg, and M. R. Bendall, *J. Magn. Reson.* **40**, 591 (1980).
102. T. S. Mansour, T. C. Wong, and E. M. Kaiser, *Org. Magn. Reson.* **21**, 71 (1983).
103. R. T. C. Brownlee, J. G. Hall, and J. A. Reiss, *Org. Magn. Reson.* **21**, 544 (1983).
104. P. H. Bolton, *J. Magn. Reson.* **41**, 287 (1980).
105. B. Wrackmeyer, *J. Magn. Reson.* **54**, 174 (1983); **66**, 172 (1986).
106. H. Kessler, W. Hehlein, and R. Schuck, *J. Am. Chem. Soc.* **104**, 4534 (1982).
107. W. Städeli, P. Bigler, and W. von Philipsborn, *Org. Magn. Reson.* **16**, 170 (1981).
108. L. Kozerski and W. von Philipsborn, *Org. Magn. Reson.* **17**, 306 (1981).
109. D. H. Live, D. G. Davis, W. C. Agosta, and D. Cowburn, *J. Am. Chem. Soc.* **106**, 1939 (1984).
110. C. Brevard, G. C. van Stein, and G. van Koten, *J. Am. Chem. Soc.* **103**, 6746 (1981).
111. J. Schraml, *J. Magn. Reson.* **59**, 515 (1984).
112. C. Brevard and R. Schimpf, *J. Magn. Reson.* **47**, 528 (1982).
113. C. L. Dumoulin and E. A. Williams, *J. Magn. Reson.* **66**, 86 (1986).
114. A. Bax, *J. Magn. Reson.* **57**, 314 (1984).
115. A. Bax, C. H. Niu, and D. Live, *J. Am. Chem. Soc.* **106**, 1150 (1984).
116. D. G. Davis, D. H. Live, W. C. Agosta, and D. Cowburn, *J. Magn. Reson.* **53**, 350 (1983).
117. D. Marion, C. Garbay-Jaureguiberry, and B. P. Roques, *J. Am. Chem. Soc.* **104**, 5573 (1982).

118. J. Kowalewski and G. A. Morris, *J. Magn. Reson.* **47**, 331 (1982).
119. A. G. Avent and R. Freeman, *J. Magn. Reson.* **39**, 169 (1980).
120. G. A. Morris, *J. Magn. Reson.* **41**, 185 (1980).
121. O. W. Sørensen, R. Freeman, T. Frankiel, T. H. Mareci, and R. Schuck, *J. Magn. Reson.* **46**, 180 (1982).
122. Y. Huang, S. Macura, and R. R. Ernst, *J. Am. Chem. Soc.* **103**, 5327 (1981).
123. D. G. Davis, W. C. Agosta, and D. Cowburn, *J. Am. Chem. Soc.* **105**, 6189 (1983).
124. W. F. Reynolds, D. W. Hughes, M. Perpick-Dumont, and R. G. Enriquez, *J. Magn. Reson.* **64**, 304 (1985).
125. G. A. Pearson, *J. Magn. Reson.* **64**, 487 (1985).
126. A. Bax and R. Freeman, *J. Magn. Reson.* **44**, 542 (1981).
127. D. M. Doddrell, D. T. Pegg, and M. R. Bendall, *J. Magn. Reson.* **48**, 323 (1982).
128. D. T. Pegg, D. M. Doddrell, and M. R. Bendall, *J. Chem. Phys.* **77**, 2745 (1982).
129. M. R. Bendall and D. T. Pegg, *J. Magn. Reson.* **52**, 164 (1983).
130. D. T. Pegg and M. R. Bendall, *J. Magn. Reson.* **55**, 51 (1983).
131. M. R. Bendall, D. T. Pegg, and D. M. Doddrell, *J.C.S. Chem. Commun.* p. 872 (1982).
132. M. R. Bendall, D. T. Pegg, D. M. Doddrell, and D. H. Williams, *J. Org. Chem.* **47**, 3021 (1982).
133. M. R. Bendall, D. T. Pegg, and D. M. Doddrell, *J. Magn. Reson.* **52**, 81 (1983).
134. T. T. Nakashima, B. K. John, and R. E. D. McClung, *J. Magn. Reson.* **57**, 149 (1984).
135. T. T. Nakashima, B. K. John, and R. E. D. McClung, *J. Magn. Reson.* **59**, 124 (1984).
136. B. Coxon, *J. Magn. Reson.* **66**, 230 (1986).
137. O. W. Sørensen and R. R. Ernst, *J. Magn. Reson.* **51**, 477 (1983).
138. M. R. Bendall and D. T. Pegg, *J. Magn. Reson.* **53**, 272 (1983).
139. M. R. Bendall and D. T. Pegg, *J. Magn. Reson.* **59**, 237 (1984).
140. D. T. Pegg and M. R. Bendall, *J. Magn. Reson.* **63**, 556 (1985).
141. H. J. Jakobsen, U. B. Sørensen, H. Bildsøe, and O. W. Sørensen, *J. Magn. Reson.* **63**, 601 (1985).
142. U. B. Sørensen, H. Bildsøe, H. J. Jakobsen, and O. W. Sørensen, *J. Magn. Reson.* **65**, 222 (1985).
143. S. W. Sparks and P. D. Ellis, *J. Magn. Reson.* **62**, 1 (1985).
144. M. R. Bendall and D. T. Pegg, *J. Magn. Reson.* **53**, 144 (1983).
145. D. T. Pegg and M. R. Bendall, *J. Magn. Reson.* **55**, 114 (1983).
146. D. M. Doddrell, W. Brooks, J. Field, and R. M. Lynden-Bell, *J. Magn. Reson.* **59**, 384 (1984).
147. D. M. Doddrell, W. Brooks, J. Field, and R. M. Lynden-Bell, *J. Am. Chem. Soc.* **105**, 6973 (1983).
148. M. R. Bendall, D. T. Pegg, D. M. Doddrell, and J. Field, *J. Magn. Reson.* **51**, 520 (1983).
149. A. J. Shaka and R. Freeman, *J. Magn. Reson.* **50**, 502 (1982).
150. D. M. Doddrell, D. T. Pegg, M. R. Bendall, W. M. Brooks, and D. M. Thomas, *J. Magn. Reson.* **41**, 492 (1980).
151. D. T. Pegg, D. M. Doddrell, W. M. Brooks, and M. R. Bendall, *J. Magn. Reson.* **44**, 32 (1981).
152. D. T. Pegg and M. R. Bendall, *J. Magn. Reson.* **58**, 14 (1984).
153. N. Chandrakumar, *J. Magn. Reson.* **60**, 28 (1984).
154. P. L. Rinaldi and N. J. Baldwin, *J. Magn. Reson.* **61**, 165 (1985).
155. P. L. Rinaldi and N. J. Baldwin, *J. Am. Chem. Soc.* **104**, 5791 (1982).
156. P. L. Rinaldi and N. J. Baldwin, *J. Am. Chem. Soc.* **105**, 7523 (1983).
157. T. T. Nakashima, R. E. D. McClung, and B. K. John, *J. Magn. Reson.* **58**, 27 (1984).
158. M. R. Bendall, D. T. Pegg, G. M. Tyburn, and C. Brevard, *J. Magn. Reson.* **55**, 322 (1983).
159. D. M. Doddrell, J. Staunton, and E. D. Laue, *J.C.S. Chem. Commun.* p. 602 (1983).

160. M. R. Bendall, D. T. Pegg, J. R. Wesener, and H. Günther, *J. Magn. Reson.* **59**, 223 (1984).
161. K. V. Schenker and W. von Philipsborn, *J. Magn. Reson.* **61**, 294 (1985).
162. K. V. Schenker and W. von Philipsborn, *J. Magn. Reson.* **66**, 219 (1986).
163. A. Bax and S. K. Sarkar, *J. Magn. Reson.* **60**, 170 (1984).
164. G. A. Morris, *in* "Topics in Carbon-13 NMR Spectroscopy" (G. C. Levy, ed.), Vol. 4, Chap. 7. Wiley, New York, 1984.
165. M. H. Levitt, *Prog. NMR Spectrosc.* **18**, 61 (1986).
166. M. H. Levitt and R. Freeman, *J. Magn. Reson.* **33**, 473 (1979).
167. M. H. Levitt and R. Freeman, *J. Magn. Reson.* **43**, 65 (1981).
168. M. H. Levitt, *J. Magn. Reson.* **50**, 95 (1982).
169. R. Freeman, S. P. Kempsell, and M. H. Levitt, *J. Magn. Reson.* **38**, 453 (1980).
170. A. J. Shaka, J. Keeler, and R. Freeman, *J. Magn. Reson.* **53**, 313 (1983).
171. M. H. Levitt and R. R. Ernst, *J. Magn. Reson.* **55**, 247 (1983).
172. A. J. Shaka and R. Freeman, *J. Magn. Reson.* **55**, 487 (1983).
173. Z. Starčuk and V. Sklenář, *J. Magn. Reson.* **62**, 113 (1985).
174. R. Freeman, T. Frenkiel, and M. H. Levitt, *J. Magn. Reson.* **44**, 409 (1981).
175. R. Tycko, H. M. Cho, E. Schneider, and A. Pines, *J. Magn. Reson.* **61**, 90 (1985).
176. A. J. Shaka, *Chem. Phys. Lett.* **120**, 201 (1985).
177. A. J. Shaka, P. B. Barker, and R. Freeman, *J. Magn. Reson.* **67**, 580 (1986).
178. R. Tycko and A. Pines, *Chem. Phys. Lett.* **111**, 462 (1984).
179. R. Tycko, *Phys. Rev. Lett.* **51**, 775 (1983).
180. D. J. Lurie, *Magn. Reson. Imaging* **3**, 235 (1985).
181. D. J. Lurie, *J. Magn. Reson.* **70**, 11 (1986).
182. M. H. Levitt and R. R. Ernst, *Mol. Phys.* **50**, 1109 (1983).
183. R. Freeman and J. Keeler, *J. Magn. Reson.* **43**, 484 (1981).
184. H. P. Hetherington and D. L. Rothman, *J. Magn. Reson.* **65**, 348 (1985).
185. H. P. Hetherington, D. Wishart, S. M. Fitzpatrick, P. Cole, and R. G. Shulman, *J. Magn. Reson.* **66**, 313 (1986).
186. A. J. Shaka and R. Freeman, *J. Magn. Reson.* **59**, 169 (1984).
187. J. Frahm and W. Hänicke, *J. Magn. Reson.* **60**, 320 (1984).
188. R. Freeman and S. Wittekoek, *J. Magn. Reson.* **1**, 238 (1969).
189. G. Bodenhausen, R. Freeman, and G. A. Morris, *J. Magn. Reson.* **23**, 171 (1976).
190. G. A. Morris and R. Freeman, *J. Magn. Reson.* **29**, 433 (1978).
191. W. S. Warren, *J. Chem. Phys.* **81**, 5437 (1984).
192. A. J. Temps and C. F. Brewer, *J. Magn. Reson.* **56**, 355 (1984).
193. J. W. Carlson, *J. Magn. Reson.* **67**, 551 (1986).
194. J. Mao, T. H. Mareci, K. N. Scott, and E. R. Andrew, *J. Magn. Reson.* **70**, 310 (1986).
195. C. Bauer, R. Freeman, T. Frenkiel, J. Keeler, and A. J. Shaka, *J. Magn. Reson.* **58**, 442 (1984).
196. M. A. McCoy and W. S. Warren, *J. Magn. Reson.* **65**, 178 (1985).
197. M. S. Silver, R. I. Joseph, and D. I. Hoult, *J. Magn. Reson.* **59**, 347 (1984).
198. J. Baum, R. Tycko, and A. Pines, *J. Chem. Phys.* **79**, 4643 (1983).
199. B. L. Tomlinson and H. D. W. Hill, *J. Chem. Phys.* **59**, 1775 (1973).
200. R. Brandes and D. R. Kearns, *J. Magn. Reson.* **64**, 506 (1985).
201. R. Brandes and D. R. Kearns, *J. Magn. Reson.* **67**, 14 (1986).
202. T. Cosgrove, S. Neck, and N. A. Finch, *J. Magn. Reson.* **62**, 309 (1985).
203. M. R. Morrow, *J. Magn. Reson.* **69**, 501 (1986).
204. J. Boyd, R. Porteous, C. Redfield, and N. Soffe, *J. Magn. Reson.* **63**, 392 (1985).
205. G. Bodenhausen, *J. Magn. Reson.* **34**, 357 (1979).
206. M. Hintermann, L. Braunschweiler, G. Bodenhausen, and R. R. Ernst, *J. Magn. Reson.* **50**, 316 (1982).

1. BASIC METHODS AND SIMPLE PULSED EXPERIMENTS

207. T. Frenkiel and J. Keeler, *J. Magn. Reson.* **50**, 479 (1982).
208. M. Decorps, J. P. Albrand, P. Blondet, F. Devreux, and M. F. Foray, *J. Magn. Reson.* **66**, 364 (1986).
209. E. Guittet, D. Piveteau, M.-A. Delsuc, and J.-Y. Lallemand, *J. Magn. Reson.* **62**, 336 (1985).
210. E. O. Stejskal and J. Schaefer, *J. Magn. Reson.* **13**, 249 (1974).
211. E. O. Stejskal and J. Schaefer, *J. Magn. Reson.* **14**, 160 (1974).
212. D. I. Hoult and R. E. Richards, *Proc. R. Soc. London, Ser. A* **344**, 311 (1975).
213. G. Bodenhausen, R. Freeman, and D. L. Turner, *J. Magn. Reson.* **27**, 511 (1977).
214. A. D. Bain, *J. Magn. Reson.* **56**, 418 (1984).
215. G. Bodenhausen, H. Kogler, and R. R. Ernst, *J. Magn. Reson.* **58**, 370 (1984).
216. T. T. Nakashima and R. E. D. McClung, *J. Magn. Reson.* **70**, 187 (1986).
217. A. Bax, P. G. de Jong, A. F. Mehlkopf, and J. Smidt, *Chem. Phys. Lett.* **69**, 567 (1980).
218. P. Barker and R. Freeman, *J. Magn. Reson.* **64**, 334 (1985).
219. C. J. R. Counsell, M. H. Levitt, and R. R. Ernst, *J. Magn. Reson.* **64**, 470 (1985).
220. J. R. Garbow, D. P. Weitekamp, and A. Pines, *Chem. Phys. Lett.* **93**, 504 (1982).
221. S. Wimperis and R. Freeman, *J. Magn. Reson.* **58**, 348 (1984).
222. S. Wimperis and R. Freeman, *J. Magn. Reson.* **62**, 147 (1985).
223. K. G. R. Pachler, *J. Magn. Reson.* **7**, 442 (1972).
224. A. Bax, *J. Magn. Reson.* **52**, 76 (1983).
225. J. Bernassau, *J. Magn. Reson.* **62**, 533 (1985).
226. N. C. Nielsen, H. Bildsøe, H. J. Jakobsen, and O. W. Sørensen, *J. Magn. Reson.* **66**, 456 (1986).
227. V. Sklenář, K. Nejezchleb, and Z. Starčuk, *J. Magn. Reson.* **69**, 144 (1986).
228. S. D. Simova, *J. Magn. Reson.* **63**, 583 (1985).
229. T. T. Nakashima, R. E. D. McClung, and B. K. John, *J. Magn. Reson.* **56**, 262 (1984).
230. N. Chandrakumar, *J. Magn. Reson.* **63**, 174 (1985).

2

Density-Operator Theory of Pulses and Precession

MALCOLM H. LEVITT*
LABORATORIUM FÜR PHYSIKALISCHE CHEMIE
EIDGENÖSSISCHE TECHNISCHE HOCHSCHULE
CH-8092 ZÜRICH, SWITZERLAND

I.	Introduction	111
II.	Overview	113
	A. Statistical Behavior on the Level of Individual Spins	113
	B. Spin Ensembles—Density Operator	115
	C. Cartesian Operators	116
	D. Product Operators—Single-Element Basis	118
	E. Cartesian Product Basis	121
III.	Theory and Use of Product Operators	125
	A. Quantization	125
	B. Density Operator	126
	C. Base Operators	127
	D. Thermal Equilibrium	128
	E. Time Evolution—Free Precession	129
	F. Time Evolution—Strong Pulses	133
	G. Selective Pulses	134
	H. Orders of Coherence—Phase Cycling	136
IV.	A Typical Two-Dimensional Experiment—Double-Quantum Spectroscopy of Two Coupled Spins-$\frac{1}{2}$	140
V.	Summary	145
	References	145

I. Introduction

In this chapter we will present an operator formalism suitable for analyzing the spin dynamics of weakly coupled systems. We will be interested in showing how the formalism arises from first principles, and how

* Present address: MIT Building NW14-5122, Cambridge, Massachusetts 02139.

it may be used to treat fundamental concepts in 2D spectroscopy like coherence transfer, multiple-quantum coherence, phase cycling, and the like. Finally, we will treat a single application in detail for illustrative purposes.

The reason for choosing this formalism in preference to some alternatives is that it takes advantage of the simplifying features of weakly coupled spin-$\frac{1}{2}$ systems. A universal approach to treating spin dynamics is by density matrix calculation (1, 2). Although this is quite well suited to computer simulation, in application to systems with more than a few energy levels it is cumbersome to use, and it provides very little insight. The same is essentially true for the use of single-transition operators (3, 4), which concentrate attention on pairs of energy levels, and hence can be useful in cases where a pulse only perturbs appreciably a single transition—but this is rarely the case in isotropic liquids, unless special precautions are taken. A different approach which may be of use for systems of spins of high angular momentum is the use of spherical tensor operators (5). However, once again, except in the simplest cases, the use of tensor operators is often extremely cumbersome, and better adapted to computer work than to give insight into how an experiment works or can be made to work. In fact, to produce a universal operator theory which works and yet which is simple enough to provide physical insight seems to be a hopeless task.

However, by setting our sights somewhat lower, it is possible to go some way to satisfy these two requirements at the same time. Almost always in carbon-13 NMR, and to an increasing extent in proton NMR, high-resolution liquid-state NMR spectra can be assumed to be first order. This assumption may be expected to grow in scope and in accuracy as static magnetic fields continue to increase. Now the dynamics of weakly coupled systems are relatively simple. Each spin is affected by its neighbors' spin states only by a slight increase or decrease of the local magnetic field, causing its conventional NMR spectrum (the single-quantum spectrum) to split up into a characteristic multiplet. Apart from this effect, each spin rotates in its own three-dimensional space quite independently from its neighbors. There are no cooperative effects in a weakly coupled system: if one spin is flipped by a pulse, there is no tendency for neighboring spins to turn as well. Hence there is no need to use a mathematical formalism which explicitly treats the weakly coupled system as a single entity. The system behaves more as a loose bag of independent spins. Mathematically we may express this by using operators which are simple *products* of operators associated with the individual spins, each of which has an independent "motion."

Product operator formalisms have been used before in the literature (6), but it is only recently that several workers have tried to bring together the rather scattered threads and weave them into a coherent pattern. Recently, the formalism that we will call the Cartesian product operator formalism has received extensive discussion by Sørensen et al. (7). Independent work has also been done by van de Ven and Hilbers (8), Packer and Wright (9), and Lynden-Bell et al. (10). In this chapter we will try to show how the Cartesian product operators arise from first principles and how they are related to perhaps more fundamental product operators—the "single-element basis," introduced to NMR by Banwell and Primas (11) and used for spin dynamics by Warren (12) and Weitekamp et al. (13, 14).

First, we will try to present a motivation for using a density-operator formalism by examining quite closely the quantum-mechanical properties of spins-$\frac{1}{2}$ on the microscopic level and then on the macroscopic or ensemble level. The use of density operators will be seen to be forced because spin behavior is at the microscopic level where quantum-mechanical consequences are inescapable. In this section we will also try to give an overview of the use of product operators, in two different manifestations, for the dynamics of weakly coupled spins-$\frac{1}{2}$. In the following section density operators will be considered in more formal terms, allowing discussion of spin dynamics in the basic building blocks of most 2D experiments: strong pulses, free precession, selective pulses, and phase-cycling schemes. Finally, the simple example of two coupled spins-$\frac{1}{2}$, of importance in the INADEQUATE experiment, will be treated in more detail.

II. Overview

We consider first the simple case of an ensemble of isolated spins-$\frac{1}{2}$ in a static magnetic field B_0 along the z axis. Each spin has two "stationary states," or eigenfunctions, $|\alpha\rangle$ and $|\beta\rangle$, in which the spin is either "up" or "down" with respect to B_0. In general the state of an individual spin $|\psi\rangle$ may always be represented as a linear superposition of these two eigenstates:

$$|\psi\rangle = c_\alpha |\alpha\rangle + c_\beta |\beta\rangle \qquad [1]$$

A. Statistical Behavior on the Level of Individual Spins

One often receives the impression that spins-$\frac{1}{2}$ are "classical" objects in the sense that quantum mechanics is not necessary to predict their behavior. This is a false impression. To emphasize this, imagine that we could measure the x or y components of magnetization for the individual spins. After a $\pi/2$ pulse, a classical treatment, such as the Bloch equations (15), would lead us to expect a sinusoidal oscillation of the magnetization components as the spin precesses around B_0 at the Larmor frequency ω_0:

$$M_x(t) = M_0 \sin \omega_0 t$$
$$M_y(t) = -M_0 \cos \omega_0 t, \qquad [2]$$

where, for an individual spin, $M_0 = (\gamma/2)\hbar$. This would *not* be the observed behavior. In fact, every time we would measure the component of magnetization, we would only get one of the two answers, $+(\gamma/2)\hbar$ or $-(\gamma/2)\hbar$, and if we repeated the experiment, we would not always get the same answer. This is because the only possible results of an observation are the eigenvalues of the observable operator,* which in this case are $\pm(\gamma/2)\hbar$ (*16*). On the level of individual spins, the famous statistical nature of reality implicit in quantum mechanics intrudes, and, for example, it becomes impossible to remove the statistical spread in a measured value of M_y without introducing a broad scatter in measurements of M_x. This is not classical behavior. However, the *expectation values*, that is, the averages of M_x and M_y over a large number of measurements, do obey Eq. [2], and so behave "classically." In the usual NMR experiment, the signals observed originate from the whole ensemble of a large number of spins, so the measured quantities approach very closely the classical expectation values of Eq. [2], leading to a general feeling that an ensemble of isolated spins-$\frac{1}{2}$ is a classical system.

As an aside, an experiment in which the magnetization of individual spins is detected is not completely hypothetical. There is a strong relationship to the renowned Stern–Gerlach experiment (*16*), but a nice example very closely related to NMR is the detection of muon resonance (*16, 17*). Muons are particles similar to the electron except that their mass is 207 times greater and that they are unstable, decaying into a positron and two neutrinos. Like the electron, the muon has spin $\frac{1}{2}$. Muons may be created in a spin-polarized beam and may be captured by a chemical target in which they usually exist as a muon–electron pair, chemically an isotope of hydrogen, which is known as muonium. It is of interest to measure the muonium frequencies in various chemical states. This is done by placing a homogeneous magnetic field perpendicular to the direction of the incident muon beam, and thus, as in NMR, perpendicular to the spin polarization. After the individual muon has precessed in this field, it decays, releasing the positron along the angular momentum direction, which may therefore be experimentally determined. The relevance to the present discussion is that the apparent Larmor pre-

---

* This assumes that the NMR experiment provides a measurement of spin angular momentum in the same sense that, for example, the Stern–Gerlach experiment (*16*) does. This is actually a matter of controversy. It may be that this particular quantum-mechanical postulate does not apply to continuous "nondemolition" measurements as occur in NMR.

2. DENSITY-OPERATOR THEORY OF PULSES AND PRECESSION

cession in the magnetic field is found to be a precession of the *expectation value* only and not of the individual spins, so that the "free induction decay" is only revealed after averaging millions of measurements. In NMR, one typically observes 10^{17} spins at a time, so that statistical averaging of experiments is unnecessary.

B. Spin Ensembles—Density Operator

It is clear from the above discussion that ensemble averaging may introduce a deceptive simplicity into the properties of spin systems. The *density operator* formalism is a means to exploit the simplifying properties of ensemble averaging in a rigorous way. It allows one to specify the state of the ensemble *as a whole* in a quantum-mechanical way without reference to the state $|\psi\rangle$ of each individual spin. The state of an esemble of spins-$\frac{1}{2}$, for example, is represented by a *density operator* σ, which may be described as a linear superposition of the four *shift operators* I^α, I^β, I^+, and I^-:

$$\sigma = \sigma_{\alpha\alpha} I^\alpha + \sigma_{\alpha\beta} I^+ + \sigma_{\beta\alpha} I^- + \sigma_{\beta\beta} I^\beta \qquad [3]$$

where

$$I^\alpha = |\alpha\rangle\langle\alpha| \qquad I^+ = |\alpha\rangle\langle\beta|$$
$$I^\beta = |\beta\rangle\langle\beta| \qquad I^- = |\beta\rangle\langle\alpha| \qquad [4]$$

This "ket-bra" notation defines the matrix elements of the operators, through, for example, $\langle\alpha|I^+|\beta\rangle = \langle\alpha|\alpha\rangle\langle\beta|\beta\rangle = 1 \cdot 1 = 1$, $\langle\alpha|I^+|\alpha\rangle = \langle\alpha|\alpha\rangle\langle\beta|\alpha\rangle = 0$, and so on.

Exercise: Using Eqs. [4], show that the shift operators obey the equations $I^+|\alpha\rangle = 0$, $I^-|\alpha\rangle = |\beta\rangle$, $I^\alpha|\alpha\rangle = |\alpha\rangle$, $I^\beta|\alpha\rangle = 0$, $I^+|\beta\rangle = |\alpha\rangle$, $I^-|\beta\rangle = 0$, $I^\alpha|\beta\rangle = 0$, and $I^\beta|\beta\rangle = |\beta\rangle$.

Equation [3], which applies for an ensemble of spins-$\frac{1}{2}$, is analogous to Eq. [1], which is appropriate for an individual spin. Instead of there being two wave functions $|\alpha\rangle$ and $|\beta\rangle$ in the expansion, the ensemble density operator σ is expanded into the four basis operators I^α, I^β, I^+, and I^-. (We prefer this notation to the equivalent I_{0+}, I_{0-}, I^+, and I^- used in Refs. 11–14.)

The coefficients σ_{ij} in the expansion in Eq. [3] have some physical significance. If the coefficients are arranged in a matrix (called the *density matrix*) as

$$\begin{pmatrix} \sigma_{\alpha\alpha} & \sigma_{\alpha\beta} \\ \sigma_{\beta\alpha} & \sigma_{\beta\beta} \end{pmatrix}$$

then the diagonal elements $\sigma_{\alpha\alpha}$ and $\sigma_{\beta\beta}$ represent *ensemble populations* of states $|\alpha\rangle$ and $|\beta\rangle$, respectively. The off-diagonal elements $\sigma_{\alpha\beta}$ and $\sigma_{\beta\alpha}$ indicate *ensemble coherence* between states $|\alpha\rangle$ and $|\beta\rangle$.

But what do these terms mean? First we should realize that both population and coherence are quantum-mechanical concepts, so we should not expect to be able to "understand" either of them in macroscopic terms. However, the population of an eigenstate is such a widely used concept in all forms of spectroscopy that it excites little uneasiness nowadays. More difficulty is encountered with the concept of coherence because it is only observed in exotic circumstances in other spectroscopies than magnetic resonance. This difference exists because of the unusually long time scales and long wavelengths in NMR and ESR. To magnetic resonance spectroscopists, coherence should be as basic and familiar a concept as population.

In case of an ensemble of spins-$\frac{1}{2}$, coherence between states $|\alpha\rangle$ and $|\beta\rangle$ is associated with precessing transverse magnetization, but in more complicated systems, coherence may not give rise to directly observable quantities. Note that coherence is an ensemble property, and, in the view of the author, one can no more talk about the coherence of a single spin than about the population or temperature of a single spin. In diagrams we shall express coherence by a wavy line connecting two eigenstates, as in Fig. 2-1a.

This diagrammatic representation is useful but also potentially misleading. Coherence is certainly a property of a pair of eigenstates but should *not* be described as a "transition between" the two eigenstates. The term "transition" gives the misleading impression that there is an exchange of population between the eigenstates, as for example in optical spectroscopy in the presence of driving irradiation. Coherence is a long-lived persistent interference phenomenon associated with the two states in the absence of irradiating fields; emphatically, coherence in itself does not involve any change in the populations of the two states, although changes in population brought about by some irradiating pulse sequence may be necessary in order to *prepare* the coherence.

C. Cartesian Operators

The choice of basis operators of Eq. [4] is not unique and other choices may be used, depending on the problem. If we wish to express the dynamics of the spin ensemble in the presence of an applied rf field, it is convenient to form the operators for the Cartesian components of angular momentum:

$$I_z = \tfrac{1}{2}(I^\alpha - I^\beta)$$
$$I_x = \tfrac{1}{2}(I^+ + I^-) \qquad [5]$$
$$I_y = \frac{1}{2i}(I^+ - I^-)$$

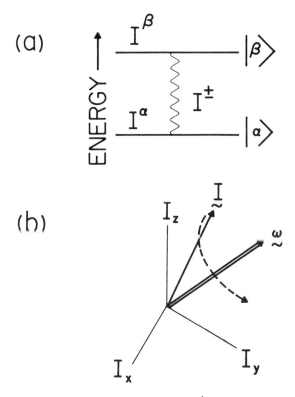

Fig. 2-1. Dynamics of an ensemble of isolated spins-$\frac{1}{2}$. (a) The two energy levels of an isolated spin-$\frac{1}{2}$ in a magnetic field, the associated eigenfunctions $|\alpha\rangle$ and $|\beta\rangle$, and population operators I^α and I^β. Coherence (transverse magnetization) is associated with operators I^+ and I^-. (b) Spin dynamics are economically represented by Cartesian operators I_x, I_y, and I_z. The state of the spin ensemble is represented by a vector **I** which rotates in a three-dimensional space around a vector **ω**.

A scalar operator equal to one-half of the unity operator completes the set:

$$\tfrac{1}{2}E = \tfrac{1}{2}(I^\alpha + I^\beta) \qquad [6]$$

The unity operator E is invariant to rotations, while the three Cartesian operators form a three-dimensional space in which the density operator is represented by a vector **I**, rotated at $|\omega|$ rad sec^{-1} about a vector **ω** with Cartesian components (Fig. 2-1b):

$$\omega_x = \omega_1 \cos \phi$$
$$\omega_y = \omega_1 \sin \phi \qquad [7]$$
$$\omega_z = \Omega$$

Here Ω is the resonance offset expressed in rad sec^{-1}, ϕ is the phase of an applied rf pulse, and $\omega_1 = -\gamma B_1$, where γ is the magnetogyric ratio and B_1 is the radio-frequency field strength. All quantities are expressed in the usual rotating frame. Thus, longitudinal magnetization I_z is converted into transverse magnetization $-I_y$ by the application of an on-resonance pulse with $\phi = 0$ for a duration t_p given by $\omega_1 t_p = \pi/2$, a "90°_x pulse." Application of the pulse twice this duration, a "180° pulse" or "π pulse," inverts the longitudinal magnetization. Notice that we shall continue to use the convention for the sign of rotation introduced in Chapter 1, in which a positive rotation about the x axis by $\pi/2$ radians converts z magnetization into $-y$ magnetization, $-y$ magnetization into $-z$ magnetization, and so on.

We want to emphasize again the alternative ways to decompose the density operator of an ensemble of isolated spins-$\frac{1}{2}$ in terms of either (a) the *single-element operators* of Eq. [4], which allow us to represent individual coherences and populations, or (b) the *Cartesian operators* of Eq. [5], in which the individual coherences and populations are not separately represented, but which allow a ready physical prediction of spin dynamics in terms of rotations in a three-dimensional space. Before a more formal discussion we proceed briefly to the interesting case of a system composed of two coupled spins-$\frac{1}{2}$ and show in the next two sections how the two types of product operators are developed for this spin system.

D. Product Operators—Single-Element Basis

The extension of the concept of single-element operators and Cartesian operators to an ensemble of weakly coupled pairs of spins-$\frac{1}{2}$ is quite simple, but immediately introduces some new features which are of central importance in two-dimensional NMR. Each pair of spins now has four eigenstates, $|\alpha_1\alpha_2\rangle$, $|\alpha_1\beta_2\rangle$, $|\beta_1\alpha_2\rangle$, and $|\beta_1\beta_2\rangle$, which are simple products of the two eigenstates of each of the two spins. The ensemble population of each eigenstate may be associated with a population operator, for example, $|\alpha_1\alpha_2\rangle\langle\alpha_1\alpha_2|$. These operators may be expressed as simple products of the population operators for the individual spins, thus:

$$\begin{aligned} |\alpha_1\alpha_2\rangle\langle\alpha_1\alpha_2| &= I_1^\alpha I_2^\alpha \\ |\alpha_1\beta_2\rangle\langle\alpha_1\beta_2| &= I_1^\alpha I_2^\beta \\ |\beta_1\alpha_2\rangle\langle\beta_1\alpha_2| &= I_1^\beta I_2^\alpha \\ |\beta_1\beta_2\rangle\langle\beta_1\beta_2| &= I_1^\beta I_2^\beta \end{aligned} \quad [8]$$

There are also four single-quantum transitions, each associated with a pair of coherence operators:

2. DENSITY-OPERATOR THEORY OF PULSES AND PRECESSION

$$\left.\begin{array}{l}|\alpha_1\alpha_2\rangle\langle\alpha_1\beta_2| = I_1^\alpha I_2^+ \\ |\alpha_1\beta_2\rangle\langle\alpha_1\alpha_2| = I_1^\alpha I_2^- \end{array}\right\}$$
$$\left.\begin{array}{l}|\beta_1\alpha_2\rangle\langle\beta_1\beta_2| = I_1^\beta I_2^+ \\ |\beta_1\beta_2\rangle\langle\beta_1\alpha_2| = I_1^\beta I_2^- \end{array}\right\}$$
$$\left.\begin{array}{l}|\alpha_1\alpha_2\rangle\langle\beta_1\alpha_2| = I_1^+ I_2^\alpha \\ |\beta_1\alpha_2\rangle\langle\alpha_1\alpha_2| = I_1^- I_2^\alpha \end{array}\right\}$$
$$\left.\begin{array}{l}|\alpha_1\beta_2\rangle\langle\beta_1\beta_2| = I_1^+ I_2^\beta \\ |\beta_1\beta_2\rangle\langle\alpha_1\beta_2| = I_1^- I_2^\beta \end{array}\right\}$$

[9]

As an illustration of the meaning of these operators, consider the first pair. These two operators are associated with two eigenstates which differ in the quantum number of spin 2, spin 1 always remaining in state $|\alpha\rangle$. *Very loosely* we may say that this coherence operator refers to the single-quantum transition of spin 2, spin 1 remaining polarized "up," but always reminding ourselves that this is only a convenient way of speaking, and that actually no transition is taking place. Another question is why two operators are required for this one "transition." The answer to this is that the coherence needs a direction, or phase, in the rotating frame, and two operators are required to supply this. The two operators given have particularly convenient rotational properties and it is often useful to treat them as independent objects when analyzing two-dimensional experiments, as will be shown below.

We also find the possibility of coherence involving both spins simultaneously, such as "double-quantum transitions" associated with the operators

$$|\alpha_1\alpha_2\rangle\langle\beta_1\beta_2| = I_1^+ I_2^+$$
$$|\beta_1\beta_2\rangle\langle\alpha_1\alpha_2| = I_1^- I_2^-$$

[10]

and "zero-quantum transitions" associated with the operators

$$|\alpha_1\beta_2\rangle\langle\beta_1\alpha_2| = I_1^+ I_2^-$$
$$|\beta_1\alpha_2\rangle\langle\alpha_1\beta_2| = I_1^- I_2^+$$

[11]

The relationship between the 16 operators and the energy-level diagram for two coupled spins-$\frac{1}{2}$ is given in Fig. 2-2. Each product operator has a straightforward "physical interpretation" in terms of the energy levels or the transitions connecting them. We call the basis set consisting of the 16 operators listed above the *single-element basis*, following Warren (12).

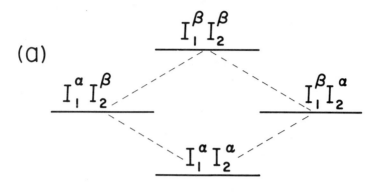
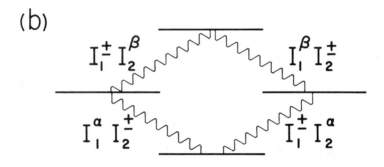
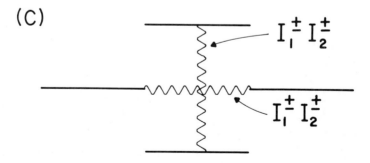

Fig. 2-2. Operators for two weakly coupled spins-$\frac{1}{2}$: (a) populations, (b) single-quantum coherences, (c) multiple-quantum coherences.

E. Cartesian Product Basis

Just as was explained previously for the case of an ensemble of isolated spins-$\frac{1}{2}$, we may form Cartesian components in order to describe the dynamics under a nonselective radio-frequency pulse. For an ensemble of pairs of coupled spins-$\frac{1}{2}$, we replace the 16 single-element operators of Eqs. [8] to [11] by the 16 possible binary products of the four operators of Eqs. [5] and [6]. Some examples are:

$$\tfrac{1}{2}E = 2(\tfrac{1}{2}E_1)(\tfrac{1}{2}E_2) = \tfrac{1}{2}(I_1^\alpha + I_1^\beta)(I_2^\alpha + I_2^\beta)$$

$$I_{1z} = 2(I_{1z})(\tfrac{1}{2}E_2) = \tfrac{1}{2}(I_1^\alpha - I_1^\beta)(I_2^\alpha + I_2^\beta)$$

$$I_{2z} = 2(\tfrac{1}{2}E_1)(I_{2z}) = \tfrac{1}{2}(I_1^\alpha + I_1^\beta)(I_2^\alpha - I_2^\beta)$$

$$I_{2y} = 2(\tfrac{1}{2}E_1)(I_{2x}) = \tfrac{1}{2}(I_1^\alpha + I_1^\beta)(I_2^+ + I_2^-) \quad [12]$$

$$2(I_{1z})(I_{2z}) = \tfrac{1}{2}(I_1^\alpha - I_1^\beta)(I_2^\alpha - I_2^\beta)$$

$$2(I_{1z})(I_{2y}) = \frac{1}{2i}(I_1^\alpha - I_1^\beta)(I_2^+ - I_2^-)$$

$$2(I_{1x})(I_{2y}) = \frac{1}{2i}(I_1^+ + I_1^-)(I_2^+ - I_2^-)$$

Note the introduction of a numerical factor 2 in the last three equations to give all operators the same norm.

Before considering the implications of the rotations of this Cartesian product base, let us examine the physical significance of some of the typical terms expressed in Eq. [12]. For example, the first operator in Eq. [12], $\tfrac{1}{2}E$, is an operator associated with equal populations of all four energy levels in the two-spin-$\frac{1}{2}$ system. This is clearly seen from the expanded form:

$$\tfrac{1}{2}E = \tfrac{1}{2}(I_1^\alpha I_2^\alpha + I_1^\alpha I_2^\beta + I_1^\beta I_2^\alpha + I_1^\beta I_2^\beta) \quad [13]$$

All four energy-level populations appear with equal weight.

The significance of three more operators from Eq. [12] is represented in Fig. 2-3:

1. The operator I_{1z} is associated with equal polarizations across the two single-quantum transitions of the first spin. In its explicit expansion,

$$I_{1z} = \tfrac{1}{2}(I_1^\alpha I_2^\alpha + I_1^\alpha I_2^\beta - I_1^\beta I_2^\alpha - I_1^\beta I_2^\beta) \quad [14]$$

2. The operator I_{2z} is associated with equal polarizations across the two single-quantum transitions of the second spin.

3. The operator $2I_{1z}I_{2z}$ is associated with a non-Boltzmann distribution of populations in which the levels $|\alpha_1\alpha_2\rangle$ and $|\beta_1\beta_2\rangle$ have higher populations

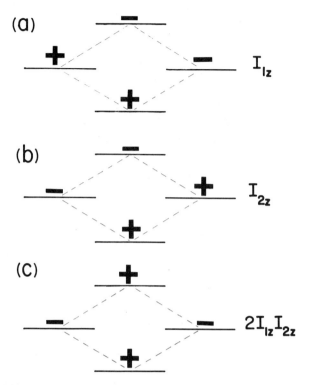

Fig. 2-3. Physical interpretation of diagonal Cartesian product operators for two spins-$\frac{1}{2}$. Plus signs represent population excess from a completely saturated state; minus signs represent population depletion.

than the levels $|\alpha_1\beta_2\rangle$ and $|\beta_1\alpha_2\rangle$. This type of population distribution is called "two-spin order":

$$2I_{1z}I_{2z} = \tfrac{1}{2}(I_1^\alpha I_2^\alpha - I_1^\alpha I_2^\beta - I_1^\beta I_2^\alpha + I_1^\beta I_2^\beta) \qquad [15]$$

The relaxation of such a population difference is examined in the Jeener–Broekaert experiment (18).

The physical significance of the operators involving x and y Cartesian components is also apparent by expansion. Thus the operator I_{2x} becomes

$$I_{2x} = \tfrac{1}{2}(I_1^\alpha I_2^+ + I_1^\alpha I_2^-) + \tfrac{1}{2}(I_1^\beta I_2^+ + I_1^\beta I_2^-) \qquad [16]$$

This is the sum of the x magnetizations contributed from the two single-quantum "transitions" of the second spin; it is thus the "in-phase" component of the x magnetization. In contrast, the operator $2I_{1z}I_{2x}$ implies opposite contributions from the two "transitions" of the second spin. The doublet should be "antiphase" to give a contribution to this type of coherence, hence the minus sign:

$$2I_{1z}I_{2x} = \tfrac{1}{2}(I_1^\alpha I_2^+ + I_1^\alpha I_2^-) - \tfrac{1}{2}(I_1^\beta I_2^+ + I_1^\beta I_2^-) \qquad [17]$$

The physical significance of some terms of this type involving a single transverse operator in the product for a two-spin-$\tfrac{1}{2}$ system is illustrated in Fig. 2-4.

Operators with *transverse* components for *both* spins involve double- and zero-quantum coherences. Again this is most clearly understood by expansion, as for the example

$$2I_{1x}I_{2y} = \frac{1}{2i}(I_1^+ I_2^+ - I_1^- I_2^-) - \frac{1}{2i}(I_1^+ I_2^- - I_1^- I_2^+) \qquad [18]$$

The expression of the effect of an rf pulse in terms of these Cartesian product operators is exceedingly straightforward because each component of the product is rotated independently in a three-dimensional space by a nonselective pulse. For example, let us consider the transformation induced

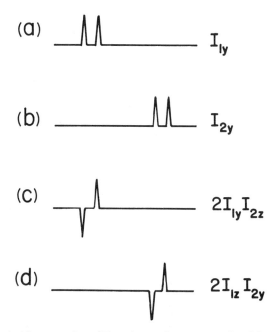

Fig. 2-4. Physical interpretation of Cartesian product operators involving transverse terms. The one-dimensional spectra which would be generated by terms I_{1y}, I_{2y}, $2I_{1y}I_{2z}$, and $2I_{1z}I_{2y}$ are represented.

by a strong $(\pi/2)_x$ pulse, assumed to rotate both spins equally as in a homonuclear system:

1. The operator $\frac{1}{2}E$ is of course unchanged by the pulse. This is consistent with the fact that the sum of all populations is a constant of the motion.

2. An operator $I_{1z} = 2I_{1z}\frac{1}{2}E_2$ is transformed into $-2I_{1y}\frac{1}{2}E_2 = -I_{1y}$ by the $(\pi/2)_x$ pulse. Thus polarizations of the type depicted in Fig. 2-3a are transformed into in-phase magnetization of the first spin by the nonselective pulse, giving rise to a spectrum as in Fig. 2-4a.

3. Two-spin order $2I_{1z}I_{2z}$ is rotated into the term $2I_{1y}I_{2y}$ by the $\pi/2$ pulse. This is a superposition of zero- and double-quantum coherence, and contributes no observable signal. A pulse of flip angle other than $\pi/2$ is necessary to convert two-spin order into observable signal (18).

4. An antiphase component such as $2I_{1x}I_{2z}$ is rotated into $-2I_{1x}I_{2y}$ by the pulse. This also represents the creation of zero- and double-quantum coherence. However, an antiphase component $2I_{1y}I_{2z}$ is rotated into $-2I_{1z}I_{2y}$. This represents a coherence-transfer process—a transformation of antiphase magnetization of the first spin into antiphase magnetization of the second spin by the pulse. Such coherence-transfer processes are extremely important in two-dimensional spectroscopy.

5. A multiple-quantum term such as $2I_{1x}I_{2y}$ is rotated by the pulse into $2I_{1x}I_{2z}$, once more representing antiphase magnetization of the first spin.

These simple examples are sufficient to illustrate the simplicity of this Cartesian product approach in describing the dynamics of weakly coupled spins-$\frac{1}{2}$. In fact, points 1 to 5 above sum up processes which are at the root of almost every experiment in two-dimensional spectroscopy in isotropic liquids. Thus the conversion of longitudinal magnetization I_{1z} into $-I_{1y}$ is the usual starting point for most experiments. The preparation of multiple-quantum coherence by rotation of an antiphase component like $2I_{1x}I_{2z}$ into $-2I_{1x}I_{2y}$, and its reconversion for observation, is a standard part of multiple-quantum spectroscopy (Chapter 4), while the coherence transfer implicit in the rotation of the antiphase term $2I_{1y}I_{2z}$ into $-2I_{1z}I_{2y}$ is used in two-dimensional correlation spectroscopy (Chapter 5). More detailed discussion of some examples may also be found in Sørensen et al. (7) and in Chapter 3 of this book.

Exercise: Work out the results of applying a nonselective $(\pi/2)_y$ pulse to each of the product operators I_{1z}, $2I_{1y}I_{2x}$, $2I_{1x}I_{2z}$, and $2I_{1z}I_{2z}$. Using geometric pictures, work out the transformations of the same operators under $(\pi/4)_y$ pulses, and under a $\pi/4$ rotation about the z axis. Compare the answers with the result of three successive rotations by $\pi/2$ about the y axis, $\pi/4$ about the x axis, and $\pi/2$ about the $-y$ axis, in turn. Then consult Ref. 46.

III. Theory and Use of Product Operators

In this section we will expand the concepts of the previous section in a somewhat more formal way. The aim is also to present some basic principles of quantum mechanics in a form specifically related to the problem of spin dynamics.

A. QUANTIZATION

We consider first an isolated spin **I** in a magnetic field **B**$_0$, in an arbitrary state $|\psi\rangle$. The wave function $|\psi\rangle$ obeys the Schrödinger equation

$$\mathcal{H}_0^{\text{lab}}|\psi\rangle = i\frac{\partial}{\partial t}|\psi\rangle \qquad [19]$$

The factor $\hbar$ has been omitted, and the Hamiltonian operator $\mathcal{H}_0^{\text{lab}}$ expresses the interaction of the spin magnetic moment $\gamma\mathbf{I}$ with the static field **B**$_0$, assumed to be along the z axis:

$$\mathcal{H}_0^{\text{lab}} = -\gamma\mathbf{I}\cdot\mathbf{B}_0 = \omega_0 I_z \qquad [20]$$

where

$$\omega_0 = -\gamma B_0 \qquad [21]$$

Again the particular sign convention employed throughout should be noted.

Of fundamental importance are the eigenfunctions $|m\rangle$ of the Hamiltonian, defined by the quantum numbers m:

$$I_z|m\rangle = m|m\rangle \qquad [22]$$

From the theory of angular momentum (16, 19, 20), m may take any one of $2I+1$ values, from $-I$ to $+I$, in integer steps, I being the quantum number for the total angular momentum of the spin. Accordingly, Eq. [22] allows an equivalent definition of the longitudinal angular momentum operator I_z:

$$I_z = \sum_{m=-I}^{+I} m|m\rangle\langle m| \qquad [23]$$

where the last factor $|m\rangle\langle m|$ corresponds to the population operator of the state labeled by the quantum number m. This equation follows from the orthonormality of the spin eigenfunctions:

$$\langle m|m'\rangle = \delta_{mm'} \qquad [24]$$

For $I = \frac{1}{2}$, Eq. [23] becomes the first equation of the set Eq. [5], with the definitions

$$|\alpha\rangle = |+\tfrac{1}{2}\rangle$$
$$|\beta\rangle = |-\tfrac{1}{2}\rangle \qquad [25]$$

B. Density Operator

An arbitrary wave function $|\psi\rangle$ of the isolated spin may be expressed as an expansion in terms of the orthonormal eigenfunctions $|m\rangle$:

$$|\psi\rangle = \sum_m |m\rangle\langle m|\psi\rangle \qquad [26]$$

Again from standard quantum-mechanical theory, the expectation value $\langle Q \rangle$ of an observable may then be derived as follows:

$$\langle Q\rangle = \langle\psi|Q|\psi\rangle = \sum_{mm'} \langle\psi|m\rangle\langle m|Q|m'\rangle\langle m'|\psi\rangle \qquad [27]$$

This equation is adequate for assessing the expectation value of observables derived from an ensemble of spins prepared in identical states $|\psi\rangle$. However this is rarely the case. In more general instances in which $|\psi\rangle$ varies from member to member in the ensemble, an ensemble average of the full expression must be taken:

$$\overline{\langle Q\rangle} = \sum_{mm'} \overline{\langle\psi|m\rangle\langle m|Q|m'\rangle\langle m'|\psi\rangle} \qquad [28]$$

Since $|\psi\rangle$ appears quadratically in the expression, replacing $|\psi\rangle$ by an ensemble average $\overline{|\psi\rangle}$ would give incorrect results. However, rearrangement of Eq. [28] yields

$$\overline{\langle Q\rangle} = \sum_{mm'} \langle m|Q|m'\rangle\overline{\langle m'|\psi\rangle\langle\psi|m\rangle} \qquad [29]$$

The ensemble averaging may be restricted to the quantity $\overline{|\psi\rangle\langle\psi|}$, which has been introduced above as the density operator σ.

$$\overline{\langle Q\rangle} = \sum_{mm'} \langle m|Q|m'\rangle\langle m'|\sigma|m\rangle = (Q|\sigma) = \mathrm{Tr}(Q\sigma) \qquad [30]$$

All notations in Eq. [30] are equivalent; "Tr" denotes the trace of a matrix.

It is clear that consideration of the density operator $\sigma = \overline{|\psi\rangle\langle\psi|}$ rather than the wave functions $|\psi\rangle$ represents an efficient method of analyzing cases in which ensemble averaging is necessary. The averaging may be done once and for all at the beginning of the calculation, and it is implicit in the use of the operator σ. Note that, providing we are only interested in ensemble observables, there is no loss of information involved in consulting σ rather than the states $|\psi\rangle$ of individual spins.

C. Base Operators

The type of density operator treatment we present here requires the density operator at any given time to be expanded as a weighted combination of orthogonal base operators, which we call in general B_l. The coefficients of the expansion are b_l

$$\sigma = \sum_l b_l B_l \quad [31]$$

There are an infinite number of possible orthogonal basis operator sets. In the section above we emphasized two possibilities, the single-element product operators and the Cartesian product operators, both of which are very useful for weakly coupled spins-$\frac{1}{2}$. For each of these operator sets, every base operator B_l is a product of N operators $B_i^{(l)}$, each of which concerns only the ith spin of a many-spin system:

$$B_l = a \prod_i B_i^{(l)} \quad [32]$$

In the single-element basis, $B_i^{(l)}$ is either I_i^α, I_i^β, I_i^+, or I_i^-. In the Cartesian product basis, $B_i^{(l)}$ may be one of $\frac{1}{2}E_i$, I_{ix}, I_{iy}, or I_{iz}; in this case a factor $a = 2^{N-1}$ is usually also introduced for convenience (7). Note also that this latter base is orthogonal, but not normalized. For N coupled spins-$\frac{1}{2}$, full operator sets contain 2^{2N} members. However, many of these do not develop during an ordinary experiment.

It will often be found necessary to transform from the single-element basis, which concerns individual populations and coherences, into the Cartesian product basis, concerning multiplet configurations, and vice versa. The product-operator nature of the basis makes this transformation easy, since it is only necessary to write each $B_i^{(l)}$ in the alternative representation and then multiply all operators out. Thus in going from the single element base to the Cartesian base, we make the substitutions

$$\begin{aligned} I_i^\alpha &= \tfrac{1}{2}E_i + I_{iz} \\ I_i^\beta &= \tfrac{1}{2}E_i - I_{iz} \\ I_i^+ &= I_{ix} + iI_{iy} \\ I_i^- &= I_{ix} - iI_{iy} \end{aligned} \quad [33]$$

To go in the opposite direction we use

$$\begin{aligned} \tfrac{1}{2}E_i &= \tfrac{1}{2}(I_i^\alpha + I_i^\beta) \\ I_{iz} &= \tfrac{1}{2}(I_i^\alpha - I_i^\beta) \\ I_{ix} &= \tfrac{1}{2}(I_i^+ + I_i^-) \\ I_{iy} &= \frac{1}{2i}(I_i^+ - I_i^-) \end{aligned} \quad [34]$$

In both directions, one term splits up into 2^N on transformation into the opposite base. For example, in a two-spin system, an operator for pure double-quantum coherence $I_1^+ I_2^+$ is expressed in Cartesian operators as follows:

$$I_1^+ I_2^+ = (I_{1x} + iI_{1y})(I_{2x} + iI_{2y})$$
$$= \tfrac{1}{2}(2I_{1x}I_{2x} - 2I_{1y}I_{2y}) + \tfrac{1}{2}i(2I_{1x}I_{2y} + 2I_{1y}I_{2x}) \quad [35]$$

The presence of a factor i in Eq. [35] should not worry us too much since the single-element base is not Hermitian, and may have complex eigenvalues.

D. Thermal Equilibrium

The Hamiltonian $\mathcal{H}_0^{\text{lab}}$ for a system of N weakly coupled spins-$\tfrac{1}{2}$, unperturbed by an applied rf field, is given by

$$\mathcal{H}_0^{\text{lab}} = \sum_i \omega_{0i} I_{iz} + \sum_{i<j} 2\pi J_{ij} I_{iz} I_{jz} \quad [36]$$

where ω_{0i} is the Larmor frequency of spin I_i.

In thermal equilibrium it is assumed that the energy levels are populated according to the Boltzmann distribution, and that coherences have died out [the "spin-temperature" hypothesis (21)]. The equilibrium density operator may then be written as

$$\sigma_{\text{eq}} = \frac{\exp(-\hbar \mathcal{H}_0^{\text{lab}}/kT)}{\text{Tr}\{\exp(-\hbar \mathcal{H}_0^{\text{lab}}/kT)\}} \quad [37]$$

Making the high-temperature approximation ($\hbar \omega_{0i} \ll kT$) and assuming the couplings are weak compared to the Zeeman interactions, the result simplifies to

$$\sigma_{\text{eq}} \approx \left(E - \frac{\hbar}{kT} \sum_i \omega_{0i} I_{iz} \right) \Big/ (\text{Tr } E) \quad [38]$$

It is usual to drop the unity operator, which is invariant to evolution and contributes to no interesting observable. The other constant terms may be gathered together. We write

$$\sigma_{\text{eq}} = \sum_i \beta_i I_{iz} \quad [39]$$

where the Boltzmann factor $\beta_i = -(\text{Tr } E)^{-1} \hbar \omega_{0i}/kT$. For nuclei of the same isotopic species, all β_i are identical and this factor is sometimes also dropped. Equation [39] is already in the form of a weighted sum of Cartesian operators, as is required by Eq. [31].

E. Time Evolution—Free Precession

The Schrödinger equation (Eq. [19]), with the definition of the density operator $\sigma = \overline{|\psi\rangle\langle\psi|}$, leads immediately to the equation of motion of the density operator:

$$\frac{\partial \sigma}{\partial t} = \frac{\partial}{\partial t}\{\overline{|\psi\rangle\langle\psi|}\} = \overline{\left\{\frac{\partial}{\partial t}|\psi\rangle\right\}\langle\psi| + |\psi\rangle\left\{\frac{\partial}{\partial t}\langle\psi|\right\}}$$

$$= -i\mathcal{H}\overline{|\psi\rangle\langle\psi|} + i\overline{|\psi\rangle\langle\psi|}\mathcal{H} = -i[\mathcal{H}, \sigma] \quad [40]$$

where the commutator $[\mathcal{H}, \sigma]$ denotes $\mathcal{H}\sigma - \sigma\mathcal{H}$. It is assumed that the Hamiltonian contains no cross terms between members of the ensemble, or terms which vary from member to member of the ensemble, both of which are responsible for relaxation. Treatment of these terms normally requires a superoperator formalism (11, 22). With the inclusion of a relaxation superoperator, Eq. [40] is known as the Liouville–von Neumann equation. In this chapter, it will be sufficient to introduce relaxation terms phenomenologically after completion of the spin dynamics.

We will always be concerned with Hamiltonians which are, at least over short, finite intervals, time-independent, or which are made time-independent by going into a rotating frame and neglecting as usual the "counter-rotating" terms. Any sequence of rectangular pulses can be treated in this way. For time-independent Hamiltonians, the solution to Eq. [40] is

$$\sigma(t) = \exp(-i\mathcal{H}t)\sigma(0)\exp(i\mathcal{H}t) \quad [41]$$

This may easily be verified by straightforward differentiation, treating the operators exactly like ordinary algebraic functions.

Let us turn at first to free precession, for which the Hamiltonian in the rotating frame is

$$\mathcal{H}_0 = \sum_i \Omega_i I_{iz} + \sum_{i<j} \pi J_{ij}(2I_{iz}I_{jz}) \quad [42]$$

Resonance offsets Ω_i are defined by $\Omega_i = \omega_{0i} - \omega_{rf}$, where ω_{rf} is the rf reference frequency, all in angular frequency units. The coupling term has been written in an explicitly Cartesian product-like way (compare Eq. [15]).

We will consider the evolution of the density operator by examining the evolution of the base operators B_l individually. For free precession, each type of base operator has some advantages. The single-element operators precess according to the frequency difference of the transition they represent, those representing populations not evolving at all. For a weakly coupled Hamiltonian, as given by Eq. [42], it is fairly simple to calculate the evolution frequency of a given single-element product operator. Let us express the product operator B_l as

$$B_l = \prod_i B_i^{(l)} = \prod_i |m_i\rangle\langle m_i'| \qquad [43]$$

where each $m_i = +\frac{1}{2}$ or $-\frac{1}{2}$ and we use the definitions of Eq. [4]. The operator under free precession evolves as

$$B_l \xrightarrow{\mathcal{H}_0 t} B_l \exp\left\{ it\left[\sum_i \Omega_i(m_i' - m_i) + \sum_{i<j} 2\pi J_{ij}(m_i' m_j' - m_i m_j) \right] \right\} \qquad [44]$$

The arrow notation is shorthand for the transformation represented in Eq. [41]. For example, the coherence $I_1^- I_2^\alpha$ in a two-spin system, which has $m_1 = -\frac{1}{2}$, $m_2 = m_1' = m_2' = +\frac{1}{2}$, evolves as

$$I_1^- I_2^\alpha \xrightarrow{\mathcal{H}_0 t} I_1^- I_2^\alpha \exp[it(\Omega_1 + \pi J_{12})]$$

while the zero-quantum coherence $I_1^+ I_2^-$ in the same system, which has $m_1 = m_2' = +\frac{1}{2}$ and $m_1' = m_2 = -\frac{1}{2}$, evolves as

$$I_1^+ I_2^- \xrightarrow{\mathcal{H}_0 t} I_1^+ I_2^- \exp[it(-\Omega_1 + \Omega_2)]$$

Equation [44] expresses the situation that the single-element operators do not split up or mix under the free precession Hamiltonian but simply rotate at their transition frequency, represented by the term in square brackets. This makes single-element operators very useful during the evolution or detection periods of 2D experiments. Equation [44] may be verified by applying the Hamiltonian of Eq. [42] to the kets and bras ($|m_i\rangle$ and $\langle m_i'|$) of Eq. [43] individually, using Eq. [41].

Despite the simplicity of the free evolution of the single-element basis operators, it is sometimes more convenient to calculate the evolution in the Cartesian product base, since, as we shall see, this is the natural one to use as far as pulses are concerned. Evolution of the operators in this base is not as clearcut as with the single element base because each operator splits into several terms as the evolution proceeds. To simplify this process conceptually, it is attractive to take each term in the Hamiltonian, Eq. [42], individually. This can be done in any order, since all terms in the weakly coupled Hamiltonian commute.

The *chemical shift* terms in Eq. [42] affect each spin individually, rotating each term about the z axis at Ω_i radians per second. Thus a term I_{iz} is unaffected, while simple trigonometric relations show that I_{ix} and I_{iy} develop into $I_{ix} \cos \Omega_i t + I_{iy} \sin \Omega_i t$ and $I_{iy} \cos \Omega_i t - I_{ix} \sin \Omega_i t$, respectively, as represented in Fig. 2-5a.

2. DENSITY-OPERATOR THEORY OF PULSES AND PRECESSION

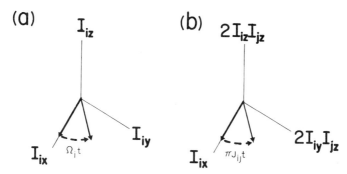

Fig. 2-5. Evolution of some Cartesian product terms under free precession. (a) Evolution of the in-phase term I_{ix} under the chemical shift term $\Omega_i I_{iz}$. (b) Evolution of the in-phase term I_{ix} under the coupling term $\pi J_{ij} 2 I_{iz} I_{jz}$.

The *spin–spin coupling* terms act in a rather more complicated way, since they involve pairs of terms in the product of Cartesian operators. Each coupling J_{ij} involves operators $B_i^{(l)}$ and $B_j^{(l)}$, with other spin operators being unaffected. As each $B_i^{(l)}$ has four possible forms, it might seem that we have to worry about the transformations of $4^2 = 16$ possible types of base operator, but fortunately most of these remain invariant for coupled spins-$\frac{1}{2}$. The only types of Cartesian product operator which do evolve under the coupling are of the form

$$
\begin{aligned}
I_{ix} = 2 I_{ix} \tfrac{1}{2} E_j &\xrightarrow{\pi J_{ij} t 2 I_{iz} I_{jz}} I_{ix} \cos(\pi J_{ij} t) + 2 I_{iy} I_{jz} \sin(\pi J_{ij} t) \\
I_{iy} = 2 I_{iy} \tfrac{1}{2} E_j &\longrightarrow I_{iy} \cos(\pi J_{ij} t) - 2 I_{ix} I_{jz} \sin(\pi J_{ij} t) \\
2 I_{ix} I_{jz} &\longrightarrow 2 I_{ix} I_{jz} \cos(\pi J_{ij} t) + I_{iy} \sin(\pi J_{ij} t) \\
2 I_{iy} I_{jz} &\longrightarrow 2 I_{iy} I_{jz} \cos(\pi J_{ij} t) - I_{ix} \sin(\pi J_{ij} t)
\end{aligned}
\qquad [45]
$$

These represent the splitting of in-phase components into antiphase components, as shown in Fig. 2-5b, and their subsequent reconvergence. Other Cartesian product operators, for example $2 I_{ix} I_{jy}$, are unaffected by evolution under J_{ij}. However, they are of course affected by other couplings J_{ik}.

By the time evolution under each term in Eqs. [45] has been calculated, a large number of terms will have been generated. This is one of the disadvantages of the Cartesian-product approach. However, it turns out to be relatively easy to work out the terms for the evolution of an arbitrarily large weakly coupled system. For example, consider a three-spin-$\frac{1}{2}$ system evolving under a coupling term

$$
\begin{aligned}
\mathcal{H}_{II} &= \mathcal{H}_{12} + \mathcal{H}_{13} + \mathcal{H}_{23} \\
&= 2\pi (J_{12} I_{1z} I_{2z} + J_{13} I_{1z} I_{3z} + J_{23} I_{2z} I_{3z})
\end{aligned}
$$

An initial operator such as I_{1x} splits up, according to Eq. [45], into four terms under such a Hamiltonian, as is best represented by a *tree* diagram, in which the effects of each term in the Hamiltonian are considered, term by term but in arbitrary order:

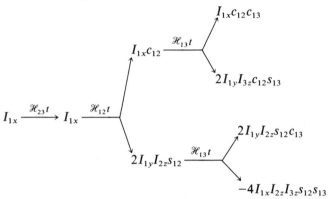

where $c_{ij} = \cos \pi J_{ij} t$ and $s_{ij} = \sin \pi J_{ij} t$.

In fact, no product operator term in a three-spin system splits up into more than four terms under the coupling Hamiltonian, as can be verified by use of Eq. [45]. As another example, consider the more general term $4I_{1x}I_{2y}I_{3z}$, which represents zero- and double-quantum coherence between spins I_1 and I_2, antiphase with respect to spin I_3 (7). This time the tree diagram takes the form

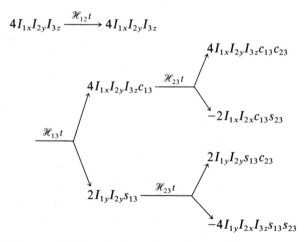

Note the different choice of the order of operations, so as to use the commutation most efficiently. The term $2I_{1x}I_{2y}$, representing a superposition of zero- and double-quantum coherence between spins 1 and 2, does not

evolve under the coupling between those spins, as mentioned above. This is because the spin states at the two ends of the "transition" have in both cases the same amount of spin–spin coupling energy.

Many of the coupling terms in a many-spin system may often be left out of the Hamiltonian. Thus couplings between chemically equivalent spins may be omitted in the weak-coupling case. Small couplings may also often be ignored if the evolution is calculated over a short enough time scale. The remaining terms are often conveniently represented as a coupling network diagram, such as

for AX_3 and A_2X_2 spin systems, respectively (23). In some cases the evolution of Cartesian product operators over quite complicated pulse sequences may be calculated by inspection of these diagrams. This is especially true if all nonvanishing couplings are assumed to have the same value and evolution is allowed to proceed over times that are integer multiples of $1/2J$. In this case, one Cartesian product operator evolves into only one other.

As an example, evolution of a three-spin Cartesian product operator $4I_{1x}I_{2y}I_{3z}$ in the A_2X_2 spin network over a time $(2J)^{-1}$ may be calculated as follows, omitting chemical shift evolution:

$$\underset{z}{\overset{x}{\diagdown}}\hspace{-2pt}\underset{}{\overset{y}{\diagup}} \xrightarrow{12} \underset{z}{\overset{x}{\diagdown}}\hspace{-2pt}\underset{}{\overset{y}{\diagup}} \xrightarrow{34} \underset{z}{\overset{x}{\diagdown}}\hspace{-2pt}\underset{}{\overset{y}{\diagup}} \xrightarrow{13} \underset{z}{\overset{y}{\diagdown}}\hspace{-2pt}\underset{}{\overset{y}{\diagup}} \xrightarrow{24} \underset{z}{\overset{y}{\diagdown}}\hspace{-2pt}\underset{}{\overset{\bar{x}}{\diagup}}$$

or

$$4I_{1x}I_{2y}I_{3z} \xrightarrow{(2J)^{-1}} -4I_{1y}I_{2x}I_{4z}$$

In these diagrams, the coupling terms relevant to the subsequent transformation step have been represented by thickened lines for clarity. Analysis of this type allows construction of procedures for selection of spin responses on the basis of the topology of their coupling networks (23, 24). It should be emphasized that the analysis described above is only valid for infinitely weakly coupled spin systems with uniform transverse relaxation.

F. TIME EVOLUTION—STRONG PULSES

The great strength of the Cartesian product basis is that the calculation of the evolution under pulses is very easy provided that the pulses are short and nonselective.

The Hamiltonian during a pulse, expressed in the rotating frame, is

$$\mathcal{H} = \mathcal{H}_0 + \omega_1 \sum_i (I_{ix} \cos \phi + I_{iy} \sin \phi) \qquad [46]$$

It is assumed here for convenience that all nuclei are of the same isotopic species. The rotation frequency induced by the rf field is ω_1 in angular frequency units. If the pulse is very strong, $\mathcal{H}_0$ may be ignored, and then $\mathcal{H}$ induces rotations of each spin operator around axes in the equatorial plane of the rotating frame. For each individual spin, the usual three-dimensional representation (Bloch picture) may be used to describe the transformations of $\frac{1}{2}E$ (invariant), I_{iz}, I_{ix}, and I_{iy}. Even if $\mathcal{H}_0$ is not small, it may be included in the picture as causing off-resonance effects, which may be represented by tilting of the effective field, as long as the coupling terms $\Sigma_{i<j} \pi J_{ij} 2 I_{iz} I_{jz}$ may still be ignored during the pulse, as discussed in the next section.

Some examples of transformations of Cartesian product operators by short strong pulses were given in Section II,E. As another example, consider a $(\pi/2)_y$ pulse applied to the Cartesian product operator given above:

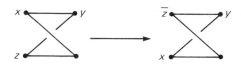

or

$$4 I_{1x} I_{2y} I_{3z} \longrightarrow -4 I_{1z} I_{2y} I_{3x}$$

G. Selective Pulses

So far we have assumed that the rf pulses are infinitely strong and short so that the dynamics are described by instantaneous rotations about axes in the equatorial plane of the rotating frame. This assumption is valid providing that $|\omega_1| \gg |\Omega_i|, |\pi J_{jk}|$, where Ω_i and J_{jk} are the resonance offsets and couplings, respectively. In practice, this condition is often violated, and we distinguish four classes of selective pulse, which we discuss in order of decreasing ω_1:

1. In heteronuclear spin systems, it is usual that the pulse is only resonant for spins I_i of a single isotopic species. This is easily accounted for by restricting the rotations of the Cartesian product operators to those spins alone.

2. In weakly coupled systems, we may often meet a situation where $|\omega_1| \gg |\pi J_{jk}|$ but $|\omega_1| \approx |\Omega_i|$. In this case it is valid to ignore the effect of spin couplings J_{jk} during the pulse and to take into account the offsets Ω_i by

performing rotations of the Cartesian operators around tilted effective fields in the rotating frame (7, 25). Calculations of spin dynamics in this case can be quite complicated in detail, but the principles are the same as with infinitely strong pulses. Undesirable effects caused by tilting of effective fields may be compensated in this range of ω_1 by using composite pulses (25-27). It may sometimes be arranged that $|\Omega_i| \ll |\omega_1|$ for a given spin I_i while $|\Omega_j| \gg |\omega_j|$ for a set of remaining spins I_j. In this case the pulse may be regarded as selective on just one spin in a multispin system, and may be treated as in case 1 above.

3. The case where $|\omega_1| \approx |\pi J_{jk}|$ is normally too complicated to be solved analytically, although solutions exist for fairly simple spin systems in the context of "spin tickling" and decoupling experiments (28, 29). Use of an operator formalism is not normally to be preferred over computer calculations involving numerical diagonalization in this case.

4. If the pulse is so weak that $|\omega_1| \ll |\pi J_{jk}|, |\Omega_i|$, and is arranged so as to be closely resonant to a single well-resolved line in the spectrum, the dynamics may be analyzed by single-transition operators (3, 4). In weakly coupled systems, the single-element product operator representation of these can be instructive. For example, consider irradiation of the allowed transition $|\alpha\alpha\rangle \leftrightarrow |\beta\alpha\rangle$ of a two-spin-$\frac{1}{2}$ system (30-32). Neglecting far-off-resonant terms, the Hamiltonian may be expressed, with the aid of Eqs. [33] and [34], as:

$$\mathcal{H} = \tfrac{1}{2}\omega_1(I_1^+ I_2^\alpha + I_1^- I_2^\alpha)$$
$$= \tfrac{1}{2}\omega_1(I_1^+ + I_1^-)I_2^\alpha$$
$$= \tfrac{1}{2}\omega_1 I_{1x} + \tfrac{1}{2}\omega_1 2 I_{1x} I_{2z} \qquad [47]$$

If this Hamiltonian is applied to an initial condition $\sigma(0) = I_{2x}$ (transverse magnetization of the *other* spin), the first term in Eq. [47] commutes and may be ignored. In accordance with Eq. [41], the second term leads to the evolution

$$\exp(-i\tfrac{1}{2}\omega_1 t 2 I_{1x} I_{2z}) I_{2x} \exp(i\tfrac{1}{2}\omega_1 t 2 I_{1x} I_{2z})$$
$$= I_{2x} \cos \tfrac{1}{2}\omega_1 t + 2 I_{1x} I_{2y} \sin \tfrac{1}{2}\omega_1 t \qquad [48]$$

Examination of this equation shows that the spin coherence is transformed periodically into zero- and double-quantum coherence. Note that a 2π pulse ($\omega_1 t = 2\pi$) in this case only *inverts* the transverse magnetization I_{2x}, a pulse of duration given by $\omega_1 t = 4\pi$ being necessary to return the system to its initial condition. This 4π periodicity is a feature of half-integral spin and has been termed "spinor" behavior (30-33). One might imagine that, since the pulse only perturbs one of the two allowed transitions of the spin, the rf field is only half as effective as when applied nonselectively.

H. Orders of Coherence—Phase Cycling

If the Cartesian operators are well suited for describing pulses, the single-element operators are good at expressing the various orders of coherence and the effect of suppressing some terms by phase cycling procedures (34, 35). As both pulses and phase cycles are employed in most two-dimensional experiments, it is advisable to be conversant with both bases and to be able to transform one into the other (Section III,C).

The order of coherence represented by a given single-element product operator B_l is indicated by the quantity $p_l = \Sigma_i(m_i - m_i')$ (see the definition given in Eq. [43]). In practice, this simply means adding 1 for every factor I^+ that appears, subtracting 1 for every factor I^-, and discounting all the factors I^α or I^β. Thus $I_1^+ I_2^+$ is a (+2)-quantum operator and $I_1^- I_2^\alpha$ is a (−1)-quantum operator. It is important in a general analysis of two-dimensional spectroscopy to retain a strong distinction between (+p)-quantum and (−p)-quantum processes, since these contribute to quite different parts of a 2D spectrum and may sometimes behave quite differently. They are distinguishable because the quadrature detection process selects exclusively (−1)-quantum operators from the density operator (35). This can be seen by writing the quadrature signal as a function of time, $S(t)$, as $S_x(t) + iS_y(t)$:

$$\langle S(t) \rangle \propto \gamma \, \text{Tr}\left\{ \rho(t) \sum_i (I_{ix} + iI_{iy}) \right\}$$

$$= \gamma \, \text{Tr}\left\{ \rho(t) \sum_i I_i^+ \right\} \quad [49]$$

and

$$\text{Tr}(I_i^+ I_i^-) = \tfrac{1}{2} \text{Tr } E, \quad \text{Tr}(I_i^+ I_i^+) = 0 \quad [50]$$

This exclusive detection of the (−1)-quantum coherences by quadrature detection lends an overall "handedness" to the two-dimensional experiment which explains many otherwise unexpected features. For example, when transferred to observable coherences, (+p)-quantum coherences form echoes, whereas (−p)-quantum coherences do not, causing characteristic selective line-narrowing effects (36–38).

One is reminded of the reactions of racemic mixtures in organic chemistry. The left- and right-handed molecules [(+p)- and (−p)-quantum coherences] appear in equal numbers (i.e., the density operator is Hermitian). In reaction with a second racemic reagent (the pulse sequence, which is a unitary transformation, also without preferential "handedness"), each enantiomer gives two chemically distinct compounds ("echo" and "antiecho" components). If only the right-handed molecules of these are somehow chosen (quadrature detection of the final signal), the reaction products from the two enantiomers of the starting reagent might be quite different. It may sometimes be helpful to remind oneself of this chemical analogy when confronted with some of the more confusing aspects of quadrature detection in two-dimensional spectroscopy.

Different orders of coherence are usually separated by phase cycling, although other methods such as field-gradient pulses may also be used (38). Phase cycling schemes may be formulated in a number of different ways. An elegant formulation of the problem of filtering out coherences of a given order or set of orders is given in a paper by Bodenhausen *et al.* (35). Consider the general scheme for a two-dimensional experiment shown in Fig. 2-6a. We assume that the experimental procedure may be divided into four periods: (a) the preparation of the desired coherences by a pulse sequence U of phase ϕ_U, (b) their evolution for a time t_1, (c) their conversion into observable signal by a transformation V of phase ϕ_V, and (d) detection, during t_2, of the signal with respect to a reference frequency of phase ϕ_R. For suppression of all signals except those deriving from coherences of order p in the evolution period, a series of experiments is performed for which the following condition is always satisfied, with n as an arbitrary integer:

$$p\phi_U - (p+1)\phi_V + \phi_R + \phi_C = 2n\pi \qquad [51]$$

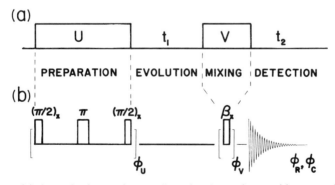

Fig. 2-6. (a) General scheme of a two-dimensional experiment with preparation period (pulse sequence U), evolution period t_1, mixing period V, and detection period t_2. (b) Pulse sequence used for double-quantum spectroscopy of a two-spin-$\frac{1}{2}$ system. The three pulses in the excitation sequence have phase ϕ_U, the mixing pulse has phase ϕ_V, and the phase-sensitive detector is fed with a reference frequency of phase ϕ_R. These phases are described numerically in relation to x as zero. After analog-to-digital conversion, the computer may also apply an arbitrary phase shift ϕ_C before addition to the accumulated signal. The optimum flip angle β of the mixing pulse is a matter of controversy and depends on the magnet homogeneity and the available size of data storage (see text).

Here a phase ϕ_C has been included for the possibility of phase manipulation by the computer after the analog-to-digital conversion. For example, alternate addition and subtraction of data may be represented by having $\phi_C = 0, \pi, 0, \pi$, etc. on a series of experiments, whereas the CYCLOPS scheme for suppressing quadrature artefacts (39) involves $\phi_C = 0, \pi/2, \pi, 3\pi/2$, etc.

The selection of signals derived from coherences of order p during the evolution period is accomplished by incrementing at least one of the independent phase variables, ϕ_U or ϕ_V, in steps of $2\pi/m$ radians over a multiple of m experiments, where m is an integer, and adjusting ϕ_R or ϕ_C each time so as to fulfill Eq. [51]. Optimal suppression is achieved if all possible independent combinations of phases are run through in the series of experiments. The increment $2\pi/m$ is chosen according to the desired *order-selectivity* of the phase-cycling scheme, as discussed in detail in Bodenhausen *et al.* (35). In general, not only signals arising from coherences of order p survive, but also those from all orders $p \pm qm$, where q is an integer. For example, if $m = 4$ (corresponding to phase increments in units of $\pi/2$), and $p = 2$, generating all the combinations in Eq. [51] requires a cycle of $4^3 = 64$ steps. This allows signals arising from both +2Q and −2Q coherences to pass (40), as well as those derived from ±6Q, etc. Some representative phase cycles are given in Tables 2-1 to 2-3. These tables do not include CYCLOPS, which could be implemented by incrementing ϕ_C by $\pi/2$ on each repeat, adjusting ϕ_R according to the analytic formula.

Sometimes allowing both $(+p)$- and $(-p)$-quantum coherences to pass is not desirable. This is true, for example, in a two-dimensional double-quantum experiment (Chapter 4), where allowing both signs of coherence order to appear in the spectrum may lead to confusion, since these evolve with opposite signs of frequency. The ambiguity may be overcome by

TABLE 2-1

BASIC PHASE CYCLES TO SELECT (+1)-QUANTUM COHERENCES, REJECTING ORDERS 0 AND −1; ANALYTIC FORMULA: $\phi_U + \phi_R + \phi_C = 2n\pi$

(a) Three-Step Cycle: $m = 3$					(b) Four-Step Cycle: $m = 4$				
Step	ϕ_U	ϕ_V	ϕ_R	ϕ_C	Step	ϕ_U	ϕ_V	ϕ_R	ϕ_C
(1)	0	0	0	0	(1)	0	0	0	0
(2)	$2\pi/3$	0	$4\pi/3$	0	(2)	$\pi/2$	0	$3\pi/2$	0
(3)	$4\pi/3$	0	$2\pi/3$	0	(3)	π	0	π	0
(1)	0	$2\pi/3$	0	0	(4)	$3\pi/2$	0	$\pi/2$	0
(2)	$2\pi/3$	$2\pi/3$	$4\pi/3$	0	(1)	0	$\pi//2$	0	0
⋮					⋮				
(27 combinations in all)					(64 combinations in all)				

TABLE 2-2

FOUR-STEP PHASE CYCLE TO SELECT
(± 2)-QUANTUM COHERENCES, REJECTING
0 AND ± 1; ANALYTIC FORMULA:
$2\phi_U - 3\phi_V + \phi_R + \phi_C = 2n\pi$; $m = 4$

Step	ϕ_U	ϕ_V	ϕ_R	ϕ_C
(1)	0	0	0	0
(2)	$\pi/2$	0	π	0
(3)	π	0	0	0
(4)	$3\pi/2$	0	π	0
(1)	0	$\pi/2$	$3\pi/2$	0
		$\vdots$		

(64 combinations in all)

positioning the carrier at one end of the spectrum, or more elegantly by using *time-proportional phase incrementation* (TPPI) (41, 42) to give the plus and minus double-quantum coherences an equal and opposite frequency shift (35). In this method, ϕ_U is advanced in steps of $\pi/(2p)$ between each t_1 increment, over and above the phase cycling on each increment. The effect of this is to shift the apparent multiple-quantum frequencies by plus or minus half the spectral width in the ω_1 dimension.

However, both of these options require more data storage space, leading often to the preferred use of phase-cycling schemes which retain only the $(+2)$-quantum coherences, rejecting the (-2)-quantum peaks and so removing the ambiguity in the assignment of peaks to positive or negative orders.

TABLE 2-3

FIVE-STEP PHASE CYCLE TO SELECT
$(+2)$-QUANTUM COHERENCE, REJECTING
0, ± 1, AND -2; ANALYTIC FORMULA:
$2\phi_U - 3\phi_V + \phi_R + \phi_C = 2n\pi$; $m = 5$

Step	ϕ_U	ϕ_V	ϕ_R	ϕ_C
(1)	0	0	0	0
(2)	$2\pi/5$	0	$6\pi/5$	0
(3)	$4\pi/5$	0	$2\pi/5$	0
(4)	$6\pi/5$	0	$8\pi/5$	0
(5)	$8\pi/5$	0	$4\pi/5$	0
(1)	0	$2\pi/5$	$6\pi/5$	0
		$\vdots$		

(125 combinations in all)

All that is required in order to do this is to set the phase increment $2\pi/m$ to less than $\pi/2$. The first cycle of this type reported had $m = 8$ (43), which was clearly an overkill. The most efficient would have $m = 5$ as shown in Table 2-3; a cycle with $m = 6$ has been demonstrated (35). Keeping m as small as possible reduces the time required for the complete experiment.

The use of analytical expressions such as Eq. [51] together with the recipe of Bodenhausen *et al.* (35) for the necessary basic increment in phase allows the phase cycle to be calculated easily by the computer without reference to cumbersome phase tables. In the pulse programming system in use in our laboratory, just such a software calculation of phases has proven to be indispensable for implementing the more complicated phase cycles.

The modern view of phase cycling summarized here makes it clear that it is necessary to have phase shifts available as any fraction of 2π radians. Although the technology for producing such shifts is surprisingly straightforward (44, 45), few spectrometers are as yet equipped with accurate and versatile digital phase shifters. On a standard instrument, providing pulse power, stability, and rf homogeneity are good, composite z pulses (46), which employ combinations of pulses which have conventional quadrature phases, may be used as a stop-gap (43). Here a phase shift β of the reference frequency is replaced by an active "z rotation" $(-\beta)_z$ of the density operator, conveniently implemented by a composite pulse sequence such as $(\pi/2)_x(\beta)_y(\pi/2)_{-x}$.

The suppression of certain orders of coherence by a phase-cycling procedure or by use of field gradients may be taken into account in the density operator simply by deleting all single element terms not representing coherences of the desired order or orders. This necessarily requires use of the single-element basis at this stage in the calculation, rather than the Cartesian-product base, since each term in the latter base includes a mixture of coherences of different orders.

IV. A Typical Two-Dimensional Experiment— Double-Quantum Spectroscopy of Two Coupled Spins-$\frac{1}{2}$

The use of the operators and concepts given above will be exemplified by a brief treatment of homonuclear double-quantum spectroscopy of two coupled spins-$\frac{1}{2}$, employed, for example, in the so-called INADEQUATE experiment (40, 43), for which the spins are typically ^{13}C nuclei.

The pulse sequence used for double-quantum spectroscopy is shown in Fig. 2-6b. The preparation sequence U is a pulse sequence

$$(\pi/2)_{\phi_U}-\tau/2-(\pi)_{\phi_U}-\tau/2-(\pi/2)_{\phi_U} \qquad [52]$$

2. DENSITY-OPERATOR THEORY OF PULSES AND PRECESSION

For optimum creation of double quantum coherence, the periods $\tau/2$ are adjusted to fulfill the condition (6, 40)

$$\sin \pi J\tau = \pm 1 \qquad [53]$$

We evaluate the action of the pulse sequence U assuming at first that $\phi_U = 0$, corresponding to phase x.

The equilibrium density operator for the two-spin-$\frac{1}{2}$ system, applicable at the start of the pulse sequence, is given by

$$\sigma_{eq} = \beta_I (I_{1z} + I_{2z}) \qquad [54]$$

where β_I is the Boltzmann factor introduced in Eq. [39]. For simplicity, we shall omit this factor in the following equations. We may follow the transformations of this operator through the preparation sequence using Cartesian product operators. The first pulse $(\pi/2)_x$ rotates the magnetization into the $-y$ direction, converting the density operator into

$$\sigma = -(I_{1y} + I_{2y}) \qquad [55]$$

There now follows a $\tau/2$-$(\pi)_x$-$\tau/2$ period. The refocusing pulse removes the effects of chemical shift evolution so we have to be concerned only with the effects of couplings during this period. These are easily obtained from Eqs. [45], so that at the end of the τ period, the density operator is

$$\sigma = (I_{1y} + I_{2y}) \cos(\pi J\tau) - (2I_{1x}I_{2z} + 2I_{1z}I_{2x}) \sin(\pi J\tau) \qquad [56]$$

The $(\pi)_x$ rotation under the refocusing pulse has also been included. Assuming $\tau = 1/2J$, the cosine factor is zero and the sine factor is unity. Considering the effect of the next $(\pi/2)_x$ pulse, we find

$$\sigma = -(2I_{1x}I_{2y} + 2I_{1y}I_{2x}) \qquad [57]$$

This we may interpret by expanding the Cartesian operators into the single-element base by equations [34]. If we do this, we find much cancellation, and the resultant density operator contains only $(+2)$-quantum and (-2)-quantum terms:

$$\sigma = i(I_1^+ I_2^+ - I_1^- I_2^-) \qquad [58]$$

At this stage phase cycling according to Table 2-2 would remove any zero- and single-quantum terms contributed either by this spin system or by the more abundant systems of isolated spins-$\frac{1}{2}$. Also, use of a phase cycle with increments in phase $2\pi/m$ of less than $\pi/2$ may lead to suppression of either the $I_1^+ I_2^+$ or the $I_1^- I_2^-$ term in Eq. [58]. We shall return to this point later, but for the moment we assume that both terms survive the phase cycling.

The coherences are now allowed to evolve for a time t_1, and the density operator at the end of this interval may be obtained from Eq. [44]:

$$\sigma(t_1) = i \exp[-i(\Omega_1 + \Omega_2)t_1 - \lambda_1 t_1]I_1^+ I_2^+$$
$$- i \exp[i(\Omega_1 + \Omega_2)t_1 - \lambda_1 t_1]I_1^- I_2^- \qquad [59]$$

where relaxation of the double-quantum coherences at the rate λ_1 has been included. We note that the (+2)-quantum coherence appears to evolve in the sense opposite to that of the (−2)-quantum coherence, making it imperative that the latter is suppressed by a phase cycle or displaced by TPPI (time-proportional phase incrementation), if confusion of the resultant spectrum is to be avoided. Note also that the double-quantum coherences are not affected by the coupling between the two spins; this is a general property of coherences involving flips of all spins in a system.

Before the detection period, the multiple-quantum coherences are converted into observable signal by application of a pulse of angle θ. This can be analyzed by transforming back into the Cartesian-product representation. Thus from Eq. [35], the (+2)-quantum term becomes

$$I_1^+ I_2^+ = (I_{1x} + iI_{1y})(I_{2x} + iI_{2y}) \qquad [60]$$

To this we may apply a rotation by angle θ around the x axis, giving

$$\exp\left(-i\theta \sum_i I_{ix}\right) I_1^+ I_2^+ \exp\left(i\theta \sum_i I_{ix}\right)$$
$$= (I_{1x} + iI_{1y} \cos\theta + iI_{1z} \sin\theta)(I_{2x} + iI_{2y} \cos\theta + iI_{2z} \sin\theta) \qquad [61]$$

In the case of quadrature detection, only the (−1)-quantum terms of this are observable. These may be picked out by transforming back into the single-element basis by the standard procedure and omitting all but the (−1)-quantum terms. This gives

$$\left[\exp\left(-i\theta \sum_i I_{ix}\right) I_1^+ I_2^+ \exp\left(i\theta \sum_i I_{ix}\right)\right]_{-1Q}$$
$$= -\frac{1}{4i} \sin\theta(1 - \cos\theta)(I_1^- I_2^\alpha - I_1^- I_2^\beta + I_1^\alpha I_2^- - I_1^\beta I_2^-) \qquad [62]$$

Some (−1)-quantum terms are also derived from the (−2)-quantum terms involving $I_1^- I_2^-$:

$$\left[\exp\left(-i\theta \sum_i I_{ix}\right) I_1^- I_2^- \exp\left(i\theta \sum_i I_{ix}\right)\right]_{-1Q}$$
$$= \frac{1}{4i} \sin\theta(1 + \cos\theta)(I_1^- I_2^\alpha - I_1^- I_2^\beta + I_1^\alpha I_2^- - I_1^\beta I_2^-) \qquad [63]$$

Several points of interest appear here. First, observe how the final (−1)-quantum coherences appear in antiphase, with $I_1^- I_2^\alpha$ of opposite sign to $I_1^- I_2^\beta$. This is a very typical feature of spectra involving coherence transfer. Second, we observe the functional dependences of the (±2)-quantum signals on the flip angle θ. These have been observed by Mareci and Freeman (47) and are manifestations of more general rules for the intensity dependence of multiple-quantum spectra. We note that the "echo" component [the (+2)-quantum signal] is at a maximum at $\theta = 2\pi/3$, for which $\sin\theta(1 - \cos\theta) = (3/4)\sqrt{3}$. Although this component is stronger than the "antiecho" component (−2Q coherence), it is still best to suppress the latter by using a phase cycle as given above, requiring phase shifts in steps of less than $\pi/2$ radians.

However, the optimum strategy at this stage is somewhat controversial. The modification (47) with $\theta = 2\pi/3$ certainly does lead to good sensitivity and small data matrices, since only the (+2)-quantum terms appear. On the other hand, it does seem that the rejection of the (−2)-quantum coherences comprises a loss of useful signal energy, particularly in a homogeneous magnetic field, in which case the two "mirror-image" components have equal linewidths. Since the *combined* signal energy from the (±2)-quantum coherences is at an optimum for $\theta = \pi/2$, one suspects that sensitivity is in fact optimized by using that value of θ and retaining both signals, using a phase cycle with steps no smaller than $\pi/2$.

If a technique such as TPPI is used to remove overlap between the two components in frequency space as described above, folding of the spectrum about $\omega_1 = 0$ can be used to superimpose the two sets of signals. This can be done by using a real rather than a complex transform in the t_1 dimension (35). Another advantage of this approach is that it removes dispersion-mode contributions to the 2D lineshapes, providing enhanced resolution. An undesirable consequence is that the size of the data matrix required to properly represent both the (±2)-quantum signals is double that required when only one component is retained.

Let us now assume that the (−2)-quantum peaks are suppressed. This done, the coherence is allowed to evolve in the detection period, during which the four terms in Eq. [62] rotate at their four individual frequencies representing the four lines of the AX quartet. Including a relaxation rate λ_2, the final expansion for the phase-cycled 2D signal is

$$\overline{s(t_1, t_2)} = -\tfrac{1}{4}\sin\theta(1-\cos\theta)\exp[-i(\Omega_1+\Omega_2)t_1 - \lambda_1 t_1]$$
$$\times \{\exp[i(\Omega_1 + \pi J)t_2 - \lambda_2 t_2]$$
$$-\exp[i(\Omega_1 - \pi J)t_2 - \lambda_2 t_2]$$
$$+\exp[i(\Omega_2 + \pi J)t_2 - \lambda_2 t_2]$$
$$-\exp[i(\Omega_2 - \pi J)t_2 - \lambda_2 t_2]\} \qquad [64]$$

Double Fourier transformation, first with respect to t_2 and then with respect to t_1, yields four 2D peaks as shown schematically in Fig. 2-7. These peaks are shown each with the characteristic "phase-twist" lineshape (48) involving a mixture of absorption-mode and dispersion-mode Lorentzians. It should be realized that this lineshape is as fundamental a feature of 2D spectroscopy as absorption-mode and dispersion-mode Lorentzians are of 1D Fourier spectroscopy. If the absorption Lorentzian $a(\omega)$ and dispersion Lorentzian $d(\omega)$ are defined as

$$a(\omega) + id(\omega) = \int_{-\infty}^{\infty} \exp(-\lambda t) \exp(i\omega t) \, dt \qquad [65]$$

where both $a(\omega)$ and $d(\omega)$ are real, then the phase-twist lineshape has an equally simple definition in two Fourier dimensions:

$$S(\omega_1, \omega_2) = \int_{-\infty}^{\infty} \int_{-\infty}^{\infty} \exp(-\lambda_1 t_1) \exp(-\lambda_2 t_2) \exp(i\omega_1 t_1) \exp(i\omega_2 t_2) \, dt_1 \, dt_2$$
$$[66]$$

The real and imaginary parts of the lineshape are superpositions of absorption- and dispersion-mode Lorentzians:

$$\text{Re } S(\omega_1, \omega_2) = a(\omega_1)a(\omega_2) - d(\omega_1)d(\omega_2)$$
$$\text{Im } S(\omega_1, \omega_2) = a(\omega_1)d(\omega_2) + a(\omega_2)d(\omega_1) \qquad [67]$$

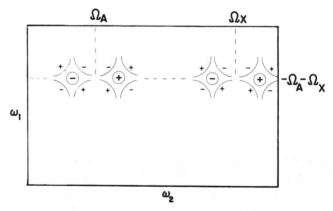

Fig. 2-7. Schematic representation of a double-quantum spectrum of a two-spin-$\frac{1}{2}$ system, assuming that (-2)-quantum peaks are rejected by phase cycling or by a proper choice of β. The peaks arise solely from $(+2)$-quantum coherences which evolve with frequency $-\Omega_A - \Omega_X$ in the ω_1 dimension. The four lines are shown as schematic representations of the typical 2D "phase-twist" lineshapes, with signs indicated. Note the antiphase character.

2. DENSITY-OPERATOR THEORY OF PULSES AND PRECESSION

We may regard the phase-twist lineshapes as another manifestation of the fundamental "handedness" of the quadrature 2D experiment. It is sometimes possible to defeat it by a deliberate superposition of opposite "hands," that is, superposition of signals arising from coherences of opposite signs of coherence order in t_1 (35, 49–52). As mentioned above, this normally requires more data storage space, but may improve sensitivity as well as resolution. Unfortunately, the superposition of opposite orders in this way does not seem to be applicable to all 2D experiments. Those lacking a mixing period, such as 2D homonuclear J-spectroscopy, do not seem amenable to this type of manipulation.

V. Summary

A great number of other systems may be treated by product operators; I have not yet considered here spins greater than $\frac{1}{2}$, for example. Some brief examples are to be found in Sørensen et al. (7). However, I hope that I have been successful in showing that the product-operator formalisms can be useful in clearing up such topics as phase cycling, discrimination of sign of coherence order, 2D lineshapes, and the like, particularly when the single-element basis is used. In mathematical terminology, the nice features of this basis arise because the operators are eigenoperators of the Hamiltonian superoperator (22) and so bring out the rotational properties around the static magnetic field. On the other hand, the Cartesian product base is well adapted for calculating the effects of strong pulses and will be found to be most convenient for many applications, although there are some pulse sequences for which a Cartesian-operator treatment is inadequate and use of the single-element operators is much more efficient (23, 24). Also, anything concerning the "handedness" of the 2D experiment, such as coherence-transfer echoes, is rather difficult to visualize using Cartesian operators but becomes clearer on working in the "natural" single-element base.

Acknowledgments

I would like to thank Prof. R. R. Ernst for support during the writing of this chapter. Much of its contents grew directly out of discussions with him and his group, including Ole Sørensen, Pablo Caravatti, Lukas Braunschweiler, Dieter Suter, and Alex Wokaun. For particular thanks I would like to single out Geoffrey Bodenhausen for continuous assistance and illuminating discussions.

References

1. W. P. Aue, E. Bartholdi, and R. R. Ernst, *J. Chem. Phys.* **64**, 2229 (1976).
2. A. Bax, "Two Dimensional Nuclear Magnetic Resonance in Liquids." Delft Univ. Press (Reidel), Dordrecht, Netherlands, 1982.
3. A. Wokaun and R. R. Ernst, *J. Chem. Phys.* **67**, 1753 (1977).

4. S. Vega, *J. Chem. Phys.* **68**, 5518 (1978).
5. B. C. Sanctuary, *Mol. Phys.* **48**, 1155 (1983).
6. L. Müller, *J. Am. Chem. Soc.* **101**, 4481 (1979).
7. O. W. Sørensen, G. W. Eich, M. H. Levitt, G. Bodenhausen, and R. R. Ernst, *Prog. NMR Spectrosc.* **16**, 163 (1983).
8. F. J. M. van de Ven and C. W. Hilbers, *J. Magn. Reson.* **54**, 512 (1983).
9. K. Packer and K. Wright, *Mol. Phys.* **50**, 797 (1983).
10. R. M. Lynden-Bell, J. M. Bulsing, and D. M. Doddrell, *J. Magn. Reson.* **55**, 128 (1983).
11. C. N. Banwell and H. Primas, *Mol. Phys.* **6**, 225 (1963).
12. W. S. Warren, Ph.D. Thesis, Univ. of California, Berkeley, 1980.
13. D. P. Weitekamp, J. R. Garbow, and A. Pines, *J. Chem. Phys.* **77**, 2870 (1982).
14. D. P. Weitekamp, *Adv. Magn. Reson.* **11**, 111 (1983).
15. C. P. Slichter, "Principles of Magnetic Resonance," 2nd Ed. Springer-Verlag, Berlin and New York, 1980.
16. C. Cohen-Tannoudji, B. Diu, and F. Laloe, "Quantum Mechanics," 2nd Ed. Wiley, New York, 1977.
17. V. W. Hughes, *Sci. Am.* **214** (Apr.), 93 (1966).
18. J. Jeener and P. Broekaert, *Phys. Rev.* **157**, 232 (1967).
19. A. R. Edmonds, "Angular Momentum in Quantum Mechanics." Princeton Univ. Press, Princeton, New Jersey, 1957.
20. P. W. Atkins, "Molecular Quantum Mechanics," Chap. 6. Oxford Univ. Press (Clarendon), London and New York, 1970.
21. M. Goldman, "Spin Temperature and NMR in Solids." Oxford Univ. Press (Clarendon), London and New York, 1970.
22. J. Jeener, *Adv. Magn. Reson.* **10**, 1 (1983).
23. M. H. Levitt and R. R. Ernst, *Chem. Phys. Lett.* **100**, 119 (1983).
24. M. H. Levitt and R. R. Ernst, *J. Chem. Phys.* **83**, 3297 (1985).
25. M. H. Levitt and R. R. Ernst, *Mol. Phys.* **50**, 1109 (1983).
26. M. H. Levitt and R. Freeman, *J. Magn. Reson.* **33**, 473 (1979).
27. M. H. Levitt, *Prog. NMR Spectrosc.* **18**, 61 (1986).
28. W. A. Anderson and R. Freeman, *J. Chem. Phys.* **37**, 85 (1962).
29. J. S. Waugh, *J. Magn. Reson.* **50**, 30 (1982).
30. P. M. Henrichs and L. J. Schwartz, *J. Magn. Reson.* **28**, 477 (1979).
31. M. E. Stoll, A. J. Vega, and R. W. Vaughan, *Phys. Rev. A* **16**, 1521 (1977).
32. R. Kaiser, *J. Magn. Reson.* **40**, 439 (1980).
33. M. Mehring, E. K. Wolff, and M. E. Stoll, *J. Magn. Reson.* **37**, 475 (1980).
34. A. Wokaun and R. R. Ernst, *Chem. Phys. Lett.* **52**, 407 (1977).
35. G. Bodenhausen, H. Kogler, and R. R. Ernst, *J. Magn. Reson.* **58**, 370 (1984).
36. A. A. Maudsley, A. Wokaun, and R. R. Ernst, *Chem. Phys. Lett.* **55**, 9 (1978).
37. D. P. Weitekamp, J. R. Garbow, J. B. Murdoch, and A. Pines, *J. Am. Chem. Soc.* **103**, 3578 (1981).
38. A. Bax, P. G. de Jong, A. F. Mehlkopf, and J. Smidt, *Chem. Phys. Lett.* **69**, 567 (1980).
39. D. I. Hoult and R. E. Richards, *Proc. R. Soc. London, Ser. A* **344**, 311 (1975).
40. A. Bax, R. Freeman, and S. P. Kempsell, *J. Am. Chem. Soc.* **102**, 4849 (1980).
41. G. Drobny, A. Pines, S. Sinton, D. P. Weitekamp, and D. Wemmer, *Faraday Symp. Chem. Soc.* **13**, 49 (1979).
42. G. Bodenhausen, R. L. Vold, and R. R. Vold, *J. Magn. Reson.* **37**, 93 (1980).
43. A. Bax, R. Freeman, T. Frenkiel, and M. H. Levitt, *J. Magn. Reson.* **43**, 478 (1981).
44. M. Hintermann, L. Braunschweiler, G. Bodenhausen, and R. R. Ernst, *J. Magn. Reson.* **50**, 316 (1982).

45. T. Frenkiel and J. Keeler, *J. Magn. Reson.* **50**, 479 (1982).
46. R. Freeman, T. Frenkiel, and M. H. Levitt, *J. Magn. Reson.* **44**, 409 (1981).
47. T. H. Mareci and R. Freeman, *J. Magn. Reson.* **48**, 158 (1982).
48. R. Freeman, *Proc. R. Soc. London, Ser. A* **373**, 149 (1980).
49. P. Bachmann, W. P. Aue, L. Müller, and R. R. Ernst, *J. Magn. Reson.* **28**, 29 (1977).
50. R. Freeman, S. P. Kempsell, and M. H. Levitt, *J. Magn. Reson.* **34**, 663 (1979).
51. D. J. States, R. A. Haberkorn, and D. J. Ruben, *J. Magn. Reson.* **48**, 286 (1982).
52. D. Marion and K. Wüthrich, *Biochem. Biophys. Res. Commun.* **113**, 967 (1983).

PULSE METHODS IN 1D AND 2D LIQUID-PHASE NMR

3

Polarization Transfer and Editing Techniques*

OLE W. SØRENSEN

LABORATORIUM FÜR PHYSIKALISCHE CHEMIE
EIDGENÖSSISCHE TECHNISCHE HOCHSCHULE
CH-8092 ZÜRICH, SWITZERLAND

HANS J. JAKOBSEN

DEPARTMENT OF CHEMISTRY
UNIVERSITY OF AARHUS
DK-8000 AARHUS C, DENMARK

I. Introduction	150
II. Theory	152
A. The Density Operator	152
B. Product Operators	153
C. Transformations of Product Operators	159
D. Multiple-Quantum Coherence	164
E. Some Shortcuts in Density-Operator Calculations	166
III. Polarization Transfer and Purging	169
A. Introduction	169
B. Purging	169
C. Heteronuclear Polarization Transfer	175
D. Heteronuclear Chemical Shift Correlation from One-Dimensional Polarization-Transfer Experiments	185
E. Polarization Transfer and Relative Signs of Coupling Constants	192
IV. Heteronuclear Spectral Editing Techniques	207
A. Introduction	207
B. APT (*J*-Modulated Spin Echo) and INEPT	207
C. SEMUT and DEPT	209
D. Magnitude of *J* Cross Talk in Edited Subspectra	211
E. SEMUT GL and DEPT GL	214

* Based in part on Chapters 1-3, O. W. Sørensen, "Modern Pulse Techniques in Liquid State Nuclear Magnetic Resonance Spectroscopy," ETH Dissertation 7658, Zürich (1984).

Copyright © 1988 by Academic Press, Inc.
All rights of reproduction in any form reserved.

F. Practical Aspects of Subspectral Editing ... 220
G. Editing of Coupled Spectra ... 227
H. SEMUT Editing of INADEQUATE Spectra ... 229
I. Proton-Coupling Networks from Extended DEPT GL Editing ... 242
J. Editing of 2D Spectra ... 245
Appendixes ... 248
   A. Basic Quantum Mechanics of Commutators and Exponential Operators ... 248
   B. Pulses with Arbitrary Phase ... 250
   C. Pulses with Tilted rf Axes ... 251
   D. Operator Analysis of DEPT for I_2S System ... 252
   E. Operator Analysis of Editing by GL^+ Sequences ... 253
References ... 257

I. Introduction

Heteronuclear polarization transfer (PT) and editing methods are very important experimental NMR techniques which in recent years have found widespread applications for solving chemical problems and have been utilized in a number of pulse sequences. It is likely that future implementations and applications in different fields of chemistry and biochemistry will increase rapidly.

Selective population transfer, described in Chapter 1, is induced by applying a selective π pulse to one or more of the transitions in the hydrogen satellite spectrum for a heteronuclear spin system, and represents the first heteronuclear polarization transfer experiment reported for liquids. This experiment, introduced independently by two research groups in the early 1970s (1, 2), is the predecessor of today's family of modern nonselective PT experiments. The development of the SPT experiment was inspired by the homonuclear ^1H-{^1H} transitory selective irradiation (TSI) cw NMR experiment devised some years before by Hoffman, Gestblom and Forsen (3). These authors also suggested a π-pulse version of the TSI experiment (4), which is the cw analog of the SPT Fourier-transform experiment. With the introduction of gated decoupling capability on commercial FT spectrometers, heteronuclear selective population transfer became experimentally feasible. Selective polarization transfer enhancement proved advantageous in comparison with conventional methods in that, unlike nuclear Overhauser enhancement, the effect does not depend on relaxation mechanisms but has a magnitude equal to γ_H/γ_X, where X is the observed nucleus. An additional advantage of the proton PT experiment is that the repetition rate is determined by the usually faster relaxation of the protons rather than by the often slower relaxation of the observed nucleus. It thus is of significant importance for the observation of nuclei with low magnetogyric ratios and long spin–lattice relaxation times, and particularly for nuclei such as silicon-29 and nitrogen-15, which have negative γ values.

Disadvantages of the SPT method are that it requires a preknowledge of proton chemical shifts and that generally only a single spin may be enhanced at one time. To a large extent these problems were eliminated with the development by Bertrand *et al.* (5) in 1978 of J cross polarization for liquids, the first one-dimensional nonselective PT method. In spite of the ingenuity of the liquid cross-polarization experiment, this technique has not been widely used, probably because of its sensitivity to a mismatch for the Hartman–Hahn condition, $\gamma_I B_{1I} = \gamma_S B_{1S}$, as well as the subsequent development of other experimentally less critical nonselective methods, such as INEPT, DEPT, and SINEPT (6, 7). Following their publication in the period since 1979, these methods and a variety of extensions of them have been applied for sensitivity enhancement, determination of proton multiplicities in carbon-13 and nitrogen-15 NMR, and one-dimensional heteronuclear chemical shift correlation. It is the pulse sequences for these PT experiments and several of their extensions and modifications which will form the basis for the discussion of nonselective PT methods in this chapter.

Spectral editing is a valuable method for simplification of complex NMR spectra. Pegg, Doddrell, and Bendall (8) introduced the term *subspectral editing* in connection with the construction of spectra containing the resonances of only CH, only CH_2, and only CH_3 carbons, respectively, by use of the DEPT technique. In its broadest sense, "subspectral editing" means sorting according to multiplicity—that is, conventional spectra are decomposed into independent subspectra according to proton multiplicities. In many applications, spectral editing may be considered as an alternative to two-dimensional methods. In this chapter spectral editing will be applied principally to various types of carbon-13 spectra; however, the techniques described apply, of course, although with some limitations, to other nuclei as well.

Several extensions of INEPT have been used to retrieve proton multiplicities from decoupled carbon-13 spectra. However, the fact that INEPT requires phase-shifting capabilities which were not found on many NMR spectrometers led to the design of multiplicity determination techniques based on *J modulation* of *spin echoes*, as described in Chapter 1, Section I,D. The earlier investigations aimed to differentiate the proton multiplicities on the basis of intensity variations of the carbon-13 resonance and were mainly concerned with a distinction according to whether the number of attached protons was even or odd. However, later studies demonstrated the feasibility of decomposing the spectra into four subspectra according to the individual proton multiplicities. These developments include the DEPT sequence with the improved DEPT GL (9) as well as the SEMUT (10) and SEMUT GL (9) sequences, to be described in this chapter.

Historically, it is interesting that the first application of J modulation of spin echoes in spectral editing was published in 1975 by Campbell *et al.* (11) for the case of first-order homonuclear coupled proton spectra. This work was concerned with a distinction of proton resonance according to whether the number of coupling partners is even or odd. An improvement of this method, which copes with the problem of a spread in J_{HH} values by use of a z filter, has recently been described (12).

Spectral editing techniques may be applied not only to decoupled carbon-13 spectra but also, with minor but important modifications of the pulse sequences, to proton-coupled spectra (13). The two cases will be discussed separately in two subsections. It has recently been demonstrated that editing methods serve as useful tools when incorporated into other pulse sequences. For example, spectrum simplification in the INADEQUATE experiment can be achieved by incorporating SEMUT editing techniques to give the SEMINA experiments (14, 15). Furthermore, in certain cases subspectral editing of decoupled spectra can be extended to yield information not only on the number of directly bonded protons, but also on the number of homonuclear coupling partners of these protons on contiguous atoms—that is, relayed spectral editing (16). Finally, incorporation of editing techniques into 2D carbon-13 spectroscopy has also proved possible (17, 18). These more recent extensions and applications of spectral editing will be covered in the final subsections.

II. Theory

A. The Density Operator

The state of a spin system at any one time may be described (19, 20) by the density operator $\sigma(t)$. In discussing the density operator in this chapter, we will neglect relaxation effects since they are not important in understanding the principles of the pulse sequences. As represented in Eq. [41] of Chapter 2, the time evolution of this operator during interval n under a time-independent Hamiltonian $\mathcal{H}_n$ can be expressed by a unitary transformation with the *propagator* $\exp(-i\mathcal{H}_n \tau_n)$. For a sequence of intervals for each of which a time-independent average Hamiltonian $\mathcal{H}_n$ can be defined,

$$\sigma(\tau_1 + \tau_2 + \tau_3) = \exp(-i\mathcal{H}_3\tau_3)\exp(-i\mathcal{H}_2\tau_2)\exp(-i\mathcal{H}_1\tau_1)\sigma(0)$$
$$\times \exp(i\mathcal{H}_1\tau_1)\exp(i\mathcal{H}_2\tau_2)\exp(i\mathcal{H}_3\tau_3) \quad [1]$$

An alternative way of representing the evolution corresponding to Eq. [1] is by the symbolic notation

$$\sigma(0) \xrightarrow{\mathcal{H}_1\tau_1} \sigma(\tau_1) \xrightarrow{\mathcal{H}_2\tau_2} \sigma(\tau_1+\tau_2) \xrightarrow{\mathcal{H}_3\tau_3} \sigma(\tau_1+\tau_2+\tau_3) \quad [2]$$

where the signs and chronological sequence follow the order of arguments of the exponential operators on the right-hand side of Eq. [1]. This is consistent with the convention that the operators in equations such as Eq. [1] are applied in order from the innermost to the outermost.

From the density operator calculated at any particular time during an experiment, the observable magnetization can be computed using the trace relations presented in Eq. [30] of Chapter 2:

$$M_x = N\gamma\hbar \operatorname{Tr}(F_x \sigma) \qquad [3a]$$

$$M_y = N\gamma\hbar \operatorname{Tr}(F_y \sigma) \qquad [3b]$$

where N is the number of nuclei per unit volume and the operators F_x and F_y corresponding to observables represent sums over the operators I_x and I_y, respectively, of all spins of a particular nuclear species, such as hydrogen-1 or carbon-13 or nitrogen-15:

$$F_x = \sum_k I_{kx} \qquad [4a]$$

$$F_y = \sum_k I_{ky} \qquad [4b]$$

Because the product operators are orthogonal, only I_x contributes to Eq. [3a] and only I_y to Eq. [3b].

In this chapter, we shall use the Schrödinger representation in which the time dependence is carried by the density operator $\sigma(t)$, and the operators such as F_x and F_y, associated with particular properties of the system, remain constant.

B. PRODUCT OPERATORS

As described in Eq. [31] of Chapter 2, the density operator is expressed as a linear combination of certain base operators B_l, with time-dependent coefficients b_l:

$$\sigma(t) = \sum_l b_l(t) B_l \qquad [5]$$

We shall express the base operators in turn as products in which each factor I_{kq} refers to one of the nuclei in the spin system:

$$B_l = 2^{(n-1)} \prod_{k=1}^{N} (I_{kq})^{a_{lk}} \qquad [6]$$

In this equation, k labels a particular nucleus and q is one of the Cartesian coordinates, x, y, or z; $a_{lk} = 1$ for n nuclei and $a_{lk} = 0$ for the remaining nuclei. Use of Cartesian base operators is particularly convenient in treating the effects of pulses, although alternative bases may be employed as described in Chapter 2. The number of operators appearing in the product in Eq. [6] may be any number from one up to the total number N of spin-$\frac{1}{2}$ nuclei in the system. For each spin system, there is also a zero-spin base operator, equal to one-half the unity operator E.

For a system with N spin-$\frac{1}{2}$ nuclei, there are 2^{2N} different product operators. If there are three nuclei in the system, the operators may be designated zero-spin, one-spin, two-spin, and three-spin. For a two-spin system consisting of spins labeled 1 and 2, the Cartesian product operators are

$$n = 0 \quad \tfrac{1}{2}E$$

$$n = 1 \quad I_{1x}, I_{1y}, I_{1z}, I_{2x}, I_{2y}, I_{2z}$$

$$n = 2 \quad 2I_{1x}I_{2x}, 2I_{1x}I_{2y}, 2I_{1x}I_{2z},$$
$$2I_{1y}I_{2x}, 2I_{1y}I_{2y}, 2I_{1y}I_{2z},$$
$$2I_{1z}I_{2x}, 2I_{1z}I_{2y}, 2I_{1z}I_{2z}$$

A one-spin operator represents the normal in-phase multiplet of the nucleus to which it corresponds. Thus, I_{kz} represents equal longitudinal or z polarization across all transitions of spin k. The transverse operators, I_{kx} and I_{ky}, represent in-phase x magnetization and in-phase y magnetization, respectively, of spin k. These magnetizations may be depicted in a semiclassical vector model by parallel vectors along the x and y axes, respectively, of the rotating frame, as shown in Fig. 3-1. Alternatively, these magnetizations may be represented as in the lower part of Fig. 3-1 by lines which are wavy in order to indicate coherence, rather than straight arrows, which would indicate population transfer. A solid wavy arrow designates the x component, and a dashed wavy arrow designates the y component.

Considering two-spin product operators, we recall from Chapter 2 the significance of various types:

1. $2I_{kx}I_{lz}$, $2I_{ky}I_{lz}$ correspond to magnetizations of spin k, respectively in the x and y directions, each of which is antiphase with respect to the splitting caused by coupling to spin l, of magnitude J_{kl}. Antiphase magnetization denotes a multiplet with opposite phases for the individual lines in the two halves of the multiplet; for a doublet, one component is positive and the other is negative. Thus an antiphase multiplet has a total integrated intensity of zero. Figure 3-1 includes a pictorial representation of $2I_{kx}I_{lz}$ with oppositely directed vectors along the $+x$ and $-x$ axes of the rotating frame for the two transitions in the vector diagram, and oppositely directed wavy arrows for the two transitions in the energy-level diagram, the anti-

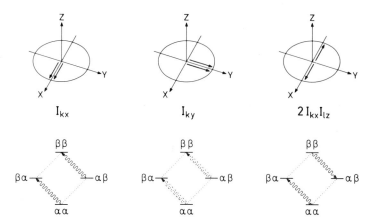

Fig. 3-1. Pictorial representations of product operators representing transverse and longitudinal magnetization for two coupled spins, k and l, with $I = \frac{1}{2}$. The oscillating x and y magnetization components are represented by the usual vectors in the xy plane of the rotating frame or by wavy lines in the energy-level diagram (dashed lines for y components). Wavy lines are used to distinguish coherence from a flow of population.

parallel arrangement in each diagram indicating antiphase magnetization. The operators $2I_{kz}I_{lx}$ and $2I_{kz}I_{ly}$ correspond to analogous magnetizations of spin l, antiphase with respect to the splitting produced by spin k.

2. The operators which have two transverse components, $2I_{kx}I_{lx}$, $2I_{ky}I_{ly}$, $2I_{kx}I_{ly}$, and $2I_{ky}I_{lx}$, correspond to *two-spin coherence* of spins k and l. These are superpositions of zero- and double-quantum coherence and will be discussed further in several later sections. Linear combinations representing pure zero- and double-quantum coherence are represented graphically in Fig. 3-2.

3. The operator $2I_{kz}I_{lz}$ represents *longitudinal two-spin order*, a situation in which there is spin-correlated population of energy levels without net polarization and without observable magnetization; this was depicted in Fig. 2-3.

For spin systems composed of three or more nuclei, there may appear, in addition to one- or two-spin operators such as we have listed for a two-spin system, three-spin operators of forms and significance such as the following:

1. $4I_{ky}I_{lz}I_{mz}$ represents y magnetization of spin k, antiphase with respect to the splitting produced by spin l and also antiphase with respect to the splitting produced by spin m.

2. $4I_{kz}I_{lx}I_{mx}$ represents antiphase two-spin coherence of spins l and m, a superposition of zero- and double-quantum coherence. The multiplet

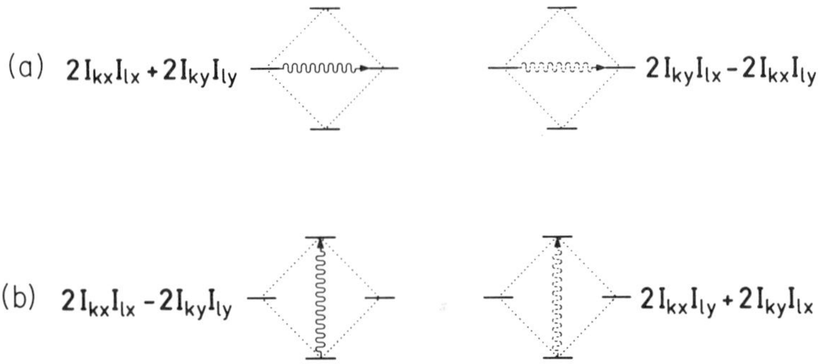

(a) $2I_{kx}I_{lx}+2I_{ky}I_{ly}$ ⟶〰️〰️⟶   ⟶〰️〰️⟶ $2I_{ky}I_{lx}-2I_{kx}I_{ly}$

(b) $2I_{kx}I_{lx}-2I_{ky}I_{ly}$ ⟶⦵⦵⦵⟶   ⟶⦵⦵⦵⟶ $2I_{kx}I_{ly}+2I_{ky}I_{lx}$

Fig. 3-2. Pictorial representations of the linear combinations of product operators representing (a) *pure* zero-quantum and (b) *pure* double-quantum coherence for two coupled spins, k and l, with $I=\tfrac{1}{2}$. The solid and dashed wavy arrows are indicative of the different phase relations for the coherence (x and y components, respectively). Single products never represent *pure* p-quantum coherence; for example, $2I_{kx}I_{lx}$ is represented by a superposition of two of the diagrams.

components have opposite phases depending on the polarization I_{kz} of "passive" spin k.

3. $4I_{ky}I_{ly}I_{my}$, with three transverse components, represents *three-spin coherence*, a combination of single-quantum and triple-quantum coherence.

4. $4I_{kz}I_{lz}I_{mz}$ represents *longitudinal three-spin order*.

Exercise: Describe the state of a three-spin system under each of the following operators: $4I_{kx}I_{ly}I_{mz}$; $4I_{kx}I_{lx}I_{mx}$; $4I_{kx}I_{lz}I_{mx}$; $4I_{kz}I_{ly}I_{mz}$.

Some typical product operators that may occur in the course of evaluation of the density operator for a three-spin system are represented in Fig. 3-3. They include in-phase magnetization (I_{kx}), antiphase magnetization with respect to one or two other nuclei coupled to nucleus k ($2I_{kx}I_{lz}, 4I_{kx}I_{lz}I_{mz}$), pure in-phase double-quantum coherence involving two of the three spins ($2I_{kx}I_{lx}-2I_{ky}I_{ly}$), and pure double-quantum coherence in antiphase with respect to the third spin, a "passive" spin m ($4I_{kx}I_{lx}I_{mz}-4I_{ky}I_{ly}I_{mz}$).

For systems with all spins of $I=\tfrac{1}{2}$, only the one-spin operators I_{kx} and I_{ky} correspond to directly observable magnetization. Product operators representing antiphase magnetization may evolve during the detection period into operators which *do* correspond to observable magnetization. For example, although $2I_{kx}I_{lz}$ is not, strictly speaking, directly observable, because it does not contribute to the observable signal in Eq. [3], it evolves under the effect of resolved spin–spin coupling between k and l:

$$2I_{kx}I_{lz} \to 2I_{kx}I_{lz}\cos\pi J_{kl}\tau + I_{ky}\sin\pi J_{kl}\tau \qquad [7]$$

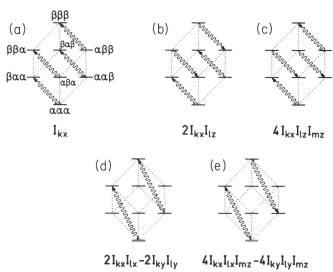

Fig. 3-3. Pictorial representations of some typical product operators for a system of three coupled spins, k, l, and m, with $I = \frac{1}{2}$. (a) In-phase single quantum coherence for one of the spins. (b and c) Antiphase single-quantum magnetization with respect to one and two of the three coupled nuclei, respectively. (d) Pure in-phase double quantum coherence for two of the three spins. (e) Double-quantum coherence in antiphase with respect to the third spin. The arrows indicate parallel and antiparallel coherence components. Each operator term represents an entire multiplet rather than a single transition.

Subsequent Fourier transformation yields, from the term $I_{ky} \sin \pi J_{kl}\tau$, a doublet centered at the chemical shift of nucleus k. The fact that the term includes the sine function, which is an odd function, is the clue to the antiphase nature of the resulting doublet. The first term, that involving $2I_{kx}I_{lz}$, remains unobserved. Any product operator which contains a single transverse component with the remaining components longitudinal, such as $2I_{kx}I_{lz}$ and $4I_{kx}I_{lz}I_{mz}$, evolves to produce observable magnetization provided that all couplings J_{kl}, J_{km}, ... are resolved.

The relative amplitudes and phases of the spectral lines can be derived immediately from the form of the product operators. Figure 3-4 illustrates some one-dimensional spectra of three-spin systems obtained after Fourier transformation of FIDs induced by typical product operators. Given a particular product operator, one can derive the corresponding spectrum by carrying out, as described below in Section II,C, a transformation like Eq. [7] for each I_z factor in the product. An alternative approach is to apply a "tree" procedure. In using this method, we require a parameter defined by the expression $a_{kl} = \gamma_k \gamma_l J_{kl} / |\gamma_k \gamma_l J_{kl}|$, where J_{kl} is the spin–spin coupling constant between k and l and γ_k and γ_l are the magnetogyric ratios of these spins. Suppose, for example, that the spins are ^{15}N and ^{29}Si, for both of

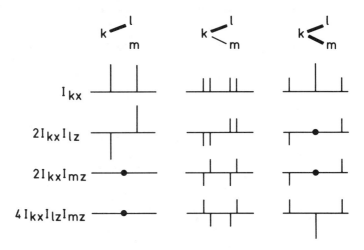

Fig. 3-4. Stick-plot spectra corresponding to some typical product operators involving single-quantum coherence of nucleus k in a system of three spins, k, l, and m, with $I = \frac{1}{2}$. For $J(km) = 0$, signals arising from product operators with I_k magnetization in antiphase with respect to I_m cancel. The observation of single transitions within a multiplet corresponds to a linear combination of product operators. The spectra are drawn assuming that all coupling constants are positive, that all magnetogyric ratios have the same sign, and that the receiver reference phase is x. The positive direction of the frequency axis is from right to left.

which the magnetogyric ratio is negative, and that the sign of J is negative. The numerator of a_{kl} is therefore negative, and, since the denominator is always positive and of the same magnitude as the numerator, the result is that $a_{kl} = -1$. Multiplication of a spectrum by this parameter a_{kl} means that the phases of all lines are changed by 180° if the value is -1 and remain unchanged if the value is $+1$.

Referring to Fig. 3-5, consider the product operator $8I_{1x}I_{2z}I_{3z}I_{4z}$ with $|J_{12}| > |J_{13}| > |J_{14}|$. This corresponds to magnetization of spin 1, and we begin with I_{1x} as a singlet. Then the singlet is split into a doublet with separation J_{12}, the phase of the high-frequency component is changed by 180°, and the spectrum is multiplied by a_{12} as symbolized in the third row in the diagram. Now each of the two lines is treated as a singlet, and the above procedure is repeated to introduce I_{3z}. This yields four lines which, when each is considered as a singlet, make up the starting point for the introduction of I_{4z}. Again each line is doubled, the high-frequency line of each new doublet is inverted, and the spectrum is multiplied by a_{14}. If there were further couplings to spin 1 not represented in the product operator, each of the eight lines of $8I_{1x}I_{2z}I_{3z}I_{4z}$ would now be split in the normal manner without the phase-inversion step.

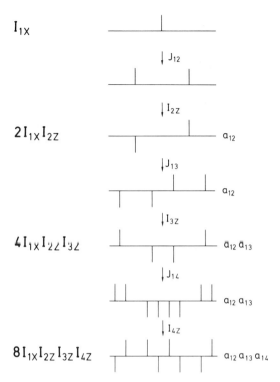

Fig. 3-5. "Tree" procedure for the visualization of product operators in terms of the spectra to which they give rise. The principles of the procedure are explained in the text. There is no significance to relative line intensities between the several stick-plot spectra. Each spectrum is to be multiplied by the product of the coefficients a_{ij} indicated at the right.

With some practice, the "tree" procedure can be carried out in one step starting from the entire in-phase multiplet. The phase of the line at highest frequency is inverted n times, where n is the number of I_z factors in the product operator. The phases of the other lines follow from the fact that the two lines separated by a given J have opposite phase if the spin responsible for the splitting is represented by an I_z operator in the product operator to be interpreted; inspection, in the final diagram in Fig. 3-5, of any pair of lines separated by J_{12}, J_{13}, or J_{14} shows the two lines of the pair to be opposite in phase. Finally, the spectrum is multiplied by the product of the a_{kl} factors.

C. Transformations of Product Operators

We now consider the time evolution of product operators, first during periods of free precession, such as the interval between pulses, and then under the influence of rf pulses. The effects of free precession or rf pulses on a particular product operator B_l can be represented by a series of transformations, corresponding to Eq. [41] of Chapter 2 and to Eqs. [A.15] and [A.18] in the appendix, of the form

$$\exp(-i\eta B_r) B_l \exp(i\eta B_r) = \sum_t b_{tl}(r, \eta) B_t \quad [8]$$

This equation implies that the initial product operator B_l is converted into a linear combination of product operators B_t with coefficients b_{tl}. For an rf pulse of rotation angle θ about axis ϕ applied to nucleus k, ηB_r is of the form $\theta I_{k\phi}$. Using the Hamiltonian for free precession given by Eq. [42] of Chapter 2, the form of ηB_r for chemical shift evolution is $(\Omega_k \tau) I_{kz}$ and for evolution under weak scalar coupling between two spin-$\frac{1}{2}$ nuclei is $(\pi J_{kl} \tau) 2 I_{kz} I_{lz}$. Since for weakly coupled systems the terms in the Hamiltonian commute, the effects of chemical shift and spin-spin coupling may be calculated separately and in any order. In this respect, there is some analogy to the concept of "pulse cascades" introduced by Bodenhausen and Freeman (21).

1. *Evolution During Free Precession*

We consider first the effect of *chemical shift* on the density operator. The result for spin k of chemical shift precession with an angular frequency of Ω_k in the rotating frame during a period of length τ is given by the relations on page 130 of Chapter 2:

$$I_{kx} \xrightarrow{\Omega_k \tau I_{kz}} I_{kx} \cos \Omega_k \tau + I_{ky} \sin \Omega_k \tau \quad [9]$$

$$I_{ky} \xrightarrow{\Omega_k \tau I_{kz}} I_{ky} \cos \Omega_k \tau - I_{kx} \sin \Omega_k \tau \quad [10]$$

These transformations rotate the magnetization around the z axis, in the sense that the x component goes into $+y$ and y component goes into $-x$, and have the same effect as would pulses along the z axis, represented diagramatically in Fig. 3-6a. The I_z component of magnetization is unaffected by chemical shift evolution.

As an example, we may describe the result of chemical shift evolution for a product operator representing two-spin coherence by the following transformation:

$$2 I_{ky} I_{ly} \xrightarrow{\Omega_k \tau I_{kz}} \xrightarrow{\Omega_l \tau I_{lz}} 2(I_{ky} \cos \Omega_k \tau - I_{kx} \sin \Omega_k \tau)$$
$$\times (I_{ly} \cos \Omega_l \tau - I_{lx} \sin \Omega_l \tau) \quad [11]$$

This illustrates the result that $\Omega_k \tau I_{kz}$ affects only I_{ky} and $\Omega_l \tau I_{lz}$ affects only I_{ly}.

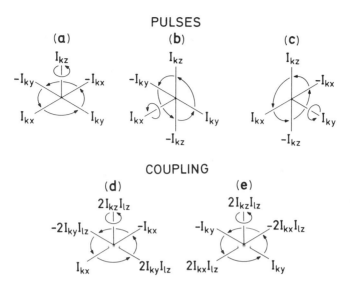

Fig. 3-6. The effects on product operators of chemical shift precession (a), of on-resonance rf pulses applied from the z axis (a), from the x axis (b), and from the y axis (c), and of positive scalar J-coupling evolution (d, e). The sense of each rotation indicated corresponds to a positive angular displacement.

Chemical shift evolution does not change the number of single-spin operators in each product operator. An operator with n transverse components (I_{kx} or I_{ky}) is transformed into a linear combination of no more than 2^n terms, while longitudinal components I_{kz}, as mentioned above, remain unchanged.

The evolution of in-phase magnetization under *spin–spin coupling* between spin-$\frac{1}{2}$ nuclei k and l results in conversion to antiphase magnetization according to Eq. [45] in Chapter 2:

$$I_{kx} \to I_{kx} \cos \pi J_{kl}\tau + 2I_{ky}I_{lz} \sin \pi J_{kl}\tau \qquad [12]$$

$$I_{ky} \to I_{ky} \cos \pi J_{kl}\tau - 2I_{kx}I_{lz} \sin \pi J_{kl}\tau \qquad [13]$$

Conversely, if the initial magnetization is antiphase, evolution under spin–spin coupling produces in-phase magnetization:

$$2I_{kx}I_{lz} \to 2I_{kx}I_{lz} \cos \pi J_{kl}\tau + I_{ky} \sin \pi J_{kl}\tau \qquad [14]$$

$$2I_{ky}I_{lz} \to 2I_{ky}I_{lz} \cos \pi J_{kl}\tau - I_{kx} \sin \pi J_{kl}\tau \qquad [15]$$

For simplicity, in these equations we have omitted the explicit statement of the factor corresponding to ηB_r in Eq. [8] for spin–spin coupling evolution, which in each case is $\pi J_{kl}\tau 2I_{kz}I_{lz}$. It is also to be understood that in each term of equations presented here, all factors following "cos" or "sin" are included in the arguments of these functions. The transformations corresponding to these equations are represented geometrically in Figs. 3-6d and 3-6e. Equations [12]–[15] are restricted to nuclei l with $I = \frac{1}{2}$, but apply if I_k is greater than $\frac{1}{2}$, as well.

2. Effects of Radio-Frequency Pulses

The effect on Cartesian product operators of nonselective pulses with $|\omega_1| = |\gamma B_1| \gg |\Omega_k|$ and $|\omega_1| \gg |\pi J_{kl}|$, so that the rf strength exceeds any chemical shift offset or coupling constant in the spin system, is now considered. For pulses with phase angle ϕ and flip angle of θ, the transformation of $\sigma(t_1)$ into $\sigma(t_2)$ is given by

$$\sigma(t_2) = \exp\left(-i\theta \sum_k I_{k\phi}\right) \sigma(t_1) \exp\left(+i\theta \sum_k I_{k\phi}\right) \qquad [16]$$

The sum includes all spins affected by the pulses; usually for heteronuclear systems this is limited to either only the I species or only the S species. The various nuclei in the spin system may be treated in any order.

For a pulse about the x axis with rotation angle θ affecting nucleus l, the transformations are given by

$$I_{lz} \rightarrow I_{lz} \cos \theta - I_{ly} \sin \theta \qquad [17]$$

$$I_{ly} \rightarrow I_{ly} \cos \theta + I_{lz} \sin \theta \qquad [18]$$

For a similar pulse about the y axis, the transformations are

$$I_{lz} \rightarrow I_{lz} \cos \theta + I_{lx} \sin \theta \qquad [19]$$

$$I_{lx} \rightarrow I_{lx} \cos \theta - I_{lz} \sin \theta \qquad [20]$$

These equations can be obtained by simple trigonometric considerations: the cosine term gives the component remaining along the initial direction and the sine term follows the rule for rotation about the pulse axis; for example, for an x pulse, $+z$ goes into $-y$ and $+y$ goes into $+z$.

The effects of an rf pulse on each single component of a product operator may be treated separately. For example, a 90° nonselective pulse about the x axis converts antiphase magnetization of spin l into antiphase magnetization of spin k:

$$2I_{kz}I_{ly} \xrightarrow{90°(I_{kx}+I_{lx})} -2I_{ky}I_{lz} \qquad [21]$$

A transfer of this type is involved in many polarization transfer experiments, such as INEPT. Use of a 90° pulse along the same axis as one transverse

3. POLARIZATION TRANSFER AND EDITING TECHNIQUES 163

component of the existing antiphase magnetization leads to two-spin coherence:

$$2I_{kz}I_{ly} \xrightarrow{90°(I_{ky}+I_{ly})} 2I_{kx}I_{ly} \qquad [22]$$

Here the y component of spin l is unchanged by the pulse along the y axis. This method is used in the generation of zero- and double-quantum coherence, as in DEPT, SEMUT, and INADEQUATE experiments.

3. *Selective Pulses*

Frequently it is desirable to generate pulses which cover a limited frequency range and to analyze their effects. To describe the transformation effected by a *semiselective* pulse, one which affects uniformly all components of the multiplet for a given spin, the sum in Eq. [16] is simply restricted to a single spin.

A *selective* pulse, which acts only on one line as described in Chapter 1 for selective population transfer, may be represented by a suitable linear combination of related product operators. As an example, it may be observed from Fig. 3-4a that each of the lines in the doublet for spin k can be constructed by a linear combination of spectra corresponding to the operators I_{kx} and $2I_{kx}I_{lz}$. Addition gives the low-frequency line and subtraction the high-frequency line. A selective pulse on one of these lines can then be described by the transformation

$$\sigma(t_1) \xrightarrow{\theta\frac{1}{2}(I_{kx}\pm 2I_{kx}I_{lz})} \sigma(t_2) \qquad [23]$$

That this transformation is equivalent to the following sequence of rotations can be demonstrated by using Eq. [A.22] of the appendix with U in that equation equal to $\exp[i(\pi/2)I_{ky}]$:

$$\sigma(t_1) \xrightarrow{-\frac{\pi}{2}I_{ky}} \xrightarrow{\frac{\theta}{2}I_{kz}} \xrightarrow{\pm\frac{\theta}{2}2I_{kz}I_{lz}} \xrightarrow{\frac{\pi}{2}I_{ky}} \sigma(t_2) \qquad [24]$$

Thus a selective pulse can be represented or, indeed, experimentally generated, by a series of semiselective pulses surrounding a free precession period (22).

For a three-spin system comprising nuclei k, l, and m, appropriate combinations of operators may be derived by inspection of Fig. 3-4b. Thus a selective x pulse acting on the k-spin transition at lowest frequency can be described by the transformation

$$\sigma(t_1) \xrightarrow{\theta\frac{1}{4}(I_{kx}+2I_{kx}I_{lz}+2I_{kx}I_{mz}+4I_{kx}I_{lz}I_{mz})} \sigma(t_2) \qquad [25]$$

If the a values, defined in Section II,B, are not +1 for the three-spin system, then the signs of the terms in this transformation must be appropriately modified; however, the correct combination of spectra can easily be established by inspection of diagrams like those in Figs. 3-4 and 3-5.

When evaluating the effect of the transformation in Eq. [25] in terms of rotations, as was done above for Eq. [23], one must expand terms containing three operators. Thus the last term in Eq. [25] represents a transformation that may be written $\exp(-i\theta\frac{1}{4}4I_{kz}I_{lz}I_{mz})I_{kx}\exp(i\theta\frac{1}{4}4I_{kz}I_{lz}I_{mz})$. Based on Eqs. [A.15], [A.18], and [A.7], this is given by

$$I_{kx} \xrightarrow{\theta\frac{1}{4}4I_{kz}I_{lz}I_{mz}} I_{kx}\cos(\theta/4) + 4I_{ky}I_{lz}I_{mz}\sin(\theta/4) \quad [26]$$

D. Multiple-Quantum Coherence

Coherences involving several spins have been introduced in previous chapters as well as in Section II,B of this chapter. To understand their properties, it is instructive to consider the chemical shift evolution of operators such as $2I_{ky}I_{ly}$. By multiplying out the right-hand side of Eq. [11] and using standard formulas for the sine and cosine products in terms of the sine and cosine of sums and differences of angles, it is possible to show that

$$2I_{ky}I_{ly} \xrightarrow{\Omega_k\tau I_{kz}} \xrightarrow{\Omega_l\tau I_{lz}} (-I_{kx}I_{lx} + I_{ky}I_{ly})\cos(\Omega_k+\Omega_l)\tau$$
$$+ (-I_{kx}I_{ly} - I_{ky}I_{lx})\sin(\Omega_k+\Omega_l)\tau + (I_{kx}I_{lx} + I_{ky}I_{ly})\cos(\Omega_k-\Omega_l)\tau$$
$$+ (-I_{kx}I_{ly} + I_{ky}I_{lx})\sin(\Omega_k-\Omega_l)\tau \quad [27]$$

The term evolving with the sum frequency is double-quantum coherence between the two spins, and the term with the difference frequency represents zero-quantum coherence.

Any multiple-quantum coherence (MQC) can be represented in terms of the shift operators described in Section II,B of Chapter 2. For example, $I_k^+I_l^+$ represents the +2QC between spins k and l and $I_k^-I_l^-$ the corresponding −2QC. An MQC may be a combination of some spins flipping up (I^+) and others flipping down (I^-). MQCs with both up and down flips are associated with *combination lines*. The order of coherence p is given by the net flip; thus $I_k^+I_l^+I_m^-I_n^+$ represents a +2 quantum coherence in a four-spin system. The free precession frequency of this coherence—the effective chemical shift—is $\Omega_k+\Omega_l-\Omega_m+\Omega_n$. Product operators containing even or odd numbers of transverse operators or shift operators (I_x, I_y or I^+, I^-) represent even- or odd-order coherences, respectively.

Consider a phase shift by an angle ϕ, which corresponds to a rotation of angle ϕ about the z axis. The effects on the shift operators are given by

3. POLARIZATION TRANSFER AND EDITING TECHNIQUES

the equations

$$I^+ \xrightarrow{\phi I_z} e^{-i\phi}I^+ \qquad [28]$$

$$I^- \xrightarrow{\phi I_z} e^{+i\phi}I^- \qquad [29]$$

Exercise: Show that these equations are correct by using the expansions of I^+ and I^- in Cartesian operators.

When several operators appear in a product, the effects of the phase shift are cumulative and therefore, *compared to single-quantum coherence, coherence of order p has a p-fold sensitivity to phase shifts of rf pulses and to static field inhomogeneity.* This is expressed in the equation

$$(pQC) \xrightarrow{\phi F_z} e^{-ip\phi}(pQC) \qquad [30]$$

where F_z is the sum over the I_z operators for all the spins in the system, and p is a positive or negative integer or zero.

The positive and negative components of a given MQC correspond to the two possible directions of precession of the coherence. They may be combined according to the equations following, which are analogous to the equations relating the Cartesian operators to the shift operators:

$$(q\text{-spin } pQC)_x = \frac{1}{2}[(q\text{-spin } +pQC) + (q\text{-spin } -pQC)] \qquad [31]$$

$$(q\text{-spin } pQC)_y = -\frac{i}{2}[(q\text{-spin } +pQC) + (q\text{-spin } -pQC)] \qquad [32]$$

A particular example would be the following:

$$(4\text{-spin 2QC})_x = \tfrac{1}{2}(I_k^+ I_l^+ I_m^- I_n^+ + I_k^- I_l^- I_m^+ I_n^-) \qquad [33]$$

For an MQC, there may be defined an effective chemical shift (23):

$$\Omega_{\text{eff}} = \sum_k \Delta m_k \Omega_k \qquad [34]$$

The frequency defined by this equation is equal to the frequency of the transition with which we associate the coherence. The equations for chemical shift precession, Eqs. [9] and [10], may now be written in more general form:

$$(q\text{-spin } pQC)_x \xrightarrow{\Sigma_k \Omega_k I_{kz} \tau} (q\text{-spin } pQC)_x \cos(\Omega_{\text{eff}}\tau)$$
$$+ (q\text{-spin } pQC)_y \sin(\Omega_{\text{eff}}\tau) \qquad [35]$$

$$(q\text{-spin } pQC)_y \xrightarrow{\Sigma_k \Omega_k I_{kz} \tau} (q\text{-spin } pQC)_y \cos(\Omega_{\text{eff}}\tau)$$
$$- (q\text{-spin } pQC)_x \sin(\Omega_{\text{eff}}\tau) \qquad [36]$$

The evolution of multiple-quantum coherence in spin-$\frac{1}{2}$ systems is independent of coupling between the nuclei directly involved in the coherence. Multiplet structure results only from coupling to "passive" spins. For a q-spin coherence of order p, an effective coupling constant to spin m can be defined in a way parallel to the definition of effective chemical shift:

$$J_m^{\text{eff}} = \sum_k \Delta m_k J_{km} \qquad [37]$$

The evolution of an MQC as a result of this effective coupling is described in full analogy to Eqs. [12]–[15]. For example,

$$(q\text{-spin } pQC)_x \xrightarrow{\sum_k \pi J_{km} \tau 2 I_{kz} I_{mz}} (q\text{-spin } pQC)_x \cos(\pi J_m^{\text{eff}} \tau)$$
$$+ 2 I_{mz} (q\text{-spin } pQC)_y \sin(\pi J_m^{\text{eff}} \tau) \qquad [38]$$

E. Some Shortcuts in Density-Operator Calculations

The transformations discussed in the previous sections are sufficient to analyze any pulse sequence applied to a weakly coupled spin system. However, it is often unnecessary to perform each individual step in a series of transformations. In this section there are described some simplifications which may be used in the application of the product operator formalism to analysis of pulse experiments. Although examples are given specifically for a heteronuclear two-spin system, with nuclei of the two species represented by I and S, respectively, the conclusions are valid for any system.

1. Spin-Echo Variants

A common fragment in pulse sequences is a refocusing segment consisting of a 180° pulse preceded and followed by periods of free precession, as used in spin-echo experiments and for refocusing of antiphase magnetization, for example, in the INEPTR experiment: $-\tau-180°-\tau-$.

If a 180° pulse is applied only to the I species, the density operator after the period of 2τ is determined by the following transformations:

$$\sigma_1 \xrightarrow{\pi J \tau 2 I_z S_z} \xrightarrow{\Omega_S \tau S_z} \xrightarrow{\Omega_I \tau I_z} \xrightarrow{\pi I_\phi} \xrightarrow{\Omega_S \tau S_z} \xrightarrow{\Omega_I \tau I_z} \xrightarrow{\pi J \tau 2 I_z S_z} \sigma_2 \qquad [39]$$

Using the results of the Appendix (e.g., Eq. [A.24]), one can formulate the following rule: a 180° pulse can be moved toward the beginning of a sequence of transformations, provided that chemical shift evolution or spin–spin coupling evolution involving the nucleus to which the pulse is applied is changed in sign whenever the 180° pulse is moved ahead of it. Thus, in Eq. [39], moving the 180° pulse on I to the beginning of the sequence changes the first and third transformations to $-\pi J\tau 2I_zS_z$ and $-\Omega_I\tau I_z$, respectively. These now exactly cancel the seventh and sixth terms in the sequence. Further, the second and fifth terms in Eq. [39] can now be combined, yielding $\Omega_S 2\tau S_z$, which means that S chemical shift evolution is effective for the full length of the period 2τ. The result is thus a condensed form, requiring explicit consideration of only two transformation steps:

$$\sigma_1 \xrightarrow{\pi I_\phi} \xrightarrow{\Omega_S 2\tau S_z} \sigma_2 \quad [40]$$

If 180° refocusing pulses are applied to both I and S species, the full sequence of transformations becomes

$$\sigma_1 \xrightarrow{\pi J\tau 2I_zS_z} \xrightarrow{\Omega_S \tau S_z} \xrightarrow{\Omega_I \tau I_z}$$
$$\xrightarrow{\pi I_\phi} \xrightarrow{\pi S_\psi} \xrightarrow{\Omega_S \tau S_z} \xrightarrow{\Omega_I \tau I_z} \xrightarrow{\pi J\tau 2I_zS_z} \sigma_2 \quad [41]$$

Moving both 180° pulses to the beginning and simplifying as in the preceding example, we obtain the condensed equivalent of Eq. [41]:

$$\sigma_1 \xrightarrow{\pi I_\phi} \xrightarrow{\pi S_\psi} \xrightarrow{\pi J 2\tau I_zS_z} \sigma_2 \quad [42]$$

The result of this sequence can thus be seen to be refocusing of both chemical shifts but *retention* of the effects of spin–spin coupling, in contrast to the sequence described above with only a 180° pulse on I, for which S chemical shifts are retained but spin–spin coupling effects are refocused. A vector model for this distinction was presented in Chapter 1, and experiments of this type are of basic importance.

If there are several coupled I spins in the system, there may be effects of homonuclear spin–spin interactions to be considered. In experiments where it is desirable to reduce, or "scale," the effects of homonuclear and heteronuclear interactions to different extents, by allowing them to be effective for different time intervals, we encounter another variant. With total length τ, this, in its most general form, is

$$-\tau_1-180°_\phi(I)-\tau_2-180°_\psi(S)-(\tau-\tau_1-\tau_2)$$

Here the condensed series appears as

$$\sigma_1 \xrightarrow{\pi I_\phi} \xrightarrow{\pi S_\psi} \xrightarrow{\pi J(\tau-2\tau_2)2I_zS_z} \xrightarrow{\Omega_I(\tau-2\tau_1)I_z} \xrightarrow{\Omega_S[\tau-2(\tau_1+\tau_2)]S_z} \sigma_2 \quad [43]$$

The scaling of the heteronuclear coupling is determined by the time between the two 180° pulses. The effective I-spin chemical shift evolution has opposite signs for $\tau_1 < \tau/2$ and for $\tau_1 > \tau/2$ and is zero for $\tau_1 = \tau/2$.

Exercise: Write out the full series of transformations for the experiment just described and show how Eq. [43] is derived from it. Formulate a rule for the sign of S-spin chemical evolution in this experiment.

We will also encounter heteronuclear pulse experiments where the refocusing periods are displaced with respect to one another. An example is the sequence

$$-\tau-\theta_{1,\alpha}(S),\ 180^\circ_\phi(I)-\tau-180^\circ_\psi(S),\ \theta_{2,\beta}(I)-\tau-$$

The symbol $\theta_{1,\alpha}$ represents a pulse of flip angle θ_1 and phase α. This gives the following series of propagators, where the refocusing effects of the 180° pulses on the chemical shifts of I and S have already been taken into account:

$$\sigma_1 \xrightarrow{\Omega_S\tau S_z} \xrightarrow{\pi J\tau 2I_zS_z} \xrightarrow{\theta_1 S_\alpha} \xrightarrow{\pi I_\phi}$$
$$\xrightarrow{\pi J\tau 2I_zS_z} \xrightarrow{\pi S_\psi} \xrightarrow{\theta_2 I_\beta} \xrightarrow{\Omega_1\tau I_z} \xrightarrow{\pi J\tau 2I_zS_z} \sigma_2 \qquad [44]$$

Again the 180° pulses can be moved to the beginning of the respective precession periods, with corresponding sign changes, yielding

$$\sigma_1 \xrightarrow{\pi I_\phi} \xrightarrow{\Omega_S\tau S_z} \xrightarrow{-\pi J\tau 2I_zS_z} \xrightarrow{\theta_1 S_\alpha}$$
$$\xrightarrow{\pi S_\psi} \xrightarrow{-\pi J\tau 2I_zS_z} \xrightarrow{\theta_2 I_\beta} \xrightarrow{\Omega_1\tau I_z} \xrightarrow{\pi J\tau 2I_zS_z} \sigma_2 \qquad [45]$$

If the 180° pulses are omitted from the preceding sequence, yielding

$$-\tau-\theta_{1,\alpha}(S)-\tau-\theta_{2,\beta}(I)-\tau-$$

the series of transformations is

$$\sigma_1 \xrightarrow{\Omega_1 2\tau I_z} \xrightarrow{\Omega_S\tau S_z} \xrightarrow{\pi J\tau 2I_zS_z} \xrightarrow{\theta_1 S_\alpha}$$
$$\xrightarrow{\Omega_S 2\tau S_z} \xrightarrow{\pi J\tau 2I_zS_z} \xrightarrow{\theta_2 I_\beta} \xrightarrow{\Omega_1\tau I_z} \xrightarrow{\pi J\tau 2I_zS_z} \sigma_2 \qquad [46]$$

Analysis of the possibilities of simplification, as illustrated by the examples given above, shows that the evolution of the transverse operators for the I and S spins can to some extent be computed separately. The chemical shift evolution of one of the spin types over a period which contains no pulses on that spin, except possibly 180° pulses, can be condensed to a single transformation, no matter what happens on the other pulse channel. However, the evolution under heteronuclear J couplings cannot be condensed across pulses of flip angles other than 180°.

2. Magnetic Equivalence

When magnetically equivalent nuclei are present, there is little to be gained by invoking symmetry considerations for purposes of simplifying the pulse sequence analysis. Weakly coupled systems are best handled by simply omitting terms for coupling between equivalent nuclei, justifiable since the Hamiltonian terms for such coupling commute with all propagators and observable operators.

III. Polarization Transfer and Purging

A. Introduction

In this section, one-dimensional techniques which include variants of the polarization transfer process are analyzed. This process involves transfer of antiphase magnetization of a spin k to antiphase magnetization of another spin l. This is not the only possibility for polarization transfer in isotropic liquids, but it is the most important and the one with which we shall be concerned here.

In the first subsection are introduced what we shall term "purging techniques," which can be added to many pulse sequences to eliminate the phase and multiplet distortions which frequently occur, particularly when refocusing of antiphase magnetization is employed. We then show how the performance of INEPT and DEPT sequences can be improved by incorporating purging methods. Next the design and application of polarization transfer pulse sequences for one-dimensional heteronuclear chemical shift correlation spectroscopy will be discussed. Finally, the following question is addressed: under what circumstances can a polarization transfer experiment give information about relative signs of coupling constants and how should possible pulse sequences for this purpose be designed?

B. Purging

Purging techniques are designed to provide selective removal of antiphase or dispersive magnetization contributions prior to acquisition of FIDs. Nonselective removal of magnetization is very easy, but since we want to retain the absorptive in-phase components, purging must involve some kind of coherence transfer. The clue to most of the available techniques is transformation of antiphase magnetization into unobservable multiple-quantum coherence.

1. Purging in Heteronuclear Experiments

Purging in heteronuclear experiments with observation of a rare isotope such as carbon-13, nitrogen-15, or silicon-29, is especially attractive, because a very high degree of perfection in elimination of distortions may be achieved. The reason for this will be explained and illustrated later. However, we shall first demonstrate the connection between coherence transfer and purging by a simple experiment in which magnetization is transferred within carbon-13 multiplets which are split by one-bond $^{13}C-^{1}H$ coupling. All product operators observable for the S spin in IS, I_2S, and I_3S spin systems are shown in Fig. 3-7, together with stick diagrams of the spectra they give rise to. The line intensities and phases may be verified by using the procedures outlined above in Section II,B.

The experiment we are describing employs selective excitation by a DANTE sequence, described in Chapter 1, of one of the lines in the carbon-13 multiplets, followed by a nonselective hydrogen pulse of variable flip angle θ and then data acquisition for carbon-13. Experimental spectra for a sample of 1,3-dibromobutane are shown in Fig. 3-8; part (a) is for the CH group, parts (b) and (c) for a CH_2 group, and parts (d) and (e) for the methyl group. The spectra in the first column were recorded without a hydrogen pulse, that is, with θ of 0°, and their quality reflects the selectivity of the DANTE technique. To show the basis for obtaining the spectra in the second and third columns, Table 3-1 gives the product operator representations for the $\theta = 0°$ spectra of Fig. 3-8. The table also illustrates how, as described in Section II,C,4, single absorption lines within spin multiplets

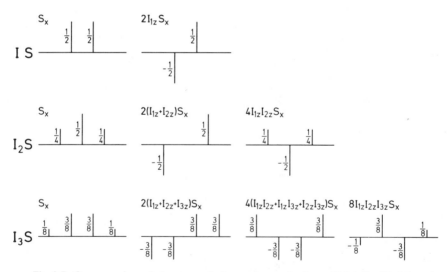

Fig. 3-7. Correspondence between product operators and spin multiplets for IS, I_2S, and I_3S spin systems. It is assumed that the receiver reference phase is x and that the a_{ij} factors described in Section II,B are equal to +1.

3. POLARIZATION TRANSFER AND EDITING TECHNIQUES

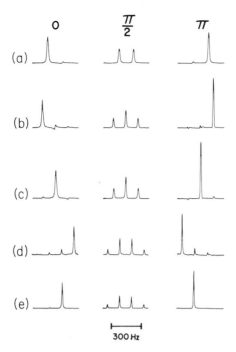

Fig. 3-8. The effects of applying nonselective proton pulses of variable tip angles to nonequilibrium carbon-13 magnetization. In the first column, one line from each of several multiplets has been selectively excited by a DANTE sequence, consisting of (a) 15 or (b–e) 30 pulses separated by 1 msec. The second and third columns show the effects of applying a 90° and a 180° pulse, respectively, immediately after the DANTE excitation. The excited lines are (a) high-frequency line of a CH doublet; (b, c) high-frequency and center lines of a CH_2 triplet; (d, e) low-frequency outer and inner lines of a CH_3 quartet. The molecule is 1,3-dibromobutane, and the spectra were recorded at 75 MHz on a Bruker CXP-300 spectrometer.

can be considered as superpositions of full in-phase multiplets and various antiphase contributions, such as represented in Fig. 3-7.

The trick to remove the antiphase components is to apply a nonselective 90°(H) pulse which transfers all such terms to heteronuclear MQC. For example a $90°_x(H)$ pulse transforms $2I_zS_x$ into $-2I_yS_x$. Only the pure S_x operators are unaffected, and the normal multiplets are reproduced in the second column of Fig. 3-8 for θ of 90°. If, instead, a nonselective 180°(H) pulse is applied, all product operators with an odd number of I_z components change their signs, and the magnetizations are transferred to the symmetrically related transitions as shown in the $\theta = \pi$ column in Fig. 3-8. The lines from this procedure appear sharper because of refocusing of long-range J_{CH} interactions.

TABLE 3-1

PRODUCT OPERATOR REPRESENTATIONS OF THE SPECTRA FROM
1,3-DIBROMOBUTANE SHOWN IN THE FIRST COLUMN ($\theta=0$)
OF FIG. 3-8

Figure part	Representation
a	$\frac{1}{2}\{S_x - 2I_zS_x\}$
b	$\frac{1}{4}\{S_x - [2(I_{1z} + I_{2z}) - 4I_{1z}I_{2z}]S_x\}$
c	$\frac{1}{2}\{S_x - 4I_{1z}I_{2z}S_x\}$
d	$\frac{1}{8}\{S_x + [2(I_{1z} + I_{2z} + I_{3z}) + 4(I_{1z}I_{2z} + I_{1z}I_{3z} + I_{2z}I_{3z}) + 8I_{1z}I_{2z}I_{3z}]S_x\}$
e	$\frac{3}{8}\{S_x + [\frac{1}{3}2(I_{1z} + I_{2z} + I_{3z}) - \frac{1}{3}4(I_{1z}I_{2z} + I_{1z}I_{3z} + I_{2z}I_{3z}) - 8I_{1z}I_{2z}I_{3z}]S_x\}$

2. Purging in Homonuclear Experiments

The simplicity and perfection of purging in heteronuclear experiments is attributable to the possibility of pulsing the observed and the nonobserved isotopes independently of one another. Thus the antiphase magnetization of spin S represented by the term $2I_zS_x$ is converted by a $90°_y(I)$ pulse into multiple-quantum coherence $2I_xS_x$. However, in homonuclear systems, a nonselective $90°_y$ pulse affects both spins, converting $2I_zS_x$ into $-2I_xS_z$. Thus the undesired antiphase magnetization is merely transferred from one spin to another, and a 90° pulse alone is not suitable for purging in homonuclear experiments. A phase-shifted purging pulse would affect the desired in-phase components I_y, S_y and cannot be applied to improve the situation.

A suitable method involves the application of a 45° pulse, which, according to Eqs. [19] and [20], transfers part of the antiphase magnetization to multiple-quantum coherence and part to longitudinal two-spin order:

$$2I_zS_x \xrightarrow{(\pi/4)(I_y + S_y)}$$

$$2\left(I_z \cos\frac{\pi}{4} + I_x \sin\frac{\pi}{4}\right)\left(S_x \cos\frac{\pi}{4} - S_z \sin\frac{\pi}{4}\right)$$

$$= 2I_zS_x(1/\sqrt{2})^2 - 2I_zS_z(1/\sqrt{2})^2 + 2I_xS_x(1/\sqrt{2})^2 - 2I_xS_z(1/\sqrt{2})^2$$

$$= \tfrac{1}{2}(2I_zS_x - 2I_xS_z) + \tfrac{1}{2}(2I_xS_x - 2I_zS_z) \qquad [47]$$

The antiphase magnetization is distributed equally between the I and S spins. If the corresponding term for the I spin, $2I_xS_z$, is present with the same amplitude, the 45° purging pulse converts it to $\tfrac{1}{2}(2I_xS_z - 2I_zS_x) + \tfrac{1}{2}(2I_xS_x - 2I_zS_z)$. Since the antiphase terms in these two expressions cancel, the net result is the transformation of all undesired magnetization to unobserved two-spin order and two-spin coherence. In this experiment, the result of a scan recorded without purging plus a scan with a 90° pulse is the same as the result of two scans each with a 45° purging pulse. Normally the 45° pulse is preferred, because it is more effective in purging without requiring any more scans. Applications to homonuclear experiments of purging with a single pulse will be found in Section IV.

A more efficient technique for purging in homonuclear experiments is the z filter (12), which consists of two 90° pulses separated by a delay τ_z. The z filter can select the in-phase magnetization along any desired direction in the xy plane simply by adjusting the phase of the first pulse in the filter. Thus if the pulse phase is $90° - \phi$, the in-phase components along the directions $\pm \phi$ are selected. Consider the product operators representing the various possible in- and antiphase magnetizations of a spin I_1:

x family: $I_{1x}, 2I_{1y}I_{iz}, -4I_{1x}I_{iz}I_{jz}, -8I_{1y}I_{iz}I_{jz}I_{kz}, \ldots$

y family: $I_{1y}, -2I_{1x}I_{iz}, -4I_{1y}I_{iz}I_{jz}, 8I_{1x}I_{iz}I_{jz}I_{kz}, \ldots$

[48]

The labels i, j, and k denote any spins coupled to I_1. To select the magnetization along the y axis, a nonselective $90°_x$ pulse is applied and the operators in the two families transform to the following, with y going to $+z$ and z going to $-y$:

x family: $I_{1x}, -2I_{1z}I_{iy}, -4I_{1x}I_{iy}I_{jy}, 8I_{1z}I_{iy}I_{jy}I_{ky}, \ldots$

y family: $I_{1z}, 2I_{1x}I_{iy}, -4I_{1z}I_{iy}I_{jy}, -8I_{1x}I_{iy}I_{jy}I_{ky}, \ldots$

[49]

The coherences in the x family are all of odd quantum orders and the coherences in the y family are all of even orders, as indicated by the number of transverse components in the product operators.

The basis for the z filter is that, in the course of the free precession period following the first pulse, all components oscillate except for the z magnetization, I_{1z}, and can be suppressed by coadding results of experiments with a series of different values of τ_z. The second 90° pulse, perhaps of phase $-x$, transforms the retained z components, which comprise the desired signals, into transverse magnetization I_{1y} suitable for acquisition. Longitudinal spin order terms such as $I_{1z}I_{iz}$ are also invariant during the interval τ_z, but the final 90° pulse transforms these terms into multiple quantum coherence, which remains unobservable when no further pulses are applied before observation.

To improve the efficiency of filtering, it is advisable to phase-cycle the first pulse of the filter together with all previous pulses in the sequence or the second pulse together with the following pulses and the receiver reference phase, in order to eliminate all coherence components except zero-quantum coherence, which is suppressed by the variation of τ_z. It is usually sufficient to use a four-step phase cycle with 90° increments equivalent to CYCLOPS. This also eliminates spin terms of even order.

A z filter as described works properly when inserted immediately before observation without any pulse following or when inserted after a single pulse followed by a free precession period. In the rare circumstance where a z filter must be applied in the middle of a longer pulse sequence, it is necessary to suppress the multiple quantum coherence which might otherwise remain following it, by exploiting the characteristic dependence on the flip angle of the filter pulses.

The z-filtering procedure is related to the elimination of coherence during the mixing process in NOESY experiments (24, 25), and all schemes designed to eliminate the zero-quantum coherence in NOESY (26) are also applicable for the z filter. However, the effectiveness of the z filter is limited by the impossibility of suppressing low-frequency zero-quantum coherences. For applications to biomolecules, it must be kept in mind that undesired magnetization transfer through cross relaxation may occur for long τ_z values such as might be needed for suppression of low-frequency coherences.

For illustration, we have inserted a z filter into the spin-echo multiplet selection technique described in Chapter 1, which is quite sensitive to the proper setting of the time interval. The sequence is shown in Fig. 3-9. The in-phase magnetization at time τ for a given spin I_1 with $n-1$ coupling partners I_i is given by

$$\sigma_{I_1} = \prod_{i=2}^{n} I_{1y} \cos \pi J_{1i}\tau \qquad [50]$$

A prerequisite for the multiplet selection techique is that the modulation factor, $\cos \pi J_{1i}\tau$, in Eq. [50] be equal, to a good approximation, to $(-1)^{n-1}$; this condition is met for $\tau = 1/J$. For optimum timing, negative absorption signals arise for those spins having an even number of coupling partners and positive absorption signals for those with an odd number of coupling partners.

An experimental test was carried out on an approximately equimolar mixture of 2-methyl-2-propanol, 2-propanol, 2-butanone, and 2-butanol, which has a range of vicinal coupling constants from 6.1 to 7.4 Hz. The result of the simple spin-echo experiment is shown in Fig. 3-10a. The spectrum clearly exhibits phase and multiplet distortions arising because all multiplets cannot be refocused with a single value of τ. In Fig. 3-10b is presented the spin-echo spectrum obtained under the same conditions, but with a z filter inserted prior to acquisition, as in Fig. 3-9b, showing that the phase errors present in the spectrum of Fig. 3-10a have been removed by the filter.

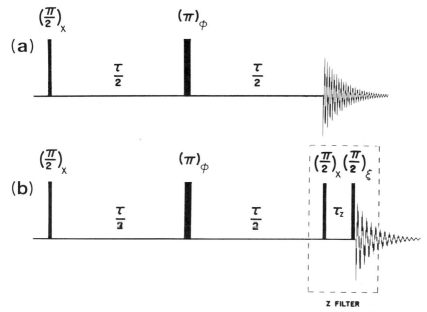

Fig. 3-9. Pulse sequences for spin-echo multiplet selection techniques. (a) Simple spin-echo sequence; (b) spin-echo pulse sequence with a z filter inserted prior to acquisition to retain only the in-phase magnetization. The period τ_z is varied within a sequence of coadded experiments to average out the unwanted coherences. The final 90° pulse and the receiver reference phase are taken through a CYCLOPS cycle. Cycling of the 180° pulse according to Exorcycle eliminates the effects from the nonideality of the 180° pulse.

Having obtained a clean spin-echo spectrum, we can now combine this with a standard one-pulse spectrum such as Fig. 3-11a. Properly weighted addition, taking into account relaxation effects, gives a subspectrum of only those spins having an odd number of coupling partners, shown in Fig. 3-11b, while subtraction produces a subspectrum of spins with even numbers of coupling partners, as in Fig. 3-11c.

Although the separation for the mixture investigated here turned out quite well, less perfect performance may be encountered in practice when the spread of J couplings is larger, as in conjugated and aromatic systems, or if there is a large range of relaxation times, since the elimination by subtraction presupposes equally scaled responses of all multiplets.

C. Heteronuclear Polarization Transfer

1. *Product-Operator Analysis of INEPT and DEPT*

We begin this section with a general product-operator analysis of these two experiments, which were introduced in Chapter 1. Heteronuclear

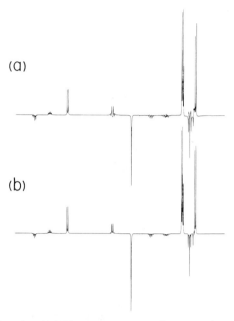

Fig. 3-10. (a) Spin-echo 300-MHz proton spectrum of an approximately equimolar mixture of 2-methyl-1-propanol, 2-isopropanol, 2-butanone, and 2-butanol, obtained by the sequence of Fig. 3-9a with $\tau = 156$ msec. (b) Same as (a) but with a z filter inserted prior to acquisition (sequence of Fig. 3-3b). Ten τ_z values were used (0, 18, 36, ... msec) for the z filter. Note the disappearance of dispersive contributions, for example, for the negative triplet on the right. A Bruker CXP-300 spectrometer was used. The spectra cover a range from 0.4 ppm on the right to 4.5 ppm on the left.

polarization transfer techniques require a minimum of three pulses of flip angles between 0° and 180°. The first pulse is applied to the source nuclei to generate transverse magnetization, which is then allowed to precess for a certain time before the actual transfer is effected by the last two pulses, one on each isotope. Usually 180° pulses are inserted in the free precession periods between the initial excitation and the polarization transfer pulses, as well as after the polarization transfer pulses, to refocus precession induced by chemical shifts.

Consider an I_nS system including n equivalent protons, each labeled by an index i, and one rare spin, denoted S. The density operator at equilibrium is given by

$$\sigma_0 = m_I \sum_{i=1}^{n} I_{iz} + m_S S_z \qquad [51]$$

The m quantities are the magnetization of the respective spins and m_I/m_S

3. POLARIZATION TRANSFER AND EDITING TECHNIQUES 177

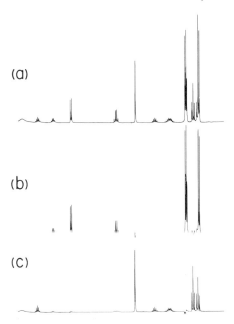

Fig. 3-11. (a) Normal one-pulse proton spectrum of the same mixture as in Fig. 3.10. (b, c) Spectra obtained by combining the spectrum in (a) with the z-filtered spin-echo spectrum in Fig. 3-10b. Additive combination yields a spectrum exclusively of protons having an odd number of coupling partners (b), and subtractive combination results in a spectrum containing only signals from those protons with an even number of coupling partners (c). Note the clean appearance of the low-frequency triplet in (c).

is equal to the ratio of magnetogyric ratios of the two species; for convenience we will, however, disregard these factors. If the experiment starts with a $90°_x(I)$ pulse, the S_z operator is unaffected and all the I_z operators are transformed to $-I_y$. A $90°_{-x}(I)$ would take the I_z operators to I_y, so subtractive combination of two such experiments adds the I magnetization and eliminates the native S-spin magnetization, provided the following parts of the two pulse sequences are identical. The same result can alternatively be obtained by introducing a phase shift of another pulse in the sequence. In any event, the initial S-spin magnetization in Eq. [51] can be assumed to be eliminated and we define a new σ_0:

$$\sigma_0 = \sum_{i=1}^{n} I_{iz} \qquad [52]$$

The $90°_x(I)$ pulse turns the I magnetization into the $-y$ direction, giving

$$\sigma_1 = -\sum_{i=1}^{n} I_{iy} \qquad [53]$$

If the chemical shift evolutions are refocused in the following period τ, only evolution under the heteronuclear couplings need be considered. From the equations in Section II,C,2,

$$\sigma_2 = -\cos(\pi J\tau) \sum_{i=1}^{n} I_{iy} + \sin(\pi J\tau)\left(2 \sum_{i=1}^{n} I_{ix}\right) S_z \qquad [54]$$

Only the terms with S_z can lead to polarization transfer, so it is important that these terms be "frozen" at a time near where $\tau = 1/2J$ at which $\sin(\pi J\tau)$ is at a maximum and the useless in-phase magnetization I_{iy} is at a minimum. The freezing may be done with a 90°(S) pulse. If this is of phase x, then

$$\sigma_3 = -\left(2 \sum_{i=1}^{n} I_{ix}\right) S_y \qquad [55]$$

The I_y operators in Eq. [54] have no influence on the S spectrum and can be disregarded. The quantity $2I_{ix}S_y$ represents heteronuclear two-spin coherence and is frozen in the sense that the transverse S operator cannot, except by relaxation, disappear during free precession, unlike S_z operators.

In the INEPT experiment, as shown in Fig. 3-12a, a $90°_y(I)$ pulse is applied at the same time as the $90°_x(S)$ pulse and the polarization transfer is completed:

$$\sigma_4^{\text{INEPT}} = \left(2 \sum_{i=1}^{n} I_{iz}\right) S_y \qquad [56]$$

The DEPT experiment (Fig. 3-12d) includes a further $\tau = 1/2J$ precession period after the state described in Eq. [55]. Here the couplings between S and the $(n-1)$ other spins which were not represented in the bilinear terms in σ_3 are effective, and for ideal timing they generate

$$\sigma_4^{\text{DEPT}} = 2^n \sum_{i=1}^{n} I_{ix} \prod_{j \neq i} I_{jz} S_{x,y} \qquad [57]$$

In this equation, the sign has been disregarded and the subscript on S is x when n is even and y when n is odd. To create observable S magnetization, a proton pulse must now rotate the I_{ix} operators back to I_{iz} while leaving the I_{jz} operators unaffected. The first effect is optimum for a flip angle θ of 90° and the second for 0°, so for n greater than one, an intermediate value of the angle is required for polarization transfer. A θ_y pulse gives the following observable magnetization:

$$\sigma_5^{\text{DEPT}} = n \sin\theta \cos^{n-1}\theta\, 2^n \prod_{i=1}^{n} I_{iz} S_{x,y} \qquad [58]$$

3. POLARIZATION TRANSFER AND EDITING TECHNIQUES 179

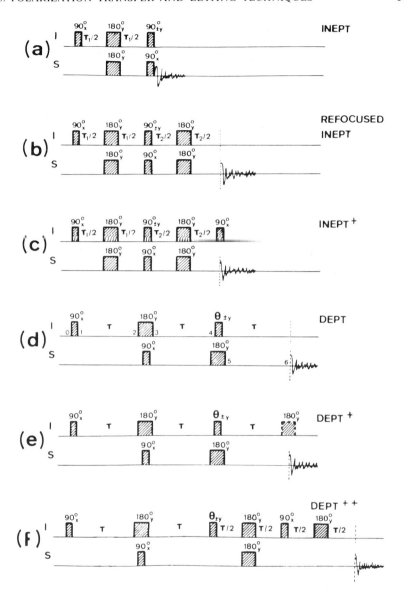

Fig. 3-12. The six polarization transfer pulse techniques discussed in the text. The polarization transfer proceeds from I to S spins. The native S-spin magnetization is eliminated by phase alternating, for example, the I-spin transfer pulse together with subtractive combination of the resulting FIDs. In DEPT[+] every second experiment uses an additional 180° pulse for the compensation of phase anomalies. The phases of the S-spin pulses and all 180° pulses are of no practical importance and have been selected solely for convenience in computation. Artifacts arising from imperfect 180° pulses can be eliminated by Exorcycle phase cycling.

This density operator represents states of maximum antiphase character, and usually a third $\tau = 1/2J$ period is included to allow refocusing to undistorted multiplets:

$$\sigma_6^{\text{DEPT}} = n \sin \theta \cos^{n-1} \theta \, S_x \qquad [59]$$

An explicit density operator calculation of DEPT applied to an I_2S system can be found in the Appendix. A more general calculation of the responses for an I_nS system is presented in Section IV on subspectral editing.

2. Elimination of Spectral Distortions in Heteronuclear Polarization Transfer Experiments

A spectrum may be defined to be distorted if it is impossible to phase all lines to positive absorption or if the relative line intensities within multiplets deviate from those encountered in a standard acquisition experiment employing a single 90° pulse. We have seen that undistorted multiplets are represented by the S_x (or S_y) operator; spectral distortions correspond to antiphase product operators ($2I_{1z}S_y$, $4I_{1z}I_{2z}S_x$, and so on) remaining after chemical shift evolution is refocused. Those with an odd number of I_z operators correspond to dispersive contributions to the S_x spectrum and are *phase anomaly* terms. Those with an even number of I_z operators are called *multiplet anomaly* terms, because they disturb only the relative line intensities within a multiplet but not the phasing of the spectrum. Both types of anomalies can be removed by purging techniques.

Figure 3-12 shows the pulse sequences for the INEPT and DEPT polarization transfer techniques and the three improved versions, INEPT$^+$, DEPT$^+$, and DEPT^{++} (19). The improved sequences are significant only for proton-coupled spectra, since the distortions, because of their antiphase character, disappear when proton decoupling is applied during acquisition in the conventional INEPTR and DEPT sequences. Table 3-2 lists the density operators for IS, I_2S, and I_3S spin systems immediately before data acquisition, for each of the six sequences in Fig. 3-12.

The INEPT sequence leads to multiplet-distorted spectra. All multiplets appear in antiphase, but without phase anomalies, and this is acceptable for many applications, or even desirable because of the enhancement of the outer lines in quartets or resonances of greater multiplicity. In contrast, refocused coupled INEPT spectra exhibit both types of anomalies, and, unlike DEPT, the anomalies do not disappear for any value of τ_2 except in an IS system. To purge the antiphase components, a 90°(H) pulse is applied at the end of the τ_2 period giving the INEPT$^+$ sequence. This purging pulse transfers all antiphase terms to unobservable multiple-quantum coherence; this corresponds to converting all the I_{iz} factors in the product operators

TABLE 3-2

Observable Components of the Density Operators Immediately Before Data Acquisition for IS, I$_2$S, and I$_3$S Spin Systems and for the Six Polarization Transfer Techniques of Fig. 3-12

INEPT

I_nS: $\quad s2\left(\sum_i^n I_{iz}\right)S_y$

Refocused INEPT

IS: $\quad s(-s_1 S_x + c_1 2 I_z S_y)$

I_2S: $\quad s[-s_2(S_x + 4I_{1z}I_{2z}S_x) + c_2 2(I_{1z} + I_{2z})S_y]$

I_3S: $\quad s[-\tfrac{3}{4}(s_3 + s_1)S_x - \tfrac{1}{4}(3s_3 - s_1)4(I_{1z}I_{2z} + I_{1z}I_{3z} + I_{2z}I_{3z})S_x$
$\qquad + \tfrac{1}{4}(3c_3 + c_1)2(I_{1z} + I_{2z} + I_{3z})S_y + \tfrac{3}{4}(c_3 - c_1)8I_{1z}I_{2z}I_{3z}S_y]$

INEPT$^+$

I_nS: $\quad -nss_1 c_1^{n-1} S_x$

DEPT^a

IS: $\quad s^2 s_\theta S_x - scs_\theta 2 I_z S_y$

I_2S: $\quad (s^4 s_{2\theta} + 2s^2 c^2 s_\theta)S_x - (s^2 c^2 s_{2\theta} - 2s^2 c^2 s_\theta)4I_{1z}I_{2z}S_x$
$\qquad -(s^3 cs_{2\theta} + sc^3 s_\theta - s^3 cs_\theta)2(I_{1z} + I_{2z})S_y$

I_3S: $\quad 3[\tfrac{1}{4}(s_{3\theta} + s_\theta)s^6 + s_{2\theta}s^4 c^2 + s_\theta s^2 c^4]S_x$
$\qquad -[(\tfrac{3}{4}s_{3\theta} + 2s_{2\theta} + \tfrac{7}{4}s_\theta)s^4 c^2 + (s_{2\theta} - 2s_\theta)s^2 c^4](I_{1z}I_{2z} + I_{1z}I_{3z} + I_{2z}I_{3z})S_x$
$\qquad -[(\tfrac{3}{4}s_{3\theta} - s_{2\theta} + \tfrac{3}{4}s_\theta)s^5 c + 2(s_{2\theta} - s_\theta)s^3 c^3 + s_\theta sc^5](I_{1z} + I_{2z} + I_{3z})S_y$
$\qquad + 3(\tfrac{1}{4}s_{3\theta} - s_{2\theta} + \tfrac{5}{4}s_\theta)s^3 c^3 8 I_{1z}I_{2z}I_{3z}S_y$

DEPT$^+$

IS: $\quad s^2 s_\theta S_x$

I_2S: $\quad (s^4 s_{2\theta} + 2s^2 c^2 s_\theta)S_x - (s_{2\theta} - s_\theta)s^2 c^2 4 I_{1z}I_{2z}S_x$

I_3S: $\quad [\tfrac{3}{4}(s_{3\theta} + s_\theta)s^6 + 3s_{2\theta}s^4 c^2 + 3s_\theta s^2 c^4]S_x$
$\qquad -[(\tfrac{3}{4}s_{3\theta} - 2s_{2\theta} + \tfrac{7}{4}s_\theta)s^4 c^2 + (s_{2\theta} - 2s_\theta)s^2 c^4](I_{1z}I_{2z} + I_{1z}I_{3z} + I_{2z}I_{3z})S_x$

DEPT^{++}

I_nS: $\quad ns^2 s_\theta (c^2 + s^2 c_\theta)^{n-1} S_x$

---

^{a} For DEPT techniques $\tau = \tau_1$. Abbreviations: $s = \sin \pi J\tau_1$, $c = \cos \pi J\tau_1$, $s_p = \sin p\pi J\tau_2$, $c_p = \cos p\pi J\tau_2$, $s_{p\theta} = \sin p\theta$, $c_{p\theta} = \cos p\theta$.

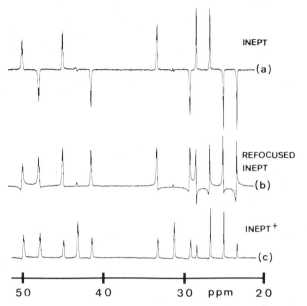

Fig. 3-13. Proton-coupled ^{13}C spectra of 1,3-dibromobutane obtained on a Bruker CXP-300 spectrometer under comparable conditions. (a) INEPT spectrum; (b) refocused INEPT spectrum; (c) INEPT$^+$ spectrum with $\tau_1 = 3.58$ msec and $\tau_2 = 2.2$ msec. In the INEPT$^+$ experiment, the multiplet and phase anomalies are eliminated irrespective of the value of J. Note the appearance of the center lines in the triplets.

in Table 3-2 to $-I_{iy}$ operators. Spectra from the various versions of INEPT are shown in Fig. 3-13.

Similar straightforward artifact elimination is not possible for DEPT because of the presence of heteronuclear multiple-quantum coherence, an unobservable portion not included in Eq. [58], which could be transformed by the purging pulse into observable S-spin magnetization. By applying a 180°(I) pulse before data acquisition in every second experiment and co-adding the FIDs, the phase anomaly terms—those with an odd number of I_z operators—are eliminated. However, this improved technique, termed DEPT$^+$ and depicted in Fig. 3-12e, leaves the multiplet anomaly terms unaffected.

For more complete purging in DEPT, it is necessary to modify further or to extend the basic sequence. One possibility, called DEPT^{++}, is represented in Fig. 3-12f; another possibility has also been proposed (27). The crucial point is to refocus fully I-spin chemical shifts so that the transverse I-spin operators of the multiple quantum coherence are aligned along the x axis before a $90°_x(I)$ purging pulse is applied at time 3τ. This guarantees that no undesired MQC is made observable by the purging pulse. The

3. POLARIZATION TRANSFER AND EDITING TECHNIQUES

refocusing pulses inserted at time $\frac{5}{2}\tau$ leave the S-spin chemical shifts defocused at time 3τ, and an additional period of length τ is required to refocus the S-spin vectors at time 4τ. During this last period, the I-S interactions are suppressed by a further 180°(I) pulse at time $\frac{7}{2}\tau$. The remaining observable product operators for DEPT^{++} are collected in Table 3-2. No phase or multiplet anomalies remain, irrespective of the value of J.

The degree of artifact suppression depends on the qualities of the 90° or 180° purging pulses. Results can be improved by use of composite pulses as described in Chapter 1. For example, the 90° pulse in INEPT$^+$ can be replaced by a $90°_x 90°_y$ composite.

Carbon-13 spectra of 1,3-dibromobutane obtained by the various DEPT versions are shown in Fig. 3-14. The delay time τ of 3.58 msec has been selected to be nearly optimum for Fig. 3-14a, but it has been decreased for the other spectra by a factor of 2 to simulate the influence of a variation of J and to demonstrate the effectiveness of the purging technique. Comparison of Fig. 3-14c with Fig. 3-14b, recorded under the same conditions, demonstrates that DEPT$^+$ efficiently corrects phase anomalies. However, the multiplet anomalies remain and are responsible for the multiplet-distorted triplet and quartet patterns. Finally, in Fig. 3-14d, recorded with DEPT^{++}, the phase and multiplet anomalies have been eliminated completely. This spectrum is qualitatively comparable to that recorded with INEPT$^+$, shown in Fig. 3-13c.

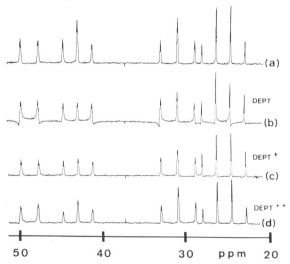

Fig. 3-14. Proton-coupled ^{13}C spectra of 1,3-dibromobutane obtained on a Bruker CXP-300 spectrometer. (a) DEPT spectrum with $\tau = 3.58$ msec; (b) DEPT spectrum with $\tau = 1.79$ msec; (c) DEPT$^+$ spectrum with $\tau = 1.79$ msec; (d) DEPT^{++} spectrum with $\tau = 1.79$ msec. The pulse angle θ was set to 45° for spectra (a)-(c) and to 135° for spectrum (d).

3. Relative Benefits of INEPT and DEPT Techniques

After a description of the principles of the six polarization transfer sequences in Fig. 3-12, it is appropriate to discuss briefly their relative importance for practical applications. The aim in applying these methods can be either sensitivity enhancement through polarization transfer or decomposition of the spectrum into independent subspectra according to the number of attached protons or both. In Section IV it will be shown that for subspectral editing, DEPT techniques are always superior to their INEPT analogs. However, when only enhancement by polarization transfer is required, INEPT techniques are recommended. They always yield the shortest pulse sequences, because DEPT involves an "inactive" precession period with two-spin coherence.

When antiphase multiplets are acceptable, the basic INEPT experiment is the sequence of choice. Undistorted coupled spectra are conveniently obtained through INEPT$^+$ and proton-decoupled spectra through INEPTR. To enhance the uniformity of excitation over larger J ranges, experiments recorded using different τ_1 or τ_2 values can be coadded. As long as the variation in these interval lengths is not too large, such a procedure does not result in cancellations.

As an example of a situation often encountered in practical applications, Fig. 3-15 illustrates the nitrogen-15 INEPT and INEPT$^+$ spectra for the nonprotonated nitrogen atom in the cyclic hydrazide shown in the figure. For the spectra, polarization was transferred via ^{15}N-^{1}H long-range couplings ($\tau_1 = 20$ msec and $\tau_2 = 42.0$ msec). From the structural formula, number of lines, and complexity of the experimental spectra in Fig. 3-15, we conclude that the multiplet results from severe overlap of a doublet of triplets of quartets. The simulated spectra shown in Figs. 3-15b and 3-15c are obtained by ordinary spectrum analysis using the following combinations of long-range coupling constants:

Figure	$^2J(^{15}$N-H$)$	$^3J(^{15}$N-C$H_2)$	$^3J(^{15}$N-C$H_3)$
3-15b	7.95 Hz	1.91 Hz	3.88 Hz
3-15c	9.89 Hz	3.83 Hz	1.97 Hz

The assignment in Fig. 3-15b apparently represents the correct choice for the couplings of the nonprotonated nitrogen to hydrogen in this molecule. This assignment would have been very difficult to obtain from the basic INEPT spectrum (Fig. 3-15a, lower row), although the simulated INEPT spectra (b and c, lower row) also favor the assignment of Fig. 3-15b. The

3. POLARIZATION TRANSFER AND EDITING TECHNIQUES

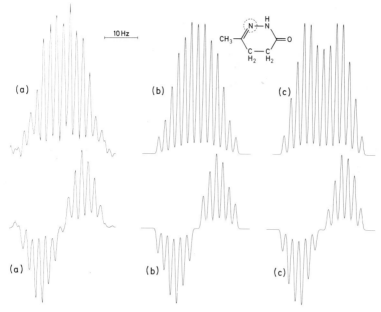

Fig. 3-15. (Above) Proton-coupled INEPT$^+$ ^{15}N NMR spectra for the nonprotonated nitrogen atom in the cyclic hydrazide shown. (a) Experimental spectrum obtained with $\tau_1 = 20.0$ msec and $\tau_2 = 42.0$ msec; (b, c) simulated spectra corresponding to the ^{15}N-^{1}H coupling constant combinations b and c given in the text. (Below) Proton-coupled INEPT ^{15}N spectra for the same nitrogen atom. (a) Experimental spectrum obtained with $\tau_1 = 20.0$ msec; (b, c) simulated INEPT spectra corresponding to the ^{15}N-^{1}H coupling constant combinations b and c given in the text.

ordinary INEPT spectra were simulated using the INEPT intensity expression in Table 3-2. We also find that it would have been impossible to extract the ^{15}N-^{1}H coupling information from an ordinary DEPT spectrum because of the expected severe spectral distortions.

D. Heteronuclear Chemical Shift Correlation from One-Dimensional Polarization-Transfer Experiments

In the heteronuclear polarization-transfer experiments discussed so far, the hydrogen chemical shift evolution is exactly refocused at the point of transfer by the 180°(H) pulse. Thus, assuming the same one-bond ^{13}C-^{1}H coupling constant for all C-H fragments, uniform polarization transfer is achieved from all chemically shifted hydrogens in the spectrum. However, if the refocusing pulses are omitted, the polarization transfer is no longer independent of the chemical shift. If no phase shifts are applied, the polarization transfer becomes a sine function of the chemical shift evolution in the rotating hydrogen frame. Examples of experiments leading to such *sine*-dependent *p*olarization *t*ransfer are shown in Fig. 3-16 for the three pulse sequences SINEPT-1, SINEPT-2, and SINEPT-3 (6, 7). These sequences are convenient and simple one-dimensional methods for correlation of heteronuclear chemical shifts for samples where only a few correlations are required. Thus, in several cases they may be considered alternatives to the more general methods of 2D heteronuclear correlation spectroscopy. Furthermore, the SINEPT sequences also serve as convenient methods of obtaining spectra enhanced by polarization transfer on older spectrometers not equipped with phase-shifting hardware for the proton channel, such as is required for the INEPT and DEPT experiments.

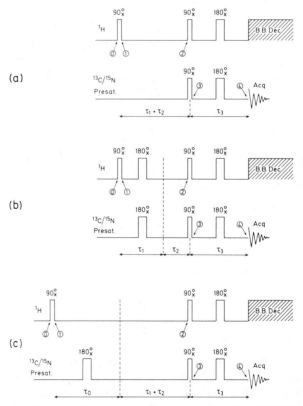

Fig. 3-16. Pulse sequences (a) SINEPT-1, (b) SINEPT-2, and (c) SINEPT-3 employed for 1D heteronuclear chemical shift correlation spectroscopy with polarization transfer. The sequences may easily be modified for a DEPT-type transfer.

3. POLARIZATION TRANSFER AND EDITING TECHNIQUES

For polarization transfer via one-bond couplings, the delays τ_1 and τ_2 in the sequences shown in Fig. 3-16 are chosen to fulfill the condition $\tau_1 + \tau_2 = 1/2J$. A refocusing period is present in each sequence to allow broadband decoupling during acquisition. The polarization transfer part of SINEPT-1 (Fig. 3-16a) is a 1D version of the original 2D heteronuclear shift correlation experiment (28) and is closely related to INEPT. It differs from INEPT in that the two 180° refocusing pulses applied to both nuclei at the center of the τ_2 period have been omitted; thus, in addition to the J evolution, proton chemical shift evolution is effective during the interval τ_2. In the sequences shown in Figs. 3-16b and 3-16c, proton chemical shift evolution is effective during the periods τ_2 and $\tau_0 + \tau_1 + \tau_2$, respectively.

As an example, we shall analyze the SINEPT-2 pulse sequence (Fig. 3-16b) in terms of product operators for a general spin system. Consider a system of n magnetically equivalent protons I_i, one heteronucleus S, and $m - n$ additional passive protons. Only terms relevant for the polarization transfer from the n equivalent protons to the S spin are included in the calculation. The analysis begins as follows:

$$\sigma_0 = \sum_{i=1}^{n} I_{iz} \quad [60]$$

$$\sigma_1 = -\sum_{i=1}^{n} I_{iy} \quad [61]$$

Before continuing, it is important to note that the evolution as a result of spin–spin coupling extends over the entire period $\tau_1 + \tau_2$, because of the presence of simultaneous 180° pulses, but that chemical-shift evolution is only occurring during τ_2. The total evolution can be calculated as two separate steps, one for the spin–spin effects and the other for chemical shift effects:

$$\sigma_2(J \text{ evol.}) = -\left\{ \sin[\pi J_{IS}(\tau_1 + \tau_2)] \sum_{i=1}^{n} 2I_{ix}S_z \right\}$$

$$\times \prod_{j=n+1}^{m} \cos[\pi J_{1j}(\tau_1 + \tau_2)] \quad [62a]$$

$$\sigma_2(J \text{ evol.} + \omega \text{ evol.}) = -\left\{ \sin(\omega_1 \tau_2) \sin[\pi J_{IS}(\tau_1 + \tau_2)] \sum_{i=1}^{n} 2I_{iy}S_z \right\}$$

$$\times \prod_{j=n+1}^{m} \cos[\pi J_{1j}(\tau_1 + \tau_2)] \quad [62b]$$

The $90°_x$ pulses convert I_{iy} to I_{iz} and S_z to $-S_y$, yielding

$$\sigma_3 = \left\{ (\sin \omega_1 \tau_2) \sin[\pi J_{IS}(\tau_1 + \tau_2)] \sum_{i=1}^{n} 2 I_{iz} S_y \right\}$$
$$\times \prod_{j=n+1}^{m} \cos[\pi J_{1j}(\tau_1 + \tau_2)] \qquad [63]$$

At this point, the antiphase S-spin signal $2I_{iz}S_y$ could be observed, or, as in the diagram, a refocusing period of length τ_3 can be included to establish in-phase magnetization at the start of detection. If broadband decoupling is applied during acquisition, the antiphase terms automatically disappear. For the observation of coupled spectra, these terms can be eliminated by a proton $90°_x$ purging pulse. Either alternative leads to the same expression for σ_4:

$$\sigma_4 = -n \sin(\omega_1 \tau_2) \sin[\pi J_{IS}(\tau_1 + \tau_2)] \prod_{j=n+1}^{m} \cos\{\pi J_{1j}(\tau_1 + \tau_2)\}$$
$$\times \sin(\pi J_{IS} \tau_3)[\cos(\pi J_{IS} \tau_3)]^{n-1} \prod_{j=n+1}^{m} \cos(\pi J_{Sj} \tau_3) S_x \qquad [64]$$

Using the relatively large one-bond coupling constants for polarization transfer, the functions in Eqs. [63] and [64] reduce to those for INEPT (Table 3-2) multiplied by $\sin(\omega_1 \tau_2)$. For transfer via heteronuclear long-range couplings, however, Eqs. [63] and [64] have important implications for sequences of SINEPT type as well as for INEPT, because all three types of coupling constants J_{IS}, J_{1j}, and J_{Sj} are now of comparable magnitude, and the polarization transfer functions do not exhibit clear maxima. The standard INEPT setting $\tau_1 + \tau_2 = 1/2J$ will usually give a signal which is very weak as a result of the many cosine factors. In general, the larger the number of "passive" protons, the shorter should be the delays. It should also be kept in mind that several different types of protons can transfer polarization to the S spin and thereby give different types of multiplet distortions in coupled spectra.

For $\tau_1 = 0$, SINEPT-2 becomes identical to SINEPT-1; consequently Eqs. [62] to [64] are applicable for SINEPT-1 if $\tau_1 = 0$ is substituted and all signs are reversed. In SINEPT-3 (Fig. 3-16c) the heteronuclear coupling J_{IS} is ineffective during τ_0. The calculation is similar to the one given above for SINEPT-2, and here we merely list the results:

$$\sigma_3 = -\left\{ \sin[\omega_1(\tau_0 + \tau_1 + \tau_2)] \sin[\pi J_{IS}(\tau_1 + \tau_2)] \right.$$
$$\left. \times \prod_{j=n+1}^{m} \cos[\pi J_{1j}(\tau_0 + \tau_1 + \tau_2)] \right\} \sum_{i=1}^{n} 2 I_{iz} S_y \qquad [65]$$

$$\sigma_4 = n \sin[\omega_1(\tau_0 + \tau_1 + \tau_2)] \sin[\pi J_{IS}(\tau_1 + \tau_2)]$$

$$\times \prod_{j=n+1}^{m} \cos[\pi J_{1j}(\tau_0 + \tau_1 + \tau_2)] \sin(\pi J_{IS}\tau_3)[\cos(\pi J_{IS}\tau_3)]^{n-1}$$

$$\times \prod_{j=n+1}^{m} \cos(\pi J_{Sj}\tau_3) S_x \qquad [66]$$

It is not possible to give general guidelines for experiments in which long-range couplings are used for polarization transfer, because of the strong dependence of the transfer function on the spin system. Therefore, in the remainder of this section we will be solely concerned with transfer via one-bond couplings; for this case, $\tau_1 + \tau_2$ can be optimized to $1/2J$.

The applicability of the SINEPT pulse sequences for heteronuclear chemical shift correlation will now be discussed. The resolution in the proton dimension, that is, the magnitude of the smallest frequency differences that can be distinguished, is determined by the length of the period during which the chemical shifts are active. This makes SINEPT-3 the sequence of choice when very high resolution is required. SINEPT-2 gives the lowest resolution and, in the limit as τ_2 approaches 0, the results become independent of the proton chemical shifts. This makes SINEPT-2 a useful method for obtaining sensitivity enhancement, which is very important in, for example, nitrogen-15 spectroscopy.

If τ_2 is set to zero and a 90° phase shift for the second 90° proton pulse is introduced, SINEPT-2 becomes equivalent to INEPT. Since $\sin(0) = 0$, one must, in the absence of a proton phase shifter, choose a different option: to obtain uniform intensity enhancements from hydrogen nuclei throughout a particular spectral region, the hydrogen transmitter frequency must be offset from the center of that region by an amount equal to $1/4\tau_2$.

As an example, suppose that a minimum PT enhancement of 80% is required from a hydrogen spectral region covering approximately 800 Hz. For SINEPT-2 these conditions are fulfilled with τ_2 of 0.25 msec and an offset for the ^{1}H transmitter of 1000 Hz from the center of the 800-Hz ^{1}H spectral region. The sine dependence of intensity enhancement by the SINEPT-2 experiment using these parameters is illustrated in Fig. 3-17, curve a. From curve b, it may be seen that complete polarization transfer from two ^{1}H resonances separated by 800 Hz can be achieved by placing the ^{1}H transmitter at the mean of the two frequencies and using τ_2 of $1/(4 \times 400)$ sec or 0.63 msec. Minimum polarization transfer is then obtained for protons in the center region.

The versatility of the SINEPT pulse sequence for fast one-dimensional heteronuclear chemical-shift correlations is a consequence of the freedom for the NMR instrument operator to "shape" the polarization transfer

Fig. 3-17. Illustrations for two examples of the sine-dependent polarization transfer that may be obtained using SINEPT-2. Curve a, a minimum of 80% polarization transfer is obtained from a ^{1}H spectral region covering approximately 800 Hz with a ^{1}H transmitter offset $\Delta\nu_H = 1000$ Hz and $\tau_2 = 0.25$ msec. Curve b, 100% polarization transfer is obtained from two ^{1}H resonances separated by 800 Hz using $\tau_2 = 0.63$ msec with the ^{1}H transmitter placed at the center of the two resonances or at multiples of 800 Hz from that position. For polarization transfer via one-bond couplings, $\tau_1 = 1/2J - \tau_2$ (see text).

function. Changing the ^{1}H transmitter frequency can be understood as causing a phase shift of the polarization transfer sine function, and variation of the chemical-shift labeling delay as equivalent to a change of its frequency.

As an illustrative example, we have used SINEPT-2 to correlate the ^{15}N and ^{1}H chemical shifts for the four NH groups in bilirubin IX-α (**1**).

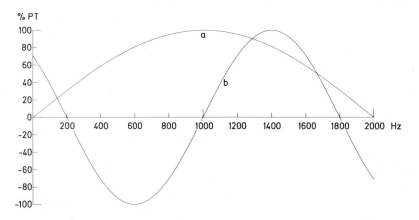

1

The proton-decoupled refocused INEPT or SINEPT-2 spectrum, obtained by polarization transfer via $^1J(^{15}\text{N}-^1\text{H})$, shows four resolved resonances for the four nitrogens (Fig. 3-18). Correlation of the ^{15}N chemical shifts with the four NH resonances observed in the ^{1}H spectrum follows

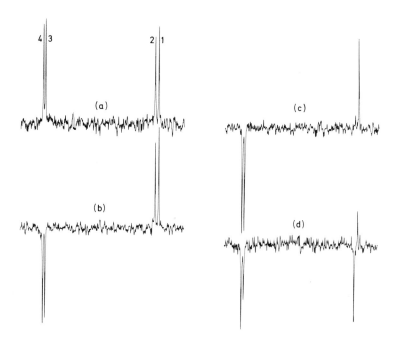

Fig. 3-18. Proton-decoupled natural abundance ^{15}N NMR spectra of bilirubin IX-α [0.034 M in (CD$_3$)$_2$SO] obtained using polarization transfer techniques on a Varian XL-300 spectrometer; the shift scale is relative to external CH$_3$NO$_2$. (a) Normal INEPT spectrum, $\tau_1 = \tau_3 = 5.55$ msec. (b)-(d) SINEPT-2 spectra obtained using the following settings for the ^{1}H transmitter (in hertz from TMS) and τ_2 values: (b) 2772.1 Hz and 3.39 msec, corresponding to the sine-dependent polarization transfer curve b in Fig. 3-19; (c) 2864.8 Hz and 4.55 msec, corresponding to curve c in Fig. 3-19; (d) 2889.8 Hz and 4.55 msec, corresponding to curve d in Fig. 3-19. For all experiments, $\tau_1 + \tau_2 = 1/2J = 5.55$ msec and $\tau_3 = 5.55$ msec. The heteronuclear shift correlation follows from Fig. 3-19 (see text). The number of scans varies from 10,000 to 15,000.

from Fig. 3-19, which shows some sine polarization transfer functions superimposed on a stick plot of the ^{1}H spectrum, along with the corresponding SINEPT-2 ^{15}N spectra shown in Fig. 3-18. The enhancements observed for the four ^{15}N signals in the spectra of Fig. 3-18 show that the ^{15}N and ^{1}H chemical shifts (both sets have been numbered 1-4 from lower to higher frequency) have the following correlations: N1-H2, N2-H1, N3-H3, and N4-H4. It is also seen that, by proper design of the polarization transfer sine function, a complete correlation may be obtained from only two experiments (spectra b and d in Fig. 3-18). Experiment d also gives an impression of the resolution achievable with SINEPT-2, since the separation of the two hydrogen resonances at highest frequency is only 14 Hz.

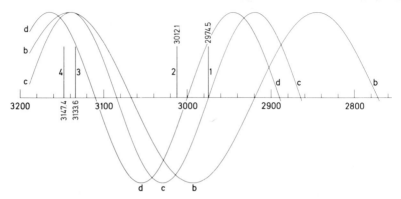

Fig. 3-19. The sine-dependent polarization transfer function (Eq. [64]) used in the SINEPT-2 experiments of Fig. 3-18 superimposed on a stick plot of the ^{1}H spectrum for the four NH protons of bilirubin IX-α. The frequency scale is in hertz relative to TMS, at a spectrometer frequency of 300 MHz. Curves b-d correspond to the experimental conditions for spectra (b)-(d) in Fig. 3-18.

E. POLARIZATION TRANSFER AND RELATIVE SIGNS OF COUPLING CONSTANTS

1. *Introduction*

The determination of the signs of scalar coupling constants is of interest in relation to the understanding of spin–spin interactions and their dependence on molecular structure. The absolute signs of some coupling constants have been established by their determination relative to the sign of the direct dipole–dipole interaction in studies of molecules oriented in liquid crystals (29). However, ordinary experiments are limited to the determination of the sign of one J value relative to another; the starting point for many such determinations is the positive sign invariably observed for the one-bond ^{13}C-^{1}H coupling constant. For strongly coupled systems, relative sign information can often be extracted from the spectrum by, for example, computer simulations, but this is not possible for weakly coupled systems, for which specially designed experiments are required (30–32).

The establishment of the relative sign for J_{12} and J_{23} in a three-spin system, consisting of I_1, I_2, and I_3, comes about by comparing the influence of the I_2 spin state on the energies of the spin states of I_1 and I_3, respectively. At least three coupled spins are required for a determination of relative signs of coupling constants. It is normally required that all three couplings be resolved, and only techniques related to this situation will be discussed here, although a triple-resonance experiment has been described where only two couplings need be resolved (30).

Double-resonance experiments involving simultaneous irradiation at several frequencies have been extensively employed to determine relative signs of coupling constants (31, 32). The essence of these methods is the identification of I_1 and I_3 transitions that correspond to just one of the two possible spin states of I_2. In practice this is accomplished by perturbing, in the I_1 spectrum, one line corresponding to a specific I_2 spin state while observing the I_3 spectrum. Closely parallel in approach are the selective population transfer techniques which were described in Chapter 1, and which will be analyzed in the next section in terms of product operators, although these methods involve sequential rather than simultaneous irradiation. Multiple-quantum experiments also give information about relative signs. As an example, consider double-quantum coherence between spin-$\frac{1}{2}$ nuclei I_1 and I_3. According to Eqs. [37] and [38], this coherence is a doublet with splitting $|J_{12} + J_{23}|$, so that, if the magnitudes of J_{12} and J_{23} are known from a normal 1D spectrum, the relative signs of J_{12} and J_{23} can be established.

2. Selective Population Transfer

The SPT experiment has been applied mostly to heteronuclear spin systems, and here we discuss the simplest possibility of one S spin coupled to I_1 and I_2. One of the four I_1 transitions is inverted by a selective 180° pulse, to create a nonequilibrium state of the first kind, one in which the diagonal elements of the density matrix describe nonequilibrium populations but the off-diagonal elements are all equal to zero (33). This is followed by a nonselective pulse on S and acquisition of the S-spin FID. The information about the relative signs of the coupling constants follows from the fact that the transition of I_1 inverted corresponds to a specific spin state of the passive spin I_2 and that only the two transitions of S corresponding to the same I_2 spin state are influenced by this particular SPT experiment. Suppose the I_1 transition at highest frequency is the one which is inverted; we may represent its x component by a linear combination of operators such as were given in Fig. 3-4:

$$M_x = \tfrac{1}{4}(I_{1x} - a_{12}2I_{1x}I_{2z} - a_{1S}2I_{1x}S_z + a_{12}a_{1S}4I_{1x}I_{2z}S_z) \qquad [67]$$

The a terms were defined on page 157, and, for simplicity, we have omitted the symbols I in the labels on them. This represents magnetization in the xy plane; if a $90°_{-y}(I_1)$ pulse were applied, it would be rotated to the z axis, and the resulting equilibrium magnetization would be represented by

$$M_z = \tfrac{1}{4}(I_{1z} - a_{12}2I_{1z}I_{2z} - a_{1S}2I_{1z}S_z + a_{12}a_{1S}4I_{1z}I_{2z}S_z) \qquad [68]$$

Assuming that the native S-spin magnetization is suppressed by phase cycling, we can ignore that contribution and start the SPT analysis with the following density operator:

$$\sigma_0 = I_{1z} + I_{2z} \quad [69]$$

It is convenient to divide I_{1z} into two parts, expressing it as the sum of the high-frequency component given by Eq. [68] and all the remaining components:

$$\sigma_0 = \tfrac{1}{4}(I_{1z} - a_{12}2I_{1z}I_{2z} - a_{1S}2I_{1z}S_z + a_{12}a_{1S}4I_{1z}I_{2z}S_z)$$
$$+ \tfrac{1}{4}(3I_{1z} + a_{12}2I_{1z}I_{2z} + a_{1S}2I_{1z}S_z - a_{12}a_{1S}4I_{1z}I_{2z}S_z) + I_{2z} \quad [70]$$

A selective 180° pulse on the high-frequency component of I_1 changes the signs of terms in the first set of parentheses, permitting Eq. [70] to be simplified:

$$\sigma_1 = \tfrac{1}{2}(I_{1z} + a_{12}2I_{1z}I_{2z} + a_{1S}2I_{1z}S_z - a_{12}a_{1S}4I_{1z}I_{2z}S_z) + I_{2z} \quad [71]$$

A 90°(S) pulse, which for illustration may have phase y, generates S-spin magnetization, leading to the following results for the products involving S-spin operators:

$$\sigma_2 = \tfrac{1}{2}(a_{1S}2I_{1z}S_x - a_{12}a_{1S}4I_{1z}I_{2z}S_x) \quad [72]$$

If we assume that the magnitude of J_{1S} is larger than that of J_{2S}, this results in the following signals:

Each multiplet is to be multiplied by the indicated function of the a values. If a_{12} and a_{2S} are equal (+1 or −1), these two add to give

If a_{12} and a_{2S} are of opposite sign, subtraction of the second multiplet from the first gives

By using, in place of Cartesian operators, the polarization or shift operators which were introduced in Chapter 2, Section II,B, we can express nicely the situation that the selective 180° pulse affects only I_1 transitions corresponding to a specific spin state of the passive nucleus I_2, and that only S transitions associated with the same I_2 spin state are polarized. The polarization operators specify the state, α or β, of the spin they represent:

$$I^\alpha = \tfrac{1}{2}(E + 2I_z) \quad [73]$$

$$I^\beta = \tfrac{1}{2}(E - 2I_z) \quad [74]$$

Equation [72] for the observable S-spin magnetization may be factored to give a form including the expression for one of these operators:

$$\sigma'_2 = \tfrac{1}{2}a_{1S}2I_{1z}S_x(E_2 - a_{12}2I_{2z}). \quad [75]$$

Thus the I_2 polarization operator corresponds to either I_2^α or I_2^β, depending upon the sign of a_{12}. The relative sign of J_{12} and J_{2S} can then be determined by observing which pair of lines in the S spectrum has been polarized and taking into account the relative signs of the I and S spin magnetogyric ratios.

3. Sign-Labeled Polarization Transfer Using Nonselective Pulses

What appears to be a general principle for techniques capable of determining relative signs of coupling constants in a spin system consisting of I_1, I_2, and I_3 is the following: *to determine the relative signs of* J_{12} *and* J_{23}, *the operator* I_{2z} *for the common spin must be introduced into the density operator via* J_{12} *and removed via* J_{23} *or vice versa.* SPT experiments as well as multiple-quantum experiments can be interpreted in terms of this principle, but it is illustrated specifically by the following two experiments.

a. *Homonuclear Spin-Labeled Polarization Transfer.* The pulse sequence which may be used in this connection is

$$90°_{\pm x} - \frac{\tau}{2} - 180°_y - \frac{\tau}{2} - \theta_x - \text{Acquire}(\pm)$$

As usual, the phase of the refocusing pulse is not important, but it is specified so that specific equations may be written out. If this phase is y, the pulse commutes with the density operator immediately after the first pulse. If we consider only the magnetization on spin I_1 in a three-spin system, the initial density operator σ_0 is I_{1z}, which becomes $-I_{1y}$ after the initial $90°_x(I)$ pulse. In the following τ period, the chemical shift is refocused, but, because the 180° pulse affects both partners in each coupled pair of nuclei, the density operator evolves from σ_1 to σ_2 under the effects of two couplings:

$$\sigma_2 = -c_{12}c_{13}I_{1y} + c_{12}s_{13}2I_{1x}I_{3z} + s_{12}c_{13}2I_{1x}I_{2z} + s_{12}s_{13}4I_{1y}I_{2z}I_{3z} \quad [76]$$

where the abbreviations $s_{ij} = \sin \pi J_{ij}\tau$ and $c_{ij} = \cos \pi J_{ij}\tau$ have been used.

From the general principle for relative sign determination introduced above, the significant term in the density operator must include an I_z factor for the spin I_3 to which we wish to transfer polarization and an I_z operator for the passive spin I_2; this term is evidently $s_{12}s_{13}4I_{1y}I_{2z}I_{3z}$. To avoid a total transfer into multiple-quantum coherence, the pulse θ must be of flip angle other than 90°. The relevant part of the density operator after this pulse is

$$\sigma_3 = -\sin^2\theta \cos\theta\, s_{12}s_{13}(4I_{1z}I_{2y}I_{3z} + 4I_{1z}I_{2z}I_{3y}) \qquad [77]$$

These are the only terms describing transfer to I_2 and I_3. For I_1, parts of all four terms in Eq. [76] remain; thus, the practical way to perform this experiment is to make the first 90° pulse semiselective for one spin such as I_1, and focus attention on the resulting spectra of its coupling partners. From the I_2 spectrum, the sign of $J_{13} \times J_{23}$ can be determined and from the I_3 spectrum, the sign of $J_{12} \times J_{23}$. Both spectra show four lines in antiphase, and, for a correct interpretation of the multiplet patterns, it is crucial to set the receiver reference phase properly. This is done by phase-correcting a normal one-pulse experiment, using the same carrier frequency as in the subsequent experiment, to pure positive absorption.

After this setup experiment, the four-line antiphase patterns for I_2 and I_3 can be characterized by

multiplied, respectively, by $s_{12}s_{13}a_{12}a_{23}$ and $s_{12}s_{13}a_{13}a_{23}$. From τ and the magnitude of J_{12}, the sign of $s_{12}a_{12} = \sin \pi |J_{12}|\tau$ is known. The sign of the term $s_{13}a_{23}$ is the same as the sign of $J_{13} \times J_{23} \sin \pi |J_{13}|\tau$.

An experimental test has been carried out on a sample of 2,3-dibromopropionic acid, a much-studied three-spin system. Figure 3-20a shows the setup spectrum and Fig. 3-20b the spectrum resulting from semiselective excitation of the center multiplet (spin I_2) by an on-resonance DANTE sequence and a δ filter. The δ filter involved coaddition of two experiments, the second of which included an extra delay after the DANTE excitation to allow for a 180° precession of the high-frequency multiplet. In Fig. 3-20c, a large amount of coherence corresponding to $4I_{1z}I_{2y}I_{3z}$ has been created for $\tau = 38$ msec, which corresponds to the first maximum of the product $s_{12}s_{23}$. The optimum setting of the flip angle θ is the "magic angle," and Fig. 3-20d shows the spectrum resulting when $\theta = 55°$. Numbering the three spins I_1, I_2, and I_3 from high to low frequency, the I_1 spectrum shows that $s_{12}s_{23}a_{12}a_{13} < 0$, and the I_3 spectrum shows that $s_{12}s_{23}a_{13}a_{23} > 0$.

3. POLARIZATION TRANSFER AND EDITING TECHNIQUES 197

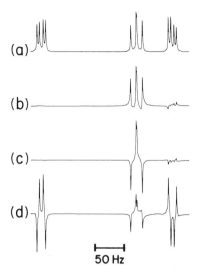

Fig. 3-20. Spectra illustrating sign-labeled polarization transfer between proton spins. (a) Setup spectrum for 2,3-dibromopropionic acid in CDCl$_3$; (b) selective excitation of the center multiplet (six-pulse DANTE sequence with 2.6 msec pulse spacing and δ filter with 2.8 msec delay); (c) same as (b) but with a waiting time of 38 msec prior to acquisition; (d) the effect of applying a nonselective pulse ($\beta = 55°$) to the state shown in (c). All spectra were obtained with the same phase correction, on a Bruker CXP-300 spectrometer. The results in (d) show that $J_{13} \times J_{23} < 0$, $J_{12} \times J_{13} > 0$, and therefore $J_{12} \times J_{23} < 0$. The 180° pulse indicated in Section III,E,3,b was omitted to avoid strong coupling effects.

This yields $J_{13} \times J_{23} < 0$ and $J_{12} \times J_{13} > 0$. From these two relations, it follows that $J_{12} \times J_{23} < 0$.

The technique described above can be considered the basis for the two-dimensional COSY-45 experiment (90°—t_1—45°—t_2) of Aue (34) and Bax (35). The three-operator terms such as $4I_{1y}I_{2z}I_{3z}$, plus possibly higher terms, are superimposed on two-operator terms such as $2I_{1y}I_{2z}$ to give characteristic tilts of the cross peaks in the two-dimensional spectrum. The directions of the tilts depend on the relative signs of the coupling constants. The 1D experiment described above as well as COSY-45 can easily be extended to heteronuclear spin systems, but this will not be discussed further here.

b. *Sign Determination of ^{13}C-^{13}C Coupling Constants.* Carbon–carbon scalar couplings as a probe for molecular structure have been used increasingly during the last decade (36–38). A major breakthrough in the observation of ^{13}C-^{13}C coupling constants at natural abundance was achieved by Bax and co-workers with the introduction of the INADEQUATE technique (39), which was described in Chapter 2 and will be discussed further in

Chapter 4. This method provides efficient suppression of the signals of singly labeled molecules and is convenient for measuring magnitudes of carbon–carbon couplings, but it does not allow the determination of their signs. Sign determination may be achieved by double selective polarization transfer, which uses selective inversion of ^{13}C satellites in the ^{1}H spectrum while the pair of ^{13}C resonances is observed with a nonselective pulse (40). The disadvantages of this technique are that only couplings involving a particular ^{13}C nucleus can be investigated in one experiment, and that sensitivity is reduced because the spectra are recorded without proton decoupling.

A general method, called SLAP for *sign-la*beled *p*olarization transfer, for the determination of signs of carbon–carbon couplings does not require selective pulses and is only weakly dependent on the magnitudes of the couplings involved (41, 42). In terms of the general principle for sign determination stated at the beginning of Section III,E,3, the essence of the experiment is readily described for an H_1—C_1—C_2 fragment (denoted by I_1—S_1—S_2). The quantity S_{2z} is introduced via the long-range C_2–H_1 coupling, the proton magnetization of H_1 is transferred to the directly bonded carbon C_1, and S_{2z} can be removed via $J(C_1C_2)$. Thus the experiment provides the relative sign of $^nJ(CC)$ and $^{n+1}J(CH)$. Relating the sign of $^nJ(CC)$ to that of $^1J(CH)$ would require a longer and more complicated pulse sequence than that for the SLAP sequence shown in Fig. 3-21, which is designed to give optimum sensitivity for the ^{13}C-^{13}C satellites and optimum suppression of the signals from singly ^{13}C-labeled molecules. It includes a mixing period, τ_2, in which antiphase ^{13}C-^{13}C homonuclear two-spin coherence rephases before the decoupler is switched on. A 90°(C) pulse then transfers the MQC to observable antiphase doublets for detection.

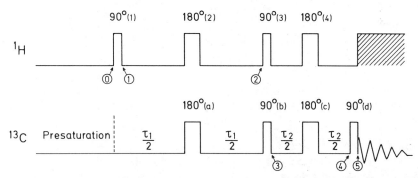

Fig. 3-21. The SLAP pulse sequence for the determination of the signs of ^{13}C-^{13}C coupling constants. The setting of the delays is discussed in the text and a recommended phase cycling scheme is given in Table 3-3. The states of the system referred to in the text are numbered 1, 2, 3, 4, 5.

3. POLARIZATION TRANSFER AND EDITING TECHNIQUES

We first analyze the response of the three-spin system I_1—S_1—S_2 and later generalize the result to arbitrary spin systems with two ^{13}C nuclei. In its simplest form, the sequence uses pulses of the same phase, in the following set equal to x. We will neglect product operators which do not finally lead to observable magnetization. The native S-spin (^{13}C) magnetization is eliminated by presaturation, so that the experiment starts with pure I-spin (^{1}H) magnetization.

From rules presented in Section II, the relevant terms of the density operators at the points indicated in Fig. 3-21 can be written as follows:

$$\sigma_0 = I_{1z} \qquad [78]$$

$$\sigma_1 = -I_{1y} \qquad [79]$$

$$\sigma_2 = -\sin[\pi J(C_1H_1)\tau_1]\sin[\pi J(C_2H_1)\tau_1]4I_{1y}S_{1z}S_{2z} \qquad [80]$$

$$\sigma_3 = -\sin[\pi J(C_1H_1)\tau_1]\sin[\pi J(C_2H_1)\tau_1]4I_{1z}S_{1y}S_{2y} \qquad [81]$$

$$\sigma_4 = -\sin[\pi J(C_1H_1)\tau_1]\sin[\pi J(C_2H_1)\tau_1]\sin[\pi J(C_1H_1)\tau_2]2S_{1x}S_{2y} \qquad [82]$$

The τ_2 delay is made sufficiently short so that only one-bond carbon-hydrogen couplings are effective. The last 90° pulse transforms the carbon MQC into observable magnetization. If the phase of this pulse is x, we find

$$\sigma_5 = -\sin[\pi J(C_1H_1)\tau_1]\sin[\pi J(C_2H_1)\tau_1]\sin[\pi J(C_1H_1)\tau_2]2S_{1x}S_{2z} \qquad [83a]$$

If the phase is y, the density operator becomes

$$\sigma_5 = \sin[\pi J(C_1H_1)\tau_1]\sin[\pi J(C_2H_1)\tau_1]\sin[\pi J(C_1H_1)\tau_2]2S_{1z}S_{2y} \qquad [83b]$$

Thus magnetization originating from H_1 is found on either C_1 or C_2, depending on the phase of the detection pulse. Figure 3-22 shows these features for a sample of doubly ^{13}C-enriched acetic acid. Trace (b) was obtained with an x detection pulse, causing antiphase magnetization of the methyl carbon, while the y detection pulse for trace (c) produces almost exclusively antiphase magnetization of the *complementary* carboxyl carbon in dispersion mode. Thus the $90°_y$ experiment has been dubbed COSLAP (*co*mplementary *SLAP*). A 90° phase correction of trace (c) leads to the absorption mode signals in trace (d). The low-intensity doublets observed for the carboxyl carbon in b and the methyl carbon in c and d result from refocusing via the long-range C_2-H_1 coupling during the τ_2 period:

$$-4I_{1z}S_{1y}S_{2y} \longrightarrow -\cos[\pi J(C_1H_1)\tau_2]\sin[\pi J(C_2H_1)\tau_2]2S_{1y}S_{2x} \qquad [84]$$

The value of τ_2 is chosen to be less than $1/{}^1J(CH)$ so that $\sin[\pi J(C_1H_1)\tau_2]$ is always positive and has no influence on the relative sign determination.

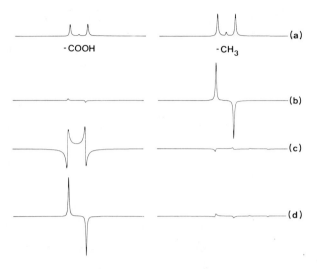

Fig. 3-22. Proton-decoupled ^{13}C spectra of doubly ^{13}C-enriched acetic acid recorded on a Bruker CXP-300 spectrometer. Each spectrum was obtained from a total of 16 scans. (a) Setup experiment with a 90° pulse of phase $-y$ (no NOE). (b) Spectrum obtained with the SLAP sequence and the phase cycle in Table 3-3, $\tau_1 = 73.3$ msec and $\tau_2 = 1.5$ msec. Note that the magnetization is almost exclusively on the methyl carbon. (c) Same as (b) but with the last 90° pulse phase shifted by 90°. The magnetization appears almost exclusively on the carboxyl carbon and in dispersion. (d) Spectrum obtained with a 90° phase correction of the spectrum in (c).

Deletion of this factor from Eq. [83a] leads to the simplified expression:

$$\sigma_5 = -\sin[\pi J(C_1 H_1)\tau_1] \sin[\pi J(C_2 H_1)\tau_1] 2S_{1x} S_{2z} \qquad [85]$$

The evolution of σ_5 during detection leads to an antiphase multiplet in the spectrum, with polarity determined by the signs of the coupling constants. The product operator $2S_{1x}S_{2z}$ itself produces an antiphase doublet in which the line at frequency $+J(CC)/2$ relative to the chemical shift frequency is negative. Taking into account the trigonometric functions in Eq. [85] and assuming that τ_1 is less than $1/J(C_2H_1)$, we arrive at the following interpretation. If the high-frequency component of the $^{13}C_1$ doublet is positive,

$$\sin[\pi J(C_1 H_1)\tau_1] \times J(C_2 H_1) \times J(C_1 C_2) > 0 \qquad [86]$$

If the high-frequency component is negative, the expression in Eq. [86] is negative. Knowledge of the sign of $J(C_2 H_1)$ and the magnitude of $J(C_1 H_1)$ thus allows the determination of the sign for $J(C_1 C_2)$.

Although straightforward, density-operator calculations on large spin systems can be extremely lengthy if all transformations are carried out in sequential order and all terms retained throughout. Fortunately, once the response has been calculated for a small system, it can normally be easily generalized by simply including additional cosine factors, reflecting the condition that certain couplings can not be allowed to cause dephasing if a desired response is to be obtained. In some cases, care must be taken if additional pathways can lead to observable signals for larger spin systems, but this is not true for SLAP.

The general spin system is a fragment consisting of two ^{13}C nuclei with n and m directly bonded protons, respectively:

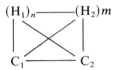

Since τ_2 is less than $1/{}^1J(CH)$, negligible polarization is transferred from a given proton to the non-directly bonded carbon using a $90°_x$ detection pulse, and we focus on the transfer from H_1 to C_1. If the molecule has a total of N protons, there can be coupling between $(H_1)_n$ and the remaining $N-n$ protons during τ_1. During τ_2, one $^1J(CH)$ coupling should rephase and the other $n+m-1$ one-bond couplings should stay in phase. The general function for the transfer of polarization from H_1 to C_1 may thus be written, inserting cosine functions into the coefficients of Eq. [83]:

$$n \sin[\pi J(C_1H_1)\tau_1] \sin[\pi J(C_2H_1)\tau_1] \prod_{i=n+1}^{N} \cos[\pi J(H_1H_i)\tau_1]$$

$$\times \sin[\pi J(C_1H_1)\tau_2] \cos^{n-1}[\pi J(C_1H_1)\tau_2] \cos^m[\pi J(C_2H_2)\tau_2] \quad [87]$$

If τ_1 and τ_2 are sufficiently short, so that $\tau_1 < 1/2J(HH)$ and $\tau_2 < 1/{}^1J(CH)$, then $\cos[\pi J(HH)\tau_1]$ and all trigonometric functions depending on τ_2 remain positive and the sign of the expression [87] is determined by the first two factors. The interpretations in Eq. [86] and its converse are thus valid for any spin system.

The suppression of the signals of the much more abundant singly ^{13}C-labeled molecules is a prerequisite for the observation of the resonances of doubly labeled molecules. For the magnetization originating from proton I_1, the density operator after the first $90°_x(H)$ pulse is $-I_{1y}$. During τ_1, both the ^{13}C-^{1}H and the ^{1}H-^{1}H couplings can lead to dephasing of the multiplet components of spin I_1, and the density operator contains the following types of products at the end of τ_1:

$$\sigma_2: \quad I_{1y}, 2I_{1x}S_z, 2I_{1x}I_{iz}, 4I_{1y}I_{iz}S_z, 4I_{1y}I_{iz}I_{jz}, \ldots$$

where i and j denote any protons having nonvanishing couplings to I_1. The simultaneous ^{1}H and ^{13}C $90°_x$ pulses at time τ_1 create the following operators:

$$\sigma_3: \quad I_{1z}, 2I_{1x}S_y, 2I_{1x}I_{iy}, 4I_{1z}I_{iy}S_y, 4I_{1z}I_{iy}I_{jy}, \ldots$$

Except for I_{1z}, these product operators describe heteronuclear and proton homonuclear MQC and are unobservable. This indicates that the signals of singly labeled molecules should be well suppressed in the SLAP experiment. In addition, they may be further suppressed by presaturation of the native ^{13}C magnetization.

c. *Practical Aspects of the SLAP Technique.* The determination of the sign of a carbon–carbon coupling constant requires knowledge of the signs of the coupling constants of the two carbon spins to a proton directly bonded to one of them. The sign of $^1J(CH)$ is known to be invariably positive, while the signs of the long-range couplings between carbon and hydrogen should be determined in advance.

For selection of τ_1, during which sign-labeling occurs, three factors must be considered:

1. Dephasing of the ^{1}H-^{1}H couplings during τ_1 causes intensity loss, and each hydrogen–hydrogen spin–spin interaction contributes a cosine factor to the finally observed ^{13}C signal. The value of τ_1 should not be larger than the reciprocal of twice the largest ^{1}H-^{1}H coupling constant; a value equal to $1/2J(HH)$ gives zero polarization transfer.

2. Further restrictions are imposed by the heteronuclear long-range couplings. The cosine modulations caused by the hydrogen–hydrogen couplings are decreasing functions of τ_1, whereas the sine modulations caused by the long-range carbon–hydrogen couplings are increasing functions of τ_1 in the applicable τ_1 range. The first maximum of the sine function is at

$$\tau_1 = 1/2\,{}^nJ(CH) \qquad [88]$$

Maxima for longer τ_1 values are normally not of interest because of the $J(HH)$ limitation.

3. The selection must also be matched to the rapidly oscillating sine functions, caused by the one-bond ^{13}C-^{1}H couplings, with extrema at

$$\tau_1 = \frac{2p+1}{2\,{}^1J(CH)}, \qquad p = 0, 1, 2, \ldots \qquad [89]$$

For optimum sensitivity of the experiment, τ_1 is selected to fulfill Eq. [89] for a value of p which allows appropriate dephasing of the heteronuclear long-range couplings, taking into account the upper limit imposed by the ^{1}H-^{1}H couplings. It is normally possible to find conditions acceptable for many carbon sites at the same time. The only critical adjustment is with regard to the $^1J(CH)$ values, which are easily determined.

3. POLARIZATION TRANSFER AND EDITING TECHNIQUES

For the optimization of τ_2, which is needed to rephase the ^{13}C-^1H interactions in order to permit decoupling during acquisition, only one-bond carbon–hydrogen coupling constants need be considered. The maxima for the part of Eq. [87] which depends on τ_2 are given in Table 1-6 for various values of $n+m$. A good compromise, valid for all combinations of n and m, is $\tau_2 = 1/5\ ^1J(CH)$.

Although the ideal experiment is insensitive to singly labeled molecules, pulse imperfections may produce spurious response from such molecules. A suitable 16-step phase cycle which compensates possible pulse errors is outlined in Table 3-3. It is based on the different phase properties of MQC compared to single-quantum coherence. Depending on the required degree of suppression and the number of scans required for natural abundance studies, Table 3-3 may be extended in the following order: (*i*) incrementation of 90°(1) and 90°(3) in 90° steps; (*ii*) Exorcycling of 180°(2); (*iii*) Exorcycling of 180°(c); (*iv*) phase alternation of 180°(a). Each of the extensions (*i*)–(*iii*) multiplies by four, and (*iv*) by two, the number of experiments in the scheme. It is our experience that good suppression of unwanted parent signals is obtained using this phase cycle scheme.

For correct interpretation of the positive/negative doublet line intensities, the receiver phase must be adjusted in advance. A setup experiment with

TABLE 3-3

BASIC 16-STEP PHASE CYCLING SCHEME EMPLOYED IN THE SLAP EXPERIMENT[a]

1	2	3	4	a	b	c	d	Receiver
x	x	x	x	x	x	x	x	x
					−x	−x	−x	−x
					−x	x	x	x
					x	−x	−x	−x
−x					x	x	x	−x
					−x	−x	−x	x
					−x	x	x	−x
					x	−x	−x	x
x					x	x	−x	−x
					−x	−x	x	x
					−x	x	−x	−x
					x	−x	x	x
−x					x	x	−x	x
					−x	−x	x	−x
					−x	x	−x	x
					x	−x	x	−x

[a] In order to compensate possible phase errors it can be advantageous to repeat the basic 16 steps with 90° added to the b, d, and receiver phases.

a single 90°C pulse of phase y or $-y$ and broadband decoupling is performed, and the singlet lines of the spectrum are phase-corrected to positive absorption.

The proper choice of the sign of the y phase depends on the sense of rotation used in the definition of the x and y phase, which may differ from spectrometer to spectrometer. By performing two test experiments, one may easily establish the proper phase for a particular instrument. To begin, a single-pulse experiment, $90°_y$ followed by acquisition, is phase-corrected to give signals in positive absorption. Then, with the same phase correction, the experiment is repeated with a $90°_{-x}$ pulse. All signals now appear in dispersion mode. If the high-frequency tails of the dispersion signals are negative, the setup experiment should be performed with phase y. If the high-frequency tails are positive, phase $-y$ should be used.

The features of the SLAP and COSLAP techniques have been demonstrated for three molecules, doubly ^{13}C-enriched acetic acid, representing an $(I_1)_3 S_1 S_2$ spin system with $^1J(I_1 S_1) = 129.7$ Hz and $^2J(I_1 S_2) = -6.8$ Hz, and natural-abundance 2-bromothiazole and 2,3-dibromothiophene. The values of $J(CH)$ and $J(HH)$ for the two latter compounds, determined from conventional proton-coupled spectra (43), are given in Tables 3-4 and 3-5.

From the sequence of positive/negative intensities in Fig. 3-22b for acetic acid, we conclude that

$$\sin[\pi {}^1J(CH)\tau_1] \times {}^2J(CH) \times {}^1J(CC) > 0 \qquad [90]$$

Since, for the selected τ_1 value,

$$\sin[\pi {}^1J(CH)\tau_1] < 0 \qquad [91]$$

this means that $^2J(CH)$ and $J(CC)$ have opposite signs; since $J(CC)$ is positive, we deduce that $^2J(CH)$ is negative.

To demonstrate the feasibility of the SLAP and COSLAP transfer techniques for natural abundance samples, the signs of the two long-range

TABLE 3-4

COUPLING CONSTANTS FOR 2-BROMOTHIAZOLE[a]

J	H4	H5
C2	19.48	8.81
C4	191.0	5.79
C5	15.31	192.0
H4		3.63

[a] From •I. Hansen (43). All couplings are positive and in hertz.

TABLE 3-5
^{13}C-^{1}H AND ^{1}H-^{1}H COUPLING CONSTANTS IN HERTZ FOR 2,3-DIBROMOTHIOPHENE[a]

J	H4	H5
C2	11.18	6.46
C3	2.52	11.91
C4	176.6	3.70
C5	5.72	191.30
H4	—	5.75

[a] From M. Hansen (43). All couplings are positive.

^{13}C-^{13}C coupling constants in 2-bromothiazole as well as the assignment and signs of some ^{13}C-^{13}C couplings in 2,3-dibromothiophene have been determined. Before these experiments no information was available, either on the signs or on the magnitudes of these constants. Figures 3-23 and 3-24 show the experimental results. Following the setup experiment for 2-bromothiazole, the SLAP spectrum shown in Fig. 3-23a was recorded with

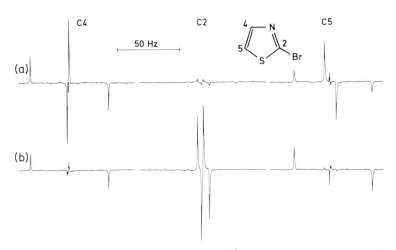

Fig. 3-23. Proton-decoupled natural abundance ^{13}C NMR spectra of 2-bromothiazole obtained on a Varian XL-300 spectrometer. (a) SLAP spectrum obtained with $\tau_1 = 23.5$ msec and $\tau_2 = 1.74$ msec; 4096 scans. (b) COSLAP spectrum with the same conditions as for (a) but with phase y for the last 90° ^{13}C pulse. Note the opposite polarization of the doublet from the long-range coupling across the nitrogen in the C4 spectrum. The spectra show J(C2C4) to be negative and J(C2C5) to be positive.

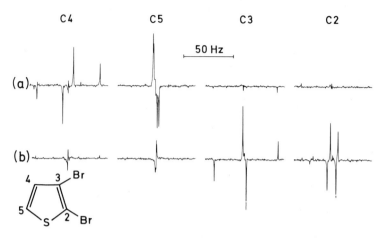

Fig. 3-24. Proton-decoupled natural-abundance ^{13}C spectra of 2,3-dibromothiophene obtained on a Varian XL-300 spectrometer. (a) SLAP spectrum obtained with $\tau_1 = 54.9$ msec and $\tau_2 = 2.61$ msec; 2048 scans. (b) COSLAP spectrum obtained with the same conditions as in (a). Peaks from a minor impurity appear in the spectra. The spectra give both the assignments and signs of the long-range ^{13}C-^{13}C couplings (see text).

τ_1 of 23.5 msec and τ_2 of 1.74 msec. The τ_2 value was selected as a compromise allowing all the ^{13}C-^{13}C couplings to appear. According to Eq. [86], we conclude that $^1J(C4C5) = +61.8$ Hz, $J(C2C4) = -1.65$ Hz, and $J(C2C5) = +9.35$ Hz.

The corresponding COSLAP spectrum shown in Fig. 3-23b for 2-bromothiazole gives no additional information for this molecule. It simply serves as a test for the COSLAP technique with natural abundance samples and also confirms the assignment of coupling constants. In judging the intensities for the spectra in Fig. 3-23, one should note that the lines for the C5 resonances are somewhat broader than for the C4 and C2 resonances. As an example of the SLAP/COSLAP methods for assignment of ^{13}C-^{13}C couplings in organic molecules, the two types of spectra for 2,3-dibromothiophene are shown in Fig. 3-24. First, the two experiments unambiguously assign the four ^{13}C resonances as C4, C5, C3, and C2, in order from low to high field. More importantly, the two overlapping doublets observed for the long-range ^{13}C-^{13}C couplings in the C5 SLAP spectrum are unambiguously assigned to their respective C5-C3 and C5-C2 couplings from the COSLAP spectrum. In addition, from Eq. [86] and these spectra, we deduce that $^1J(C3C4) = +64.6$ Hz, $^2J(C2C4) = +11.25$ Hz, $^2J(C2C5) = +5.90$ Hz, and $^2J(C3C5) = +3.60$ Hz.

Finally, the SLAP method, in addition to providing relative sign information, can also overcome a limitation of INADEQUATE, which, for observation of very small two-bond ^{13}C-^{13}C coupling constants, requires a long preparation delay, equal to $1/2J(CC)$, with accompanying sensitivity loss by relaxation. However, in SLAP, the transfer proceeds via the three-bond carbon–hydrogen coupling, which is often considerably larger than the corresponding two-bond carbon–carbon coupling.

IV. Heteronuclear Spectral Editing Techniques

A. INTRODUCTION

Heteronuclear spectral editing, described in Chapter 1 and introduced by Doddrell, Bendall, and Pegg in a series of papers, first with INEPT, then with DEPT (8), allows decomposition of conventional ^{13}C spectra into four subspectra according to the number of directly attached protons, and this aids in the simplification of complex spectra. Recently, spectral editing methods have been introduced into a number of new pulse sequences.

This section describes editing of ^{13}C spectra with a single example of application to ^{15}N spectroscopy. Various approaches to editing decoupled spectra will first be described, and then modifications required for coupled spectra will be explained. To cope with a spread in the values of one-bond carbon–hydrogen coupling constants, there are three levels of performance: (1) APT or INEPT; (2) SEMUT or DEPT; and (3) SEMUT GL or DEPT GL. The second experiment in each pair utilizes polarization transfer, while the first relies entirely on nuclear Overhauser enhancement to increase sensitivity. These methods will be analyzed in order. SEMINA editing of natural abundance INADEQUATE spectra is described, and a final section deals with "extended" editing of ^{13}C and ^{15}N spectra to obtain information about structural relationships more remote than one bond.

B. APT (J-MODULATED SPIN ECHO) AND INEPT

For the J-modulated spin-echo experiment, the two alternatives were shown in Figs. 1-6 and 1-7. For INEPT, we shall consider the refocused version in Fig. 3-12b with decoupling during acquisition. For these experiments and the improvements to them given in the following subsections, we shall determine the general response of an I_nS spin system.

The pulse sequences which rely on NOE for sensitivity enhancement are in each instance identical to the last part, starting with the 90°(C) pulse, of their polarization transfer analogs, so that it is convenient to analyze the NOE versions first.

For APT the density operator after the first carbon pulse is given by

$$\sigma_1 = -S_y \quad [92]$$

Introducing a sign change to allow for the effect of the refocusing pulses, we can write the essential transformations describing the effect of spin–spin coupling to n I nuclei:

$$S_y \xrightarrow{\sum_{i=1}^{n} \pi J\tau 2 I_{iz} S_z} \quad [93]$$

The n transformations represented by the summation lead to 2^n terms in the density operator, but, since the decoupler is on during detection, only the in-phase part is observable. The propagators in Eq. [93] all commute with each other, and the in-phase part after each is proportional to $\cos(\pi J\tau)$. Thus the intensity expression for an $I_n S$ group is

$$\text{Int}(I_n S) = \cos^n(\pi J\tau) \quad [94]$$

The calculation of the density operator evolution up to the 90°(C) pulse in PT sequences is easily performed because the proton spectrum of an $I_n S$ system is a doublet, whatever the value of n. Each of the n protons gives rise to the same amount of polarization transfer to the carbon spin. Therefore the total transfer can be simply calculated by multiplying the transfer from one of them, say I_1, by the factor n.

INEPT was discussed in Section III,C,1, and the applicable transformations in the APT part are

$$-\sin(\pi J\tau_1) 2 I_{1z} S_y \xrightarrow{\sum \pi J\tau_2 2 I_{iz} S_z} \sin(\pi J\tau_1) \sin(\pi J\tau_2) \cos^{n-1}(\pi J\tau_2) S_x \quad [95]$$

This transformation sequence has achieved the desired result of removing the operator I_{1z}, which would lead to antiphase character and is converted to the factor $\sin(\pi J\tau_2)$, and allowing the remaining couplings to stay in-phase, according to the factor $\cos^{n-1}(\pi J\tau_2)$. There results an INEPT intensity expression for the system with n equivalent I spins of the following form:

$$\text{Int}(I_n S) = n \sin(\pi J\tau_1) \sin(\pi J\tau_2) \cos^{n-1}(\pi J\tau_2) \quad [96]$$

The $I_n S$ spin systems with different values of n can in principle be distinguished by their different functional dependences on τ_2, as given in Eq. [94] for APT and in Eq. [96] for INEPT. However, these techniques require accurate knowledge of J and are very sensitive to variations in J.

Experimental modulated spin-echo or APT spectra of β-pinene are shown in Fig. 1-12. In Fig. 1-12b, $\pi J\tau$ was adjusted to the "magic angle" of 54.7°, aimed at distinguishing CH and CH_3 resonances. Ideally, the CH and CH_3 signals should be attenuated by 42% and 81%, respectively, compared to a normal spectrum. This method of distinction is obviously unsafe for molecules with wide ranges of J. Figure 1-12c shows the odd–even distinction, which is reliable over very wide ranges of J, but problems may arise if resonances with odd and even n overlap, as illustrated by the appearance

3. POLARIZATION TRANSFER AND EDITING TECHNIQUES

of C5 and C6 as one negative line in Fig. 1-12c. Figure 1-12d represents an attempt to null the signals from all protonated carbons by setting $\tau = 1/2J$, but we see feedthrough from two CH groups and minor feedthrough in the low-frequency part of the spectrum.

When INEPT is used, the quaternary carbons are always suppressed and the odd/even distinction is possible with $\tau_2 = 3/4J$. A value of τ_2 of $1/2J$ favors the CH groups.

C. SEMUT AND DEPT

With the next level of accuracy, represented by the SEMUT and DEPT pulse sequences, it becomes feasible to generate independent ^{13}C subspectra for molecules with moderate ranges of one-bond carbon–hydrogen couplings.

The pulse sequence for the SEMUT technique (10) is shown in Fig. 3-25. After the creation of states with maximum antiphase character at point 2 (that is, product operators containing n I_z operators for I_nS groups), a proton pulse of variable flip angle θ acts as a multiple quantum trap in the sense that it partially transfers the magnetization to unobservable multiple quantum coherence. The propagator between the points 1 and 4 in Fig. 3-25 is parallel to the last half of the propagator in Eq. [45], aside from the I-spin chemical shift evolution:

$$\sigma_1 \xrightarrow{\pi S_x} \xrightarrow{-\Sigma \pi J \tau 2 I_{iz} S_z} \xrightarrow{\Sigma \theta I_{ix}} \xrightarrow{\Sigma \pi J \tau 2 I_{iz} S_z} \sigma_4 \qquad [97]$$

Since operators for different spins commute, we can indicate the first transformation to $+S_y$ and then carry out the remaining sequence of transformations on each of the nuclei, one by one, followed by summing over all

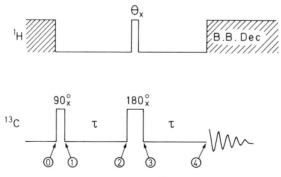

Fig. 3-25. Pulse sequence SEMUT employed for ^{13}C subspectral editing. With the modified order of transformations indicated in Eq. [98], the numbers 0 to 4 refer to corresponding density operators at these points in the sequence as discussed in the text.

the nuclei, leading to a form similar to Eq. [93]:

$$S_y \xrightarrow{\Sigma(-\pi J\tau 2I_{iz}S_z)} \sigma_2 \xrightarrow{\theta I_{ix}} \sigma_3 \xrightarrow{\pi J\tau 2I_{iz}S_z} \sigma_4 \quad [98]$$

The density-operator calculation for an IS spin system using Eq. [98] goes as follows for the various steps, including only observable terms:

$$\sigma_1 = S_y \quad [99]$$

$$\sigma_2 = S_y \cos \pi J\tau + 2I_zS_x \sin \pi J\tau \quad [100]$$

$$\sigma_3 = S_y \cos \pi J\tau + 2I_zS_x \sin \pi J\tau \cos \theta \quad [101]$$

$$\sigma_4 = S_y(\cos^2 \pi J\tau + \sin^2 \pi J\tau \cos \theta) \quad [102]$$

The general intensity expression for SEMUT is thus

$$\text{Int}(I_nS) = (\cos^2 \pi J\tau + \sin^2 \pi J\tau \cos \theta)^n \quad [103]$$

If τ is exactly equal to $1/2J$, so that $\cos^2 \pi J\tau = 0$ and $\sin^2 \pi J\tau = 1$, this becomes

$$I = \cos^n \theta$$

The propagator for the DEPT experiment (Fig. 3-12d) is given by

$$\sigma_0 \xrightarrow{\Sigma \frac{\pi}{2}I_{ix}} \sigma_1 \xrightarrow{-\Sigma \pi J\tau 2I_{iz}S_z} \sigma_2 \xrightarrow{\frac{\pi}{2}S_x}$$

$$\sigma_3 \xrightarrow{-\Sigma \pi J\tau 2I_{iz}S_z} \sigma_4 \xrightarrow{\Sigma \theta I_{iy}} \sigma_5 \xrightarrow{\Sigma \pi J\tau 2I_{iz}S_z} \sigma_6 \quad [104]$$

From this sequence of steps, the stages of the magnetization from I_1 in an I_nS system can be calculated:

$$\sigma_0 = I_{1z} \quad [105]$$

$$\sigma_1 = -I_{1y} \quad [106]$$

$$\sigma_2 = -2I_{1x}S_z \sin \pi J\tau \quad [107]$$

$$\sigma_3 = 2I_{1x}S_y \sin \pi J\tau \quad [108]$$

The remaining part of the sequence is identical to SEMUT, and apart from a phase shift of the θ pulse, the propagator is given in Eq. [98].

It is convenient to distinguish I_1 and the $n-1$ passive protons. For I_1, the following transformation applies:

$$\sigma_3 = 2I_{1x}S_y \sin \pi J\tau \rightarrow \sigma_6' = S_x \sin^2 \pi J\tau \sin \theta \quad [109]$$

The operator σ_6' is proportional to the density operator at point 1 of SEMUT, except for an unimportant 90° phase shift, and the $n-1$ passive I spins give

3. POLARIZATION TRANSFER AND EDITING TECHNIQUES

a modulation equivalent to Eq. [103], leading to the general DEPT response:

$$\text{Int}(I_n S) = n \sin^2(\pi J\tau) \sin \theta [\cos^2(\pi J\tau) + \sin^2(\pi J\tau) \cos \theta]^{(n-1)} \quad [110]$$

For ideal conditions, with $\tau = 1/2J$, Eqs. [103] and [110] simplify to

$$\text{SEMUT:} \qquad \text{Int}(I_n S) = \cos^n \theta \qquad [111]$$

$$\text{DEPT:} \qquad \text{Int}(I_n S) = n \sin \theta \cos^{n-1} \theta \qquad [112]$$

Recording four ^{13}C SEMUT spectra with different settings of θ (0°, 60°, 120°, and 180°), followed by taking appropriate linear combinations of these four experiments, yields four subspectra, one for each multiplicity ($n = 0$, 1, 2, 3 in CH_n). With DEPT, the quaternary carbons are suppressed, but experiments for three different θ values (such as 38°, 90°, and 142°) yield edited CH, CH_2, and CH_3 subspectra. These three subspectra may be supplemented with a SEMUT experiment at $\theta = 90°$ to give a quaternary carbon subspectrum.

One source of cross talk between subspectra is imperfections of pulses. However, a more important and general source is variation of $^1J(CH)$ values in the sample, which is the subject of the next section. Figure 3-26 illustrates SEMUT editing of ^{13}C spectra for a limited range of values of J (10).

D. MAGNITUDE OF J CROSS TALK IN EDITED SUBSPECTRA

Expansion of the right-hand side of Eq. [103] or [110] shows that, if $\cos \pi J\tau$ is not exactly zero, there are terms in the response of an $I_n S$ group proportional to $\cos^m \theta$ for SEMUT and $\sin \theta \cos^{m-1} \theta$ for DEPT, where m is less than n. These terms will lead to responses which appear in the subspectrum for the $I_m S$ group. However, since m is less than n, J cross talk can only occur in the downward direction. For example, methyl groups will give rise to extraneous peaks in a methylene spectrum, but not vice versa. The intensity expressions for cross talk from $I_n S$ to $I_m S$ are obtained by collecting terms proportional to $\cos^m \theta$ and $\sin \theta \cos^{m-1} \theta$ for SEMUT and DEPT, respectively:

$$\text{SEMUT:} \qquad \binom{n}{m}(\sin^{2m} \pi J\tau)(\cos^{2(n-m)} \pi J\tau) \cos^m \theta \qquad [113]$$

$$\text{DEPT:} \qquad n\binom{n-1}{m-1}(\sin^{2m} \pi J\tau)(\cos^{2(n-m)} \pi J\tau) \sin \theta \cos^{m-1} \theta \qquad [114]$$

with the binomial coefficients

$$\binom{n}{m} = \frac{n!}{(n-m)!\,m!}$$

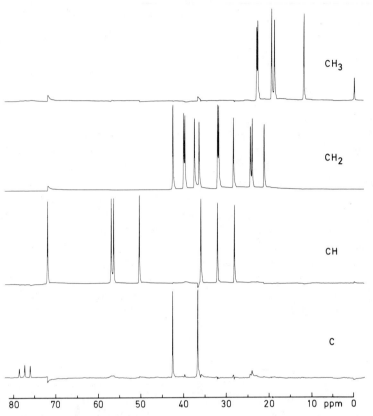

Fig. 3-26. Edited ^{13}C SEMUT spectra for the aliphatic region of cholesterol ($\tau = 3.9$ msec) recorded on a Varian XL-100 spectrometer at 25 MHz. The values of θ employed were 0°, 60°, 120°, and 180° (10).

The ratio of cross talk from I_nS in the I_mS subspectrum to the ideal I_mS response, calculated by comparing Eqs. [113] and [111] and Eqs. [114] and [112], turns out to be the same for SEMUT as for DEPT:

$$\frac{\text{Int}(I_nS \to I_mS)}{\text{Int}(I_mS)} = \binom{n}{m}(\sin^{2m} \pi J\tau)(\cos^{2(n-m)} \pi J\tau) \qquad [115]$$

Thus the two techniques have the same editing performance with respect to J cross talk. Based on this expression, the magnitudes of J cross talk have been calculated for the variation in J of up to 25% and are plotted in Fig. 3-27. It is apparent that the cross talk is most severe when m is one less than n.

The presence of J cross talk is especially critical in editing spectra of mixtures of compounds present in different concentrations. For example,

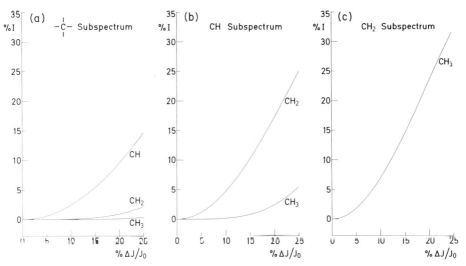

Fig. 3-27. Magnitudes of J cross talk as given by Eq. [115] for edited C, CH, and CH_2 subspectra obtained via SEMUT or DEPT (CH and CH_2 subspectra only), as a function of the percentage deviation of J (10). (a) CH_3, CH_2, and CH cross talk into the quaternary carbon subspectrum; (b) CH_3 and CH_2 cross talk into the CH subspectrum; (c) CH_3 cross talk into the CH_2 subspectrum.

a small signal in the CH_2 subspectrum at the frequency of a CH_3 resonance could be either a genuine CH_2 signal or an artifact from the CH_3 resonance. Figure 3-27 shows that editing by DEPT or SEMUT can only be carried out with acceptable accuracy for a variation of J within $\pm 10\%$. Similar calculations have been carried out for the amount of J cross talk resulting in linear combinations of carbon-13 APT experiments carried out with settings of τ corresponding to $\pi J_0 \tau = 0°, 60°, 120°,$ and $180°$ (44). The results, plotted in Fig. 3-28, show that APT is extremely susceptible to cross talk and that the magnitude of cross talk is not symmetrical in the variation of the coupling constant from the ideal value J_0 used in calculating the delays.

The performance of SEMUT and DEPT can be improved by setting the two time intervals τ somewhat unequal. The effect on Eq. [115] is to introduce separate sine and cosine factors for each of the two intervals, which we shall now designate τ_1 and τ_3:

$$\frac{\text{Int}(I_n S \to I_m S)}{\text{Int}(I_m S)}$$

$$= \binom{n}{m}(\sin^m \pi J \tau_1)(\sin^m \pi J \tau_3)(\cos^{(n-m)} \pi J \tau_1)(\cos^{(n-m)} \pi J \tau_3) \quad [116]$$

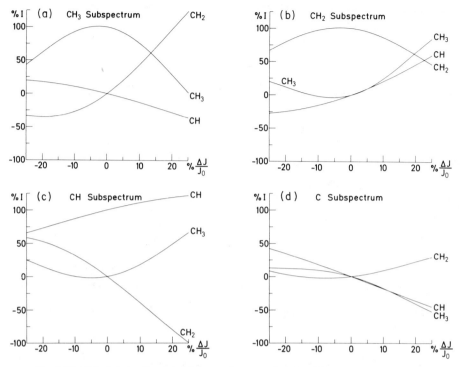

Fig. 3-28. Calculated ratio of the J cross talk to genuine signal intensities for edited CH_3, CH_2, and CH subspectra obtained via the APT experiment, as a function of the percentage deviation of J from the ideal coupling constant J_0 used in setting the τ delay (44). (a) CH_2 and CH cross talk into the CH_3 subspectrum; (b) CH_3 and CH cross talk into the CH_2 subspectrum; (c) CH_3 and CH_2 cross talk into the CH subspectrum; (d) CH_3, CH_2, and CH cross talk into the quaternary carbon subspectrum.

The product $(\cos \pi J\tau_1)(\cos \pi J\tau_3)$ can be made "small" over a larger range of J than $\cos^2 \pi J\tau$.

E. SEMUT GL AND DEPT GL

To find a means of further suppressing J cross talk in SEMUT and DEPT spectral editing, we investigate the density operator just prior to the editing step, the θ pulse. For ideal timing, the SEMUT density operator for an I_nS group consists of just one product which contains n I_z operators, such as $8I_{1z}I_{2z}I_{3z}S_x$ for an I_3S spin system. However, for nonideal timing, the density operator also contains products with fewer than n I_z operators, terms which cause J cross talk.

In the GL versions of SEMUT and DEPT, shown in Fig. 3-29, cross talk is suppressed by insertion, just prior to the θ pulse, of a sequence of pulses

3. POLARIZATION TRANSFER AND EDITING TECHNIQUES   215

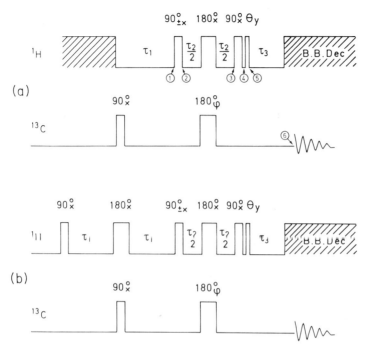

Fig. 3-29. Pulse sequences (a) SEMUT GL and (b) DEPT GL for accurate editing of ^{13}C NMR spectra. For $\tau_1 > \tau_3$, both sequences require delayed acquisition. To allow for a faster repetition rate in SEMUT GL, the first ^{13}C pulse length may be set larger than 90°.

corresponding to a propagator which commutes with the $n\text{-}I_z$-product-operator terms but not with the other terms. This sequence, called a *purging sandwich*, is

$$\text{I} \quad 90°_x - \frac{\tau_2}{2} - 180°_x - \frac{\tau_2}{2} - 90°_x$$

$$\text{S} \qquad\qquad\qquad 180°_x$$

The value of τ_2 is selected for the requirements of a particular experiment, as discussed below. The propagator corresponding to the purging sandwich is

$$\sigma_1 \xrightarrow{\pi S_x} \xrightarrow{-\sum_{i=1}^n \frac{\pi}{2}I_{ix}} \xrightarrow{\sum_{i=1}^n \pi J\tau_2 2 I_{iz}S_z} \xrightarrow{\sum_{i=1}^n \frac{\pi}{2}I_{ix}} \sigma_4 \qquad [117]$$

where the indices of the density operator refer to points in the sequence in Fig. 3-29a. The 180°(S) pulse merely refocuses S-spin chemical shifts during the entire pulse sequence. The remaining terms of the propagator can be condensed to

$$\sigma_1 \xrightarrow{-\sum_{i=1}^{n} \pi J \tau_2 2 I_{iy} S_z} \sigma_4 \quad [118]$$

The components transformed by the purging propagator represent heteronuclear multiple-quantum coherences which remain unobserved if an editing pulse θ_y is applied. For compensation of possible pulse imperfections, it is useful to alternate the phase of the two 90° pulses in the purging sandwich. This changes the sign of the rotation angle of the propagator in Eq. [118], but this is unimportant for the editing procedure, because the desired terms of σ_1 commute with the purging sandwich.

The features of the purging sandwich may also be explained in the following qualitative terms. Consider the density operator for an $I_n S$ system at point 1 in Fig. 3-29a. For ideal timing it contains only the n-I_z product operators. The $90°_x(I)$ pulse takes all the I_z operators to the y axis, representing a state of heteronuclear multiple quantum coherence. The evolution of this state of $(n+1)$-spin coherence is not influenced by the couplings between the spins involved. The 180° refocusing pulses in the middle of the sandwich ensure that the state is "frozen" independent of the length of τ_2. The second $90°_x(I)$ pulse brings the I_y operators back to the z axis and the state at point 1 is reestablished at point 4.

If τ_1 is not exactly $1/2J$, the density operator at point 1 also contains products with less than n I_z operators. From the term with operators having $n-1$ I_z components, which is responsible for cross talk to the $I_{n-1}S$ subspectrum, the $90°_x$ proton pulse generates only n-spin coherence, which can be considered a doublet in the n-spin coherence space and thus evolves under the influence of J during the τ_2 delay. The in-phase part at point 3 of the sequence is proportional to $\cos \pi J \tau_2$. This product operator leads to J cross talk, which, however, compared to SEMUT and DEPT, is attenuated by the additional factor $\cos \pi J \tau_2$. The missing I_z operator is introduced, modulated by $\sin \pi J \tau_2$, at point 3. However, this operator is trapped as $-I_y$ at point 4 and thus contributes only to unobservable heteronuclear MQC.

The intensity expressions for the GL versions of SEMUT and DEPT are

SEMUT: Int(I_nS)

$= [(\cos \pi J \tau_1)(\cos \pi J \tau_2)(\cos \pi J \tau_3) \pm (\sin \pi J \tau_1)(\sin \pi J \tau_3) \cos \theta]^n$ [119]

DEPT: Int(I_nS)

$= [(\cos \pi J \tau_1)(\cos \pi J \tau_2)(\cos \pi J \tau_3) \pm (\sin \pi J \tau_1)(\sin \pi J \tau_3) \cos \theta]^{(n-1)}$

$\times n(\sin \pi J \tau_1)(\sin \pi J \tau_3) \sin \theta$ [120]

In these equations the plus sign applies when the phase of the first 90° pulse in the purging sandwich is $+x$ and the minus sign applies when the phase is $-x$. The equivalent of Eq. [115] for cross talk in the GL versions of SEMUT and DEPT is

$$\frac{\text{Int}(I_n S \to I_m S)}{\text{Int}(I_m S)}$$

$$= \binom{n}{m}[(\sin \pi J\tau_1)(\sin \pi J\tau_3)]^m[(\cos \pi J\tau_1)(\cos \pi J\tau_2)(\cos \pi J\tau_3)]^{(n-m)}$$

[121]

Suppression of cross talk over wide J ranges corresponds to minimizing the product of three cosine factors in Eq. [121]. A numerical analysis, carried out in connection with low-pass J filtering (45), showed that, with the set of τ values equal to 3.79, 2.87, and 2.30 msec, the product of the three cosine factors has a maximum value of 2.2% within the range 125–225 Hz for J. From the functions in Eq. [121] for the C, CH, and CH$_2$ subspectra using these values, the maximum residual amplitudes for cross talk are less than about 4% over this range. For unknown compounds, these values should be appropriate. However, with a J range known to be smaller, other settings yield better suppression. A convenient way of selecting τ_1, τ_2, and τ_3 to cover a desired range from $J_{\min}$ to $J_{\max}$ with nearly optimum suppression is to set one of the values equal to $1/2J$ for each of the following values of J: $J_{\min} + 0.07(J_{\max} - J_{\min})$, $0.5(J_{\max} + J_{\min})$, $J_{\max} - 0.07(J_{\max} + J_{\min})$. For the range of J from 125 to 190 Hz, the three resulting τ values are 3.86, 3.17, and 2.70 msec; the corresponding J cross-talk functions are plotted in Fig. 3-30. There is considerable improvement compared to the corresponding plots in Fig. 3-27 for the SEMUT and DEPT sequences.

The procedure outlined for suppression of cross talk results in some attenuation of the intensities of the genuine signals. Furthermore, the decrease in intensity is unevenly distributed over the range of J, thus impeding quantitative measurements. One way to compensate partially for this effect is to coadd the results of a series of experiments in which each of the three τ values in turn occupies the position of τ_2. This scrambling of τ values leads to better relative signal intensities and almost unchanged cross-talk suppression. The graphs of genuine-signal and cross-talk intensities in Fig. 3-30 were obtained by the scrambling procedure. Since Eqs. [119] and [120] are symmetrical in τ_1 and τ_3, τ_1 can always be set larger than τ_3.

Figure 3-31 shows edited SEMUT GL spectra for a mixture of brucine and 2-bromothiazole, representing a spread in $^1J_{CH}$ values from about 125 to 192 Hz, recorded at a carbon-13 frequency of 75.43 MHz. Very little cross talk is observed.

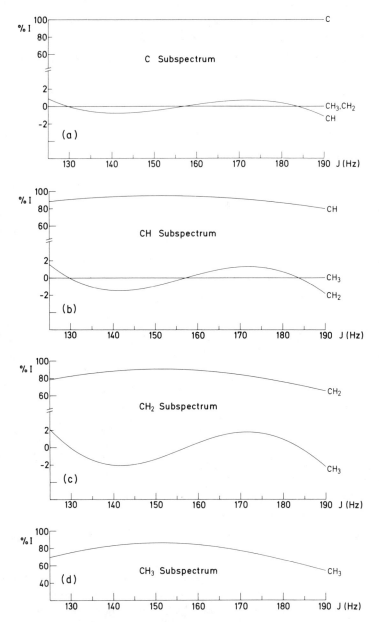

Fig. 3-30. Cross talk and genuine signal intensities for subspectra obtained via SEMUT GL or DEPT GL. The J cross talk has been minimized for the range $125 < J < 190$ Hz, using the τ values 3.86, 3.17, and 2.70 msec. (a) CH_3, CH_2, and CH cross talk into the quaternary carbon subspectrum; (b) CH_3 and CH_2 cross talk into the CH subspectrum; (c) CH_3 cross talk into the CH_2 subspectrum; (d) genuine peak signal intensity of the CH_3 subspectrum. The ranges of coupling for the genuine signal intensities in the CH_3 and CH_2 subspectra are $120 < J < 145$ Hz and $120 < J < 165$ Hz, respectively (9).

3. POLARIZATION TRANSFER AND EDITING TECHNIQUES

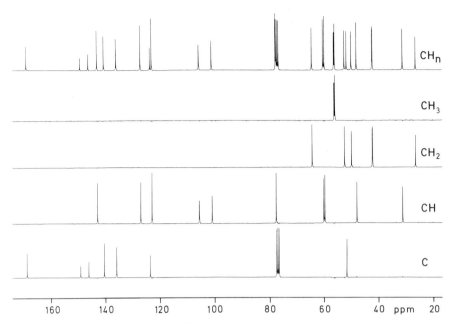

Fig. 3-31. Edited SEMUT GL ^{13}C NMR spectra for a mixture of brucine and 2-bromothiazole (~1:1) in CDCl$_3$ obtained using $\tau_1 = 3.86$, $\tau_2 = 2.70$, and $\tau_3 = 3.17$ msec. From top to bottom: CH$_n$ (all carbons), CH$_3$, CH$_2$, CH, and C spectra. The experiment was performed on a Varian XL-300 spectrometer using an automated least-squares analysis for the spectral editing (9).

The distinction between odd and even n for I$_n$S spin systems is made by alternating, in Eq. [103], [110], [119], or [120], the sign of the term containing cos θ. For SEMUT or DEPT, this requires pairs of experiments having angles θ and $\pi - \theta$. In the GL versions, the angle need not be larger than 90°, because the sign alternation can be achieved by a phase alternation of one of the 90° pulses in the purging sandwich. Consequently, editing via the GL versions is less sensitive to rf inhomogeneity.

A refinement of the two GL pulse sequences in Fig. 3-29 may be made by merging the second 90$^\circ_x$ pulse with the immediately following θ_y pulse. Applying the procedures in the appendix, one may represent the result of this combination as follows:

$$\exp\left(i\frac{\pi}{2}I_x\right)\exp(i\theta I_y) = \exp\left(i\frac{\pi}{2}I_x\right)\exp\left(-i\frac{\pi}{2}I_x\right)\exp(-i\theta I_z)\exp\left(i\frac{\pi}{2}I_x\right)$$

$$= \exp(-i\theta I_z)\exp\left(i\frac{\pi}{2}I_x\right) \qquad [122]$$

The effect of the resulting θ_{-z} pulse is to phase shift all preceding pulses by $-\theta$, which is equivalent to phase shifting the following $90°_x$ pulse by θ and leaving the phases of the preceding pulses unchanged. Thus $90°_x\theta_y$ is equivalent to $90°_\theta$, and the refined sequences, shown in Fig. 3-32, also obey Eqs. [119] and [120]. Whereas θ values larger than 90° are undesirable because of rf inhomogeneity, extended phase cycling using phase shifts of the angle θ larger than 90° performs the same function in the sequences shown in Fig. 3-32.

F. Practical Aspects of Subspectral Editing

In this section, we discuss the choice of flip angles for the editing experiments, the procedures for constructing the subspectra by linear combinations of the experiments, and the use of phase cycling.

For distinguishing between I_nS spin systems with n odd and with n even, SEMUT or DEPT requires experiments with symmetrical pairs of flip angles, θ and $180° - \theta$, and the GL versions require pairs of experiments with $180°$

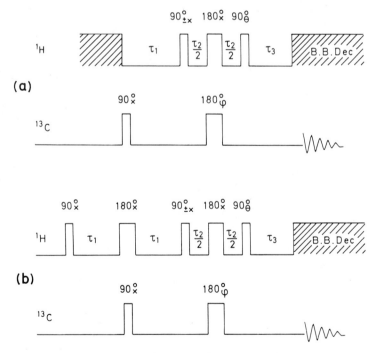

Fig. 3-32. Refined versions of the (a) SEMUT GL and (b) DEPT GL pulse sequences. As shown in the text, two consecutive pulses $90°_x\theta_y$ have the same effect as a single 90° pulse of phase θ. These versions are less sensitive to pulse imperfections than are those of Fig. 3-29.

difference in phase of one of the 90° pulses in the purging sandwich. When two experiments in such a pair are added or subtracted, we obtain a spectrum for the spin systems with n even or odd, respectively. Such partial editing may in many cases be sufficient. For $\theta = 0°$, or 180° in SEMUT GL, the sensitivity is the same as that of a normal experiment provided that $\tau_i = 1/2J$ and that the effect of long-range coupling and relaxation during the sequence can be neglected. When, in the following, we state that the sensitivities of two experiments do not significantly differ, we understand this to be valid within the same limitation.

As an example of partial editing, Fig. 3-33 shows combinations of the $\theta = 0$ and $\theta = 180°$ SEMUT carbon-13 spectra of cholesterol (10). The equivalent separation using DEPT is achieved by combining, for example, spectra with $\theta = 55°$ and 125°.

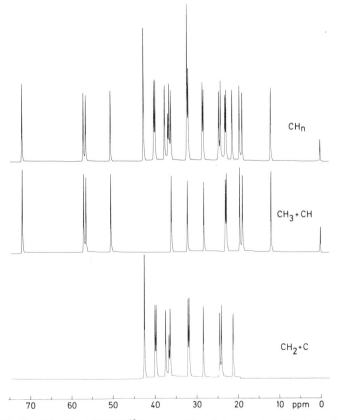

Fig. 3-33. Partially edited SEMUT ^{13}C spectra of the aliphatic region of cholesterol obtained on a Varian XL-100 spectrometer. Upper trace, all carbons; middle trace, $CH_3 + CH$ subspectrum; lower trace, $CH_2 + C$ subspectrum.

To separate CH and CH_3 or C and CH_2 resonances, it is necessary for the SEMUT experiments to perform another pair of θ and $180° - \theta$ experiments with a different setting of θ. These two experiments are combined to yield two new odd and even subspectra, which must then be combined with the partially edited $\theta = 0°$ subspectra to yield full separation into C, CH, CH_2, and CH_3 subspectra.

Two questions arise: "What is the optimum choice of the angle θ and of the relative numbers of scans for the subexperiments?" (46). If we always record symmetrical pairs of subexperiments as described above, only θ values between 0° and 90° need be considered. Thus for either SEMUT or DEPT or their GL versions, two different values of θ suffice. An angle θ of 0° is good for SEMUT because all multiplicities appear with full intensity, and $\theta = 90°$ is appropriate for DEPT because it automatically gives the CH subspectrum. We then must determine the optima for the second angles, θ_S for SEMUT and θ_D for DEPT, and the numbers of scans η, relative to the number for the 0° or 90° subexperiments. The total number of scans is normalized to unity, corresponding to $1/(1+\eta)$ scans for $\theta = 0°$ or 90° and $\eta/(1+\eta)$ scans for $\theta = \theta_S$ or θ_D. In practice, to obtain an integral number of scans, multiples of these numbers are recorded. For simplicity, we assume that τ_i is exactly $1/2J$, so that the intensities are given by Eqs. [111] or [112].

For SEMUT, the editing task is reduced to separating CH_n and CH_{n+2} resonances. The noise and signal levels in a single-scan experiment are arbitrarily set to unity. The editing procedures and the signal/noise ratios are given in Table 3-6. The maximum signal to noise is obtained for values of η of $1/(\cos^{n+2}\theta)$ for the CH_n subspectrum and $1/(\cos^n \theta)$ for the CH_{n+2} subspectrum. The optimum flip angle θ_S is given, for both CH_n and CH_{n+2}, by the equation

$$2\cos^2\theta_S(1+\cos^n\theta_S) - n\sin^2\theta_S = 0 \quad [123]$$

From Table 3-6, we see that the optimum S/N in general decreases as the number of attached protons increases. The optimum SEMUT parameters correspond to those giving the best S/N in the subspectrum for CH_3, the multiplicity which yields the least favorable results. Thus the best SEMUT or SEMUT GL performance is obtained by recording twice as many scans for $\theta = 60°$ as for $\theta = 0°$, with an average of 63% loss in sensitivity compared to a standard experiment.

Because of the elimination of quaternary carbons, the DEPT editing procedure is simpler than that for SEMUT. The S/N considerations are summarized in Table 3-7. The sensitivity improvement may be compared to that given by the parameter suggested by Doddrell et al. (8, 47). The individual subspectra can be optimized or a smooth distribution of S/N among the spectra can be pursued. If equal S/N is required in all subspectra,

TABLE 3-6

SIGNAL-TO-NOISE (S/N) RATIOS FOR SEMUT GL EDITING OF CARBON SPECTRA OF CH_n GROUPS

Group	C	CH	CH_2	CH_3
Formula for subspectrum generation[a]	$[\theta = \theta_S] - \eta c^2[\theta = 0]$	$[\theta = \theta_S] - \eta c^3[\theta = 0]$	$[\theta = 0] - [\theta = \theta_S]$	$\eta c[\theta = 0] - [\theta = \theta_S]$
Signal intensity	$\dfrac{\eta s^2}{1+\eta}$	$\dfrac{\eta s^2 c}{1+\eta}$	$\dfrac{\eta s^2}{1+\eta}$	$\dfrac{\eta s^2 c}{1+\eta}$
Noise level	$\left(\dfrac{\eta(1+\eta c^4)}{1+\eta}\right)^{1/2}$	$\left(\dfrac{\eta(1+\eta c^6)}{1+\eta}\right)^{1/2}$	$\left(\dfrac{\eta(1+\eta)}{1+\eta}\right)^{1/2}$	$\left(\dfrac{\eta(1+\eta c^2)}{1+\eta}\right)^{1/2}$
S/N	$s^2\left(\dfrac{\eta}{(1+\eta)(1+\eta c^4)}\right)^{1/2}$	$s^2 c\left(\dfrac{\eta}{(1+\eta)(1+\eta c^6)}\right)^{1/2}$	$s^2\left(\dfrac{\eta}{(1+\eta)^2}\right)^{1/2}$	$s^2 c\left(\dfrac{\eta}{(1+\eta)(1+\eta c^2)}\right)^{1/2}$
η for optimum S/N	c^{-2}	c^{-3}	1	c^{-1}
S/N for this value of η	$\dfrac{s^2}{1+c^2}$	$\dfrac{s^2 c}{1+c^3}$	$\dfrac{s^2}{2}$	$\dfrac{s^2 c}{1+c}$
S/N for optimized subspectrum				
C ($\theta_S = 90°$, $\eta = \infty$)	1	0	0	0
CH ($\theta_S = 60°$, $\eta = 8$)	0.58	0.33	0.24	0.20
CH_2 ($\theta_S = 90°$, $\eta = 1$)	0.71	—	0.50	—
CH_3 ($\theta_S = 60°$, $\eta = 2$)	0.58	0.30	0.35	0.25

[a] The square brackets indicate spectra taken with the indicated value of θ. The quantity η is the ratio of the number of scans recorded with $\theta = \theta_S$ to the number recorded with $\theta = 0$. The numbers of scans for the $\theta = \theta_S$ and $\theta = 0$ experiments are then $\eta/(1+\eta)$ and $1/(1+\eta)$, respectively. The coefficients in front of the square brackets represent the weighting factors to be used in combining these spectra to obtain a subspectrum. The S/N is expressed relative to a standard experiment with unit S/N for each carbon atom. Abbreviations: $s = \sin\theta_S$; $c = \cos\theta_S$.

the results are $\eta = 81/(4+20\sqrt{13})$ and $\sin^2 \theta_D = (7-\sqrt{13})/9$. Corresponding practical numbers are eight experiments with $\theta = 37.90°$, eight with $\theta = 37.90°$ with phase inversion of the second proton 90° pulse for DEPT GL or eight with $\theta = 180° - 37.90° = 142.10°$ for normal DEPT, and 17 experiments with $\theta = 90°$. Almost identical results are obtained with $\theta = 38.33°$ and a 1:1:2 ratio of numbers of experiments.

Since the optimum mixing coefficients in the editing procedures may diverge from theoretical values, computer programs which calculate the best subspectra by a least squares procedure are usually employed (9).

Inspection of Tables 3-6 and 3-7 shows that, in SEMUT GL as well as in DEPT GL, the optimum parameters for the CH_2 subspectrum correspond to selecting the proton double-quantum coherence present in the τ_2 period

TABLE 3-7

SIGNAL-TO-NOISE RATIOS FOR DEPT GL EDITING OF CARBON SPECTRA OF CH_n GROUPS

Group	CH	CH_2	CH_3
Formula for subspectrum generation[a]	$[\theta = 90°]$	$[\theta = \theta_D]$	$[\theta = \theta_D] - \eta s[\theta = 90°]$
Signal intensity	$\dfrac{1}{1+\eta}$	$\dfrac{\eta 2sc}{1+\eta}$	$\dfrac{\eta 3sc^2}{1+\eta}$
Noise level	$\left(\dfrac{1}{1+\eta}\right)^{1/2}$	$\left(\dfrac{\eta}{1+\eta}\right)^{1/2}$	$\left(\dfrac{\eta(1+\eta s^2)}{1+\eta}\right)^{1/2}$
S/N	$\left(\dfrac{1}{1+\eta}\right)^{1/2}$	$2sc\left(\dfrac{\eta}{1+\eta}\right)^{1/2}$	$3sc^2\left(\dfrac{\eta}{(1+\eta)(1+\eta s^2)}\right)^{1/2}$
η for optimum S/N	0	∞	s^{-1}
S/N for this value of η	1	$2sc$	$3s^2/(1+s)$
S/N for one experiment with $\theta = 54.74°$ (no editing)	0.82	0.94	0.82
S/N for optimized subspectrum CH ($\eta = 0$)	1	0	0
CH_2 ($\eta = \infty$, $\theta_D = 45°$)	—	1	—
CH_3 ($\eta = 2$, $\theta_D = 30°$)	0.58	0.71	0.75
S/N for equal intensity ($\eta = 17/16$, $\theta_D = 37.90°$)	0.70	0.70	0.70

[a] Where η is the ratio of the number of scans with $\theta = \theta_D$ to the number with $\theta = 90°$; the numbers of scans are $\eta/(1+\eta)$ and $1/(1+\eta)$, respectively. Abbreviations: $s = \sin \theta_D$; $c = \cos \theta_D$.

and those for the CH_3 subspectra correspond to filtering out the proton triple-quantum coherence existing during τ_2. The optimum quaternary carbon and CH parameters can be considered weighted combinations of different multiple-quantum-coherence phase cycles. A general statement which can be made about all the optimum procedures is that the best S/N for a given subspectrum is obtained by changing the number of scans to fit a scaling factor of unity, rather than scaling the spectrum by a certain factor prior to linear combination with another spectrum.

An important feature of DEPT editing is that, on the average, only 19% sensitivity is lost, compared to the standard nonedited polarization transfer spectrum with $\theta = 54.74°$. A quaternary subspectrum can be generated as a supplement to the three obtained with DEPT by using a special case of SEMUT GL corresponding to a phase $\theta = y$ in Fig. 3-32a. One scan of this experiment with a total of $1 + \eta$ DEPT or DEPT GL spectra, provided the NOE and PT enhancements are equal, should yield an S/N of unity in all subspectra. A number $2 + \eta$ SEMUT or SEMUT GL experiments would give S/N values of 1.01, 0.53, 0.62, and 0.44 for the C, CH, CH_2, and CH_3 subspectra, respectively.

In comparing NOE and PT enhancements, one must keep in mind that, while ideally NOE yields a factor 3 and PT a factor 4, when DEPT GL is used for a system of one carbon with n directly bonded as well as $N - n$ passive protons, the PT enhancement for genuine signals is scaled down by a factor of

$$\prod_{i=n+1}^{N} (\cos 2\pi J_{1i}\tau_1)(\cos^n \pi J_{1i}\tau_2)$$

because of homonuclear couplings. Heteronuclear long-range couplings also attenuate the PT enhancement in a similar manner, but this is also true for SEMUT GL.

Because of the greater number of pulses and the greater length of time for DEPT or DEPT GL compared to the corresponding experiments with SEMUT or SEMUT GL, we believe that equality of NOE and PT enhancements may reasonably be expected for many ^{13}C spectra. A point in favor of the NOE-based scheme in ^{13}C NMR is that, when proton decoupling is applied during acquisition, polarization transfer experiments require a relaxation delay to reestablish proton magnetization after the decoupler is switched off. Relying on NOE, one obtains the optimum S/N within a given time without a relaxation delay (48, 49), by adjusting the flip angle of the first carbon pulse of SEMUT. The value for this angle can be determined from Table 1 of Waugh (49); α from this table should be replaced by $180° - \alpha$ because of the refocusing pulse in SEMUT or SEMUT GL. By this scheme, the sensitivity of SEMUT or SEMUT GL is improved. However, only when long acquisition times are used will the performance be comparable to that of DEPT or DEPT GL. For other applications than editing of ^{13}C spectra, such as ^{15}N editing, polarization transfer can be substantially superior to NOE.

For spectrometers capable of shifting the phase of the rf in steps of less than 90°, it is recommended that ^{13}C spectra be edited via the pulse sequences in Fig. 3-32. The awkward optimum phase shift for DEPT is probably not feasible on most spectrometers, but CH, CH$_2$, and CH$_3$ S/N ratios within 0.03 of the optimum value of 0.70 can be obtained by combining two sets of experiments, one with $\eta = 2$, $\theta_D = 30°$, and one with $\eta = 3/4$, $\theta_D = 45°$.

Imperfect pulses interfere with editing, so that phase cycling schemes are necessary to eliminate artifacts resulting from this source. The carbon 180° pulse should be Exorcycled, and, if there are severe off-resonance effects, replaced by a composite pulse. This four-step cycle is sufficient for SEMUT, but for DEPT, independent phase alternation of the proton 90° pulse is also recommended to suppress the native ^{13}C magnetization.

The purging sandwiches in SEMUT GL and DEPT GL involve manipulation of high-order multiple-quantum coherence, so independent phase cycling of these pulses is critical. Table 3-8 gives the basic four-step phase cycle. Together with Exorcycling of the 180°(S) pulse, this results in a minimum of 16 steps for SEMUT GL and 32 steps for DEPT GL. When sensitivity considerations require accumulation of more scans, Table 3-8 should be extended by independent phase alternation of the 180° and the θ pulses. For DEPT GL, the first proton 180° pulse can also be Exorcycled. If the versions of Fig. 3-32 are applied, Table 3-8 can easily be modified by noting the following equivalences, which are similar to those represented in Eq. [122]:

$$90°_x \theta_{\pm y} = 90°_{\pm \theta} \qquad [124]$$

$$90°_{-x} \theta_{\pm y} = 90°_{\pi \mp \theta} \qquad [125]$$

TABLE 3-8

Basic Four-Step Phase Cycle Applied for the ^{1}H Pulses in the Purging Sandwiches of SEMUT GL and DEPT GL

$\pi/2$	π	$\pi/2$	θ_{SEMUT}	θ_{DEPT}	FID
x	x	x	y	y	+
x	y	$-x$	y	$-y$	+
$-x$	x	$-x$	y	y	+
$-x$	y	x	y	$-y$	+

G. EDITING OF COUPLED SPECTRA

In this section we evaluate the application of spectral editing methods for proton-coupled carbon-13 spectra, where the purpose is not to determine proton multiplicities but rather to obtain spectrum simplification. When a carbon-13 spectrum is acquired without proton decoupling, there can be an enormous increase in the number of resonances, so that interpretation of the crowded spectrum is difficult. Separation into subspectra is therefore useful, for example, for the measurement of heteronuclear coupling constants.

The usual pulse sequences for editing decoupled spectra cannot be directly adapted to coupled spectra by simply omitting proton decoupling, because the decoupling is responsible for removing product operators which seriously influence the editing accuracy and introduce phase distortion. Therefore the pulse sequences must be extended to include purging techniques such as those described in Sections III,B,1 and III,C,2.

APT and INEPT are suitable for the recording of coupled spectra if a 90° proton pulse is applied at the start of acquisition to give INEPT$^+$ (Fig. 3-12c) or APT$^+$. The same modifications which were described for DEPT$^+$ and DEPT^{++} in Section III,C,2, can be applied to give SEMUT$^+$ and SEMUT^{++}, which each yield coupled spectra free of phase distortions.

In SEMUT GL and DEPT GL, the heteronuclear multiple-quantum coherence present after the θ pulse is "stored" along both orthogonal directions in the rotating frame (for example, $4I_{1x}I_{2x}S_x$ and $4I_{1y}I_{2y}S_x$), and a 90° purging pulse would inevitably convert part of this back to antiphase magnetization. Therefore, only the purging scheme as shown in Fig. 3-34 with a 180° pulse in every second experiment is applicable for the GL pulse sequences (13). Since the GL procedure introduces an additional cosine factor (cos $\pi J \tau_2$) into all anomaly terms of the density operator, the "GL$^+$" scheme results in good suppression of the multiplet distortions in addition to elimination of phase distortions and excellent suppression of J cross talk. For practical applications, the multiplet distortions are unimportant because they do not disturb the measurements of coupling constants, but "clean" edited coupled ^{13}C spectra with spin multiplets without phase distortions are obtained from the GL$^+$ scheme.

Because the proton 180° purging pulse must be applied at the very start of acquisition, the GL$^+$ sequences employ "delayed coupling" ($\tau_1 < \tau_3$) instead of delayed acquisition ($\tau_1 > \tau_3$), as used in the sequences for editing decoupled spectra shown in Figs. 3-29 and 3-32. From the S/N analysis for SEMUT GL and DEPT GL subspectral editing, it may be concluded that the best sensitivity for editing coupled spectra is achieved by using DEPT GL$^+$ to edit CH, CH$_2$, and CH$_3$ resonances and SEMUT GL$^+$ to obtain (a)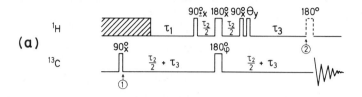

(b)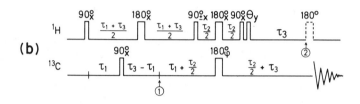

(c)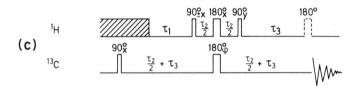

Fig. 3-34. Pulse sequences (a) SEMUT GL$^+$, (b) DEPT GL$^+$, and (c) SEMUT GL$^+$(q), for editing proton-coupled ^{13}C spectra. The phase cycling scheme for the purging sandwich and θ proton pulses is the same as for SEMUT GL and DEPT GL. Two series of experiments are coadded, one including a 180° pulse prior to acquisition and one without this pulse.

the quaternary carbon subspectrum. For the latter sequence, the two consecutive proton pulses, $90°_x 90°_y$, should be replaced by a $90°_y$ pulse, as shown in Fig. 3-32. An explicit analysis of coupled SEMUT GL and DEPT GL is given in the Appendix.

To illustrate the effect of GL$^+$ editing of a complex spectrum, Fig. 3-35 shows the result for a sample of cholesterol, a molecule with a moderate spread in $^1J(CH)$ coupling constants, obtained with ordinary SEMUT and DEPT. Figure 3-36 shows the corresponding edited spectra obtained using the GL$^+$ sequences. In Figs. 3-37 and 3-38 are shown the results of the same set of experiments performed on a mixture of two molecules, menthol and 3-methylpyridine, which have a spread in $^1J(CH)$ from 125 to 185 Hz. The spectra clearly demonstrate the benefits of using the GL$^+$ pulse sequences for editing coupled spectra with a wide range of coupling constants.

3. POLARIZATION TRANSFER AND EDITING TECHNIQUES

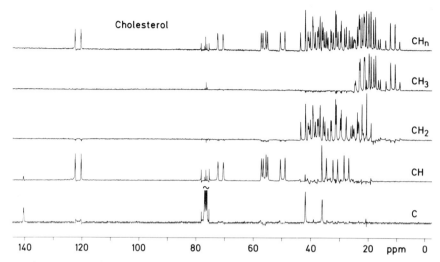

Fig. 3-35. Edited proton-coupled ^{13}C NMR spectra of cholesterol (0.015 M in CDCl$_3$) obtained using the standard DEPT and SEMUT sequences for the protonated and quaternary carbon spectra, respectively. The delay τ has been optimized corresponding to an average value of $^1J(\text{CH}) = 135$ Hz. The CH$_n$ spectrum, which contains all protonated carbons, is the $\theta = 38°$ DEPT subexperiment.

H. SEMUT EDITING OF INADEQUATE SPECTRA

In the one-dimensional version of the INADEQUATE experiment, serious overlap of ^{13}C-^{13}C doublets can make the coupling determination ambiguous, even for simple molecules. It is therefore of interest to consider incorporation of editing techniques into this experiment. The SEMUT approach is well suited for this purpose, and Fig. 3-39 shows several pulse sequences including this approach; these are termed SEMINA sequences (15). They have the SEMUT level of accuracy, but can be improved by replacing the θ pulse by purging sandwiches, although only one of the θ pulses in the general version of SEMINA-2 (Fig. 3-39b) can be so replaced.

In SEMINA-1, carbon-carbon double-quantum coherence is created under conditions of proton decoupling, but instead of immediate reconversion to observable single-quantum magnetization as in INADEQUATE, the 2QC processes for a period $2\tau_1$ before reconversion. During this period, the decoupler is switched off and a SEMUT sequence is inserted. Consider the structural unit:

$$(\text{H}_1)_n \quad\quad (\text{H}_2)_m$$
$$| \quad\quad\quad\quad |$$
$$\text{C}_1 \cdots\cdots \text{C}_2$$

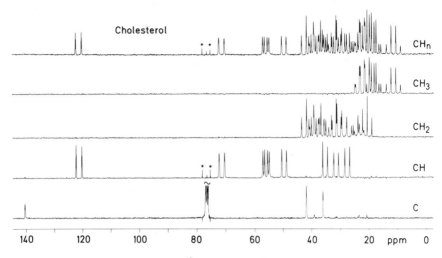

Fig. 3-36. Edited proton-coupled ^{13}C NMR spectra of cholesterol (0.015 M in CDCl$_3$) obtained on a Varian XL-300 spectrometer using the DEPT GL$^+$ and SEMUT GL$^+$(q) pulse sequences for the protonated and quaternary carbon spectra, respectively. The τ delays have been adjusted according to $J_{min} = 125$ Hz and $J_{max} = 160$ Hz (9). The signals marked with an asterisk are caused by residual CHCl$_3$ (1%) of the deuterated solvent. The CHCl$_3$ molecules cause negative J cross-talk signals in the quaternary carbon subspectrum, because ^1J(CH) (= 213 Hz) is outside the optimized J range. The CH$_n$ spectrum is from the $\theta = 38°$ DEPT GL$^+$ subexperiment.

where C$_1$ and C$_2$ can be separated by an arbitrary number of bonds. Because of the low concentration of carbon-13 nuclei, this is the only type of atomic arrangement likely to be encountered. The multiplet structure of the 2QC created by the second 90°(C) pulse is determined by the two coupling constants (see Eq. [37]):

$$J_a = J(C_1H_1) + J(C_2H_1) \quad [126]$$

$$J_b = J(C_2H_2) + J(C_1H_2) \quad [127]$$

As far as editing is concerned, there is no fundamental difference between this double-quantum coherence and the single-quantum coherence present in normal SEMUT, and the intensity expression for the in-phase 2QC at the end of the $2\tau_1$ period can be readily inferred from Eqs. [103] and [119]:

$$\text{Int}(CH_n \cdots CH_m)$$
$$= [(\cos \pi J_a \tau_1)(\cos \pi J_a \tau_3) + (\sin \pi J_a \tau_1)(\sin \pi J_a \tau_3) \cos \theta]^n$$
$$\times [(\cos \pi J_b \tau_1)(\cos \pi J_b \tau_3) + (\sin \pi J_b \tau_1)(\sin \pi J_b \tau_3) \cos \theta]^m$$
$$\times \sin \pi J_{CC} \tau_C \quad [128]$$

3. POLARIZATION TRANSFER AND EDITING TECHNIQUES

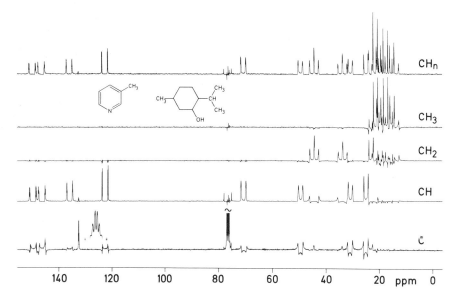

Fig. 3-37. Edited proton-coupled ^{13}C NMR spectra of a mixture of menthol (0.025 M) and 3-methylpyridine (0.030 M) in CDCl$_3$ obtained on a Varian XL-300 spectrometer using the standard DEPT and SEMUT sequences for the protonated and quaternary carbon spectra, respectively. The τ delay has been optimized corresponding to an average value of $^1J(CH) = 155$ Hz. Note the enormous J cross-talk signals and also the illustration of the fact that J cross talk only takes place in the "downward direction." The CH$_n$ spectrum is from the $\theta = 38°$ DEPT subexperiment.

When the products containing only cosine factors are made to vanish by judicious setting of the τ_1 and τ_3 delays, it follows that the θ flip angle dependence is $\cos^{n+m} \theta$, that is, like that for a hypothetical CH$_{n+m}$ group in the SEMUT experiment. Then appropriate linear combinations of experiments obtained with different θ flip angles can decompose the INADEQUATE spectrum into seven subspectra depending on the total number of hydrogens on the two carbons, $n + m$, which can vary from 0 to 6. Each subspectrum may include contributions from one-bond as well as long-range carbon–carbon pairs. Complete editing, generating seven subspectra, would lead to considerable sensitivity loss and is not recommended for general applications. On the other hand, it may often be useful to suppress the resonances corresponding to a certain value of $n + m$ or to generate two subspectra according to whether $n + m$ is odd or even. The latter option is especially useful, since it does not significantly degrade the sensitivity of the INADEQUATE experiment. In this version, the total time is divided between two subexperiments performed with $\theta = 0°$ and $\theta = 180°$,

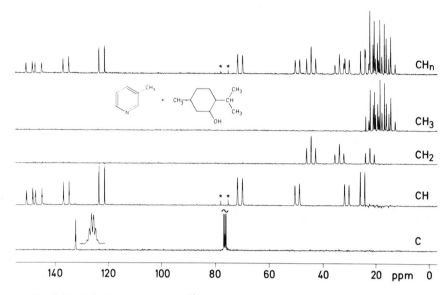

Fig. 3-38. Edited proton-coupled ^{13}C NMR spectra, of the sample used for Fig. 3-37, obtained on a Varian XL-300 spectrometer, with the DEPT GL$^+$ and SEMUT GL$^+$(q) sequences for the protonated and quaternary carbon spectra, respectively. The τ delays have been adjusted according to $J_{min} = 125$ Hz and $J_{max} = 185$ Hz. The signals marked with an asterisk are caused by residual CHCl$_3$ of the deuterated solvent. Note that the negative J cross-talk signals for CHCl$_3$ in the quaternary carbon subspectrum are suppressed in this case (cf. Fig. 3-36). The CH$_n$ spectrum is the $\theta = 38°$ DEPT GL$^+$ subexperiment.

with τ_1 and τ_3 equal. The first experiment gives a spectrum identical to the normal antiphase INADEQUATE spectrum and the second has all antiphase ^{13}CH$_n$-^{13}CH$_m$ doublets for $n + m$ odd inverted. Thus addition and subtraction of the two spectra produce one subspectrum for $n + m$ even and another for $n + m$ odd.

Although SEMINA-1 can in principle be applied to separate a conventional INADEQUATE spectrum into seven subspectra, it is limited by the inability to distinguish the individual values of n and m. SEMINA-2, shown in Fig. 3-39b, overcomes these limitations. The θ_1 pulse acts on the ^{13}C-^{13}C double-quantum coherence and again introduces a cos$^{n+m}\theta_1$ dependence. If the trapped proton MQC precesses freely during the following $2\tau_2$ delay, the second editing pulse, θ_2, may reconvert part of it to observable magnetization. A central 180° pulse avoids this result and simultaneously the ^{13}C-^{13}C double-quantum coherence is transferred to single-quantum coherence, which then undergoes a SEMUT sequence. Ideally, this gives the following flip-angle dependences for the C$_1$ and C$_2$ resonances in the

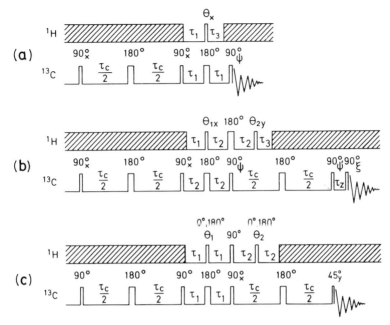

Fig. 3-39. The SEMINA pulse sequences (15) for editing INADEQUATE spectra. (a) SEMINA-1, (b) general version of SEMINA-2, and (c) $\theta = 0°$, 180° version of SEMINA-2. The double quantum excitation is optimized for $\tau_C = (2n+1)/2J(CC)$, $n = 0, 1, 2, \ldots$. The times τ_1, τ_2, and τ_3 are all of the order of $1/2\ {}^1J(CH)$. Further, the following conditions should be fulfilled: $\tau_1 > \tau_3$ in (a), $\tau_C/2 > \tau_2 + \tau_3$ in (b), and $\tau_C/2 > 2\tau_2$ in (c). If these conditions are not satisfied, two ^{13}C 180° pulses are needed in the second τ_C delay. Purging of antiphase magnetization in SEMINA-2 is done by either a z filter or a single 45° pulse, although the efficiency of the latter approach is limited (see text).

structural unit which we are considering:

$$C_1: \quad (\cos\theta_1)^{n+m}(-\cos\theta_2)^n \quad [129]$$

$$C_2: \quad (\cos\theta_1)^{n+m}(-\cos\theta_2)^m \quad [130]$$

Thus in SEMINA-2 a distinction can be made according to the individual values of n and m rather than merely their sum, as in SEMINA-1. A particularly useful version of SEMINA-2, because it preserves the sensitivity of INADEQUATE apart from losses caused by relaxation or long-range couplings, is shown in Fig. 3-39c. This version combines four experiments with $\theta_1 = 0°$ and 180°, and $\theta_2 = 0°$ and 180°.

Table 3-9 lists the $^{13}C-{}^{13}C$ coupling constants and Fig. 3-40 shows expected outcomes of the $\theta = 0°$, 180° versions of SEMINA-1 and SEMINA-2 for 1,3-dibromobutane. These predictions are fully confirmed in the

TABLE 3-9

^{13}C-^{13}C Coupling Constants^a Determined from SEMINA-1 Subspectra of 1,3-Dibromobutane

$$CH_3-CHBr-CH_2-CH_2Br$$

	A	B	C	D
$n+m$				
Even	$^1J(AB)$ 37.1		$^1J(CD)$ 37.0	
Odd	$^1J(BC)$ 36.6	$^2J(AC)$ 0.70	$^2J(BD)$ 1.45	$^3J(AD)$ 3.70

^{a} Values in hertz; uncertainty, ±0.1 Hz.

experimental spectra shown in Figs. 3-41 and 3-42. In Fig. 3-41, less intense doublets arising from the small two-bond couplings between carbons are also observed. The nature of these weaker peaks was confirmed by separate experiments tuned for the $^2J(CC)$ values.

The connectivities of carbon atoms in 1,3-dibromobutane can be unambiguously determined from the edited SEMINA-2 spectra in Fig. 3-42,

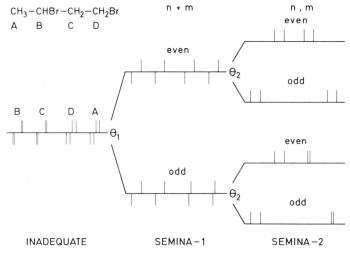

Fig. 3-40. Stick-plot INADEQUATE, (0°, 180°) SEMINA-1, and (0°, 180°) SEMINA-2 spectra of 1,3-dibromobutane taking the one- and three-bond ^{13}C-^{13}C couplings into account. The θ_1 pulse separates the INADEQUATE spectrum according to the total multiplicity $n+m$, and θ_2 separates according to the individual multiplicities n and m of the $^{13}CH_n$-$^{13}CH_m$ fragments.

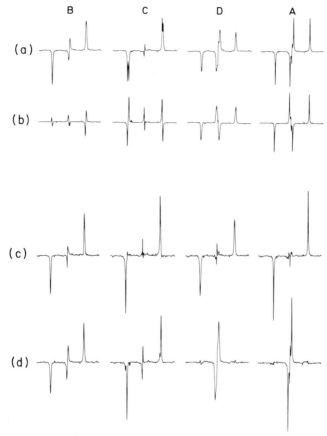

Fig. 3-41. SEMINA-1 ^{13}C NMR spectra of 1,3-dibromobutane (50% v/v in C_6D_6) recorded on a Varian XL-300 spectrometer. (a, b) Two subexperiments obtained using $\theta = 0°$ and $\theta = 180°$, respectively, and $\tau_C = 121.6$ msec. Spectrum (a) corresponds to the normal INADEQUATE spectrum. (c) The SEMINA-1, $n + m$ even, subspectrum obtained by addition of spectra (a) and (b). (d) The SEMINA-1, $n + m$ odd, subspectrum obtained as the difference between (a) and (b). All spectra are shown on the same absolute intensity scale.

without the need to measure coupling constants. The guidelines for an analysis of this sort are illustrated in the following paragraphs as applied to the SEMINA-2 spectra of menthol in Fig. 3-43.

For $n + m$ even, the two doublets for the fragment under consideration both appear in the same subspectrum, either in the even/even (b) or even/odd (c). Thus an even number of doublets is always observed in the even/even and the even/odd subspectra. For $n + m$ odd, one doublet is found in the odd/even (d) and the other in the odd/odd (e) subspectrum,

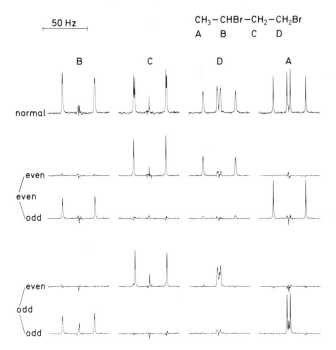

Fig. 3-42. SEMINA-2 spectra of 1,3-dibromobutane recorded on a Varian XL-300 spectrometer (15). The uppermost trace is the $\theta_1 = 0°$, $\theta_2 = 180°$ subexperiment, which corresponds to a refocused INADEQUATE spectrum. The editing into four independent subspectra by linearly combining the four θ_1, $\theta_2 = 0°$, $180°$ subexperiments is fully consistent with the prediction made in Fig. 3-40. The vertical scale in the edited subspectra has been reduced by a factor of four compared to the spectrum in the uppermost trace. Parameters: $\tau_C = 121.6$ msec and $\tau_1 = \tau_2 = 3.85$ msec.

so that the numbers of doublets in the odd/even and odd/odd subspectra may be either even or odd but will always be identical.

To ease the assignment, independent determination of proton multiplicities for the normal ^{13}C resonances is recommended before the SEMINA-2 experiments are performed. With this information, the even/even subspectrum (b) of menthol, for simplicity only optimized for one-bond carbon-carbon couplings, in Fig. 3-43 shows directly that resonances 4 and 7 constitute a $-CH_2-CH_2-$ fragment. However, proton multiplicities, both directly attached and neighboring, can be deduced from edited SEMINA-2 spectra. For example, CH_3 and CH resonances which appear only in spectra (c) and (e) are assigned as follows. CH_3 doublets will appear in either (c) or (e) with unit intensity, but not in both, since they can have no more than one carbon neighbor. Thus the (c) and (e) spectra show the presence of

3. POLARIZATION TRANSFER AND EDITING TECHNIQUES   237

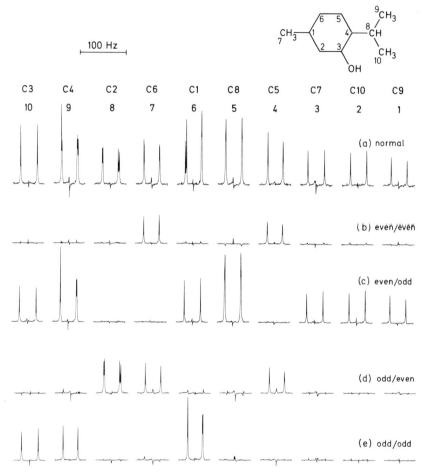

Fig. 3-43. SEMINA-2 spectra of menthol recorded on a Varian XL-300 spectrometer. For simplicity, the double-quantum excitation was optimized only for one-bond ^{13}C-^{13}C coupling constants ($\tau_C = 14.28$ msec). (a) The $\theta_1 = 0°$, $\theta_2 = 180°$ subspectrum corresponding to a refocused INADEQUATE spectrum. (b)–(e) Edited spectra with the vertical scale reduced by a factor of four compared to (a).

three CH$_3$ groups (1, 2, and 3) all directly attached to CH groups. Comparison of intensities of the remaining resonances in spectra (c) and (e) shows that 5 is a CH group with three neighboring carbons of odd proton multiplicities, since spectrum (c) must contain an even number, or ten, doublets. Further, this carbon (resonance 5) is attached to either one or two CH$_3$ groups. For the other CH resonances, it is concluded that 6 is attached to two CH$_2$ groups, one of which is 8 and the other either 4 or 7, and to one CH or CH_3 group; that 9 is attached to one CH_2 group and two CH or CH_3 groups; and that 10 is attached to one CH_2, one CH or CH_3 group, and a heteroatom substituent. In a similar manner, spectra (b) and (d) indicate the presence of three CH_2 carbons. The $-CH_2-CH_2-$ fragment, formed by resonances 4 and 7, is attached to CH carbon 6 at one end and to either the 9 or 10 CH carbon at the other. The third CH_2 carbon, 8, is flanked by two CH groups, one of which is again 6 and the other whichever one of the 9 or 10 carbons that is not attached to the $-CH_2-CH_2-$ fragment.

These qualitative considerations leave only a few alternative frameworks to be differentiated for the correct structure of menthol, for which additional information can be obtained from $^1J(CC)$ and $^3J(CC)$ values. A more challenging application of SEMINA-2 editing is shown in Fig. 3-44, the spectra of 5α-androstane, where the C8-C9-C14 fragment can be unambiguously identified with the aid of the odd/even subspectrum.

For many molecules, especially aromatic ring systems, application of the SEMINA-1 pulse sequence is sufficient and useful for (i) resolving

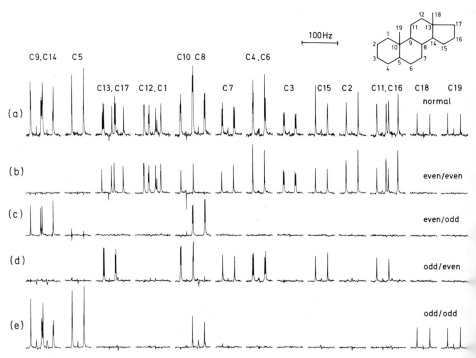

Fig. 3-44. SEMINA-2 spectra of 5α-androstane recorded on a Varian XL-300 spectrometer (15). The experiments were optimized only for one-bond $^{13}C-^{13}C$ coupling constants.

ambiguities often encountered in the assignment of ^{13}C-^{13}C couplings, particularly long-range ^{13}C-^{13}C couplings, and (ii) resolving the extensive overlap of spectra arising from long-range ^{13}C-^{13}C couplings. A particular example is shown in Figs. 3-45 and 3-46 with the SEMINA-1 spectra for acenaphthalene. These spectra show the advantage of the SEMINA experiments over the INADEQUATE method for the determination of ^{13}C-^{13}C coupling constants. The SEMINA experiments illustrated here were performed at a carbon-13 frequency of 75.43 MHz and employed phase cycles consisting of (i) a 128-step cycle (50) for the four "INADEQUATE pulses" and the receiver reference phase, and (ii) independent Exorcycling of additional carbon pulses. For each subexperiment, 2048 scans were accumulated. Proton pulses of 180° were replaced by composite $90^\circ_x 180^\circ_y 90^\circ_x$ pulses. Since, in SEMINA-2, all heteronuclear couplings are refocused for θ_1 of 0° or θ_2 of 180°, the mixing coefficients for generating subspectra may deviate from the usual optimum value of unity.

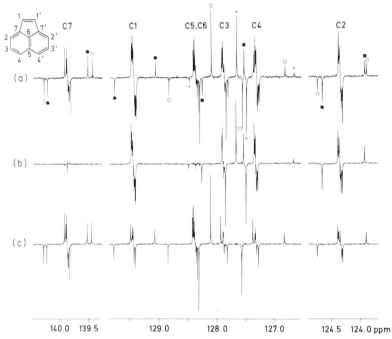

Fig. 3-45. SEMINA-1 ^{13}C spectra of acenaphthylene (2.0 M in acetone-d_6) recorded on a Varian XL-300 spectrometer using $\tau_C = 60$ msec, $\tau_1 = 3.2$ msec, and 2048 scans for each subexperiment. (a) Normal INADEQUATE spectrum. (b, c) Edited SEMINA-1 subspectra corresponding to $n + m$ even and $n + m$ odd, respectively. The ^{13}C-^{13}C satellite spectra for the one-bond ^{13}C-^{13}C couplings are marked. Expansions of the long-range ^{13}C-^{13}C satellites are shown in Fig. 3-46.

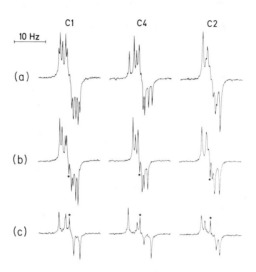

Fig. 3-46. Expansions of the ^{13}C-^{13}C satellite spectra in Fig. 3-45 of the long-range ^{13}C-^{13}C couplings for the carbons C1, C4, and C2 in acenaphthylene. (a) Spectra corresponding to the normal INADEQUATE experiment and resulting from the $\theta = 0$ SEMINA-1 subexperiment. (b, c) Edited SEMINA-1 subspectra corresponding to $n + m$ even and $n + m$ odd, respectively. Residual parent signals are indicated by an asterisk. The spectra clearly show that, in contrast to the INADEQUATE method, the SEMINA-1 experiment is very useful for assigning the ^{13}C-^{13}C couplings and resolving the extensive overlap of spectra arising from long-range coupling in aromatic molecules.

The 0°, 180° version of SEMINA-2 in Fig. 3-39c contains two purging fragments. The first is a 90°(H) pulse which purges ^{13}C-^{13}C double-quantum coherence which is antiphase with respect to heteronuclear couplings at the end of the second τ_1 delay. Thus at the beginning of the first τ_2 delay, we start with ^{13}C single-quantum coherence in antiphase with respect to $J(CC)$ but in-phase with respect to all carbon–hydrogen couplings.

The need for the second purging step arises in the following way: during the second τ_C period, the antiphase ^{13}C magnetization rephases to an extent which depends on values of $J(CC)$ and τ_C:

$$2S_{1x}S_{2z} \pm 2S_{1z}S_{2x} \xrightarrow{\pi J(CC)\tau_C 2I_{1z}S_{2z}} (2S_{1x}S_{2z} \pm 2S_{1z}S_{2x})[\cos \pi J(CC)\tau_C]$$
$$+ (S_{1y} + S_{2y})[\sin \pi J(CC)\tau_C] \qquad [131]$$

The positive sign always applies for $n + m$ even, but for $n + m$ odd holds only when θ_2 is 180°. The negative sign applies for $n + m$ odd with θ_2 of 0°. The antiphase terms in Eq. [131] give rise to phase distortions, so that purging prior to acquisition is desirable.

For this purpose, two possibilities are indicated in Figs. 3-39b and 3-39c. The z filter, in Fig. 3-39b, is applied as described in Section III,B,2, but the alternative 45° purging pulse needs some consideration. When the positive sign in Eq. [131] is applicable, the 45° pulse is perfect in purging the antiphase terms, as shown in Eq. [47]. However, when the negative sign applies, the antiphase terms are invariant to rotation about the y axis:

$$2S_{1x}S_{2z} - 2S_{1z}S_{2x} \xrightarrow{\beta(S_{1y}+S_{2y})} 2S_{1x}S_{2z} - 2S_{1z}S_{2x} \qquad [132]$$

One may describe this result as a zero-quantum coherence with respect to the y axis. This means that when purging with $45°_y$ is employed, the SEMINA-2 subspectra with $n+m$ even can always be properly phased and that phase errors in the $n+m$ odd subspectra are reduced by 50%. This is usually sufficient provided that the spread in the $J(CC)$ values is small, as when the experiment is "tuned" only for one-bond carbon–carbon couplings.

In the experiments described here, a 45° pulse was employed for purging. For the general case where observation of long-range carbon–carbon coupling constants is also of interest, it is necessary to purge by a z filter with τ_z averaging. This is illustrated in Fig. 3-47, which shows edited SEMINA-2 spectra for the carbons C1 and C3 of menthol (simultaneous observation of one-bond and long-range ^{13}C-^{13}C couplings) obtained (a) without purging at all, (b) with a 45° purging pulse, and (c) by purging using the z-filter approach. The spectra show that the 45°-pulse purging method is effective for ^{13}C-^{13}C doublets only in the even/odd subspectrum (C3) but not in the

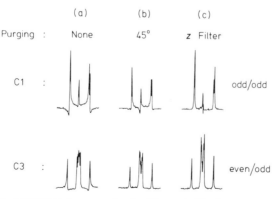

Fig. 3-47. Edited SEMINA-2 spectra for the C3 even/odd subspectrum and C1 odd/odd subspectrum of menthol (cf. Fig. 3-43), illustrating the effects of purging techniques. The experiments were optimized for both one-bond and long-range ^{13}C-^{13}C couplings, using $\tau_C = 100$ msec, and recorded with (a) no purging; (b) a $45°_y$ purging pulse; (c) a z filter with eight τ_z values (2.125, 1.625, 1.375, 1.125, 0.875, 0.625, 0.375, and 0.125 msec).

odd/odd subspectrum (C1). On the other hand, the z-filter approach works for ^{13}C resonances in both sets of subspectra. The spectra in Fig. 3-47 represent an attractive application of the use of the z filter for purging phase anomalies in homonuclear coupled spectra. It is clearly superior to purging by a 45° pulse.

The SEMINA pulse sequences may be compared with the INADEQUATE experiment. When relaxation during the evolution period can be neglected, the inherent sensitivity of 2D INADEQUATE is only insignificantly lower than that for the 1D experiment. Because of the straightforward interpretation of 2D INADEQUATE, this is in general the preferred technique if the full connectivity information of large molecules is desired. SEMINA techniques are superior in studies of small molecules, as well as for the accurate determination of coupling constants. Two-dimensional SEMINA versions could also be considered to resolve possible overlaps in 2D INADEQUATE spectra.

I. PROTON COUPLING NETWORKS FROM EXTENDED DEPT GL EDITING

Various possibilities of spectral editing techniques exist for obtaining information about structural relations more remote than those concerned with the number of directly attached protons. Here we describe a method that does not require, as does the SEMINA approach just discussed, two spin-coupled ^{13}C nuclei, but makes use of all protonated ^{13}C nuclei present. It yields CH_3, CH_2, and CH subspectra which have the additional benefit of providing information about the hydrogen–hydrogen coupling network.

This method (16) combines the homonuclear spin-echo multiplet selection technique described in Section III,B,2 with DEPT or DEPT GL spectral editing. In the combined technique the first hydrogen spin-echo period of DEPT or DEPT GL, usually taken as $2\tau = 1/2J(CH)$, is extended to $2\tau = 1/J(HH)$. At the same time, the 90°(C) pulse is shifted toward the end of this period and introduced at a time $\tau_1 = 1/2J(CH)$ before the hydrogen chemical shift and heteronuclear CH spin coupling evolutions are refocused. This pulse has the same effect as in DEPT, that is, transfer of I-spin magnetization antiphase with respect to the S spin to heteronuclear two-spin coherence. The following steps of the sequence are identical to DEPT or DEPT GL, and the complete sequence, based on DEPT GL, is shown in Fig. 3-48.

The initial homonuclear spin-echo part of the resulting sequence tests whether the number of homonuclear coupling partners is even or odd. This distinction utilizes the fact that a proton I_1, interacting with $n-1$ coupling partners in a first-order spectrum at a time 2τ, is modulated by a factor

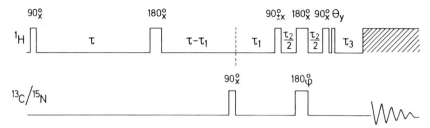

Fig. 3-48. Pulse sequence for extended DEPT GL spectra editing of decoupled ^{13}C and ^{15}N NMR spectra; $\tau = 1/2J(HH)$. The delays τ_1, τ_2, and τ_3, which are all approximately equal to $1/2\ ^1J$, are optimized as for DEPT GL. The phase-cycling scheme is the same as employed in the normal DEPT GL experiment (9). Extended spectral editing may be incorporated in a similar way into the DEPT sequence.

$\prod_{i=2}^{n} \cos \pi J_{1,i} 2\tau$. Thus for optimum sensitivity and for multiplet assignment, this modulation factor should be close to $(-1)^{n-1}$, which is true for $2\tau = 1/J(HH)$. DEPT GL editing then provides the usual three subspectra, which contain positive and negative lines depending on whether the number of homonuclear coupling partners of the protons directly attached to the ^{13}C nucleus is even or odd. If the phase is such that in the standard DEPT GL subspectra all lines are positive, a positive line in these subspectra corresponds to an even number of homonuclear coupling partners of the proton(s) attached to the ^{13}C.

Experimental spectra for the extended DEPT GL sequence shown in Fig. 3-48 are shown in Fig. 3-49 for an approximately equimolar mixture of 1,1-dibromoethane, isobutyl acetate, and diethyl ethyl malonate, and these spectra greatly facilitate the assignment of the usual carbon-13 spectra.

The spectrum shown in Fig. 3-49e illustrates a special case of the extended DEPT GL sequence using the shorter delay $2\tau = 1/2J(HH)$ rather than $2\tau = 1/J(HH)$, leading to signals only for those carbon atoms for which the resonance of the directly attached protons is a singlet in the conventional proton spectrum.

The accuracy of the editing into CH, CH$_2$, and CH$_3$ subspectra is independent of the length of the delay τ and is identical with the accuracy of standard DEPT GL editing. Within a range of about ±50%, deviations from the $J(HH)$ value used in setting the value of τ lead only to intensity reductions but do not alter the positive–negative sense of the resonances in the edited subspectra.

The extended DEPT or DEPT GL editing schemes have interesting applications in NMR analyses of peptides. From the ^{13}C spectra, the C$_\alpha$ resonances can be classified according to whether the number of β and NH protons is even or odd. From the ^{15}N spectra, there is a unique identification

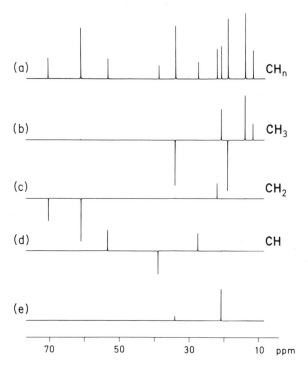

Fig. 3-49. ^{13}C DEPT GL NMR spectra recorded on a Varian XL-300 spectrometer for an approximately equimolar mixture of 1,1-dibromoethane, isobutyl acetate, and diethyl ethylmalonate (each about 0.15 M in $CDCl_3$). (a) Normal DEPT GL spectrum ($\tau = \tau_1$, $\theta = 38°$, 64 scans). (b)–(d) Extended-edited DEPT GL spectra. Positive or negative lines are associated with carbons where the directly attached protons have an even or odd number, respectively, of homonuclear coupling partners. The DEPT GL part of the pulse sequence was optimized according to $\beta_D = 38°$, $\eta = 1$, $J_{min} = 130$ Hz, and $J_{max} = 150$ Hz (9, 46). The τ delay was adjusted for a ^1H-^1H coupling constant of 6.5 Hz, and 64 scans were accumulated for each subexperiment. (e) Nonedited extended DEPT GL spectrum obtained with the special delay of $2\tau = 1/2J(HH)$, $\theta = 38°$, and 64 scans (see text).

of glycine resonances, for only this amino acid has two α protons. Such an application is shown in Fig. 3-50 with ^{15}N spectra of an approximately equimolar mixture of two dipeptides containing the amino acids alanine, glycine, valine, and phenylalanine.

Extended editing is a valuable tool for structure elucidation, although it is less general than editing based only on large one-bond couplings, where all long-range and strong coupling effects are negligible. As for most other techniques which exploit ^1H-^1H couplings, extended editing is likely to fail in the presence of strong coupling, and intensity reductions of the signals can occur because of short T_2 values and variations in the magnitudes of

3. POLARIZATION TRANSFER AND EDITING TECHNIQUES

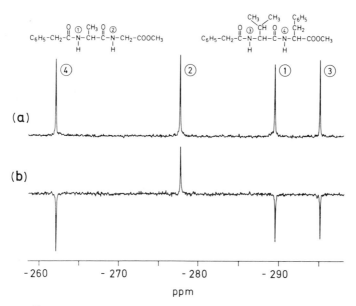

Fig. 3-50. ^{15}N DEPT GL ($\theta = 90°$) NMR spectra of an approximately equimolar mixture of the two dipeptide esters (each 0.22 M in DMSO-d_6) shown at the top. The spectra, recorded at 30.4 MHz, were optimized according to $J_{min} = 85$ Hz and $J_{max} = 95$ Hz for the one-bond ^{15}N-^{1}H coupling constants. (a) Normal DEPT GL spectrum (4096 scans). (b) Extended DEPT GL spectrum (8192 scans) optimized for $J(HH) = 7.5$ Hz. This spectrum unambiguously identifies the glycine ^{15}N resonance. The shift scale is relative to external CH_3NO_2.

$J(HH)$. The technique described in this section is just one possibility of combining homo- and heteronuclear editing methods. Other potentially useful experiments include multiple-quantum filters (51, 52) and spin pattern recognition sequences (53).

J. Editing of 2D Spectra

Editing can be incorporated into many two-dimensional ^{13}C NMR experiments, but in most cases it is not worthwhile. One-dimensional editing techniques are so fast and reliable that they can always be employed when proton multiplicities are of interest. Editing of 2D experiments is valuable when it is necessary to resolve overlapping resonances of differing multiplicities. In the 2D pulse sequences with editing, it is most convenient to have the editing step after the t_1 labeling in the evolution period. If this is done, one principle must be observed, namely, that the coherences must, at the beginning of the editing part of the pulse sequences, be in-phase with respect to the one-bond ^{13}C-^{1}H couplings.

Two-dimensional ^{13}C J spectra, to be described in Chapter 5, show the normal proton-decoupled spectrum along the ω_2 frequency axis and proton-coupled multiplets, centered at $\omega_1 = 0$, along the orthogonal ω_1 axis. The appropriate pulse sequence of the SEMUT type for editing "gated decoupler" J spectra is outlined in Fig. 3-51. The editing procedure employed is the same as for 1D SEMUT editing.

Editing of ^{13}C 2D spectra involving polarization transfer from protons can employ DEPT or DEPT GL. In the applicable 2D polarization-transfer sequences, pure proton magnetization evolves in the evolution period, and the in-phase principle stated in the first paragraph of this section is easily fulfilled by judicious placement of a proton or carbon-13 180° pulse.

Morris (54) has described an experiment for indirect 2D J spectroscopy which yields ^{13}C-decoupled proton spectra in ω_1, centered at $\omega_1 = 0$, and proton-decoupled ^{13}C spectra in ω_2. The corresponding editing sequence can be obtained from DEPT or DEPT GL by replacing the 180°(H)90°(C) pulses in the middle of the $2\tau_1$ period by $-t_1/2-180°(H)-t_1/2-90°(C)-$.

Heteronuclear 2D chemical shift correlation spectra, also described in Chapter 5, show ^{13}C-decoupled proton spectra in ω_1 and proton-decoupled ^{13}C spectra in ω_2. A CH$_n$ fragment gives rise to a peak in the 2D spectrum centered at $(\omega_1, \omega_2) = (\Omega_H, \Omega_C)$, where Ω represents a chemical shift. When the sequence $-t_1/2-180°(C)-t_1/2-$ is inserted between the first proton pulse and the first τ_1 delay in DEPT or DEPT GL, as shown in Fig. 3-52, editing of heteronuclear 2D chemical shift correlation spectra is possible.

In 2D polarization transfer sequences involving DEPT in which an editing step is included, it is recommended that the spectrum always be edited according to whether the number of directly attached protons is odd or even, since this does not significantly degrade the sensitivity. The distinction between CH + CH$_3$ and CH$_2$ resonances is obtained by adding or subtracting 2D data files recorded with, for example, $\theta = 55°$ and $125°$. In another variation (55), only one θ value of about $125°$ is employed, and then the positive and negative contour levels of the phase-sensitive 2D

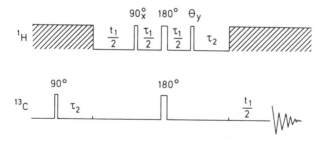

Fig. 3-51. Pulse sequence for editing of heteronuclear "gated decoupler" 2D J spectra. In the conventional 2D J experiment, $\tau_1 = \tau_2 = 0$ and no proton pulses are applied.

3. POLARIZATION TRANSFER AND EDITING TECHNIQUES

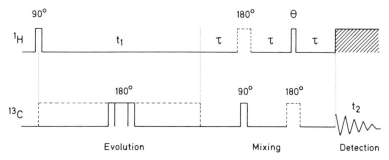

Fig. 3-52. Pulse sequence for heteronuclear chemical shift correlation employing DEPT for the heteronuclear magnetization transfer. The value of τ is chosen, as in normal DEPT, to favor strongly transfer via one-bond heteronuclear couplings, such as $\tau = 1/2J(\text{CH})$. The 180° refocusing pulses (dotted pulses) can be omitted if the spectra are represented in absolute-value mode (55).

spectrum are plotted separately. Such a spectrum of menthol is shown in Fig. 3-53. For spectra with well-separated peaks, the two approaches have the same sensitivity, but in general the first alternative is preferable because no cancellation of overlapping resonances can occur.

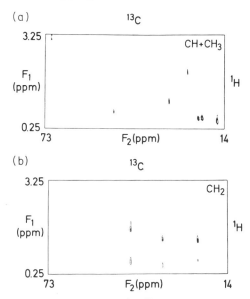

Fig. 3-53. Phase-sensitive two-dimensional ^1H-^{13}C heteronuclear chemical shift correlation spectra of menthol obtained with $\theta = 135°$ and plotted twice with (a) positive and (b) negative contour levels. The pulse sequence is shown in Fig. 3-52. The two representations show CH_2 group resonances (b) separated from CH and CH_3 group resonances (a). The CH_2 responses are split by about 1 ppm in the ω_1 dimension as a result of the nonequivalence of axial and equatorial protons. Both plots span 0.25–3.25 ppm in ω_1 (proton shifts) and 14–73 ppm in ω_2 (^{13}C shifts), relative to TMS. All fixed delays were 3.8 msec.

Appendixes

A. Basic Quantum Mechanics of Commutators and Exponential Operators

1. Commutators

Relations involving the operators A, B, C, D:

$$[A, B] = AB - BA = -[B, A] \qquad [\text{A.1}]$$

$$[AB, C] = A[B, C] + [A, C]B \qquad [\text{A.2}]$$

$$[A, BC] = [A, B]C + B[A, C] \qquad [\text{A.3}]$$

$$[AB, CD] = A[B, C]D + C[A, D]B + AC[B, D] + [A, C]DB \qquad [\text{A.4}]$$

The important commutator properties of angular momentum operators are:

$$[I_x, I_y] = iI_z \qquad [\text{A.5}]$$

$$[I_y, I_z] = iI_x \qquad [\text{A.6}]$$

$$[I_z, I_x] = iI_y \qquad [\text{A.7}]$$

$$[I_a, [I_a, I_b]] = I_b \qquad [\text{A.8}]$$

where a and b are x, y, or z, but $a \neq b$. Equation [A.8] may be verified by substituting pairs of coordinates and expanding in terms of Eqs. [A.5]–[A.7].

For $I = \tfrac{1}{2}$, the following relations are also useful:

$$I_a I_b = \frac{i}{2} I_c \qquad [\text{A.9}]$$

$$I_b I_a = -\frac{i}{2} I_c \qquad [\text{A.10}]$$

$$I_d^2 = I_d I_d = \tfrac{1}{4} E \qquad [\text{A.11}]$$

where a, b, c constitute the elements of a cyclic permutation of x, y, z, d is x, y, or z, and E is the unity operator.

2. Exponential Operators

Exponential operators are used to represent unitary transformations, and many of their properties are most easily proved by a series expansion:

$$\exp(i\phi A) = \sum_{n=0}^{\infty} \frac{1}{n!} (i\phi)^n A^n = E + i\phi A + \frac{(i\phi)^2}{2!} A^2 + \frac{(i\phi)^3}{3!} A^3 + \cdots \qquad [\text{A.12}]$$

3. POLARIZATION TRANSFER AND EDITING TECHNIQUES

The following two relations, which are valid if A and B commute, can, for example, be proved by using the expansion in Eq. [A.12]:

$$A \exp(i\phi B) = [\exp(i\phi B)]A \qquad \text{[A.13]}$$

$$\exp[i(\phi A + \psi B)] = \exp(i\phi A)\exp(i\psi B) = \exp(i\psi B)\exp(i\phi A) \qquad \text{[A.14]}$$

As a consequence of these two equations, it is, for our purpose, only necessary to evaluate a single expression of the type of Eq. [1], namely,

$$f(\phi) = [\exp(-i\phi kAS_b)]S_a \exp(i\phi kAS_b), \qquad a, b = x, y, z \qquad \text{[A.15]}$$

where S is an angular momentum operator, A is a product of angular momentum operators for other spins than S, and k is the factor $2^{(n-1)}$ in Eq. [6]. For example, $k = 2$, $A = I_z$, and $S_b = S_z$ if the evolution under J_{IS} coupling is to be calculated. For pulses and evolution under chemical shifts, kA is unity.

The derivatives of $f(\phi)$ in Eq. [A.15] with respect to ϕ are

$$\frac{df}{d\phi} = -i \exp(-i\phi kAS_b)kA[S_b, S_a]\exp(i\phi kAS_b) \qquad \text{[A.16]}$$

$$\frac{d^2f}{d\phi^2} = -\exp(-i\phi kAS_b)[S_b, [S_b, S_a]]\exp(i\phi kAS_b) \qquad \text{[A.17]}$$

For $a = b$, it follows with the aid of Eq. [A.13] that $f(\phi) = S_a$. For the more interesting case, $a \neq b$, we have by use of Eq. [A.8] that $(d^2f/d\phi^2)(\phi) = -f(\phi)$. With $f(0) = S_a$ and $(df/d\phi)(0) = -ikA[S_b, S_a]$ as boundary conditions, $f(\phi)$ can be found:

$$f(\phi) = S_a \cos\phi - ikA[S_b, S_a]\sin\phi \qquad \text{[A.18]}$$

Suppose that it is desired to determine the results of a transformation involving the product of two operators such as

$$\exp(-i\phi kAS_b)2I_cS_a \exp(i\phi kAS_b)$$

where a and c represent Cartesian coordinates. It is always possible by insertion, between the two factors of the product operator, of an expression such as

$$\exp(i\phi kAS_b)\exp(-i\phi kAS_b)$$

which is equal to unity, to separate the transformation into two parts, each of which is similar to Eq. [A.15] and which may then be evaluated by Eq. [A.18]:

$$\exp(-i\phi kAS_b)2I_cS_a\exp(i\phi kAS_b) = 2[\exp(-i\phi kAS_b)I_c\exp(i\phi kAS_b)]$$
$$\times [\exp(-i\phi kAS_b)S_a\exp(i\phi kAS_b)]$$

$$\text{[A.19]}$$

A further commutator is useful to show invariance of product operators under certain transformations:

$$[S_b I_d, S_a I_c] = I_d I_c S_b S_a - I_c I_d S_a S_b \qquad [\text{A.20}]$$

For spins 1/2 the commutator is equal to zero in two cases:

(1) $a = c, b = d$ and (2) $a \neq b, c \neq d$

For magnetically equivalent spins, such as I_k and I_l, the coupling Hamiltonian $2\pi J_{kl} \mathbf{I}_k \cdot \mathbf{I}_l$ commutes with all propagators and observable operators:

$$[(I_{ka} + I_{la}), (I_{kx}I_{lx} + I_{ky}I_{ly} + I_{kz}I_{lz})] = 0 \qquad a = x, y, z \qquad [\text{A.21}]$$

A final feature of exponential operators, which will be proved by series expansion, is important for the simplification of pulse sequence analyses. Let U denote an exponential operator, such as $\exp(i\psi B)$, describing a unitary transformation; then

$$U^{-1} \exp(i\phi A) U = U^{-1} \left[\sum_{n=0}^{\infty} \frac{(i\phi)^n}{n!} A^n \right] U$$

$$= U^{-1} \left[E + i\phi A + \frac{(i\phi)^2}{2} A^2 + \cdots \right] U$$

$$= E + i\phi U^{-1} A U + \frac{(i\phi)^2}{2} (U^{-1}AU)(U^{-1}AU) + \cdots$$

$$= \sum_{n=0}^{\infty} \frac{(i\phi)^n}{n!} (U^{-1}AU)^n$$

$$= \exp[i\phi(U^{-1}AU)] \qquad [\text{A.22}]$$

Along with its other uses this identity is employed in Section II,E in the form

$$\exp(-i\pi I_x) \exp(i\Omega_I I_z) \exp(i\pi I_x) = \exp(-i\Omega_I I_z) \qquad [\text{A.23}]$$

which on multiplying both sides from the left by $\exp(i\pi I_x)$ becomes

$$\exp(i\Omega_I I_z) \exp(i\pi I_x) = \exp(i\pi I_x) \exp(-i\Omega_I I_z) \qquad [\text{A.24}]$$

B. Pulses with Arbitrary Phase

The transformation of pulses with an arbitrary phase ϕ defined as the displacement from the x axis toward the y axis is represented by

$$\sigma(t_+) = \exp\left[-i\theta \sum_l (I_{lx} \cos\phi + I_{ly} \sin\phi)\right] \sigma(t_-)$$

$$\times \exp\left[+i\theta \sum_l (I_{lx} \cos\phi + I_{ly} \sin\phi)\right] \qquad [\text{B.1}]$$

$$= \exp\left(-i\phi \sum_l I_{lz}\right) \exp\left(-i\theta \sum_l I_{lx}\right) \exp\left(+i\phi \sum_l I_{lz}\right) \sigma(t_-)$$

$$\times \exp\left(-i\phi \sum_l I_{lz}\right) \exp\left(+i\theta \sum_l I_{lx}\right) \exp\left(+i\phi \sum_l I_{lz}\right) \quad [\text{B.2}]$$

Thus the effect of a pulse with phase ϕ and flip angle θ may be calculated in three separate steps, which in shorthand notation may be expressed, following the right-hand side of Eq. [B.2], as

$$\sigma(t_-) \xrightarrow{-\phi \sum_l I_{lz}} \xrightarrow{\theta \sum_l I_{lx}} \xrightarrow{\phi \sum_l I_{lz}} \sigma(t_+) \quad [\text{B.3}]$$

The first operation corresponds to a z pulse with rotation angle $-\phi$, the second to an x pulse with flip angle θ, and the final step to a z pulse with rotation angle ϕ. Thus Eqs. [17]–[20] can be generalized for arbitrary phase ϕ according to

$$I_{lz} \xrightarrow{\theta(I_{lx}\cos\phi + I_{ly}\sin\phi)} I_{lz}\cos\theta$$

$$+ I_{lx}\sin\theta\sin\phi - I_{ly}\sin\theta\cos\phi \quad [\text{B.4}]$$

$$I_{lx} \xrightarrow{\theta(I_{lx}\cos\phi + I_{ly}\sin\phi)} -I_{lz}\sin\theta\sin\phi$$

$$+ I_{lx}(\cos\theta\sin^2\phi + \cos^2\phi) + I_{ly}\sin^2(\theta/2)\sin 2\phi \quad [\text{B.5}]$$

$$I_{ly} \xrightarrow{\theta(I_{lx}\cos\phi + I_{ly}\sin\phi)} I_{lz}\sin\theta\cos\phi$$

$$+ I_{lx}\sin^2(\theta/2)\sin 2\phi + I_{ly}(\cos\theta\cos^2\phi + \sin^2\phi) \quad [\text{B.6}]$$

Experimentally, such transformations are achieved either by applying a single pulse shifted in phase through an angle ϕ or alternatively by using pulse sandwiches (56).

C. Pulses with Tilted rf Axes

When the frequency of a pulse is not exactly on resonance for a spin to which it is applied, rotation results about an effective rf axis that is tilted away from the xy plane. The transformation of the density operator for a pulse with tip angle θ, phase ϕ, and angle ψ away from the z axis (which is 90° for a pulse on resonance) may be written in five steps as follows:

$$\sigma(t_-) \xrightarrow{-\phi \sum_k I_{kz}} \xrightarrow{(90°-\psi)\sum_k I_{ky}} \xrightarrow{\theta \sum_k I_{kx}} \xrightarrow{-(90°-\psi)\sum_k I_{ky}} \xrightarrow{\phi \sum_k I_{kz}} \sigma(t_+)$$

$$[\text{C.1}]$$

As a result of these steps, the following general relations are obtained:

$$I_{kz} \to I_{kz}(\cos\theta \sin^2\psi + \cos^2\psi)$$
$$+ I_{kx}[\sin\theta \sin\phi \sin\psi + \sin^2(\theta/2)\cos\phi \sin 2\psi]$$
$$+ I_{ky}[-\sin\theta \cos\phi \sin\psi + \sin^2(\theta/2)\sin\phi \sin 2\psi] \quad [\text{C.2}]$$

$$I_{kx} \to I_{kz}[\tfrac{1}{2}(1-\cos\theta)\sin 2\psi \cos\phi - \sin\theta \sin\phi \sin\psi]$$
$$+ I_{kx}[\cos\theta(\sin^2\phi + \cos^2\phi \cos^2\psi) + \cos^2\phi \sin^2\psi]$$
$$+ I_{ky}[\tfrac{1}{2}(1-\cos\theta)\sin 2\phi \sin^2\psi + \sin\theta \cos\phi \sin\psi] \quad [\text{C.3}]$$

$$I_{ky} \to I_{kz}[\tfrac{1}{2}(1-\cos\theta)\sin 2\psi \sin\phi + \sin\theta \cos\phi \sin\psi]$$
$$+ I_{kx}[\tfrac{1}{2}(1-\cos\theta)\sin 2\phi \sin^2\psi - \sin\theta \cos\phi \sin\psi]$$
$$+ I_{ky}[\cos\theta(\cos^2\phi + \sin^2\phi \cos^2\psi) + \sin^2\phi \sin^2\psi] \quad [\text{C.4}]$$

D. Operator Analysis of DEPT for I_2S System

As an example of the simplicity of the product operator formalism, we treat the DEPT technique applied to an I_2S spin system. The calculation is based on Eq. [44]. The various times at which the density operator is evaluated have been numbered in Fig. 3-12d.

The initial state is assumed to consist of I-spin magnetization and is described by the density operator

$$\sigma_0 = I_{1z} + I_{2z} \quad [\text{D.1}]$$

omitting the magnetization factors.

The density operator σ_1 after the first $90°_x(I)$ pulse,

$$\sigma_1 = -(I_{1y} + I_{2y}) \quad [\text{D.2}]$$

develops under J_{IS} to the state

$$\sigma_2 = -c(I_{1y} + I_{2y}) + s2(I_{1x} + I_{2x})S_z \quad [\text{D.3}]$$

The abbreviated trigonometric functions c and s are defined in Table 3-2. The first term does not cause S-spin magnetization in the further course of the experiment and can be disregarded. The $180°_y(I)$ pulse changes the sign of the second term in Eq. [D.3], and the following $90°_x(S)$ pulse effects a transfer into heteronuclear two-spin coherence:

$$\sigma_3 = s2(I_{1x} + I_{2x})S_y \quad [\text{D.4}]$$

This coherence evolves under the J coupling to the passive I spin:

$$\sigma_4 = s[c2(I_{1x} + I_{2x})S_y - s4(I_{1x}I_{2z} + I_{1z}I_{2x})S_x] \quad [\text{D.5}]$$

The $180°_y(S)$ pulse changes the signs of the product operators containing S_x, and the simultaneous $\theta_y(I)$ pulse converts the two-spin coherence to S-spin magnetization:

$$\sigma_5 = -s[cs_\theta 2(I_{1z} + I_{2z})S_y + 2ss_\theta c_\theta 4I_{1z}I_{2z}S_x] \qquad [\text{D.6}]$$

Product operators containing only transverse I-spin operators represent unobservable MQC and have been disregarded. Because no further I-spin pulses are applied before detection, such terms cannot lead to S-spin magnetization.

The precession during the final τ period produces the state

$$\sigma_6 = s_\theta[2s^2c^2 S_x + (s^3c - sc^3)2(I_{1z} + I_{2z})S_y + 2s^2c^2 4I_{1z}I_{2z}S_x]$$
$$+ s_{2\theta}[s^4 S_y - s^3c 2(I_{1z} + I_{2z})S_y - s^2c^2 4I_{1z}I_{2z}S_x] \qquad [\text{D.7}]$$

The undistorted triplets are represented by S_x operators; the two-operator terms represent phase anomalies and the three-operator terms are representatives of multiplet anomalies (see Section III,C,2). The fact that a 180°(I) pulse changes the signs of phase anomaly terms is exploited in the DEPT$^+$ experiment. The terms in the square brackets proportional to $\sin\theta$ cause J cross talk in the course of subspectral editing (see Section IV). Note that for matched timing ($\tau = 1/2J$), only the "genuine I$_2$S signal," $\sin(2\theta)S_x$, has a nonvanishing amplitude.

E. Operator Analysis of Editing by GL$^+$ Sequences

This section analyzes editing of proton-coupled ^{13}C NMR spectra using SEMUT GL$^+$ or DEPT GL$^+$ [Fig. 3-34]. As in the decoupled case discussed in Sections IV,C and IV,E, it is most convenient to first calculate the SEMUT GL$^+$ response of an IS spin system. We write down the series of transformations from the point immediately after the 90°(C) pulse to the end of the delay τ_3 immediately before the dotted π pulse of SEMUT GL$^+$ (Fig. 3-34a):

$$\sigma_i \xrightarrow{\pi S_x} \xrightarrow{-\pi J\tau_1 2I_zS_z} \xrightarrow{-\frac{\pi}{2}I_x}$$
$$\xrightarrow{\pi J\tau_2 2I_zS_z} \xrightarrow{\frac{\pi}{2}I_x} \xrightarrow{\theta I_y} \xrightarrow{\pi J\tau_3 2I_zS_z} \sigma_f \qquad [\text{E.1}]$$

The phase of the first 90°(H) pulse of the purging sandwich has been set to x and $\phi = x$. The last six transformations in [E.1] are denoted by the propagator $P_S = P_S(\tau_1, \tau_2, \tau_3, \theta)$, that is,

$$\sigma_i \xrightarrow{\pi S_x} \xrightarrow{P_S} \sigma_f \qquad [\text{E.2}]$$

With the abbreviations $c_k, s_k = \cos \pi J\tau_k, \sin \pi J\tau_k$, and $c_\theta, s_\theta = \cos\theta, \sin\theta$, the following two transformations may easily be derived (only observable terms included):

$$S_x \xrightarrow{P_S} (c_1c_2c_3 + s_1s_3c_\theta)S_x + (c_1c_2s_3 - s_1c_3c_\theta)2I_zS_y \qquad [\text{E.3}]$$

$$S_y \xrightarrow{P_S} (c_1c_2c_3 + s_1s_3c_\theta)S_y - (c_1c_2s_3 - s_1c_3c_\theta)2I_zS_x \qquad [\text{E.4}]$$

To save space we abbreviate [E.3] and [E.4] in the following way:

$$S_x \xrightarrow{P_S} AS_x + B2I_zS_y \qquad [\text{E.5}]$$

$$S_y \xrightarrow{P_S} AS_y - B2I_zS_x \qquad [\text{E.6}]$$

These transformations are similar to the basic equations [12] and [13], but the reverse transformations (analogs of Eqs. [14] and [15]) have coefficients different from A and B. However, only [E.5] and [E.6] are of interest for the present purpose.

The generalization from IS to I_nS does not cause particular problems. Each I spin has its associated P_S operator, and we may write

$$-S_y \xrightarrow{\pi S_x} \xrightarrow{P_{S1}} \xrightarrow{P_{S2}} \cdots \xrightarrow{P_{Si}} \cdots \xrightarrow{P_{Sn}} \qquad [\text{E.7}]$$

in analogy to a cascade of J couplings. For an I_2S spin system, this equation would be

$$-S_y \xrightarrow{\pi S_x} S_y \xrightarrow{P_{S1}} AS_y - B2I_{1z}S_x \xrightarrow{P_{S2}}$$
$$A^2 S_y - AB2(I_{1z} + I_{2z})S_x - B^2 4 I_{1z}I_{2z}S_y \qquad [\text{E.8}]$$

In general, the term containing p I_z operators for an I_nS spin system [e.g., $p = 2$ in an I_3S system: $4(I_{1z}I_{2z} + I_{1z}I_{3z} + I_{2z}I_{3z})S_y$] will be proportional to $A^{n-p}B^p$. The sign is positive for $p = 4k$ or $p = 4k + 3$, $k = 0, 1, 2, \ldots$, and negative for the other p values.

For DEPT GL$^+$ without the purging pulse (Fig. 3-34b), $P_{D1} = P_{D1}(\tau_1, \tau_2, \tau_3, \theta)$ denotes the whole set of transformations affecting both I_1 and S in an I_nS spin system. This leads to

$$I_{1z} \xrightarrow{P_{D1}} s_1s_3s_\theta S_x - s_1c_3s_\theta 2I_{1z}S_y \qquad [\text{E.9}]$$

The evolution caused by the remaining $n - 1$ protons in the I_nS spin system is governed by the SEMUT GL operator P_S. The fate of the magnetization starting on I_1 may thus be written

$$I_{1z} \xrightarrow{P_{D1}} \xrightarrow{P_{S2}} \xrightarrow{P_{S3}} \cdots \xrightarrow{P_{Si}} \cdots \xrightarrow{P_{Sn}} \qquad [\text{E.10}]$$

The total response of I_nS spin system is obtained by summing the results of the n series of transformations of the type of [E.10]. In the total response, the term containing $p\,I_z$ operators is proportional to

$$s_1 s_\theta [(n-p)s_3 A^{n-p-1} B^p - p c_3 A^{n-p} B^{p-1}] \qquad \text{[E.11]}$$

where it is understood that negative exponents cause the factor to be zero. Positive signs are obtained for $p=4k$ or $p=4k+1$. The two terms in [E.11] arise from the two different "starting conditions" (S_x or $-2I_{iz}S_y$) after the initial DEPT GL operator P_{Di} (compare [E.9] and [E.10] for $i=1$).

We now restrict the discussion of SEMUT GL$^+$ and DEPT GL$^+$ to the cases relevant for editing of ^{13}C NMR spectra, that is, $n \leq 3$.

The ^{1}H π purging pulse applied in every second experiment eliminates product operators containing odd numbers of I_z operators. Therefore only terms with zero or two I_z operators need to be considered.

The J cross talk from I_nS to I_mS is the same for the in-phase terms as for the decoupled case (Eq. [121]), and it turns out that the J cross talk represented by the terms containing two I_z operators is also the same for SEMUT GL$^+$ and DEPT GL$^+$ editing of coupled spectra. In analogy to Eq. [121], we may write

$$\frac{I(I_nS \to I_mS)^{zz}}{I_0(I_mS)} = \sum_i \binom{2}{i}\binom{n-2}{m-i}(c_1 c_2)^{n-m}(s_1)^m (c_3)^{n-m+2i-2}(s_3)^{m+2-2i}$$

[E.12]

where $I_0(I_mS)$ again is the intensity of the ideal response of an I_mS group $[\tau_1 = \tau_2 = \tau_3 = (2J)^{-1}]$. The summation over i includes $i=2$ for $m=n$, $i=2$, 1 for $m=n-1$, and else $i=2, 1, 0$. Negative exponents or "impossible" binomial coefficients should be understood as causing the respective factors to be zero.

Table 3-10 contains the explicit SEMUT GL$^+$ and DEPT GL$^+$ responses of I_nS groups with $n \leq 3$.

From Table 3-10, it can be seen that the genuine CH signals and the J cross-talk signals in the quaternary subspectrum appear as 1:1 doublets. The CH$_2$ J cross-talk signals in the CH subspectrum appear as doublets with splittings $2J$, that is, the central lines are missing. All other CH$_2$ signals and all CH$_3$ signals are superpositions of in-phase (S_y) and doubly antiphase contributions (e.g., $I_{1z}I_{2z}S_y$), where the relative proportions depend on J and the setting of the delays in the pulse sequences.

The J cross-talk signals are, as for the decoupled case, proportional to at least three cosine factors, but the intensity expressions are no longer

TABLE 3-10

SEMUT GL$^+$ AND DEPT GL$^+$ RESPONSES OF COUPLED CH$_n$ GROUPS^{a,b}

C:	$x_0 S_y$
CH:	$x_0 c_1 c_2 c_3 S_y$
	$+ x_1 s_1 s_3 S_y$
CH$_2$:	$x_0 [c_1^2 c_2^2 (c_3^2 S_y - s_3^2 4 I_{1z} I_{2z} S_y)]$
	$+ x_1 [2 c_1 s_1 c_2 c_3 s_3 (S_y + 4 I_{1z} I_{2z} S_y)]$
	$+ x_2 [s_1^2 (s_3^2 S_y - c_3^2 4 I_{1z} I_{2z} S_y)]$
CH$_3$:	$x_0 \{c_1^3 c_2^2 c_3 [c_3^2 S_y - s_3^2 4 (I_{1z} I_{2z} + I_{1z} I_{3z} + I_{2z} I_{3z}) S_y]\}$
	$+ x_1 \{c_1^2 s_1 c_2^2 s_3 [3 c_3^2 S_y + (3 c_3^2 - 1) 4 (I_{1z} I_{2z} + I_{1z} I_{3z} + I_{2z} I_{3z}) S_y]\}$
	$+ x_2 \{c_1 s_1^2 c_2 c_3 [3 s_3^2 S_y + (3 s_3^2 - 1) 4 (I_{1z} I_{2z} + I_{1z} I_{3z} + I_{2z} I_{3z}) S_y]\}$
	$+ x_3 \{s_1^3 s_3 [s_3^2 S_y - c_3^2 4 (I_{1z} I_{2z} + I_{1z} I_{3z} + I_{2z} I_{3z}) S_y]\}$

^{a} For SEMUT GL$^+$, $x_m = \cos^m \theta$ and for DEPT GL$^+$, $x_m = m \sin \theta \cos^{m-1} \theta$. For other abbreviations see the text for appendix E.

^{b} Without practical consequences the DEPT GL$^+$ responses have been phase shifted by 90°.

symmetrical in τ_1 and τ_3. For narrow J ranges, this is obviously not a problem, but for wider ranges the asymmetry can be unfavorable with respect to J cross talk. Experimentally, the choice $\tau_1 < \tau_3$ leads to the simplest pulse sequences (Fig. 3-34). The opposite choice ($\tau_1 > \tau_3$) would require an additional ^{1}H 180° pulse at the midpoint between the purging pulse and the start of acquisition (compare DEPT^{++} in Fig. 3-12f). For these sequences delayed coupling is not necessary. When a high degree of J cross-talk suppression is required, both versions should be applied with full τ scrambling, that is, a total of six scans for each step of the applied phase cycle. This was not employed to obtain the spectra presented in Fig. 3.38, and in fact small J cross-talk signals are visible.

Acknowledgment

The work presented in this chapter has been supported by the Danish Natural Science Research Council (J.no. 11-2194, 11-3294, 11-3933, 11-5309, 511-15401) and the Swiss National Science Foundation. The Varian XL-300 spectrometer at the University of Aarhus was purchased through equipment grants from the Danish research councils SNF and STVF, Carlsbergfondet, and Direktør Ib Henriksens Fond. Last, but not least, we want to express out gratitude to the co-workers H. Bildsøe, G. Bodenhausen, G. W. Eich, R. R. Ernst, and M. H. Levitt for fruitful collaboration and stimulating discussions.

References

1. K. G. R. Pachler and P. L. Wessels, *J. Magn. Reson.* **12**, 337 (1973).
2. S. Sørensen, R. S. Hansen, and H. J. Jakobsen, *J. Magn. Reson.* **14**, 243 (1974).
3. R. A. Hoffman, B. Gestblom, and S. Forsen, *J. Chem. Phys.* **39**, 468 (1963).
4. R. A. Hoffman and S. Forsen, *Prog. NMR Spectrosc.* **1**, 15 (1966).
5. R. D. Bertrand, W. B. Moniz, A. N. Garroway, and G. C. Chingas, *J. Am. Chem. Soc.* **100**, 5227 (1978).
6. H. J. Jakobsen, O. W. Sørensen, and H. Bildsøe, *J. Magn. Reson.* **51**, 157 (1983).
7. H. J. Jakobsen, H. Bildsøe, S. Dønstrup, and O. W. Sørensen, *J. Magn. Reson.* **57**, 324 (1984).
8. D. T. Pegg, D. M. Doddrell, and M. R. Bendall, *J. Chem. Phys.* **77**, 2745 (1982).
9. O. W. Sørensen, S. Dønstrup, H. Bildsøe, and H. J. Jakobsen, *J. Magn. Reson.* **55**, 347 (1983).
10. H. Bildsøe, S. Dønstrup, H. J. Jakobsen, and O. W. Sørensen, *J. Magn. Reson.* **53**, 154 (1983).
11. I. D. Campbell, C. M. Dobson, R. J. P. Williams, and P. E. Wright, *FEBS Lett.* **57**, 96 (1975).
12. O. W. Sørensen, M. Rance, and R. R. Ernst, *J. Magn. Reson.* **56**, 527 (1984).
13. U. B. Sørensen, H. Bildsøe, H. J. Jakobsen, and O. W. Sørensen, *J. Magn. Reson.* **65**, 222 (1985).
14. O. W. Sørensen, U. B. Sørensen, and H. J. Jakobsen, *J. Magn. Reson.* **59**, 332 (1984).
15. U. B. Sørensen, H. J. Jakobsen, and O. W. Sørensen, *J. Magn. Reson.* **61**, 382 (1985).
16. H. J. Jakobsen, U. B. Sørensen, H. Bildsøe, and O. W. Sørensen, *J. Magn. Reson.* **63**, 601 (1985).
17. M. H. Levitt, O. W. Sørensen, and R. R. Ernst, *Chem. Phys. Lett.* **94**, 540 (1983).
18. D. T. Pegg and M. R. Bendall, *J. Magn. Reson.* **55**, 114 (1983).
19. O. W. Sørensen and R. R. Ernst, *J. Magn. Reson.* **51**, 477 (1983).
20. O. W. Sørensen, G. W. Eich, M. H. Levitt, G. Bodenhausen, and R. R. Ernst, *Prog. NMR Spectrosc.* **16**, 163 (1983).
21. G. Bodenhausen and R. Freeman, *J. Magn. Reson.* **36**, 221 (1979).
22. H. Hatanaka and C. S. Yannoni, *J. Magn. Reson.* **42**, 330 (1981).
23. A. Wokaun and R. R. Ernst, *Mol. Phys.* **36**, 317 (1978).
24. S. Macura and R. R. Ernst, *Mol. Phys.* **41**, 95 (1980).
25. A. Kumar, G. Wagner, R. R. Ernst, and K. Wüthrich, *J. Am. Chem. Soc.* **103**, 3654 (1981).
26. S. Macura, Y. Huang, D. Suter, and R. R. Ernst, *J. Magn. Reson.* **43**, 259 (1981).
27. U. B. Sørensen, H. Bildsøe, and H. J. Jakobsen, *J. Magn. Reson.* **58**, 517 (1984).
28. A. A. Maudsley and R. R. Ernst, *Chem. Phys. Lett.* **50**, 368 (1977).
29. J. W. Emsley and J. C. Lindon, "NMR Spectroscopy Using Liquid Crystal Solvents." Pergamon, Oxford, 1975.
30. A. D. Cohen, R. Freeman, K. A. McLauchlan, and D. H. Whiffen, *Mol. Phys.* **7**, 45 (1963).
31. R. Freeman and D. H. Whiffen, *Mol. Phys.* **4**, 321 (1961).
32. J. P. Maher and D. F. Evans, *Proc. Chem. Soc. London* p. 208 (1961).
33. S. Schäublin, A. Höhener, and R. R. Ernst, *J. Magn. Reson.* **13**, 196 (1974).
34. W. P. Aue, E. Bartholdi, and R. R. Ernst, *J. Chem. Phys.* **64**, 2229 (1976).
35. A. Bax and R. Freeman, *J. Magn. Reson.* **44**, 542 (1981).
36. P. E. Hansen, *Annu. Rep. NMR Spectrosc.* **11A**, 65 (1981).
37. V. Wray and P. E. Hansen, *Annu. Rep. NMR Spectrosc.* **11A**, 99 (1981).
38. V. Wray, L. Ernst, T. Lund, and H. J. Jakobsen, *J. Magn. Reson.* **40**, 55 (1980).
39. A. Bax, R. Freeman, and S. P. Kempsell, *J. Am. Chem. Soc.* **102**, 4849 (1980).
40. O. W. Sørensen, H. Bildsøe, and H. J. Jakobsen, *J. Magn. Reson.* **45**, 325 (1981).
41. O. W. Sørensen and R. R. Ernst, *J. Magn. Reson.* **54**, 122 (1983).
42. O. W. Sørensen and R. R. Ernst, *J. Magn. Reson.* **63**, 219 (1985).
43. M. Hansen, Thesis, Univ. of Aarhus, 1974.

44. J. C. Madsen, H. Bildsøe, H. J. Jakobsen, and O. W. Sørensen, *J. Magn. Reson.* **67**, 243 (1986).
45. H. Kogler, O. W. Sørensen, G. Bodenhausen, and R. R. Ernst, *J. Magn. Reson.* **55**, 157 (1983).
46. O. W. Sørensen, *J. Magn. Reson.* **57**, 506 (1984).
47. D. M. Doddrell, D. T. Pegg, and M. R. Bendall, *J. Magn. Reson.* **48**, 323 (1982).
48. R. R. Ernst and W. A. Anderson, *Rev. Sci. Instrum.* **37**, 93 (1966).
49. J. S. Waugh, *J. Mol. Spectrosc.* **35**, 298 (1970).
50. O. W. Sørensen, R. Freeman, T. A. Frenkiel, T. H. Mareci, and R. Schuck, *J. Magn. Reson.* **46**, 180 (1982).
51. U. Piantini, O. W. Sørensen, and R. R. Ernst, *J. Am. Chem. Soc.* **104**, 6800 (1982).
52. O. W. Sørensen, M. H. Levitt, and R. R. Ernst, *J. Magn. Reson.* **55**, 104 (1983).
53. M. H. Levitt and R. R. Ernst, *Chem. Phys. Lett.* **100**, 119 (1983).
54. G. A. Morris, *J. Magn. Reson.* **44**, 277 (1981).
55. M. H. Levitt, O. W. Sørensen, and R. R. Ernst, *Chem. Phys. Lett.* **94**, 540 (1983).
56. R. Freeman, T. A. Frenkiel, and M. H. Levitt, *J. Magn. Reson.* **44**, 409 (1981).

4

Principles of Multiple-Quantum Spectroscopy

THOMAS H. MARECI
DEPARTMENT OF RADIOLOGY AND DEPARTMENT OF PHYSICS
UNIVERSITY OF FLORIDA
GAINESVILLE, FLORIDA 32610

I. Introduction	259
II. Operator Formalism	266
III. Multiple-Quantum Coherence for a System of Two Spin-$\frac{1}{2}$ Nuclei	267
A. Excitation of Zero- and Double-Quantum Coherence	268
B. Evolution of Zero- and Double-Quantum Coherence	276
C. Indirect Detection of Zero- and Double-Quantum Coherence	280
D. Summary for the Two-Spin System and Applications	287
IV. Order-Selective Detection	292
A. Coherence-Transfer Pathways	295
B. Coherence-Transfer Echoes	305
V. Multiple-Quantum Coherences for Systems Containing Three or More Spin-$\frac{1}{2}$ Nuclei	310
A. Excitation of Multiple-Quantum Coherence	310
B. Composite Rotations	319
C. Evolution of Multiple-Quantum Coherences	324
D. Single-Pulse Mixing and Coherence-Transfer Selection Rules	326
E. Uniform Excitation and Indirect Detection of Multiple-Quantum Coherence	332
VI. Summary	337
References	338

I. Introduction

The manifold of energy levels available to a system of interacting magnetic spins contains a wealth of information. This became apparent with the discovery of chemical shifts and spin-coupling interaction in nuclear magnetic resonance spectroscopy. Many methods have been devised to perturb the spin system and measure its interactions as well as spin–lattice and spin–spin relaxation rates. Subsequent applications to the study of liquids, liquid crystals and solids have had a profound influence on progress in these areas (1).

Most nuclear magnetic resonance experiments have been designed to provide a first order perturbation to the spin system, so that the measured spectrum clearly represents populations of energy levels and the transition frequencies detectable through magnetic dipole transitions of the system of nuclei in a magnetic field. Assume that $|i\rangle$ and $|j\rangle$ are eigenstates of the unperturbed Hamiltonian which describes a system of N spins. For a static field aligned along the z axis, the Zeeman quantum number m characterizes the z component of total spin angular momentum, $F_z = \Sigma_{k=1}^{N} I_{kz}$, and is given by

$$F_z|i\rangle = m_i|i\rangle \quad [1]$$

A transition between eigenstates is represented by a change in the Zeeman quantum number,

$$p = \Delta m_{ij} = m_i - m_j \quad [2]$$

Here, p is called the order of transition. As will be discussed later, the observable magnetic dipole transitions occur between states related by $p = \pm 1$, representing changes which are single-quantum transitions. Therefore, the observed spectrum is restricted to contain measurable single-quantum transition frequencies.

For a group of interacting spins, many orders of transition other than $p = \pm 1$ are possible. A transition can be classified according to its order so that transitions with $p = 0$, ± 1, or ± 2, will be referred to as zero-, single-, or double-quantum transitions, respectively. It is not possible to observe transitions with p other than ± 1 directly with simple single-pulse excitation so high-order spectra have traditionally not been observed, but, with the advent of indirect detection schemes, there has been an increase in the study of high-order spectra. Two recent excellent reviews (2, 3) summarize work in this area. It is the intent of this chapter to expand on these reviews and act as a guide to understanding the principles of multiple-quantum spectroscopy. The analysis will be restricted to systems with multiple spin-$\frac{1}{2}$ nuclei and some applications will be introduced to illustrate the salient features of the technique. Special emphasis will be placed on the indirect detection of multiple-quantum spectra with two-dimensional Fourier transformation methods (4–7). Specific applications will be taken from liquid state spectroscopy of organic molecules.

One of the simplest and most effective descriptions of the phenomenon of nuclear magnetic resonance is given by Bloch's original vector model (8). The effect of an rf excitation pulse on a single spin is visualized as a rotation about the applied field in a fashion analogous to free precession about the static magnetic field. The effect of an applied magnetic field on the spin "magnetization vector" can be described by a classical torque equation. For the applied static field aligned along the z axis, the Bloch model adequately describes the quantum-mechanical processes for transitions with $p = \pm 1$ among energy levels for a noninteracting ensemble of spins.

Very few magnetic spin systems represent an ensemble of ideal isolated single spins with only a single transition available between two Zeeman energy levels. Naturally occurring examples of isolated single-spin systems would include natural abundance ^{13}C in graphite or at a quaternary position in an organic molecule with no long-range coupling to neighboring ^1H. This assumes that the probability that two ^{13}C spins are close enough together in natural abundance to interact is so low as to be negligible. As will be shown later, this low probability of interaction can be used to distinct advantage in ^{13}C nuclear magnetic resonance spectroscopy of organic molecules.

In fact, multispin, multiple-quantum processes are more the rule than the exception. When a group of spins interact, by definition the individual spins are no longer isolated. If one spin is perturbed, the measured result reflects the presence of the other interacting spin. The effects of multiple-quantum transitions are usually observed indirectly and combine to give the final result. A good example of such a process is relaxation, which proceeds through second-order effects. For spins interacting through dipolar or scalar coupling, zero- and double-quantum transitions contribute to the rate of relaxation through the random fluctuations in the spin–spin interaction (8). When a high-order coherence is created for multiple-quantum spectroscopy involving a group of spins, the group as a whole equally populates the particular energy levels in the same manner as does a single spin. Thus the group of spins acts as a whole when a nuclear magnetic resonance is produced.

The excitation of nuclear magnetic resonance is usually accomplished by imposing an rf magnetic field which is effective in creating transverse magnetization. This is achieved by either a low-power cw field using a swept main magnetic field or by a short intense rf pulse which provides broadband excitation at a fixed main magnetic field strength. In each case, the interaction of the imposed rf magnetic field with the spin system is first order and the measured resonance frequency and intensity is a function only of the available single-quantum transitions. Multiple-pulse methods of measuring spin–lattice (9) and spin–spin relaxation (10) normally only induce single-quantum transitions and indirectly detect multiple-quantum transitions through the measurement of relaxation. Many other experiments induce multiple-quantum transitions, which are only detected indirectly.

Perhaps the simplest example of induced multiple-quantum coherences is that of the "slow beat" patterns observed in early spin echo measurements (11). In this experiment, two $\pi/2$ pulses separated by a delay time τ induce a spin echo at time 2τ. If the spin system contains J coupling, the observed echo amplitude varies as a function of τ with a period proportional to $1/J$, because the transverse single-quantum coherence created by the first $\pi/2$ pulse is modulated by the J-coupling interaction among spins, as described in Chapter 1, Section II,B. At time τ, the second $\pi/2$ pulse transfers part of this modulated single-quantum coherence into other possible orders of coherence in the spin system. The efficiency of this type of transfer is dependent on the magnitude of the J coupling and the chosen delay time τ. Hence, the amplitude of the single-quantum coherence not transferred to other orders of coherence is periodic in τ with period proportional to $1/J$. The observed spectrum is modulated by both the J modulation of spin multiplets and the transfer to high orders of coherence. The result is the characteristic phase modulation with period proportional to $1/J$.

Experiments to explore the possibilities for indirect detection of multiple-quantum coherence can be cast in the general formulation of two-dimensional NMR experiments as shown in Fig. 4-1a. For multiple-quantum spectroscopy, the spin system initially in thermal equilibrium is prepared into a state containing the desired multiple-quantum coherences by the generalized preparation operator U. The preparation takes place during the interval τ_p. The spin system then evolves for a time t_1 under the influence of the spin interactions effective during the evolution time. At the end of t_1, multiple-quantum coherences are transferred to observable single-quantum coherences by the generalized mixing operator V. The mixing interval is denoted τ_m. The observable single-quantum coherences are observed in the usual manner during the detection time t_2. Double Fourier transformation with respect to the evolution time t_1 and detection time t_2 results in a two-dimensional frequency spectrum.

To understand the significance of the resulting two-dimensional frequency spectrum, it is instructive to follow the steps necessary to create the data matrix as a function of the evolution time t_1. For a given value of t_1, the spectrum observed during t_2 has the resonance frequencies of the conventional one-dimensional spectrum. It is the amplitude and phase of the detected resonances which reflect the evolution of the spin system during t_1 as well as the details of the preparation and mixing process. For each value of t_1, the preparation and mixing processes are usually held fixed. Therefore Fourier transformation with respect to t_1 results in a spectrum with resonance frequencies which reflect the evolution of the coherences present during the t_1 interval. The resonance positions in the full two-dimensional spectrum represent coherences in the spin system which has

4. PRINCIPLES OF MULTIPLE-QUANTUM SPECTROSCOPY

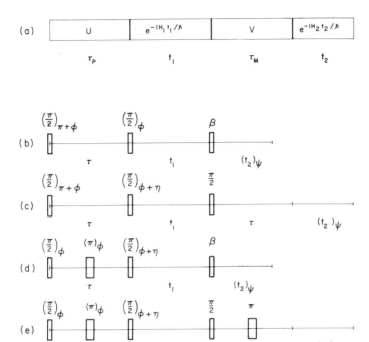

Fig. 4-1. The general formulation of the two-dimensional NMR technique and some specific examples for indirect detection of multiple-quantum spectra. (a) General scheme of the two-dimensional technique. (b–e) Specific pulse sequences discussed in the text which allow indirect detection of multiple-quantum spectroscopy. The subscripts ϕ and η represent variable rf phase and ψ represents variable receiver phase.

been prepared into a state containing some specific coherences that evolved in t_1 and have been transferred to observable single-quantum coherences, detected during t_2. The amplitude and phase of the resonances in the two-dimensional spectrum reflect the details of the preparation and mixing process.

Using the technique of two-dimensional Fourier transform spectroscopy, it is possible indirectly to study interactions in the spin system which are not directly observable by conventional one-dimensional techniques. From multiple-quantum spectroscopy, it is possible to map the details of the coupling interaction among spins for all transitions available to the system. This has opened a new "window" on the spin interactions, which provides unique information not available through other methods.

The potential information content of two-dimensional multiple-quantum spectra is indicated by Fig. 4-2. This is the ^{1}H multiple-quantum spectrum

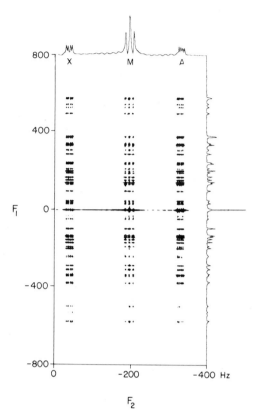

Fig. 4-2. Two-dimensional multiple-quantum spectrum of an AMX spin system 2,3-dibromopropionic acid in benzene-d_6. The spectrum was obtained with a technique of the type shown in Fig. 4-1b. In this case, β was set equal to $\pi/2$ with ϕ and ψ held fixed.

of all possible orders for an AMX spin system, 2,3-dibromopropionic acid dissolved in benzene-d_6. In general, for a system of N coupled nonequivalent spin-$\frac{1}{2}$ magnetic spins, the number of energy levels is equal to 2^N. The same system can have $2^N(2^N-1)$ possible transitions between energy levels. For the AMX system, this corresponds to eight energy levels and 56 possible transitions. Since there are 12 observable single-quantum transitions, this results in a total of 684 possible resonance positions in the two-dimensional spectrum (including those derived from the longitudinal magnetization). Clearly, the dilemma is how to deal with this much information. Fortunately, methods exist to handle this situation, and these will be discussed in the course of this chapter.

In discussing multiple-quantum spectroscopy, it is useful to develop a lexicon to allow a distinction to be drawn between the various types of multiple-quantum coherences and processes. The classification scheme for single-quantum transitions discussed by Corio (12) can be extended to multiple-quantum transitions. Thus the distinction of progessively and regressively connected transition is straightforward. The concept of a mixed transition discussed by Corio is very useful; however, it is ambiguous in the context of multiple-quantum spectroscopy. An alternative concept, partial-spin multiple-quantum coherence, will be introduced in the following lexicon.

Lexicon for Multiple-Quantum Spectroscopy

Coherence: A superposition of states of a spin or group of spins with a coherence phase relationship; in terms of energy levels, a coherence exists between levels available to a spin or group of spins when the spin or group populates each level and the phases of the coefficients for the state functions of each level have a coherent relationship. Coherences are often termed transitions, because a spin can be thought of as making transitions between energy levels in order to populate both levels.

Total-spin multiple-quantum coherence: A coherence between levels for a single spin with $I > \frac{1}{2}$ corresponding to a multiphoton process. Also a coherence between levels of an N-spin-$\frac{1}{2}$ system where all N spins are involved in the coherence. For a single-spin system this type of coherence will be denoted

$$\{p\text{-QC}\}_\varepsilon$$

where $\varepsilon = x$ or y. For an N-spin system, this type of coherence will be denoted

$$\{N\text{-spin}, p\text{-QC}\}_\varepsilon$$

When $p = N$, this will be written in the simpler form above.

Partial-spin multiple-quantum coherence: A coherence between energy levels for a group of spins such that the resonance frequency is equal to a linear combination of Larmor frequencies for a number of spins within the group. This type of coherence involves the transition of only a *portion* of the total number of coupled spins within the group. Corio termed this a mixed transition (12). The terms total-spin and partial-spin multiple-quantum coherence are used here to avoid confusion with the frequently used terms pure and mixed states of an ensemble (13). A partial-spin multiple-quantum coherence for a system of N coupled spins will be denoted as

$$\{N, q\text{-spin}, p\text{-QC}\}_\varepsilon$$

where the energy levels involved represent states of an N-spin system, the coherence is of order p, q spins make a transition, and $\varepsilon = x$ or y. It should be noted that the commonly used term "combination line coherence" can refer to either a total-spin or partial-spin multiple-quantum coherence. As used for conventional NMR spectroscopy, combination line coherences have $p = \pm 1$.

Active spin: A spin is considered active in a coherence if its magnetic quantum number m_z is changed by the transition between energy levels involved in the coherence.

Passive spin: A spin is considered passive in a coherence if its magnetic quantum number m_z is not changed by the transition between energy levels involved in the coherence.

Coherence transfer: The process of transferring coherence among the possible coherences available to the spin system. Two important types of coherence transfer can be distinguished for groups of coupled spins.

Coherence transfer to an active spin: Coherence transfer from a total-spin or partial-spin multiple-quantum coherence for a *group* of spins to a single-quantum coherence for *one* spin for which the magnetic quantum number m_z changes in the original multiple-quantum coherence. Using an AMX spin system as an example, the coherence transfer from a partial-spin AM double-quantum coherence to an active-spin single-quantum coherence, say A, could be denoted by **A**M**X** → **A**MX, where boldface letters indicate which spins are involved in the coherence.

Coherence transfer to a passive spin: Coherence transfer from a partial-spin multiple-quantum coherence of a *group* of spins to a single-quantum coherence of *one* spin that has no change in its magnetic quantum number m_z in the original coherence. Again using the AMX example, a coherence transfer from the partial-spin AM double-quantum coherence to a passive spin single-quantum coherence, say X, can be denoted by **A**M**X** → AM**X**.

These terms are used throughout this chapter, and the reader may find this lexicon useful for reference.

II. Operator Formalism

In Chapter 2, Levitt discussed the product-operator formulation for the density-operator description of NMR experiments. The product-operator formalism (14–16) is most useful and will be extended in this chapter to develop the concepts of multiple-quantum spectroscopy. This formalism provides a method of treating systems of interacting spins at a level of abstraction once removed from a consideration of the specific energy levels of the spin system associated with an observable NMR spectrum. Even in a two-dimensional multiple-quantum spectrum where one dimension rep-

4. PRINCIPLES OF MULTIPLE-QUANTUM SPECTROSCOPY

resents coherences other than single-quantum, this formalism allows the study of processes without recourse to detailed descriptions involving individual energy levels. However, on a few occasions it will be necessary to provide a detailed description of a specific spin system. Then an energy-level numbering will be assigned according to the lexicographical ordering (17) and the spin system will be described in terms of single-transition operators (18). This will allow a specific description of the final spectrum in terms of the energy-level diagram for the spin system.

III. Multiple-Quantum Coherence for a System of Two Spin-$\frac{1}{2}$ Nuclei

In the introduction a number of ways to excite the possible orders of coherence within a spin system have been outlined. Here we will concentrate on excitation schemes using short intense rf magnetic-field pulses. This will limit the discussion to broadband excitation of all the spins of a particular species; however, this will provide a description of a class of experiments that are relatively simple and easy to implement.

As a starting point, consider the two-spin-$\frac{1}{2}$, AX system. This is the simplest system which can have coherence between levels not directly observable in a conventional one-dimensional spectrum. The energy-level diagram for the AX system is shown in Fig. 4-3. There are four directly observable single-quantum coherences represented by the straight-line segments joining energy levels. The coherences of the first spin are associated with transitions from energy levels 1 to 3 and 2 to 4, while the second spin is associated with transitions from levels 1 to 2 and 3 to 4. These are the observed resonances in the conventional spectrum. There are two other possible coherences for this spin system represented in the figure by the wavy lines. The coherence associated with transition from energy level 2 to

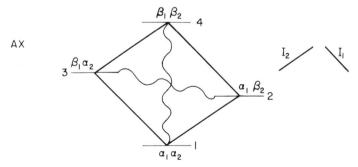

Fig. 4-3. Energy-level diagram for an AX spin system. The numbering scheme follows the lexicographical ordering (17).

3 represents a zero-quantum coherence, $p = 0$, and that associated with the transition from level 1 to 4 a double-quantum coherence, $p = \pm 2$.

A. EXCITATION OF ZERO- AND DOUBLE-QUANTUM COHERENCE

In order to see how zero- and double-quantum coherences are excited, first consider the single-element operator form of their representation. The operators that induce zero-quantum transitions can be written as $I_1^+ I_2^-$ or $I_1^- I_2^+$ and the double-quantum operator as $I_1^+ I_2^+$ or $I_1^- I_2^-$. Recall that the transverse components of angular momentum can be written as

$$I_{ix} = \tfrac{1}{2}(I_i^+ + I_i^-) \qquad [3]$$

$$I_{iy} = \frac{1}{2i}(I_i^+ - I_i^-) \qquad [4]$$

and

$$I_{iz} = \tfrac{1}{2}(I_i^\alpha - I_i^\beta) \qquad [5]$$

There are four possible products of the transverse components of these operators which represent linear combinations of zero- and double-quantum operators for a two-spin system:

$$2I_{1x}I_{2x} = \tfrac{1}{2}(I_1^+ I_2^+ + I_1^- I_2^-) + \tfrac{1}{2}(I_1^+ I_2^- + I_1^- I_2^+) \qquad [6]$$

$$2I_{1x}I_{2y} = \frac{1}{2i}(I_1^+ I_2^+ - I_1^- I_2^-) - \frac{1}{2i}(I_1^+ I_2^- - I_1^- I_2^+) \qquad [7]$$

$$2I_{1y}I_{2x} = \frac{1}{2i}(I_1^+ I_2^+ - I_1^- I_2^-) + \frac{1}{2i}(I_1^+ I_2^- - I_1^- I_2^+) \qquad [8]$$

$$2I_{1y}I_{2y} = -\tfrac{1}{2}(I_1^+ I_2^+ + I_1^- I_2^-) + \tfrac{1}{2}(I_1^+ I_2^- + I_1^- I_2^+) \qquad [9]$$

The factor of 2 is included for later convenience. The above four equations can be rearranged so that the zero- or double-quantum operators can be written in terms of the transverse angular momentum operator products. For zero-quantum coherence,

$$\tfrac{1}{2}(I_1^+ I_2^- + I_1^- I_2^+) = (I_{1x}I_{2x} + I_{1y}I_{2y}) = \{ZQC\}_x \qquad [10]$$

and

$$\frac{1}{2i}(I_1^+ I_2^- - I_1^- I_2^+) = -(I_{1x}I_{2y} - I_{1y}I_{2x}) = \{ZQC\}_y \qquad [11]$$

For double-quantum coherence,

$$\tfrac{1}{2}(I_1^+ I_2^+ + I_1^- I_2^-) = (I_{1x}I_{2x} - I_{1y}I_{2y}) = \{2QC\}_x \qquad [12]$$

4. PRINCIPLES OF MULTIPLE-QUANTUM SPECTROSCOPY

and

$$\frac{1}{2i}(I_1^+ I_2^+ - I_1^- I_2^-) = (I_{1x}I_{2y} + I_{1y}I_{2x}) = \{2QC\}_y \quad [13]$$

Expressing the zero- and double-quantum operators in this form provides some insight into the steps necessary to prepare the spin system for conversion of single-quantum coherence into other orders of coherence. Also the formal similarity of the transverse coherence operators of Eqs. [3] and [4] and the zero- and double-quantum coherence operators of Eqs. [10]-[13] can be generalized to all orders of multiple-quantum coherence (14). This analogy between transverse components of coherence allows a vectorial picture of free precession to be used to describe the evolution.

The essential step for the creation of multiple-quantum coherence is to prepare the system into a state with terms which represent products of the angular momentum operators. Recall the density-operator form for free precession in the presence of J coupling. Ignoring the effect of chemical shift precession for the moment, the density operator for free precession of the y component of angular momentum of spin 1 with J coupling to spin 2 can be written as

$$\sigma_1(t) = a[I_{1y}\cos(\pi Jt) - 2I_{1x}I_{2z}\sin(\pi Jt)] \quad [14]$$

This is the equation of motion of spin 1 following excitation of the system by a $\pi/2$ pulse about the x axis. The second term in brackets represents a product of angular momentum operators. It is terms of this form which can be converted into other orders of coherence. A second pulse of any strength, other than an integer multiple of π, applied about the x axis will convert the product term of Eq. [14] into a superposition of zero- and double-quantum terms as expressed by Eq. [7]. It is easiest to follow this conversion for a $\pi/2$ pulse about the x axis. In this case, the pulse has no effect on the I_{1x} part of the product term; however, the pulse will rotate I_{2z} to $-I_{2y}$ so that the product term becomes $2I_{1x}I_{2y}$. The combination of zero- and double-quantum coherences for this state is represented by Eq. [7]. Most efficient conversion is accomplished by a $\pi/2$ pulse.

This same type of analysis can be applied to heteronuclear spin systems where it is possible to excite separately one spin species at a time. In the case of an AX system, each individual spin can be excited separately to manipulate coherences within the spin system. In some situations it is also possible to excite separately individual spins in a homonuclear spin system with appropriate selective irradiation. In any case, the description above applies equally to any type of spin system; heteronuclear, homonuclear, or selectively irradiated homonuclear. The form of Eq. [14] is appropriate for any system and can be applied for homo- or heteronuclear coupled species.

The excitation can be considered as being applied as a single pulse to a homonuclear system or as simultaneous pulses to each species in a heteronuclear system. If Eq. [14] describes a heteronuclear spin system, the effect of a $\pi/2$ pulse to spin 2 can be developed in a straightforward manner. Only spin 2 would be affected by this pulse, so that I_{2z} would be rotated to the transverse plane and the bilinear product operator term would represent heteronuclear zero- and double-quantum coherence. Most of the development in this chapter can be treated similarly; thus, heteronuclear systems can be considered as a special case.

The inclusion of the chemical shift during free precession does not alter the basic relationship introduced in Eq. [14]: however, the amount of a particular order of coherence created by the second $\pi/2$ pulse will depend on both the chemical shifts and the values of J. The chemical shift dependence can be removed by applying a π pulse at time $\tau/2$ and delaying the second $\pi/2$ pulse by another interval $\tau/2$. The amount of coherence created will depend only on the coupling constant, and thus the choice of the optimum delay τ is simplified.

In order to be more specific, the development of the density operator for the homonuclear AX system will be followed for the $(\pi/2)_x$—$\tau/2$—$(\pi)_y$—$\tau/2$—$(\alpha)_{x+\eta}$ excitation sequence described above. The density operator following the initial pulse can be written as

$$\sigma(0) = -a(I_{1y} + I_{2y}), \qquad [15]$$

where a represents the Boltzmann polarization factor for this spin species. The effect of the free precession and of the refocusing pulse, $\tau/2$—$(\pi)_y$—$\tau/2$, can be written in a simple form for a general system of N weakly coupled spins. This procedure will be illustrated using the product operator formulation and then applied to the AX spin system.

The density operator at time τ, before the second $\pi/2$ pulse, can be written as

$$\sigma(\tau) = \exp(-i\mathcal{H}_0\tau/2) \exp(-i\pi F_y) \exp(-i\mathcal{H}_0\tau/2)\sigma(0)$$
$$\times \exp(i\mathcal{H}_0\tau/2) \exp(i\pi F_y) \exp(i\mathcal{H}_0\tau/2) \qquad [16]$$

where $F_\varepsilon = \sum_{i=1}^{N} I_{i\varepsilon}$, $\varepsilon = x$, y, or z for an N-spin system, and $\mathcal{H}_0 = \sum_{i=1}^{N} \omega_i I_{iz} + \sum_{j<k}^{N} 2\pi J_{jk} I_{jz} I_{kz}$ is the free precession Hamiltonian in frequency units for a weakly coupled spin system. The density operator can be simplified by inserting a unity operator, $E = e^{\pm i\pi F_y} e^{\mp i\pi F_y}$, into Eq. [16] before and after the density operator at $\tau = 0$, $\sigma(0)$. Then a term of the following form is encountered:

$$\exp(-i\pi F_y) \exp(-i\mathcal{H}_0\tau/2) \exp(i\pi F_y) \qquad [17]$$

4. PRINCIPLES OF MULTIPLE-QUANTUM SPECTROSCOPY

This equation has the form of a rotation of the free precession Hamiltonian about the y axis by an angle of π.

The effect of the refocusing pulse is quite important, so that it will be considered in a general way. Equation [17] can be written for any weakly coupled spin system as

$$\exp\left(-i\pi \sum_{i=1}^{N} I_{iy}\right) \exp\left[-i\left(\sum_{i=1}^{N} \omega_i I_{iz} + \sum_{j<k}^{N} 2\pi J_{jk} I_{jz} I_{kz}\right)\tau/2\right]$$

$$\times \exp\left(i\pi \sum_{i=1}^{N} I_{iy}\right) \qquad [18]$$

By expanding the exponential operator containing the Hamiltonian $\mathcal{H}_0$ and using another unity insertion step, it can be seen that the π rotation of the first term in parentheses will simply change the sign of this term, since it is effectively a rotation of the I_{iz} term by the I_{iy} operator to $-I_{iz}$. This represents an effective reversal of the sense of chemical shift precession. Since the product $I_{jz}I_{kz}$ is affected by the rotation of the individual factors separately, this product term is invariant to the π rotation. The net result is a refocusing of the chemical shift term after another delay, $\tau/2$, and a continuation of the precession in the presence of J coupling. The other π-rotation term from the unity operator has no effect on the initial state, $\sigma(0)$, given by Eq. [15]. In general, the effect of this rotation will be dependent on the form of $\sigma(0)$. However, if $\sigma(0)$ is created by a $\pi/2$ pulse applied in equilibrium, this π rotation will have no effect if the $\pi/2$ and π pulses are in quadrature and will change the sign of $\sigma(0)$ if they are applied in phase or 180° out of phase. This is a well-known result of phase shifts in spin-echo sequences (19). In general, for the pulse sequence we are considering here, Eq. [16] can be written (14)

$$\sigma(\tau) = \exp\left(-i \sum_{j<k} \pi J_{jk}\tau 2 I_{jz}I_{kz}\right) \sigma(0) \exp\left(i \sum_{j<k} \pi J_{jk}\tau 2 I_{jz}I_{kz}\right) \qquad [19]$$

where $\sigma(0)$ is given by an equation of the form of Eq. [15]. This is only exact for a weakly coupled system.

Applying this general result to an AX spin system, the density operator at time τ can be written as

$$\sigma(\tau) = -a[I_{1y}\cos(\pi J\tau) - 2I_{1x}I_{2z}\sin(\pi J\tau)$$

$$+ I_{2y}\cos(\pi J\tau) - 2I_{1z}I_{2x}\sin(\pi J\tau)]$$

$$= -a[(I_{1y} + I_{2y})\cos(\pi J\tau) - 2(I_{1x}I_{2z} + I_{1z}I_{2x})\sin(\pi J\tau)] \qquad [20]$$

Since only the product terms are transferred into zero- and double-quantum coherences by the last pulse, the delay time τ can be chosen to optimize this transfer by letting $\tau = (2J)^{-1}$. Thus Eq. [20] becomes

$$\sigma(\tau = 1/2J) = a[2(I_{1x}I_{2z} + I_{1z}I_{2x})] \quad [21]$$

To understand the significance of this state, it is perhaps easiest to relate these Cartesian-product operators to the specific number scheme for the AX spin system as defined in Fig. 4-3. First consider the expansion of these Cartesian product operators in terms of single-element operators. Such an expansion is discussed in Chapter 2 by Levitt. The two terms in Eq. [21] can be written as

$$2I_{1x}I_{2z} = \tfrac{1}{2}(I_1^+ + I_1^-)I_2^\alpha - \tfrac{1}{2}(I_1^+ + I_1^-)I_2^\beta \quad [22]$$

and

$$2I_{1z}I_{2x} = \tfrac{1}{2}I_1^\alpha(I_2^+ + I_2^-) - \tfrac{1}{2}I_1^\beta(I_2^+ + I_2^-) \quad [23]$$

The single-element operator terms in each equation above represent the two x components of single-quantum coherence for each spin. Note that the two x components have opposite sign.

This can be seen more clearly if one relates these single-element operators to the specific energy-level diagram as

$$2I_{1x}I_{2z} = \tfrac{1}{2}(|\alpha_1\alpha_2\rangle\langle\beta_1\alpha_2| + |\beta_1\alpha_2\rangle\langle\alpha_1\alpha_2|)$$
$$- \tfrac{1}{2}(|\alpha_1\beta_2\rangle\langle\beta_1\beta_2| + |\beta_1\beta_2\rangle\langle\alpha_1\beta_2|)$$
$$= \tfrac{1}{2}(|1\rangle\langle 3| + |3\rangle\langle 1|) - \tfrac{1}{2}(|2\rangle\langle 4| + |4\rangle\langle 2|) \quad [24]$$

and

$$2I_{1z}I_{2x} = \tfrac{1}{2}(|\alpha_1\alpha_2\rangle\langle\alpha_1\beta_2| + |\alpha_1\beta_2\rangle\langle\alpha_1\alpha_2|)$$
$$- \tfrac{1}{2}(|\beta_1\alpha_1\rangle\langle\beta_1\beta_2| + |\beta_1\beta_2\rangle\langle\beta_1\alpha_2|)$$
$$= \tfrac{1}{2}(|1\rangle\langle 2| + |2\rangle\langle 1|) - \tfrac{1}{2}(|3\rangle\langle 4| + |4\rangle\langle 3|) \quad [25]$$

The operator forms written in terms of specific energy levels can be recognized as representing the single-transition operators (18) appropriate for single-quantum transitions from energy levels 1 to 3, 2 to 4, 1 to 2, and 3 to 4. In analogy with the definition of the Cartesian components of angular momentum of Eqs. [3]–[5], the Cartesian components of the single-transition operators are defined as (18)

$$I_x^{rs} = \tfrac{1}{2}(|r\rangle\langle s| + |s\rangle\langle r|) \quad [26]$$

$$I_y^{rs} = \frac{1}{2i}(|r\rangle\langle s| - |s\rangle\langle r|) \quad [27]$$

and

$$I_z^{rs} = \tfrac{1}{2}(|r\rangle\langle r| - |s\rangle\langle s|) \quad [28]$$

4. PRINCIPLES OF MULTIPLE-QUANTUM SPECTROSCOPY

For the AX spin system, Eq. [21] has the simple form

$$\sigma(\tau = 1/2J) = a(I_x^{13} - I_x^{24} + I_x^{12} - I_x^{34}) \qquad [29]$$

The phase and relative amplitudes of the single-quantum coherences are seen clearly from the form of this equation [the spectrum would appear as the sum of parts (c) and (d) of Fig. 2-4].

The effect of the third pulse in the sequence is to transfer the single-quantum coherences into zero- and double-quantum coherences dependent on both the phase and magnitude of this pulse. The rotation operator for a pulse can be written in general (20) as

$$R(\alpha, \eta) = \exp(-i\eta F_z) \exp(-i\alpha F_x) \exp(i\eta F_z) \qquad [30]$$

where α is the angle of rotation and η is the phase angle of the applied field with respect to the x axis in the rotating frame of reference. Therefore the density operator immediately after the third pulse is

$$\sigma' = R(\alpha, \eta)\sigma(\tau)R^+(\alpha, \eta) \qquad [31]$$

The general situation of arbitrary α and η can be limited to the two special cases where η is equal to 0° or 90°. These two cases correspond to the situation where the third pulse is in phase or in quadrature with the first pulse.

For $\eta = 0°$, the density operator of Eq. [16] following the pulse can be written, using Eq. [21], as

$$\sigma' = a[2(I_{1x}I_{2z} + I_{1z}I_{2x}) \cos \alpha - 2(I_{1x}I_{2y} + I_{1y}I_{2x}) \sin \alpha] \qquad [32]$$

It is possible to obtain this result directly, since the rotation operator of Eq. [30] reduces to that for an α pulse about the x axis and only one operator in each product of Eq. [21] is affected by the pulse. The second term in the above equation can be identified as representing only double-quantum coherence as expressed by Eq. [13].

Experimental verification of the tip angle dependence of Eq. [32] is shown in Fig. 4-4. This figure represents the ^{1}H spectra of an AX spin system, 2,3-dibromothiophene, for the preparation sequence, $(\pi/2)_x - \tau/2 - (\pi)_y - \tau/2 - (\alpha)_x$, followed immediately by acquisition of the spectrum at each value of α. The transfer from single- to double-quantum coherence is optimum for $\alpha = n\pi/2$, where n is an odd integer.

The situation for $\eta = 90°$ is slightly more complicated. The rotation operator now reduces to that appropriate for an α pulse about the y axis. Using a pulse cascade approach (21), the rotation to each spin is applied

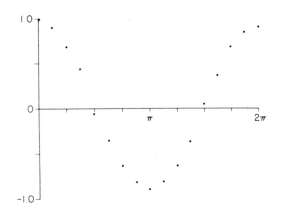

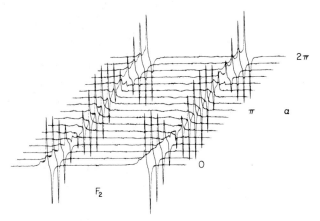

Fig. 4-4. Tip-angle dependence for the preparation of double-quantum coherence in an AX spin system 2,3-dibromothiophene in acetone-d_6. The ^{1}H spectra are shown at the bottom for the preparation sequence $(\pi/2)_x$—$\tau/2$—$(\pi)_y$—$\tau/2$—$(\alpha)_x$ followed immediately by acquisition. The graph at the top represents the normalized average peak height versus tip angle α.

separately in succession and the density operator after the pulse is then

$$\sigma' = a[2(I_{1x}I_{2z} + I_{1z}I_{2x})\cos^2\alpha$$
$$- 2(I_{1z}I_{2z} - I_{1x}I_{2x})(2\cos\alpha\sin\alpha)$$
$$- 2(I_{1z}I_{2x} + I_{1x}I_{2z})\sin^2\alpha] \qquad [33]$$

By combining terms, we can reduce this to the simpler expression,

$$\sigma' = a[2(I_{1x}I_{2z} + I_{1z}I_{2x})\cos 2\alpha - 2(I_{1z}I_{2z} - I_{1x}I_{2x})\sin 2\alpha] \qquad [34]$$

4. PRINCIPLES OF MULTIPLE-QUANTUM SPECTROSCOPY   275

In the coefficient of $\sin 2\alpha$, $I_{1x}I_{2x}$ represents a sum of zero- and double-quantum coherence as expressed by Eq. [6], and the product operator term $I_{1z}I_{2z}$ represents longitudinal two-spin order between spins 1 and 2 (14, 22).

Experimental verification of the tip-angle dependence of Eq. [34] is shown in Fig. 4-5. This figure represents the ^{1}H spectra of the same AX spin system observed in Fig. 4-4. For the spectra in Fig. 4-5, the preparation sequence $(\pi/2)_x-\tau/2-(\pi)_y-\tau/2-(\alpha)_y$, followed immediately by acquisition of the spectrum at each value of α, has been used. Transfer to the coherences represented by the second term in Eq. [34] is optimized for

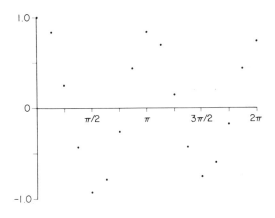

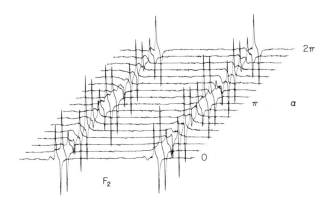

Fig. 4-5. Tip-angle dependence for the AX spin system 2,3-dibromothiophene in acetone-d$_6$. The ^{1}H spectra are shown at the bottom for the preparation sequence $(\pi/2)_x-\tau/2-(\pi)_y-\tau/2-(\alpha)_y$ followed immediately by acquisition. The graph at the top represents the normalized average peak height versus tip angle α.

$\alpha = n\pi/4$ where n is an odd integer. Note the unusual tip-angle dependence for this transfer process. A π pulse has the effect of returning the spin system to its original state before the pulse was applied. This rotation property can be interpreted in terms of rotations in a coordinate system defined with each sum of product operators and the operator for pulse rotation. This is a member of a general class of rotation properties for coupled spin systems (23).

The general properties of the preparation sequence $(\pi/2)_x$—$\tau/2$—$(\pi)_y$—$\tau/2$—$(\alpha)_{x+\eta}$ can be used to generate selectively a specific multiple-quantum coherence or combination of coherences. The evolution of these coherences can be indirectly detected following a mixing sequence which transfers the multiple-quantum coherences to observable single-quantum coherences.

B. Evolution of Zero- and Double-Quantum Coherence

Once the appropriate order of coherence has been excited, the particular special properties of that order can be exploited. The indirect effect of multiple-quantum coherences transferred to observable single-quantum coherences can be used to modify conventional one-dimensional spectra. As outlined earlier, multiple-quantum coherences can also be indirectly observed using the technique of two-dimensional spectroscopy. In this case, the spin system prepared into a state containing multiple-quantum coherences evolves for a stepped interval and then the multiple-quantum coherences are transferred to observable single-quantum coherences. Fourier transformation with respect to the evolution time results in a second frequency dimension, which represents the multiple-quantum coherences present during the evolution interval. As an example, the evolution of zero- and double-quantum coherences for the two-spin AX system will be analyzed in detail.

Consider the evolution of the Cartesian product operators representing double-quantum coherence in Eq. [32]. The evolution time will be denoted by t_1 as in a two-dimensional experiment. Then the density operator for the double-quantum coherence is

$$\sigma(t_1) = \exp(-i\mathcal{H}_0 t_1)[a2(I_{1x}I_{2y} + I_{1y}I_{2x})]\exp(i\mathcal{H}_0 t_1) \qquad [35]$$

The part of the free precession Hamiltonian which represents the J-coupling interaction does not affect these double-quantum product operators, since

$$[I_{1z}I_{2z}, I_{1\alpha}I_{2\beta}] = 0 \qquad \text{for} \quad \alpha, \beta = x, y, \text{ or } z \qquad [36]$$

Therefore the double-quantum coherence only evolves under the influence of the chemical shift interaction. The result of this evolution can be

4. PRINCIPLES OF MULTIPLE-QUANTUM SPECTROSCOPY

represented by

$$\sigma(t_1) = -a[2(I_{1x}I_{2y} + I_{1y}I_{2x})\cos(\omega_1+\omega_2)t_1$$
$$-2(I_{1x}I_{2x} - I_{1y}I_{2y})\sin(\omega_1+\omega_2)t_1] \quad [37]$$

From Eqs. [12] and [13], this can be rewritten in terms of single-element operators as

$$\sigma(t_1) = 2a\left[\frac{1}{2i}(I_1^+ I_2^+ - I_1^- I_2^-)\cos(\omega_1+\omega_2)t_1\right.$$
$$\left.-\tfrac{1}{2}(I_1^+ I_2^+ + I_1^- I_2^-)\sin(\omega_1+\omega_2)t_1\right] \quad [38]$$

The product operator terms in parentheses have the same form as the Cartesian operators defined by Eqs. [3]-[5]. Therefore, it appears that the above operators can be interpreted in terms of x and y components of double-quantum coherence.

It is most instructive to relate the single-element operators to the specific number scheme for the AX spin system as defined in Fig. 4-3. Recall the defining relations for the single-element operators in Chapter 2. With the use of these relationships, the double-quantum operators are

$$\tfrac{1}{2}(I_1^+ I_2^+ + I_1^- I_2^-) = \tfrac{1}{2}(|\alpha_1\alpha_2\rangle\langle\beta_1\beta_2| + |\beta_1\beta_2\rangle\langle\alpha_1\alpha_2|)$$
$$= \tfrac{1}{2}(|1\rangle\langle 4| + |4\rangle\langle 1|) \quad [39]$$

and

$$\frac{1}{2i}(I_1^+ I_2^+ - I_1^- I_2^-) = \frac{1}{2i}(|\alpha_1\alpha_2\rangle\langle\beta_1\beta_2| - |\beta_1\beta_2\rangle\langle\alpha_1\alpha_2|)$$
$$= \frac{1}{2i}(|1\rangle\langle 4| - |4\rangle\langle 1|) \quad [40]$$

From expressions of this form, the relationship of the double-quantum operator with a specific transition is easily seen. The operator forms in terms of specific energy levels 1 and 4 can be recognized as representing the single-transition operators between level 1 and 4. For the specific case of the AX spin system, the single-transition operators and the Cartesian operator forms of double-quantum coherence are

$$I_x^{14} = \tfrac{1}{2}(I_1^+ I_2^+ + I_1^- I_2^-) = (I_{1x}I_{2x} - I_{1y}I_{2y}) \quad [41]$$

$$I_y^{14} = \frac{1}{2i}(I_1^+ I_2^+ - I_1^- I_2^-) = (I_{1x}I_{2y} + I_{1y}I_{2x}) \quad [42]$$

and
$$I_z^{14} = \tfrac{1}{2}(I_1^\alpha I_2^\alpha - I_1^\beta I_2^\beta) = (I_{1z}E_2 + E_1 I_{2z}) \qquad [43]$$

The corresponding forms for zero-quantum coherence are
$$I_x^{23} = \tfrac{1}{2}(I_1^+ I_2^- + I_1^- I_2^+) = (I_{1x}I_{2x} + I_{1y}I_{2y}) \qquad [44]$$

$$I_y^{23} = \frac{1}{2i}(I_1^+ I_2^- - I_1^- I_2^+) = -(I_{1x}I_{2y} - I_{1y}I_{2x}) \qquad [45]$$

and
$$I_z^{23} = \tfrac{1}{2}(I_1^\alpha I_2^\beta - I_1^\beta I_2^\alpha) = (I_{1z}E_2 - E_1 I_{2z}) \qquad [46]$$

The single-transition and Cartesian forms of the operators are identical representations of the same basic coherences. It is a matter of convenience which form of the density operator is used to represent the state of the spin system. The Cartesian product operators are most convenient for calculating the effects of pulses; however, it is not always easy to interpret which coherences are represented by the operator products. Furthermore, observable magnetization is not easily calculated with the Cartesian product operators. To calculate the effect of evolution or detection, it is easiest to use the single-element operators or the single-transition operators discussed above. The effect of free precession on the single-transition operators has a direct geometric interpretation. In a fashion similar to that for the single-element operators, it can be shown that

$$\exp(-i\mathcal{H}_0 t) I_x^{rs} \exp(i\mathcal{H}_0 t) = I_x^{rs} \cos \omega_{rs} t + I_y^{rs} \sin \omega_{rs} t \qquad [47]$$

$$\exp(-i\mathcal{H}_0 t) I_y^{rs} \exp(i\mathcal{H}_0 t) = I_y^{rs} \cos \omega_{rs} t - I_x^{rs} \sin \omega_{rs} t \qquad [48]$$

and
$$\exp(-i\mathcal{H}_0 t) I_z^{rs} \exp(i\mathcal{H}_0 t) = I_z^{rs} \qquad [49]$$

where $\omega_{rs} = (E_r - E_s)/h$ and E_i is the ith energy eigenvalue of the spin system. Then the evolution of the y component of double-quantum coherences in Eq. [37] becomes

$$\sigma(t_1) = -2a(I_y^{14} \cos \omega_{14} t_1 - I_x^{14} \sin \omega_{14} t_1) \qquad [50]$$

This approach allows one to calculate directly both the phase and the resonance frequencies of the observed spectrum or the phase and the frequencies of resonance peaks in a two-dimensional spectral representation of multiple-quantum coherences.

The pulse sequence with $\alpha = \pi/2$ and $\eta = 0°$ converts single-quantum coherence into a state with only a y component of double-quantum coherence, as can be seen in Eq. [32]. The effect of the same sequence with $\alpha = \pi/4$ and $\eta = 90°$ does not have such a simple interpretation. In this case the single-quantum coherence is converted into a state which corresponds to the sum of "two-spin order" and double- and zero-quantum coherence such as given by Eq. [34]. Equation [15] in Chapter 2 represents the "two-spin order" Cartesian product in terms of the single-element operators. This in turn can be expressed as a sum of single-transition operators for a two-spin-$\frac{1}{2}$ system using Eq. [28]:

$$2I_{1z}I_{2z} = I_z^{13} - I_z^{24} = I_z^{12} - I_z^{34} \qquad [51]$$

It is interesting to note that with the expression written in terms of single-transition operators, it is easy to see the equivalence of a population inversion for one transition, here from energy level 2 to 4, of one spin and a population inversion of the progressive transition on the other spin, here from level 3 to 4. This equivalence has important implications for heteronuclear polarization transfer sequences such as INEPT (24). The $I_{1y}I_{2x}$ term in Eq. [34] represents a sum of zero- and double-quantum coherence as expressed in Eq. [6]. In terms of the appropriate single-transition operators, this is

$$2I_{1x}I_{2x} = I_x^{14} + I_x^{23} \qquad [52]$$

This indicates that the zero- and double-quantum coherences are created with equal amplitude, and that the phase of the double-quantum coherence is in quadrature with that created for the same sequence with $\alpha = \pi/2$ and $\eta = 0°$.

Evolution of the spin system following the creation pulse, $\alpha = \pi/4$ with $\eta = 90°$, is described by

$$\sigma(t_1) = \exp(-i\mathcal{H}_0 t_1)[-a2(I_{1z}I_{2z} - I_{1x}I_{2x})]\exp(i\mathcal{H}_0 t_1) \qquad [53]$$

According to Eq. [36], the part of the free precession Hamiltonian which represents the J-coupling interaction does not affect either the "two-spin order" term or the term containing zero- and double-quantum coherence, and the chemical shift interaction affects only the second term, $I_{1x}I_{2x}$. The evolution of the second term in Eq. [53] can be simplified by writing it as a sum of zero- and double-quantum product operators given in Eqs. [41] and [44]:

$$2I_{1x}I_{2x} = (I_{1x}I_{2x} + I_{1y}I_{2y}) + (I_{1x}I_{2x} - I_{1y}I_{2y}) = I_x^{23} + I_x^{14} \qquad [54]$$

Then the density operator for the system at a time t_1 is

$$\sigma(t_1) = -a[2(I_{1z}I_{2z})$$
$$+ (I_{1x}I_{2x} + I_{1y}I_{2y})\cos(\omega_1 - \omega_2)t_1 - (I_{1x}I_{2y} - I_{1y}I_{2x})\sin(\omega_1 - \omega_2)t_1$$
$$+ (I_{1x}I_{2x} - I_{1y}I_{2y})\cos(\omega_1 + \omega_2)t_1$$
$$+ (I_{1x}I_{2y} + I_{1y}I_{2x})\sin(\omega_1 + \omega_2)t_1] \qquad [55]$$

In terms of single-transition operators, this becomes

$$\sigma(t_1) = -a[(I_z^{13} - I_z^{24}) + I_x^{23} \cos \omega_{23} t_1 + I_y^{23} \sin \omega_{23} t_1$$
$$+ I_x^{14} \cos \omega_{14} t_1 + I_y^{14} \sin \omega_{14} t_1] \qquad [56]$$

The evolution of the double-quantum coherence described by Eqs. [35], [37], and [50] involves only a single coherence between the highest and lowest energy states of the AX spin system. Indirect detection would result in only *one* double-quantum frequency in the two-dimensional spectrum. However, indirect detection of the evolution described in Eqs. [53], [55], and [56] in a two-dimensional experiment results in a resonance at $F_1 = 0$ for the two-spin order term, $\omega_{23}/2\pi$ for the zero-quantum coherence, and $\omega_{14}/2\pi$ for the double-quantum coherence.

The formal description of time development can be presented in several ways, as has been illustrated for the evolution of various orders of coherence. The choice of operator forms can determine the clarity of the description. For evolution, the single-element or single-transition operators are the most informative. However, for the effect of pulse rotations, the product-operator base allows the most straightforward, illuminating description of the development of a spin system. For these reasons, it is sometimes necessary to follow evolution both in the product-operator base and in terms of single-transition elements. This is a general strategem which will be followed throughout this chapter. Both treatments are equivalent, so that the choice is only a matter of convenience.

Just as a single-quantum coherence for a single-spin system can be described as a vector rotating about the direction of the polarizing field B_0, individual multiple-quantum coherence can be described in an identical manner. The transverse multiple-quantum coherence precesses at a rate determined by the energy difference between levels involved in the transition.

C. Indirect Detection of Zero- and Double-Quantum Coherence

Since it is not possible directly to observe orders of coherence other than single-quantum coherence, one must transfer the unobservable coherence to observable coherence with a mixing scheme. The detection scheme applied during the mixing interval can have quite a general form; however, the simplest case of a single mixing pulse provides the most general transfer of coherence between connected orders. The tip angle and relative phase of the mixing pulse are very important, and single-pulse mixing will be considered for the two-spin-$\frac{1}{2}$ system.

For convenience, the mixing process will be followed in the product-operator formalism. The state of the spin system at the end of the evolution

4. PRINCIPLES OF MULTIPLE-QUANTUM SPECTROSCOPY

time will be expressed in the appropriate product-operator terms and the effect of the mixing pulse will be calculated in this operator basis. The indirect detection of the pure double-quantum coherence given in Eq. [37] will be considered first. Then the results of indirectly detecting all the coherences given by Eq. [55] will be calculated and sample spectra will be presented. For both cases, we shall assume that the mixing pulse has tip angle β and a phase of $0°$, so that the pulse is applied along the x axis in the rotating frame. The phase of the mixing pulse relative to the phase of the pulses in the preparation sequence is of importance. In this treatment the phase of the mixing pulse will be held constant at $0°$ and the preparation sequence phase will be assumed variable for the purposes of order discrimination, which will be discussed in Section IV.

First consider the double-quantum coherence of Eq. [37]. It evolves for a time t_1 in much the same way as a rotating component of transverse magnetization, as can be seen in Eq. [50]. The transfer of coherence to observable single-quantum coherence for each component of double-quantum coherence will be studied separately. The y component transfers in the same sense as described for the creation sequence in Eq. [32]. The transfer proceeds as

$$\exp(-i\beta I_x)[-a2(I_{1x}I_{2y} + I_{1y}I_{2x})] \exp(i\beta I_x)$$
$$= -a[2(I_{1x}I_{2y} + I_{1y}I_{2x})\cos\beta + 2(I_{1x}I_{2z} + I_{1z}I_{2x})\sin\beta] \quad [57]$$

The x-component transfer can be calculated in a straightforward way:

$$\exp(-i\beta I_x)[a2(I_{1x}I_{2x} - I_{1y}I_{2y})]\exp(i\beta I_x)$$
$$= a[2I_{1x}I_{2x} - 2I_{1y}I_{2y}\cos^2\beta$$
$$- 2(I_{1y}I_{2z} + I_{1z}I_{2y})\cos\beta\sin\beta - 2I_{1z}I_{2z}\sin^2\beta] \quad [58]$$

In general for quadrature detection, the detectable magnetization derived from observable single-quantum coherences is (see Chapter 2, Eq. [49])

$$M(t_1, t_2) = \gamma\hbar \operatorname{Tr}\left[\sigma(t_1, t_2) \sum_l (I_{lx} + iI_{ly})\right] \quad [59]$$

In evaluating this expression, one should note that

$$\operatorname{Tr}(I_{i\gamma}I_{j\delta}) = \begin{cases} \frac{1}{4}\operatorname{Tr} E, & i = j \text{ and } \gamma = \delta = x, y, \text{ or } z \\ 0, & i \neq j \text{ or } \gamma \neq \delta \end{cases} \quad [60]$$

Also a term of the general form

$$\operatorname{Tr}(I_{i\gamma}I_{j\delta}I_{k\varepsilon}) = 0 \quad \text{for any } i, j, k, \text{ and } \gamma, \delta, \varepsilon \quad [61]$$

is encountered for the antiphase single-quantum coherences in Eqs. [57]

and [58], which indicates that these terms are not directly observable at $t_2 = 0$. As t_2 increases, the antiphase single-quantum coherences precess in the presence of J coupling and come into phase with a periodicity of $1/J$. The explicit calculation of t_2 evolution need only be performed for two types of terms containing single-quantum coherence. These are the antiphase x and antiphase y components of single-quantum coherence in Eqs. [57] and [58]. The evolution of the general term $2I_{jx}I_{kz}$ is given by

$$\exp(-i\mathcal{H}_0 t_2)(2I_{jx}I_{kz})\exp(i\mathcal{H}_0 t_2)$$
$$= \cos \omega_j t_2 (2I_{jx}I_{kz} \cos \pi J_{jk} t_2 + I_{jy} \sin \pi J_{jk} t_2)$$
$$+ \sin \omega_j t_2 (2I_{jy}I_{kz} \cos \pi J_{jk} t_2 - I_{jx} \sin \pi J_{jk} t_2) \qquad [62]$$

The detectable magnetization, using the form of Eq. [59], is

$$\gamma\hbar \, \text{Tr}\left[\exp(-i\mathcal{H}_0 t_2)(2I_{jx}I_{kz})\exp(i\mathcal{H}_0 t_2) \sum_l (I_{lx} + iI_{ly})\right]$$
$$= \tfrac{1}{2}\gamma\hbar \sin \pi J t_2 (-\sin \omega_j t_2 + i \cos \omega_j t_2)$$
$$= \tfrac{1}{4}\gamma\hbar \{\exp[i(\omega_j - \pi J_{jk})t_2] - \exp[i(\omega_j + \pi J_{jk})t_2]\} \qquad [63]$$

The evolution of the general terms for antiphase y components, $2I_{jy}I_{kz}$, can be developed in the same manner as shown for the x component in Eqs. [62] and [63], with the result that the detected magnetization is given by

$$\gamma\hbar \, \text{Tr}\left[\exp(i\mathcal{H}_0 t_2)(2I_{jy}I_{kz})\exp(i\mathcal{H}_0 t_2) \sum_l (I_{lx} + iI_{ly})\right]$$
$$= \tfrac{1}{4}\gamma\hbar i\{\exp[i(\omega_j - \pi J_{jk})t_2] - \exp[i(\omega_j + \pi J_{jk})t_2]\} \qquad [64]$$

These generalized results can be applied to the specific case of indirect detection of double-quantum coherence given by Eqs. [57] and [58]. The detected magnetization from Eq. [59] is given as a function of t_1 and t_2 by

$$M(t_1, t_2) = -\tfrac{1}{4}\gamma\hbar a\Big\{\sin \beta \cos(\omega_1 + \omega_2)t_1\{\exp[i(\omega_1 - \pi J)t_2]$$
$$- \exp[i(\omega_1 + \pi J)t_2] + \exp[i(\omega_2 - \pi J)t_2] - \exp[i(\omega_2 + \pi J)t_2]\}$$
$$+ i \cos \beta \sin \beta \sin(\omega_1 + \omega_2)t_1\{\exp[i(\omega_1 - \pi J)t_2]$$
$$- \exp[i(\omega_1 + \pi J)t_2] + \exp[i(\omega_2 - \pi J)t_2]$$
$$- \exp[i(\omega_2 + \pi J)t_2]\Big\} \qquad [65]$$

Note that both terms have the same t_2 dependence, which can represented by

$$S(t_2) = \exp[i(\omega_1 - \pi J)t_2] - \exp[i(\omega_1 + \pi J)t_2]$$
$$+ \exp[i(\omega_2 - \pi J)t_2] - \exp[i(\omega_2 + \pi J)t_2] \quad [66]$$

Fourier transformation with respect to t_2 results in the basic form of the observed spectrum, which is modulated as a function of t_1. The form of Eq. [66] indicates that the observed spectrum at $t_1 = 0$ would consist of a pair of doublets, centered at ω_1 and ω_2, each with splitting of J, and each with the resonance lines of each doublet in antiphase. This implies that the signal acquisition time must be on the order of $1/J$; otherwise the observed signal strength will be greatly reduced. This can be thought of in terms of frequency resolution in the observed spectrum. The frequency resolution must be sufficient to resolve the J coupling, or no signal will be detected because the components of the multiplets in antiphase sum to zero (25).

With the use of the trigonometric identities

$$\sin\beta = 2\cos\frac{\beta}{2}\sin\frac{\beta}{2} \qquad \cos\beta = \cos^2\frac{\beta}{2} - \sin^2\frac{\beta}{2} \quad [67]$$

and

$$1 = \cos^2\frac{\beta}{2} + \sin^2\frac{\beta}{2}$$

the detectable magnetization given by Eq. [65] becomes

$$M(t_1, t_2) = -\tfrac{1}{2}\gamma\hbar a \left[\left(\cos^3\frac{\beta}{2}\sin\frac{\beta}{2} + \cos\frac{\beta}{2}\sin^3\frac{\beta}{2} \right) \cos(\omega_1 + \omega_2)t_1 S(t_2) \right.$$
$$\left. + i\left(\cos^3\frac{\beta}{2}\sin\frac{\beta}{2} - \cos\frac{\beta}{2}\sin^3\frac{\beta}{2} \right) \sin(\omega_1 + \omega_2)t_1 S(t_2) \right] \quad [68]$$

This can be reduced to a simpler form:

$$M(t_1, t_2) = -\tfrac{1}{2}\gamma\hbar a \left\{ \left(\cos^3\frac{\beta}{2}\sin\frac{\beta}{2} \right) \exp[i(\omega_1 + \omega_2)t_1] S(t_2) \right.$$
$$\left. + \left(\cos\frac{\beta}{2}\sin^3\frac{\beta}{2} \right) \exp[-i(\omega_1 + \omega_2)t_1] S(t_2) \right\} \quad [69]$$

Fourier transformation with respect to t_1 results in resonances in the two-dimensional spectrum symmetric about $F_1 = 0$ at $\pm(\omega_1 + \omega_2)/2\pi$. The relative intensities of these symmetric resonances depend on the tip angle of the mixing pulse as given in Eq. [69] (26).

The transfer coefficients of Eq. [69] for the mixing pulse are shown graphically in Fig. 4-6. The coefficients are equal for integer multiples of $\pi/2$. Disregarding magnet inhomogeneities for the moment, the two resonances at $F_1 = \pm(\omega_1 + \omega_2)/2\pi$ will have equal signs and magnitudes. An experimental verification of this tip-angle dependence is shown in Fig. 4-7. The graph at the top represents the average magnitude of the peak height for the four resonances in F_2 for this ^{1}H, AX spin system, in 2,3-dibromothiophene. The evolution time t_1 has been set to zero so that no precession has taken place and magnet inhomogeneity can be ignored. The effect of magnet inhomogeneities will be discussed in Section IV,B. The terms "echo" and "antiecho" refer to the sense of precession during t_1 which either opposes or adds to precession during t_2. The echo has $F_1 = -(\omega_1 + \omega_2)/2\pi$ and the antiecho has $F_1 = +(\omega_1 + \omega_2)/2\pi$. These terms have been separated by a technique involving phase cycling the preparation sequence relative to the mixing sequence. The specifics of the phase cycling technique will be treated fully in Section IV,A. There is excellent agreement between theory and experiment in this case.

Now consider the zero-quantum coherence of Eq. [55], the evolution of which is represented by the second and third terms. Note the similarity between these terms and the fourth and fifth terms, which represent the evolution of the double-quantum coherence. Except for sign differences, the overall form of the operator products is the same for the zero- and the

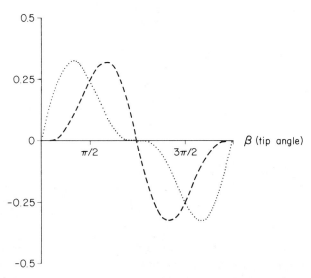

Fig. 4-6. The tip-angle effect of coherence transfer from zero- or double-quantum coherence to single-quantum coherence in an AX spin system.

4. PRINCIPLES OF MULTIPLE-QUANTUM SPECTROSCOPY   285

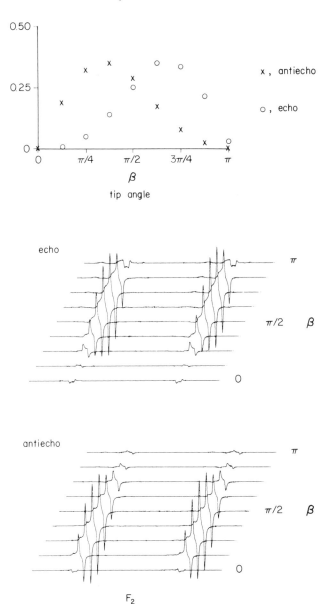

Fig. 4-7. Experimental verification of the tip-angle dependence of coherence transfer from double-quantum to observable single-quantum coherence in the AX spin system 2,3-dibromothiophene in acetone-d_6. The average peak height versus mixing-pulse tip angle is shown at the top of the figure. Peak heights have been normalized to the expected transfer-coefficient values for comparison. The evolution time t_1 was set to zero.

double-quantum coherences. The detection pulse has a similar effect on both orders of coherence, and equations of the form of Eqs. [57] and [58] can be written down directly. For the second term in Eq. [55], the coherence-transfer expression becomes

$$\exp(i\beta I_x)[-a(I_{1x}I_{2x} + I_{1y}I_{2y})]\exp(i\beta I_x)$$
$$= -a[I_{1x}I_{2x} + I_{1y}I_{2y}\cos^2\beta$$
$$+ (I_{1y}I_{2z} + I_{1z}I_{2y})\cos\beta\sin\beta + I_{1z}I_{2z}\sin^2\beta] \quad [70]$$

The third term in Eq. [55] transforms as

$$\exp(-i\beta I_x)[a(I_{1x}I_{2y} - I_{1y}I_{2x})]\exp(i\beta I_x)$$
$$= a[(I_{1x}I_{2y} - I_{1y}I_{2x})\cos\beta - (I_{1x}I_{2z} - I_{1z}I_{2x})\sin\beta] \quad [71]$$

These results are very similar to Eqs. [57] and [58] for transfer from double-quantum coherence. However, subtle differences exist which produce quite different results in the detected signal. The most important point is that the single-quantum coherence of Eq. [71] associated with spin 1 has a sign opposite to that of spin 2 as reflected in the last term (compare the sign for double-quantum transfer in Eq. [57]). This indicates that the single-quantum coherences observed for each spin are not modulated in an identical manner as was true for transfer from double-quantum coherence (see Eq. [65]). Following the procedure outlined for the indirect detection of double-quantum coherence, the observable magnetization from indirect detection of zero-quantum coherence can be written as

$$M(t_1, t_2) = -\tfrac{1}{8}\gamma\hbar a\Big\{ i\cos\beta\sin\beta\cos(\omega_1 - \omega_2)t_1\{\exp[i(\omega_1 - \pi J)t_2]$$
$$- \exp[i(\omega_1 + \pi J)t_2] + \exp[i(\omega_2 - \pi J)t_2] - \exp[i(\omega_1 + \pi J)t_2]\}$$
$$+ \sin\beta\sin(\omega_1 - \omega_2)t_1\{\exp[i(\omega_1 - \pi J)t_2] - \exp[i(\omega_1 + \pi J)t_2]$$
$$- \exp[i(\omega_2 - \pi J)t_2] + \exp[i(\omega_2 + \pi J)t_2]\}\Big\} \quad [72]$$

In order to simplify the analysis, let the t_2 dependence for each spin be

$$S_j(t_2) = \exp[i(\omega_j - \pi J)t_2] - \exp[i(\omega_j + \pi J)t_2] \quad [73]$$

Then, using the identities of Eq. [67], the detected signal of Eq. [72] is

$$M(t_1, t_2) = \tfrac{1}{8}\gamma\hbar a i\Big\{\Big(\cos\frac{\beta}{2}\sin^3\frac{\beta}{2}\Big)\exp[-i(\omega_1 - \omega_2)t_1]S_1(t_2)$$
$$- \Big(\cos^3\frac{\beta}{2}\sin\frac{\beta}{2}\Big)\exp[i(\omega_1 - \omega_2)t_1]S_1(t_2)$$

$$-\left(\cos^3\frac{\beta}{2}\sin\frac{\beta}{2}\right)\exp[-i(\omega_1-\omega_2)t_1]S_2(t_2)$$

$$+\left(\cos\frac{\beta}{2}\sin^3\frac{\beta}{2}\right)\exp[i(\omega_1-\omega_2)t_1]S_2(t_2)\Bigg\} \qquad [74]$$

Inspection of this result indicates an asymmetry in both the F_1 and F_2 directions in the two-dimensional spectrum resulting after Fourier transformation of the detected signal. The resonances symmetrically located about $F_1=0$ along the F_1 direction have relative intensities dependent on the value of the detection pulse tip angle. The functional form of the tip angle dependence is graphed in Fig. 4-6. The functional form is the same as for indirect detection of double-quantum coherence. However, the relative signs of these two zero-quantum resonances are opposite to one another, and the tip-angle dependence for spin 1 is opposite that for spin 2 (27, 28).

This effect is shown in the experimental result illustrated in Fig. 4-8. These are the two-dimensional spectra for the general sequence $(\pi/2)_x$—$\tau/2$—$(\pi)_y$—$\tau/2$—$(\pi/4)_x$—t_1—$(\beta)_x$—t_2 where the values of β have been set to $\pi/4$, $\pi/2$, and $3\pi/4$. The phase of the preparation sequence has been cycled through four steps to allow the detection of only coherence derived from longitudinal magnetization, two-spin order and zero-quantum coherence present during t_1 (see Section IV,A). The asymmetries are clearly reflected in the resulting spectra.

D. SUMMARY FOR THE TWO-SPIN SYSTEM AND APPLICATIONS

The coupled two-spin AX system is a necessary starting point for a study of multiple-quantum spectroscopy. This simplest of coupled spin systems with its six possible transition pathways between energy levels forms the basis for much of the spectroscopy of more complex systems. The treatment of excitation, evolution, and detection given here can be carried over into more complex systems where the AX system appears as a subset of the coupling network.

The study of multiple-quantum spectroscopy of a coupled two-spin system has immediate practical applications. A most interesting example is the measurement of ^{13}C-^{13}C J coupling (29-31) and ^{13}C-^{13}C connectivity through double-quantum coherence (26, 32-39) in organic molecules. With ^{1}H decoupling, the spin system of a molecule with ^{13}C in natural abundance consists effectively of an isolated single ^{13}C on roughly one molecule in a hundred. At the same time, two ^{13}C nuclei are present on roughly one molecule in ten thousand. The coupled two-spin systems in these molecules are almost all isolated pairs. Selective observation of double-quantum

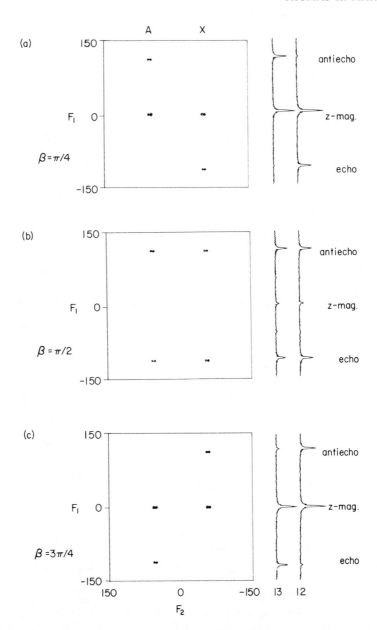

Fig. 4-8. Tip-angle effect for the indirect detection of zero-quantum coherence of the AX spin system 2,3-dibromothiophene in acetone-d_6. The two-dimensional spectra for (a) $\beta = \pi/4$, (b) $\beta = \pi/2$, and (c) $\beta = 3\pi/4$ are shown. Traces at the F_2 frequencies of coherences (1, 3) and (1, 2) are plotted at the right of the contour plots with the same vertical scale.

spectroscopy for this two-spin system allows direct measurement of coupling constants and connectivity without interference from the dominant resonance of the single ^{13}C spins. Selectivity methods will be discussed in the next section. Applications of double-quantum spectroscopy to the ^{13}C-^{13}C two-spin system illustrate two important general classes of multiple-quantum spectroscopy experiments, *filtering* and *correlation*, and these will be discussed next.

In general, one can use the special properties of a particular order of multiple-quantum coherence to *filter* out selectively all but the observable signal derived from the desired order of coherence (see next section). This allows the design of methods to observe only some special property of the spin system. For ^{13}C spectroscopy in natural abundance, molecules with two J-coupled ^{13}C spins are an effectively isolated AX or AB spin system. Since double-quantum coherence can only be created for a system of two or more coupled spins, it is possible to filter out the much more intense single-spin-system response, leaving only the response from the coupled system. As an example, the natural-abundance ^{13}C spectrum of piperidine (29) is shown in Fig. 4-9. This figure shows 50-Hz sections for the ^{1}H-decoupled spectrum of piperidine centered on the three ^{13}C resonances. There are four observable ^{13}C-^{13}C coupling constants, $^1J_{2,3} = 35.2$, $^3J_{2,3'} = 1.7$, $^2J_{2,4} = 2.6$, and $^1J_{3,4} = 33.0$ Hz. This spectrum was obtained with a pulse sequence of the type shown in Fig. 4-1d. The delay t_1 was held fixed at a minimum instrumental delay of 10 μsec to allow rf phase shifting. With t_1

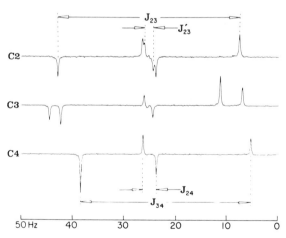

Fig. 4-9. Sections from the carbon-13 spectrum of piperidine showing ^{13}C-^{13}C direct and long-range coupling. Values of coupling constants are given in the text. [Reprinted with permission from A. Bax, R. Freeman, and S. P. Kempsell, *J. Am. Chem. Soc.* **102**, 4849 (1980). Copyright 1980 American Chemical Society.]

constant, the method is restricted to one-dimensional Fourier transformation spectrosopy. The resulting spectrum represents a filtration of the one-dimensional spectrum through double-quantum coherence by means of a basic four-step phase cycle (see next section). This cycle was augmented to reduce artifacts from imperfections in the pulse sequence (29). The observation of a particular ^{13}C-^{13}C coupling is dependent on the preparation of the appropriate double-quantum coherence. The choice of the value for the τ delay allows control over the range of couplings observed, as given in Eq. [20] for a weakly coupled spin system. The coherence generated is maximized for (40)

$$\tau = (2n+1)/(2J_{CC}) \qquad [75]$$

where n is an integer. The one-bond coupling constants are about 30 Hz, and, if the long-range couplings are assumed to be 5 Hz, both direct and long-range couplings can be observed in the same spectrum by setting $\tau = 51$ msec, which satisfies the above condition for $n = 0$ and $n = 3$. Each multiplet appears as an antiphase pair, up–down or down–up, depending on whether n is even or odd. The preparation of double-quantum coherence can be optimized for strong coupling in a straightforward manner (31) by a suitable choice of the τ delay.

Long-range ^{13}C-^{13}C coupling can be observed by modifying the pulse sequence of Fig. 4-1d to provide a two-dimensional J spectrum filtered through double-quantum coherence (31). This is accomplished by incrementing the preparation delay in order to define the J axis in the two-dimensional J spectrum. Thus small long-range couplings can be observed within the resolution limits set by the natural line width of the ^{13}C resonances.

A two-dimensional extension of the double-quantum filtering method can be used to correlate resonances of coupled ^{13}C spins. The sequence of Fig. 4-1d is used as illustrated so that the evolution of double-quantum coherences during t_1 appears in the resulting two-dimensional spectrum with double-quantum resonance frequencies parallel to the F_1 axis. Two coupled ^{13}C spins have resonance peaks in the two-dimensional spectrum at the same frequencies along F_1. Since the preparation of double-quantum coherence is dependent on the coupling constants among spins, it is possible to simplify the correlation by choosing to optimize the preparation delay τ for only directly coupled spins. In this case the two-dimensional double-quantum spectrum of directly coupled ^{13}C spins allows one to trace the connectivity of the carbons in a molecular system. The double-quantum spectrum of Fig. 4-10 is an illustration of the possibilities of connectivity mapping. This spectrum was obtained for a natural-abundance sample of panamine, an Ormosia alkaloid (26). The spectrum was run at 50.3 MHz

4. PRINCIPLES OF MULTIPLE-QUANTUM SPECTROSCOPY

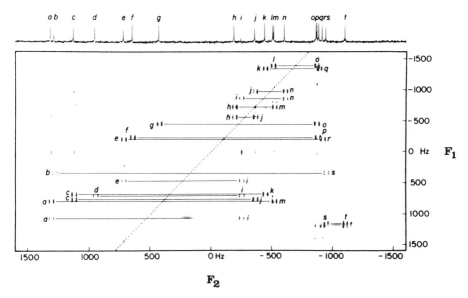

Fig. 4-10. Carbon–carbon connectivity map of panamine at 50 MHz with sign discrimination of the double-quantum frequencies. Horizontal bars indicate individual coupled pairs of ^{13}C spins. The conventional carbon-13 spectrum is shown at the top. [From Mareci and Freeman (26).]

on a Varian XL-200 spectrometer over a period of 10 hr. The tip angle of the mixing pulse was set to 135°. The tip-angle effect of the mixing pulse is given in Eq. [69] and shown in Fig. 4-6. For the tip angle used here, the responses from the antiecho are too small to appear on the contour map and it is possible to set the transmitter frequency near the center of the ^{13}C shift range without introducing ambiguity about the sign of the double-quantum frequencies.

Each pair of directly coupled carbon atoms gives rise to a four-line AX or AB spectrum in the F_2 dimension, the satellites of the conventional ^{13}C spectrum (shown along the top of the chart). The four-line subspectra have been emphasized by drawing the horizontal bars. Since the midpoint of each subspectrum is at the mean chemical shift in the F_2 dimension, all these midpoints must be on the diagonal of slope 2 passing through the origin.

The pattern of interconnections is easily determined. For example, there are just two subspectra involving carbon a at the extreme left of the diagram. The corresponding horizontal bars indicate that carbon a is directly bound to carbon i and carbon m. Examination of the vertical line representing the chemical shift of carbon i reveals three other subspectra involving carbon i, establishing that it is directly coupled to carbons d, e, and n. Panamine is known to contain three ring nitrogen atoms, and these terminate the connectivity chains, all at carbon atoms that resonate in the low-field region (a–g). The only other linkages not detected by the double-quantum method are those between p and q and between l and m. Both these pairs form strongly coupled AB spin systems which are known to give only weak signals unless the preparation delay τ is specially adjusted (31). In the general case, it is very useful to have the complementary information about ^1H-^{13}C coupling multiplicity of each carbon site, since this indicates where a carbon–carbon linkage has been "missed" or where a heteroatom is expected.

It is possible to generate a connectivity pattern in a one-dimensional double-quantum filtered spectrum for simple carbon spin systems (41). In this situation, a computer algorithm for pattern recognition can be used to interpret coupling information and provide a list of connectivity pattern possibilities. Where possible, the more direct method of two-dimensional double-quantum spectroscopy will provide an unambiguous determination of the connectivity within the limits of the technique.

IV. Order-Selective Detection

In the previous mathematical analysis it was quite easy to study separately the effects of different orders of coherence. The single-pulse detection scheme transfers multiple-quantum coherence to observable single-quantum coherence, thus making indirect detection possible. Experimentally separating the effects of different orders of coherence is a much more difficult problem. The detection process can affect all orders of coherence present during the evolution time. These coherences can be transferred to single-quantum coherence such that the observed signal is a superposition of observed single-quantum coherence derived from all orders of coherence.

Under favorable circumstances, a two-dimensional spectrum which represents the multiple-quantum coherences indirectly detected through single-quantum coherences can separate the orders of coherence, leading to an interpretable result. In general for complex spin systems, a simple approach produces a two-dimensional spectrum complicated by overlap of resonances from various orders of coherence and impossible to interpret directly.

Fortunately, it is possible to filter out selectively all but the desired orders of coherence, which are then indirectly detected. This filtering process is possible because the order of a coherence determines its characteristic response to perturbations of the spin system (42–44). In general, filtering has been accomplished by using either static field gradient pulses (3, 42, 45, 46) or rotary echo formation in an inhomogeneous rf field (47), or by combining spectra measured with the same pulse sequence but with an altered pattern of relative phases for each rf pulse, termed phase cycling (42).

Methods using either static-field gradients or rotary echoes rely on the fact that the angular velocity of precession of a coherence is proportional to both the strength of the applied field and the order of a particular coherence. A pair of "gradient" pulses can be applied symmetrically about a mixing sequence to select a particular coherence order. The relative strength and duration of the gradient pulses determine which orders are observed, and the parameters of each pulse can be modified independently to control the filtering process.

Coherence order filtering with phase cycling relies on a fundamentally different process. However, both the gradient method and phase cycling schemes can be cast in a similar formalism, which is represented as a rotation of coherences about the static field direction, in this case the z axis. Because of the similarity in representation, only the phase-cycling approach to order selection will be treated in detail.

Starting from some initial condition, a sequence of pulses and periods of free precession will place a spin system into some specific state of coherences. A preparation sequence of this type can be written as a general operator U:

$$U = R(\alpha_m, \eta_m) \exp(-i\mathcal{H}_0 \tau_{m-1}) \cdots R(\alpha_2, \eta_2) \exp(-i\mathcal{H}_0 \tau_1) R(\alpha_1, \eta_1) \quad [76]$$

Now if all the pulses are phase shifted by the same amount, ϕ, the preparation sequence can be written as

$$\begin{aligned} U(\phi) &= R(\alpha_m, \eta_m + \phi) \exp(-i\mathcal{H}_0 \tau_{m-1}) \cdots R(\alpha_2, \eta_2 + \phi) \\ &\quad \times \exp(-i\mathcal{H}_0 \tau_1) R(\alpha_1, \eta_1 + \phi) \\ &= \exp(-i\phi F_z) U \exp(i\phi F_z) \end{aligned} \quad [77]$$

Applying this phase-shifted preparation sequence to the density operator initially in a state of thermal equilibrium or nonequilibrium of the first kind (48) gives

$$\sigma' = U(\phi)\sigma_0 U^+(\phi) = \exp(-i\phi F_z) U \phi_0 U^+ \exp(i\phi F_z) \quad [78]$$

To derive this result, the pulse operator can be written in the form of Eq. [30] and the following properties can be used:

$$[F_z, \mathcal{H}_0] = 0 \quad [79]$$

and

$$[F_z, I_{jz}] = 0 \quad \text{for any spin } j \quad [80]$$

The result of Eq. [78] indicates that the coherences created by sequence U are phase shifted in a predictable manner dependent on their nature. The dependence of each coherence on the phase shift can be followed through the formalism of the product operators, single-transition operators, or single-element operators. The simplest result is derived by the use of single-element operators, but perhaps the simplest is not the most illuminating. It is worthwhile developing the effect of phase shifts in some detail using a combination of approaches.

The coherences created by sequence U can be written in a general form as product-operator terms,

$$U\phi_0 U^+ = \sum_m c_m (I_{1\delta} I_{2\partial} \cdots I_{n\varepsilon})_m, \qquad [81]$$

where δ, ∂, $\varepsilon = x$, y, or z, and n can range from 1 to the total number of spins, N, in the coupled network. Operators of the type I_{lz} in each term will not be affected by the phase shift, so only transverse coherence operators need be considered. As an example, consider a term which contains the product of two transverse coherences, $I_{jx}I_{ky}$. The phase-shifted term is then

$$\exp(-i\phi F_z)(2I_{jx}I_{ky})\exp(i\phi F_z) = 2I_{jx}I_{ky}\cos^2\phi + 2I_{jy}I_{ky}\sin\phi\cos\phi$$

$$- 2I_{jx}I_{kx}\sin\phi\cos\phi - 2I_{jy}I_{kx}\sin^2\phi \qquad [82]$$

The results of Eqs. [6]-[9] can be used to generalize the above result to give

$$\exp(-i\phi F_z)(2I_{jx}I_{ky})\exp(i\phi F_z) = \frac{1}{2i}(I_j^+ I_k^+ - I_j^- I_k^-)\cos 2\phi$$

$$-\tfrac{1}{2}(I_j^+ I_k^+ + I_j^- I_k^-)\sin 2\phi$$

$$-\frac{1}{2i}(I_j^+ I_k^- - I_j^- I_k^+),$$

or, if the I_{lz} terms are included,

$$= \{N, 2\text{-spin}, 2QC\}_y \cos 2\phi - \{N, 2\text{-spin}, 2QC\}_x \sin 2\phi$$

$$- \{N, 2\text{-spin}, ZQC\}_y \qquad [83]$$

Here $\{N, 2\text{-spin}, 2QC\}_x$ and $\{N, 2\text{-spin}, 2QC\}_y$ represent the x and y component of double-quantum coherence between spin j and k, and $\{N, 2\text{-spin}, ZQC\}_y$ represents the y component of zero-quantum coherence between spin j and k. Since product operator terms can be expressed as a sum of coherences of various orders, the above example indicated that each order of coherence is affected differently. For a particular spin system, the coherences in Eq. [83] can be related to a numbering scheme for an energy-level diagram and hence to appropriate single-transition operators.

In general for a coherence represented by a single-transition operator, the effect of the phase shift is

$$\exp(-i\phi F_z) I_\alpha^{rs} \exp(i\phi F_z) = [\exp(-i\,\Delta m_{rs}\phi I_z^{rs})] I_\alpha^{rs} [\exp(i\,\Delta m_{rs}\phi I_z^{rs})] \quad [84]$$

This result can be applied to the above example of the product operator of two transverse coherences expressed as single-transition operators. Since this product can be expressed as a sum of double- and zero-quantum coherences, Eq. [84] would predict the effect of the phase shift. The zero-quantum coherence would not be affected by the phase shift, since $\Delta m_{rs} = 0$ for this coherence, while the double-quantum coherence would be rotated by an angle twice that of the rf phase shift in the rotating frame.

As mentioned earlier, the simplest description can be derived in terms of the single-element operators. The density-matrix operator expressed in these terms will be affected by a phase shift in the following manner:

$$\exp(-i\phi F_z)|r\rangle\langle s|\exp(i\phi F_z) = \exp(-i\,\Delta m_{rs}\phi)|r\rangle\langle s|$$
$$= \exp(-ip\phi)\sigma^p \quad [85]$$

where $\sigma^p = |r\rangle\langle s|$ is a density-operator element of order $p = \Delta m_{rs} = m_r - m_s$. The single-element operators are the fundamental units which form the basis for the other operator descriptions and therefore contain the basic information. From the form of Eq. [85], it can be seen that the phase of a coherence can be inverted by a phase shift of $n\pi/p$ where n is an odd integer. It is this property which forms the basis for order-selective detection (42). A number of spectra are obtained with a variation of rf phase pattern of the pulse sequence for each spectrum; then the desired order of coherence is obtained by combining these spectra in an appropriate manner to cancel all but the desired order of coherences.

A. Coherence-Transfer Pathways

The observable magnetic resonance signal represents single-quantum coherence resulting from some preparation, evolution, and mixing sequence. The spin system has been transferred from a state of thermal equilibrium to various orders of coherences which evolved or have been transferred further to other orders of coherences. The final observable signal represents the single-quantum coherence resulting from the final transfer of various orders of coherence into observable single-quantum coherence. In general, the observed signal is the sum of all coherences transferred to observable single-quantum coherence.

The sequence of events in a pulsed NMR experiment can be represented in a "coherence-transfer map" (43, 44) as shown in Fig. 4-11 for a double-quantum preparation and mixing sequence. The orders of coherence precess

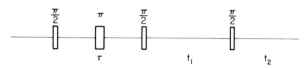

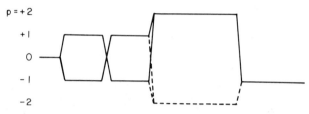

Fig. 4-11. An example of a coherence-transfer pathway map is shown at the bottom for the multiple-quantum spectroscopy technique shown at the top. The specific pathways illustrated are for the indirect detection of double-quantum coherence.

freely within the levels of the map, and pulses can produce transitions between levels. A coherence-transfer pathway is represented by the route a coherence follows during an experiment to reach its final order. This can aid in designing phase cycling sequences which provide selective indirect detection of a particular order of coherence. The pathway shown in this figure is only one of the many possible pathways available for the coherences produced by this sequence. The pathway illustrated is that appropriate for preparation and evolution of double-quantum coherence and its mixing into observable single-quantum coherence. For selective indirect detection of double-quantum coherence, it is necessary to manipulate the relative phases of the pulses in the sequence so that the illustrated pathway represents the pass route of the effective filter function produced. The particulars of the coherence transfer process will now be used to illustrate the procedure.

Starting from Eq. [59], the detectable magnetization can be written as

$$M(t_1, t_2) = \gamma\hbar \operatorname{Tr}\left[\sigma(t_1, t_2) \sum_l (I_{lx} + iI_{ly})\right]$$

$$= \gamma\hbar \operatorname{Tr}\left[\sigma(t_1, t_2) \sum_l I_l^+\right] \qquad [86]$$

The trace operation indicates that only density operator terms of the form I_k^- will be detectable and hence the coherence transfer pathway must terminate with $p = -1$.

Using the notation of Bodenhausen et al. (44), an experiment with n coherence-transfer steps can be expressed in terms of propagators $U_1, U_2, \ldots, U_n$ as

$$\sigma_0 \xrightarrow{U_1} \xrightarrow{U_2} \cdots \xrightarrow{U_n} \sigma(t) \qquad [87]$$

In general, each propagator has the form of the operator of Eq. [76]. Thus each propagator U_i causes a transfer of coherence from an initial order of coherence $\sigma^p(t_i^-)$ to the other orders $\sigma^{p'}(t_i^+)$ possible for a given propagator:

$$U_i \sigma^p(t_i^-) U_i^{-1} = \sum_{p'} \sigma^{p'}(t_i^+) \qquad [88]$$

where the time arguments refer to the density operation before and after the application of the propagator.

If the definition of the propagators is extended as in Eq. [77] to include phase ϕ_i, then

$$U_i(\phi_i) \sigma^p(t_i^-) U_i^{-1}(\phi_i) = \sum_{p'} \sigma^{p'}(t_i^+) \exp(-i\Delta p_i \phi_i) \qquad [89]$$

where $\Delta p_i = p'(t_i^+) - p(t_i^-)$ is the change in coherence order brought about by the propagator.

When all perturbations to the spin system are considered, the system will be initially in a state of equilibrium with $p = 0$. Since only coherences with $p = -1$ are observable, the sum of coherence-transfer steps must be given by

$$\sum_i \Delta p_i = -1 \qquad [90]$$

The observable density operator resulting from a general pulse sequence with n coherence-transfer steps is

$$\sigma^{p=-1}(\phi_1, \phi_2, \ldots, \phi_n, t) = \sigma^{p=-1}(\phi_1 = \phi_2 = \cdots = \phi_n = 0, t)$$
$$\times \exp[-i(\Delta p_1 \phi_1 + \Delta p_2 \phi_2 + \cdots + \Delta p_n \phi_n)]$$
$$= \sigma^{p=-1}(\boldsymbol{\phi} = 0) \exp(-i \boldsymbol{\Delta p} \cdot \boldsymbol{\phi}) \qquad [91]$$

where the coherence-order change and phase are given in the following vector notation:

$$\boldsymbol{\Delta p} = (\Delta p_1, \Delta p_2, \ldots, \Delta p_n) \qquad [92]$$

and

$$\boldsymbol{\phi} = (\phi_1, \phi_2, \ldots, \phi_n) \qquad [93]$$

The vector represented by Eq. [92] defines the steps taken along the coherence-transfer pathway. Equation [90] must be satisfied for the resulting coherence to be observable. The phase of the resulting coherence for a particular pathway with an rf phase vector as given by Eq. [93] is determined by Eq. [91]. A particular coherence-transfer pathway can be selected by adjusting the phase of the resulting coherence with the rf phase vector and combining measurements with different phase vectors to cancel all but the desired pathway.

The detected signal contribution from a particular pathway can be written as

$$s^{\Delta p}(\boldsymbol{\phi}, t) = s^{\Delta p}(\mathbf{0}, t) \exp(-i\Delta\mathbf{p} \cdot \boldsymbol{\phi}) \qquad [94]$$

Since many coherence-transfer pathways are possible, the total signal observed is the sum of the signals derived from each pathway:

$$s(\boldsymbol{\phi}, t) = \sum_{\Delta p} s^{\Delta p}(\boldsymbol{\phi}, t)$$

$$= \sum_{\Delta p} s^{\Delta p}(\mathbf{0}, t) \exp(-i\Delta\mathbf{p} \cdot \boldsymbol{\phi}) \qquad [95]$$

In practice, the selection of a particular pathway can be accomplished by performing a discrete Fourier analysis on the observed signal:

$$s^{\Delta p'}(t) = \frac{1}{N}\sum_{\boldsymbol{\phi}} s(\boldsymbol{\phi}, t) \exp(i\Delta\mathbf{p}' \cdot \boldsymbol{\phi})$$

$$= \frac{1}{N}\sum_{\boldsymbol{\phi}}\sum_{\Delta p} s^{\Delta p}(\mathbf{0}, t) \exp[i(\Delta\mathbf{p}' - \Delta\mathbf{p}) \cdot \boldsymbol{\phi}] \qquad [96]$$

The above process of discrete Fourier analysis is equivalent to summing the resulting signals for each phase vector $\boldsymbol{\phi}$, with the signal multiplied by a phase factor dependent on the desired order of coherence. The phase factor $\Delta\mathbf{p}' \cdot \boldsymbol{\phi}$ is usually derived by choosing a receiver phase value to provide this factor for the desired coherence order. For the pulse sequences of Fig. 4-1, the receiver phase Ψ is set equal to the desired value of $\Delta\mathbf{p}' \cdot \boldsymbol{\phi}$. Thus the summation can be performed during data acquisition for a given order in place of the discrete Fourier analysis. For a sequence of n propagators, this process would involve an n-dimensional sum where each dimension would range over the N_i phase values,

$$\phi_i = k_i 2\pi / N_i \qquad [97]$$

for $k_i = 0, 1, \ldots, N_i - 1$. The Fourier analysis involves a discrete sampling (according to Eq. [97]) so that, as a consequence of the sampling theorem, the resulting coherence order is not unique and is given by

$$\Delta p_i \pm nN_i, \quad \text{with} \quad n = 0, 1, 2, \ldots \qquad [98]$$

The normalization constant of Eq. [96] is given by the product of the number of discrete samples of the phase for each propagator, $N = N_1 N_2 \cdots N_n$.

The normalization constant N represents the number of times the pulse sequence must be repeated to select a particular pathway. The selection of a unique pathway can require an enormous amount of time for a sequence with a large number of coherence transfer steps. In order to discriminate a particular order, it is often only necessary to select the phase of a few specific coherence-transfer steps along a pathway. In these cases the number of phase-cycling steps can become small enough to allow the measurement of the resulting order-selective spectrum in a reasonable period of time.

The process of coherence-transfer pathway selection is effectively a filtering procedure. Only specific coherences which can exist in a given spin system will traverse a particular pathway. This leads to the possibility of eliminating for a spin system all observable subspectra which cannot have the specific coherences. If a complex spectrum contains many overlapping multiplets from spin subsystems, the spectrum can be simplified by filtering through the appropriate multiple-quantum coherences.

The applications of this filtering procedure can be divided into two general categories. The first is the use of preparation of multiple-quantum coherence as a necessary requirement for the observation of a single-quantum spectrum. Methods have been developed to separate one-dimensional spectra into component spectra from specific spin subsystems (2, 3, 29, 42, 49–61), and these ideas have been applied to filtering two-dimensional single-quantum spectroscopy (54, 62–67). The second category involves the indirect observation of the evolution of multiple-quantum coherence to create a two-dimensional multiple-quantum spectrum (2, 3, 5, 25–28, 32, 40, 42, 67–85).

An interesting illustration of the first category, the procedure of filtering complex one-dimensional spectra (55), is shown in Fig. 4-12. This figure represents ^{1}H spectra of the *Escherichia coli lac* repressor headpiece (a small globular peptide of 51 amino acid residues). The bottom spectrum was obtained by preparing the spin system into possible triple quantum coherences with a pulse sequence of the type shown in Fig. 4-1d. Here β is equal to $\pi/2$. The transmitter position is shown by the small triangle near the center of the spectrum. The bottom spectrum represents an isolated ABX multiplet from an asparagine residue. The shifts of these resonances relative to the transmitter sum to zero, so that the triple-quantum frequency will be zero in this case. This triple-quantum filtered spectrum was obtained by summing individual triple-quantum filtered spectra for a number of t_1 values. As t_1 was varied, all spectra were amplitude modulated as a function of the associated triple-quantum precession frequencies. The sum of these spectra was zero except where the triple-quantum precession frequency was

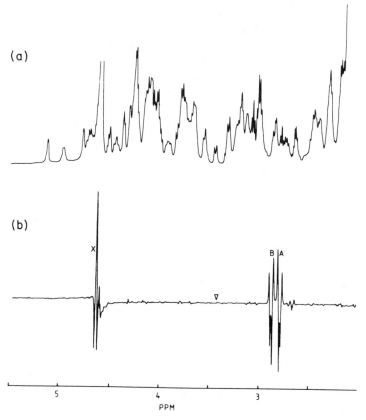

Fig. 4-12. (a) The top spectrum is the conventional 500-MHz ^{1}H NMR spectrum of 6.2 mM *lac* repressor headpiece (residues 1-51 of the complete repressor) in D$_2$O with 0.05 M phosphate, 0.4 M KCl, 0.02% (w/v) azide at pH 6.5. (b) The bottom spectrum was obtained with a pulse sequence of the type shown in Fig. 4-1d with β set equal to $\pi/2$. The preparation delay τ was 40 msec and $t_1 = 0, 1.25, 2.5, \ldots, 25$ msec. The transmitter position $[=\frac{1}{3}(\delta_A + \delta_B + \delta_c)]$ is indicated by the arrow. [From Hore, Scheek, and Kaptein (55).]

zero. Thus only the multiplet of the single αCH-βCH$_2$ spin system of the asparagine residue appears in this filtered spectrum.

Multiple-quantum filtering has proven very useful in single-quantum two-dimensional spectroscopy. Multiple-quantum filters have been introduced into homonuclear correlation spectroscopy (COSY) (54, 62), homonuclear J spectroscopy (63), and nuclear Overhauser enhancement spectroscopy (82). To illustrate the use of multiple-quantum filtering, an example of multiple-quantum filtered two-dimensional COSY (64) will next be discussed.

For a COSY spectrum, it is possible to provide a p-quantum filter which allows the observation of subspectra from spin systems which can be prepared with multiple-quantum coherences of order $\geq p$. This can greatly simplify a complex COSY spectrum such as that of a protein (81), which might contain a mixture of spin subsystems. Also the multiple-quantum filtration of COSY spectra has an important effect on the appearance of the spectra (64).

The conventional COSY experiment involves a single preparation pulse, evolution time, and a single mixing pulse. COSY spectra have two types of resonances: diagonal and cross peaks. The diagonal peaks originate from in-phase coherences present during the evolution time which are unaffected by the mixing pulse; therefore all multiplet components of a diagonal peak have the same phase. In contrast, the cross peaks originate from antiphase coherences of one spin present during the evolution time, which are transferred by the mixing pulse to the antiphase coherence of a directly coupled spin. Thus cross-peak multiplets exhibit alternating signs between components of the multiplet associated with the coupling involved in the transfer. Cross-peak multiplets tend to cancel when they are not completely resolved, but no cancellation occurs for the in-phase diagonal peaks. In a phase-sensitive COSY (86–88) spectrum where the diagonal peaks are phased to pure two-dimensional absorption lineshapes, the cross peaks have pure two-dimensional dispersive lineshapes. This can cause enhancement of diagonal peaks over cross peaks.

The multiple-quantum filtered COSY technique involves the use of a two-pulse mixing sequence. The first pulse transfers antiphase coherence for a particular spin into multiple-quantum coherence. The second mixing pulse transfers this multiple-quantum coherence to antiphase single-quantum coherence on the same spin, which appears as a diagonal peak, or transfers this multiple-quantum coherence to antiphase coherence of the active spins involved in the multiple-quantum coherence. The later coherences appear as cross peaks in the COSY spectrum. In contrast to conventional COSY, both the diagonal and cross peaks appear in antiphase relationship in the multiple-quantum filtered COSY spectra and hence have similar relative magnitudes. This allows assignment of cross peaks lying closer to the diagonal. In Fig. 4-13, the spectrum on the left is the result of a conventional phase-sensitive COSY experiment (64). Note the relatively intense resonances along the diagonal. Compare this with the double-quantum filtered COSY spectrum on the right.

In the second category of applications, the filtering procedure can be used to simplify two-dimensional multiple-quantum spectra. Filtering allows the separation of a complex spectrum into spectra from the various coherence orders available to the spin systems. The coupling patterns available

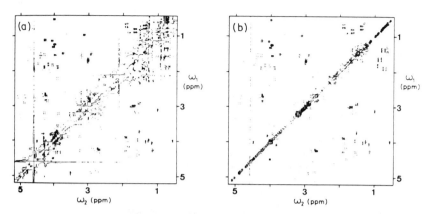

Fig. 4-13. Comparison of the spectral region from 0.5 to 5.0 ppm of (a) a conventional, phase-sensitive COSY spectrum of basic pancreatic trypsin inhibitor with (b) the corresponding region of a phase-sensitive, double-quantum filtered COSY spectrum. The data are presented as contour plots where both positive and negative levels have been plotted. [From Rance, Sørensen, Bodenhausen, Wagner, Ernst, and Wüthrich (64).]

within the spin system can be mapped by studying the two-dimensional multiple-quantum spectra of various orders.

The multiple-quantum spectrum of the relatively simple AMX system shown in Fig. 4-2 presents a complex problem in interpretation. The methods developed for order-selective detection along specific coherence-transfer pathways can be used to simplify complex spectra of this type. Specifically, rf phase cycling can be used to filter out all but the observable signal derived from the desired order of coherence. The pulse sequence for Fig. 4-1b and phase cycles of Table 4-1 can be used to observe selectively the two-dimensional multiple-quantum spectrum for a desired order of coherence. This type of preparation sequence creates both even and odd orders of coherence and is convenient to illustrate the filtering process. The tip angle of the mixing pulse is chosen to be $3\pi/4$ for later convenience. However, the mixing pulse will be effective in coherence transfer for any pulse length other than an integral multiple of π. The choice of $3\pi/4$ enhances coherence transfer to active spins (see later section). The important variables for order selection in this sequence are the relative phases of the transmitter pulses and the receiver.

The two-pulse preparation sequence can be considered as a single coherence-transfer operation. The preparation sequence can effect a transfer from the equilibrium state with order 0 to the possible coherence orders 0, ± 1, ± 2, and ± 3. A particular order will evolve during t_1 and will be transferred to observable coherence by the mixing pulse. The four-step phase cycle of the preparation sequence shown in the first column of Table 4-1, along with

4. PRINCIPLES OF MULTIPLE-QUANTUM SPECTROSCOPY

TABLE 4-1
Phase-Cycle Sequence for Multiple-Quantum Filtering

	ϕ	$\Delta p \cdot \phi$	ψ
(a) Zero-quantum coherence	0°	0°	0°
	90°	0°	0°
	180°	0°	0°
	270°	0°	0°
(b) Double-quantum coherence	0°	0°	0°
	90°	±180°	180°
	180°	±360°	0°
	270°	±540°	180°
(c) Single-quantum coherence and triple-quantum coherence	0°	0°	0°
	90°	±90°/±270°	90°
	180°	±180°/±540°	180°
	270°	±270°/±810°	270°

the receiver phase cycle of the last column, has been used for order selection. The choice of receiver phase is based on the relative phase variation of a particular coherence order with preparation phase. This is shown in the second column of the table, and the choice of receiver phase based on Eq. [96] is shown in the third column.

As shown in part (a) of the table, the zero-quantum coherence is independent of phase shifts of the preparation sequence, and zero-order selection is accomplished by holding the receiver phase constant. The resulting zero-order spectrum is shown in Fig. 4-14. As can be seen by the projection on the right-hand side, all orders except zero have been effectively suppressed by this four-step cycle. In part (b) of the table, the double-quantum order terms, ±2, are affected by the phase shift as shown in the second column. The receiver is cycled to provide double-quantum order selectivity. It is not possible to distinguish the sign of the order with this phase cycle. Additional phase cycling will be required to add unambiguous discrimination (29, 89, 90). The resulting double-quantum spectrum is shown in Fig. 4-15. The mixing pulse tip angle of $3\pi/4$ enhances +2 order (or echo) coherences, thus effectively discriminating the sign (26, 27, 71). The response of odd-order coherence for this three-spin system is shown in part (c) of the table. Two odd-orders are possible, ±1 and ±3. The ±1 order coherence can be either observable single-quantum coherence or total-spin, single-quantum coherence. The ±3 order coherence is total-spin triple-quantum coherence. Several receiver phase cycling schemes are possible. From Eq. [96], it can be seen that the one chosen here leads to a selection of the −1

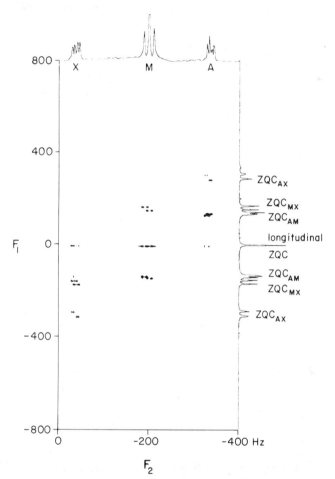

Fig. 4-14. Two-dimensional zero-quantum spectrum of the AMX spin system 2,3-dibromopropionic acid in benzene-d_6. The spectrum was acquired with the pulse sequence of Fig. 4-1b with $\beta = 3\pi/4$ using the phase cycle shown in part (a) of Table 4-1.

and +3 order coherences. The resulting two-dimensional spectrum is shown in Fig. 4-16. The phase-cycle scheme of part (c) in the table is insufficient to discriminate between first-order and third-order coherences, as can be seen by applying the results of Eqs. [97] and [98]. Assuming the receiver can be cycled through 180° phase shifts, a transmitter cycle of six steps with an increment size of 60° would be necessary to discriminate between first- and third-order coherence.

As outlined here, transmitter and receiver phase cycling provides a fundamental method of order-selective detection. This technique is used

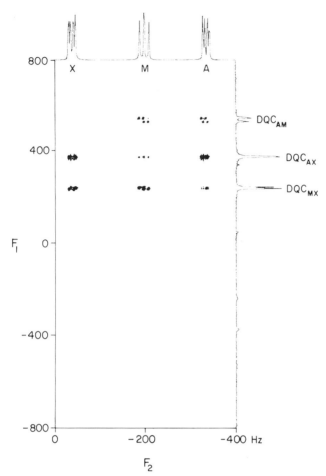

Fig. 4-15. Two-dimensional double-quantum spectrum of the AMX spin system 2,3-dibromopropionic acid in benzene-d_6. The spectrum was acquired with the pulse sequence of Fig. 4-1b with $\beta = 3\pi/4$ using the phase cycle shown in part (b) of Table 4-1.

extensively in multiple-quantum spectroscopy and other applications of multiple-pulse techniques in NMR spectroscopy. Further examples illustrating the use of multiple-quantum spectroscopy will be given in a later section.

B. Coherence-Transfer Echoes

The basic concept of the spin echo (91) can be generalized to include echoes derived from free precession of any order of coherence (92). In

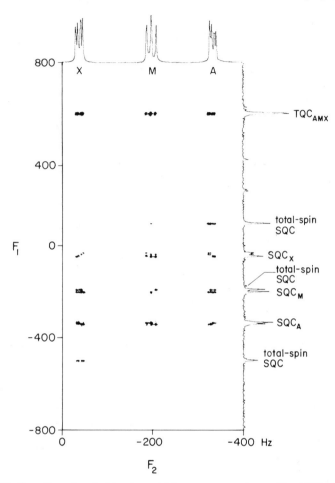

Fig. 4-16. Two-dimensional odd-order multiple-quantum spectrum of the AMX spin system 2,3-dibromopropionic acid in benzene-d_6. The spectrum was acquired with the pulse sequence of Fig. 4-1b with $\beta = 3\pi/4$ using the phase cycle shown in part (c) of Table 4-1.

other words, coherence freely precessing at one order of coherence in an inhomogeneous magnetic field will dephase and apparently lose "coherence." A transfer operation can place this coherence into another order, where the coherence can freely precess in the same inhomogeneous field, rephase, and form an echo. The process of coherence transfer echo formation is fundamental to all echo phenomena in magnetic resonance and has important effects in multiple-quantum spectroscopy. The process of echo formation can be described in the formalism of coherence-transfer pathways as follows.

Free precession in an inhomogeneous magnetic field can be represented by a term in the free precession Hamiltonian of the form

$$\hbar^{-1}\mathcal{H}_\Delta = \gamma \Delta B(\mathbf{r}) F_z \qquad [99]$$

where $\Delta B(\mathbf{r})$ represents the spatial dependence of the field variation. This assumes that only z components of the static field are present. Considering only this inhomogeneity term, we write for the free precession of a coherence of order p

$$\sigma^p(t) = \exp[-i\gamma \Delta B(\mathbf{r})tF_z]\sigma^p(0) \exp[i\gamma \Delta B(\mathbf{r})tF_z]$$
$$= \exp[-ip\gamma \Delta B(\mathbf{r})t]\sigma^p(0) \qquad [100]$$

A sum over all space of the phase factor in the equation above represents the spatially dependent coherence loss when all regions of the volume are considered. This results in an inhomogeneous line-broadening factor, which is dependent on the order of coherence (42, 92). As the order of coherence increases, the line broadening due to field inhomogeneities increases linearly. The resulting linewidths in a two-dimensional multiple-quantum spectrum will increase with increasing order (42).

In general, a coherence-transfer pathway consists of periods of free precession, τ_i, and coherence transfer steps brought about by propagators U_i. For a segment of the pathway consisting of a period of free precession τ_1, a coherence-transfer step U_1, followed by another period of free precession τ_2, the density-operator element for a particular final order of coherence is, with the aid of Eq. [100],

$$\sigma^{p_2}(\tau_2) = \exp[-ip_1\gamma \Delta B(\mathbf{r})\tau_1]\{\exp[-i\gamma \Delta B(\mathbf{r})\tau_2 F_z]U_1\sigma^{p_1}(0)U_1^{-1}$$
$$\times \exp[i\gamma \Delta B(\mathbf{r})\tau_2 F_z]\}$$
$$= \exp[-i\gamma \Delta B(\mathbf{r})(p_1\tau_1 + p_2\tau_2)]\sigma^{p_2} \qquad [101]$$

In deriving this result, it has been assumed that the inhomogeneity does not contribute during the coherence transfer step and that no diffusion is taking place. The final coherence will be independent of the field inhomogeneity at time τ_2 if the condition $p_1\tau_1 + p_2\tau_2 = 0$ is fulfilled. An echo will form at time $\tau_2 = (p_1/p_2)\tau_1$. In this example, an echo is formed only if the initial coherence is transferred to an order with opposite sign so that the above condition will be satisfied.

The effect of the field inhomogeneity during an entire coherence-transfer pathway of n steps and periods of free precession is

$$\sigma^{p_n}(\tau_n) = \exp[-i\gamma \Delta B(\mathbf{r})(p_1\tau_1 + p_2\tau_2 + \cdots + p_n\tau_n)]\sigma^{p_n}$$
$$= \exp[-i\gamma \Delta B(\mathbf{r})\mathbf{p}\cdot\boldsymbol{\tau}]\sigma^{p_n} \qquad [102]$$

where the vectors represent the ordered numbers

$$\mathbf{p} = (p_1, p_2, \ldots, p_n)$$ [103]

and

$$\boldsymbol{\tau} = (\tau_1, \tau_2, \ldots, \tau_n)$$ [104]

These two vectors define the coherence-transfer pathway of a particular coherence. A complete coherence-transfer map can be drawn using these vectors.

The general condition for echo formation at any point along the pathway is given by

$$\mathbf{p} \cdot \boldsymbol{\tau} = 0$$ [105]

This result is based on the assumptions that the field inhomogeneity is constant along the coherence-transfer pathway, the inhomogeneity has negligible effect during the coherence-transfer steps, and no diffusion takes place. If the coherence-transfer steps are assumed to occur in an infinitesimal period of time, an echo forms at any point along a coherence-transfer map where the area under a particular pathway equals zero, according to Eq. [105].

In general, the observable single-quantum coherence, here assumed to be of order -1, can be derived from any coherence-transfer pathway, as defined by the vector of Eq. [103]. If a particular pathway satisfies the echo condition, Eq. [105], the "coherence" loss due to field inhomogeneities will be recovered. Other pathways, not satisfying this condition, will continue to lose "coherence." In Fig. 4-17, the absolute-value spectra of the detected single-quantum coherences transferred from double-quantum coherence are shown as a function of double-quantum evolution time t_1. The portions of the observed signal transferred from the echoing ($+2$) and antiechoing (-2) coherences have been separated with phase cycling. The intensity pattern in the echo result is complicated by the finite acquisition time of the observed spectrum; however, the echo effect is clearly apparent. This echo effect is also apparent in the two-dimensional spectrum of Fig. 4-2. The multiple-quantum resonances in the top portion of the spectrum are derived from coherence orders with a plus sign, which lead to an echo formation during the time of acquisition of the observable single-quantum coherences (order -1). The resonances in the bottom portion of the spectrum are derived from coherence orders with a minus sign and hence continue to lose "coherence" during acquisition. As a consequence of line broadening, there is a loss of resonance peak height in the two-dimensional spectrum. The notable exceptions are zero-quantum coherences. Since these coherences are not affected by field inhomogeneities, there is no asymmetry due to echo formation for

antiecho (−2)

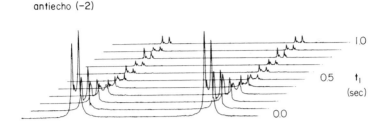

echo (+2)

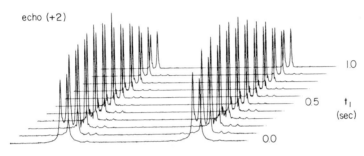

Fig. 4-17. The effect of the evolution of double-quantum coherence during t_1 in an inhomogeneous magnetic field. The absolute value of the observed magnetization for the proton AX spin system 2,3-dibromothiophene in acetone-d_6 is plotted as a function of evolution time t_1. The top portion is the antiecho magnetization and the bottom the echoing magnetization. Both stacked plots have the same vertical scale.

the zero-quantum portion of a two-dimensional spectrum, as can be seen in Fig. 4-8b.

Pulsed field gradients can be applied at various points along a coherence-transfer pathway to manipulate echo formation (42, 45–47, 93). Since echo formation depends on the orders present along a coherence transfer pathway, pulsed field gradients can be used as a filter for the desired coherence pathway and hence orders of coherence can be indirectly detected.

The effect of an inhomogeneous field on the coherence transfer pathway of a heteronuclear system of m nuclear species can be cast in the same formulation as a homonuclear system. The result for a density operator of a particular pathway is then

$$\sigma^{P_n}(\tau_n) = \exp\{-i\,\Delta B(\mathbf{r})[\gamma'\mathbf{p}' \cdot \boldsymbol{\tau} + \gamma''\mathbf{p}'' \cdot \boldsymbol{\tau} + \cdots + \gamma^{(m)}\mathbf{p}^{(m)} \cdot \boldsymbol{\tau}]\}\sigma^{P_n}, \quad [106]$$

where an echo will form at τ_n if the condition

$$\gamma'\mathbf{p}' \cdot \boldsymbol{\tau} + \gamma''\mathbf{p}'' \cdot \boldsymbol{\tau} + \cdots + \gamma^{(m)}\mathbf{p}^{(m)} \cdot \boldsymbol{\tau} = 0 \quad [107]$$

is satisfied. The definition of the coherence order vector can be generalized for a specific nuclear species i:

$$\mathbf{p}^{(i)} = (p_1^{(i)}, p_2^{(i)}, \ldots, p_n^{(i)}) \qquad [108]$$

For heteronuclear systems, it is best to develop a coherence-transfer map for nuclear species according to Eq. [108]. Using Eq. [107], it can be seen that an echo will form at a point along the coherence-transfer maps where the sum of the areas under each map weighted with the magnetogyric ratio of the specific spins equals zero.

An example of heteronuclear coherence-transfer echo formation is shown in Fig. 4-18. The static field homogeneity has been degraded to exaggerate the echo effect. A ^{1}H-^{13}C coupled spin system has been used. The system is prepared into a state of ^{1}H single-quantum coherence, where it freely precesses in the inhomogeneous field for time t_1. Then a coherence-transfer step transforms ^{1}H single-quantum coherence into ^{13}C single-quantum coherence. The echo occurs at $t_2 = 4t_1$ as predicted by Eq. [107].

V. Multiple-Quantum Coherences for Systems Containing Three or More Spin-$\frac{1}{2}$ Nuclei

The basic features of multiple-quantum-coherence excitation have been illustrated earlier with the AX spin system; however, some subtleties and complications only become apparent when considering more than two spins. In order to keep the discussion of more complex spin systems manageable, the three-spin AMX spin system will be considered as an example. This system exhibits the basic properties of more complicated systems of a larger number of nonequivalent spins and hence can be used as a basis for further generalization to a system with more spins. Also the extension of the treatment of the AMX system to the situation where two spins are equivalent is possible by setting their chemical shifts equal and simply allowing the coupling between the two spins to go to zero. The description of the resulting A_2X spin system combined with the treatment of the AMX system can act as a guide to the further extension to a more complex system.

A. Excitation of Multiple-Quantum Coherence

As illustrated previously for the AX system, the state of the spin system from which multiple-quantum coherences are excited can be represented by a product of Cartesian operators. Free precession of transverse coherence in the presence of a J-coupling interaction following a $\pi/2$ pulse will cause the system to evolve into a number of states which can be described by product operators. The explicit form containing both the chemical shift precession and J modulation is quite complicated. However, the chemical

4. PRINCIPLES OF MULTIPLE-QUANTUM SPECTROSCOPY

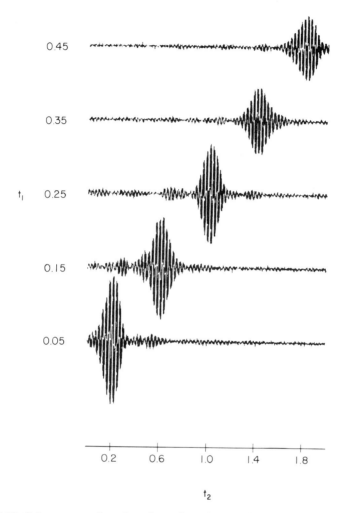

Fig. 4-18. Coherence-transfer echoes for carbon-13 magnetization after a heteronuclear chemical-shift correlation-pulse sequence. Carbon magnetization was detected for a time t_2, and proton single-quantum coherence evolved for a time t_1, before the heteronuclear coherence-transfer step. Static main magnetic-field homogeneity had been degraded to enhance the echo effect.

shift precession can be effectively removed with the spin-echo preparation sequence described previously, leaving only the terms resulting from J modulation of the transverse components of spin. Since each spin in an AMX system can be described in an identical manner, consider only the A spin and denote A, M, and X as 1, 2, and 3, respectively. The density operator for the A spin at the time τ in the spin-echo sequence is

$$\sigma_A(\tau) = -a_A[I_{1y} \cos(\pi J_{12}\tau) \cos(\pi J_{13}\tau) - 2I_{1x}I_{2z} \sin(\pi J_{12}\tau) \cos(\pi J_{13}\tau)$$
$$- 2I_{1x}I_{3z} \cos(\pi J_{12}\tau) \sin(\pi J_{13}\tau)$$
$$- 4I_{1y}I_{2z}I_{3z} \sin(\pi J_{12}\tau) \sin(\pi J_{13}\tau)] \qquad [109]$$

The terms containing products of two operators can be converted into zero- and double-quantum coherence between the two spins represented by the operator product. The J dependence is such that the amount of coherence created depends on the sine of the coupling between the two spins involved in the coherence and the cosine of all other couplings. The terms containing the product of three operators can be converted into zero-quantum, double-quantum, triple-quantum ($p = \pm 3$), or total-spin single-quantum coherence ($p = \pm 1$). The particular coherences created from these product terms depends on the relative phase and amplitude of the creation pulse. There are several basic principles of coherence transfer which can be employed in the creation of a particular order of coherence. It is worth a general discussion of these principles before applying them to the specific case of the AMX system.

In general, in a group of N nonequivalent coupled spins, the J-modulated form of the density operator for the spin echo sequence can be written for one of the spins, for example, spin 1:

$$\sigma_1(\tau) = -a \Bigg\{ I_{1y} \prod_{j=2}^{N} \cos(\pi J_{1j}\tau)$$
$$- 2I_{1x} \sum_{j \neq 1}^{N} \left[I_{jz} \sin(\pi J_{1j}\tau) \prod_{k \neq 1, j}^{N} \cos(\pi J_{1k}\tau) \right]$$
$$- 4I_{1y} \sum_{\substack{j,k \neq 1 \\ j<k}}^{N} \left[I_{jz}I_{kz} \sin(\pi J_{1j}\tau) \sin(\pi J_{1k}\tau) \prod_{l \neq 1, j, k}^{N} \cos(\pi J_{1l}\tau) \right] \cdots$$
$$\pm 2^{N-1} I_{1\varepsilon} I_{2z} \cdots I_{Nz} \sin(\pi J_{12}\tau) \cdots \sin(\pi J_{1N}\tau) \Bigg\} \qquad [110]$$

where $\varepsilon = x$ for $N =$ even integer and $\varepsilon = y$ for $N =$ odd integer. The signs of the terms in the expansion follow a pattern $+--++--++\cdots$ and consequently the sign of the last term depends on the number of coupled spins.

This form of the density operator contains product-operator terms which represent transverse magnetization of spin 1 modulated by coupling to the other spins in the coupled network. The terms can be divided into two general types as

$$I_{1x}I_{2z}\cdots I_{nz}, \quad n = \text{even integer} \quad [111]$$

and

$$I_{1y}I_{2z}\cdots I_{nz}, \quad n = \text{odd integer} \quad [112]$$

In each case the amplitude of the product-operator term is proportional to the product of the sines of the coupling constant between spin 1 and the other spins represented *in the product operator*, and to the product of the cosines of the coupling constants between spin 1 and *all the other spins in the network*.

The third pulse in the preparation sequence can transfer coherence among various orders, creating particular multiple-quantum coherences. The exact result is dependent on the amplitude and phase of this creation pulse. Consider the special case where the amplitude of the creation pulse is $\pi/2$. When the creation pulse is in phase with the first pulse in the sequence, the coherence-transfer expression for terms of the form in Eq. [111] is

$$\exp[-i(\pi/2)F_x](I_{1x}I_{2z}\cdots I_{nz})\exp[i(\pi/2)F_x]$$
$$= -I_{1x}I_{2y}\cdots I_{ny}, \quad n = \text{even integer} \quad [113]$$

For terms of the form in Eq. [112], the coherence-transfer expression is

$$\exp[-i(\pi/2)F_x](I_{1y}I_{2z}\cdots I_{nz})\exp[i(\pi/2)F_x]$$
$$= I_{1z}I_{2y}\cdots I_{ny}, \quad n = \text{odd integer} \quad [114]$$

By expanding the final product-operator terms above in terms of single-element operators, it can be shown that only even orders of coherence are created by a $(\pi/2)_x$ creation pulse in phase with the first $(\pi/2)_x$ pulse in the sequence. In general, product-operator terms with an even or odd number of transverse angular momentum operators contain even or odd orders of coherence, respectively.

Now consider a $\pi/2$ creation pulse in quadrature with the first $(\pi/2)_x$ pulse in the preparation sequence. The coherence-transfer process can be represented in a manner analogous to Eqs. [113] and [114] above as follows. For terms of the form in Eq. [111],

$$\exp[-i(\pi/2)F_y](I_{1x}I_{2z}\cdots I_{nz})\exp[i(\pi/2)F_y]$$
$$= -I_{1z}I_{2x}\cdots I_{nx}, \quad n = \text{even integer} \quad [115]$$

For terms of the form in Eq. [112],

$$\exp[-i(\pi/2)F_y](I_{1y}I_{2z}\cdots I_{nz})\exp[i(\pi/2)F_y]$$
$$= I_{1y}I_{2x}\cdots I_{nx}, \quad n = \text{odd integer} \quad [116]$$

When expanded in terms of single-element operators, the final product operators above are seen to represent only odd orders of coherence.

Using a $\pi/2$ creation pulse, the choice of the pulse phase in the preparation sequence allows selective excitation of either even or odd orders of coherence. As will be shown shortly, the amplitude of a particular odd (or even) order of coherence will be maximized for a pulse tip angle of $\pi/2$.

The situation for a creation pulse of arbitrary tip angle is slightly more complicated. However, useful relationships can be developed for product-operator terms of the general form given in Eqs. [111] and [112]. Consider a creation pulse of tip angle α in phase with the first pulse in the preparation sequence. The coherence-transfer process for terms given in Eq. [111] can be written as

$$\exp(-i\alpha F_x)(I_{1x}I_{2z}\cdots I_{nz})\exp(i\alpha F_x)$$

$$= I_{1x}I_{2z}\cdots I_{nz}\cos^{n-1}(\alpha)$$

$$+ I_{1x}\sum_{m=1}^{n-2}\sum_{j<k}^{n}(-1)^m I_{2z}\cdots I_{jy}\cdots I_{ky}\cdots I_{nz}\cos^{n-m}(\alpha)\sin^m(\alpha)$$

$$- I_{1x}I_{2y}\cdots I_{ny}\sin^{n-1}(\alpha), \qquad n = \text{even integer} \qquad [117]$$

where m equals the total number of spins of the group $j, \ldots, k$, rotated by the pulse and the inner sum is over the possible combinations of the m rotated spins. For the product-operator terms involving an odd number of spins, the coherence-transfer process for terms as given in Eq. [112] is represented by

$$\exp(-i\alpha F_x)(I_{1y}I_{2z}\cdots I_{nz})\exp(i\alpha F_x)$$

$$= I_{1y}I_{2z}\cdots I_{nz}\cos^n(\alpha)$$

$$+ I_{1y}\sum_{m=1}^{n-1}\sum_{j<k}^{n}(-1)^m I_{2z}\cdots I_{jy}\cdots I_{ky}\cdots I_{nz}\cos^{n-m}(\alpha)\sin^m(\alpha)$$

$$+ I_{1z}\sum_{m=0}^{n-2}\sum_{j<k}^{n}(-1)^m I_{2z}\cdots I_{jy}\cdots I_{ky}\cdots I_{nz}\cos^{n-m-1}(\alpha)\sin^{m+1}(\alpha)$$

$$+ I_{1z}I_{2y}\cdots I_{ny}\sin^n(\alpha), \qquad n = \text{odd integer} \qquad [118]$$

where again m equals the total number of spins rotated by the pulse and the inner sum is over the possible combinations of the m rotated spins for the second term on the right-hand side of the equation above and the $m+1$ rotated spins for the third term.

In either case, the resulting coherence transfer can create any order coherence up to and including the nth order with amplitude determined by the choice of the tip angle of the creation pulse as well as the J dependence of the pulse delays. Equations [117] and [118] indicate that even orders of coherence are created with maximum amplitude when the creation pulse is equal to $\pi/2$ in phase with the first pulse. Analogously, odd orders of coherence are maximized for a $\pi/2$ creation pulse in quadrature with the first pulse in the preparation sequence. Thus to create the highest order, N, possible in an N-spin network, one should choose a creation pulse of $\pi/2$ with appropriate phase for N even or odd. The same basic principle will apply for any desired order $n < N$ in the spin network.

The final amplitude of a particular order of coherence is dependent on the coupling constants appropriate for the development of the product-operator term. It can be seen from Eq. [110] that the amplitude of the Nth-order coherence in an N-spin network created from the transverse coherence of a particular spin is dependent on the product of the sines of each coupling constant between that spin and all others in the network. This places a restriction on the size of the coupled networks and hence the maximum order of coherence which can be excited by this method. As an example, consider a linear-chain organic molecule with a series of nonequivalent CH_2 groups. It seems quite unlikely that a coherence of greater than 10th order could be created among the protons in such a system, because the size of the coupling constants becomes vanishingly small with distance as the network size increases. The delay τ can be increased to accommodate smaller coupling constants; however, nodes in the sine dependence of larger coupling constants may occur, and also relaxation effects begin to dominate. The relaxation effects present an increasing problem particularly as the size of the molecular system increases.

Coherence of orders less than the maximum possible, which are less restricted by dependence on the coupling constants, can be created. For a coherence between a particular subgroup of spins in a coupled network, the amplitude of these coherences created from the transverse coherence of a member of this subgroup is dependent on the product of the sine of the couplings between this spin and all others in the subgroup and the cosine of the coupling constants of this spin to all others in the network. Thus it is not necessary to have a finite coupling constant between all members of a network to excite multiple-quantum coherence within a subgroup (25).

These general relationships can now be applied to the preparation of multiple-quantum coherences in an AMX spin system. The effect of a creation pulse on the spin system prepared with the spin-echo sequence can be calculated starting from Eq. [109]. This form of the density operator contains product-operator terms of the two types described in Eqs. [111] and [112]. The result of applying a creation pulse at time τ can be determined using Eqs. [113]–[118] and analogous equations for quadrature phase-shifted pulses.

In order to create double-quantum coherence with maximum efficiency, a $\pi/2$ creation pulse must be applied in phase with the first pulse as described by Eqs. [117] and [118]. The resulting form of the density operator for spin A is

$$\sigma'_A(\tau) = -a_A[I_{1z}\cos(\pi J_{12}\tau)\cos(\pi J_{13}\tau) + 2I_{1x}I_{2y}\sin(\pi J_{12}\tau)\cos(\pi J_{13}\tau)$$

$$+ 2I_{1x}I_{3y}\cos(\pi J_{12}\tau)\sin(\pi J_{13}\tau)$$

$$- 4I_{1z}I_{2y}I_{3y}\sin(\pi J_{12}\tau)\sin(\pi J_{13}\tau)] \qquad [119]$$

The second and third product-operator terms above can be expanded in terms of the appropriate single-element operators using relationships of the general form of Eq. [7]. The second term above represents a sum of y components for double- and zero-quantum coherence between spins 1 and 2. The third term represents a similar sum of y components between spins 1 and 3.

The last term in Eq. [119] can be shown to represent a sum of x components for double- and zero-quantum coherence between spins 2 and 3. This represents multiple-quantum coherence created through a passive spin (spin 1) not directly involved in the coupled interaction between spins 2 and 3. The creation of these coherences is not dependent on the presence of a finite coupling between spins 2 and 3 but only finite coupling from spins 1 to 2 and spins 1 to 3, as can be seen in Eq. [119]. This would complicate the interpretation of coupling patterns indicated by the two-dimensional representation of the double-quantum spectrum. The multiple-quantum coherence created through a passive spin is detected on the appropriate *active* spins (those actually involved in the coupled interaction) if a finite coupling exists between active spins. This coupling between active spins must be observable as a finite splitting in the detected spectrum. Even in the absence of a finite detectable splitting between active spins, a multiple-quantum coherence created through a passive spin is also detectable on the *passive* spin (see Fig. 4-22). The passive-spin multiple-quantum coherence appears at the resonance frequency of the active spins, indicating a coupled network of spins (25).

The amount of multiple-quantum coherence created can be controlled by varying the size of the tip angle of the creation pulse. This is implicit in Eqs. [117] and [118]. For the particular example of the three-spin AMX system, rotation of the second and third terms of Eq. [119] can be illustrated by considering only the second term, as follows:

$$\exp(-i\alpha F_x)(-2I_{1x}I_{2z})\exp(i\alpha F_x)$$

$$= -2I_{1x}I_{2z}\cos\alpha + 2I_{1x}I_{2y}\sin\alpha \qquad [120]$$

4. PRINCIPLES OF MULTIPLE-QUANTUM SPECTROSCOPY

The effect of tip angle of the creation pulse on the last term in Eq. [109] can be expanded as

$$\exp(-i\alpha F_x)(-4I_{1y}I_{2z}I_{3z})\exp(i\alpha F_x) = -4I_{1y}I_{2z}I_{3z}\cos^3\alpha$$
$$+4I_{1y}(I_{2z}I_{3y}+I_{2y}I_{3z})\cos^2\alpha\sin\alpha$$
$$-4I_{1y}I_{2y}I_{3y}\cos\alpha\sin^2\alpha$$
$$-4I_{1z}I_{2z}I_{3z}\cos^2\alpha\sin\alpha$$
$$+4I_{1z}(I_{2z}I_{3y}+I_{2y}I_{3z})\cos\alpha\sin^2\alpha$$
$$-4I_{1z}I_{2y}I_{3y}\sin^3\alpha \qquad [121]$$

These two equations explicitly show that the preparation of even orders of coherence in this spin system is maximized for $\alpha = n\pi/2$, where n is an odd integer. The creation-pulse tip-angle dependence is quite different for each equation. Equation [120] involves the product of two spin operators, one of which is not affected by the pulse. The result is the simple tip-angle dependence which transfers single-quantum coherence into y components of zero- and double-quantum coherence with a $\sin\alpha$ coefficient. This describes coherence transfer for both the second and third terms in Eq. [109]. The tip-angle effect on the three-spin operator term in Eq. [109], as given above in Eq. [121], is more complicated. The initial single-quantum coherence can be transferred to a number of other coherences. Considering only even orders of coherence (zero- and double-quantum), the second term above represents the x components of even-order coherence between spins 1 and 2 and 1 and 3 created through active spin 1. These coherences are created with tip-angle dependence $\cos^2\alpha\sin\alpha$. The last term represents the x components of even-order coherence between spins 2 and 3 created through passive spin 1. The tip-angle dependence in this case is given by $\sin^3\alpha$.

The dependence of the transfer coefficients on the tip angle of the creation pulse is illustrated in Fig. 4-19. The top graph represents the coefficients for Eq. [120] and the bottom graph the coefficients for Eq. [121]. Whether the multiple-quantum coherences are created from passive or active spin, single-quantum coherence can be controlled by the choice of tip angle. In particular, the amount of multiple-quantum coherence (zero and double) between passive spins 2 and 3 created from the single-quantum coherence of an active spin 1 can be manipulated. For $\alpha \leq \pi/4$, the multiple-quantum coherence between active spins predominates by at least a factor of two over that between passive spins. This tip-angle effect can be used to minimize the possibly complicating effect of the detection of passively created coherence through the passive spins. The J dependence of the multiple-quantum coherence creation will be simpler when only the active spins contribute predominantly to the indirectly detected coherence.

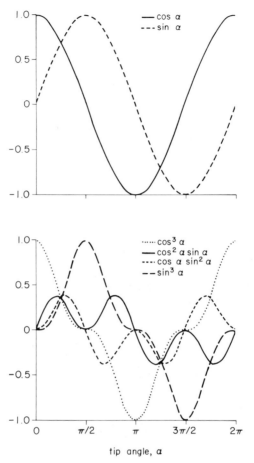

Fig. 4-19. Tip-angle dependence of various coherence-transfer coefficients for an AMX spin system. (See text, Eqs. [120] and [121].)

The analysis of the effect of a creation pulse of tip angle α applied in quadrature with the first pulse in the spin-echo sequence can be carried out in a manner analogous to the previous analysis. Starting from Eq. [109], one can derive a result for a $\pi/2$ creation pulse which produces only odd-order coherences for the AMX spin system. Similarly, the analysis for any value of tip angle α can be carried out in a straightforward manner resulting in an expression which details the tip-angle dependence for the AMX spin system.

If α is chosen to be other than an integer multiple of $\pi/2$, it is possible to create all orders of multiple-quantum coherence simultaneously. The

4. PRINCIPLES OF MULTIPLE-QUANTUM SPECTROSCOPY

relative amplitudes of the indirectly detected coherences will be dependent on the tip angle and the J coupling involved, as well as the details of the coupling network. As the number of coupled spins increases, the tip-angle dependence of the higher-order coherence becomes more sharply peaked about $\alpha = n\pi/2$, where n is an integer. This is seen clearly in Fig. 4-20. The graph represents the transfer coefficients to the highest possible multiple-quantum coherence as the number of spins in the couple network increases from 1 to 5.

Though the details of the effect of the creation-pulse tip angle can be quite complex, direct control is possible over both the order and amount of a particular multiple-quantum coherence which is created.

B Composite Rotations

The analysis of spin dynamics of a multispin system during a pulse sequence can be quite complex. It is often difficult to follow the detailed motion of each coherence as it traverses a path through a pulse sequence. Some simplification can be obtained by considering groups of pulses and precessions as a composite whole. This frequently allows a single operator to replace a group of operators, independent of the specifics of a given spin system (14, 94, 95).

This approach can be particularly useful in multiple-quantum spectroscopy (14, 95) of complex spin systems, where it leads to a simplified

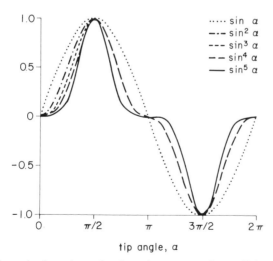

Fig. 4-20. Tip-angle dependence for the coherence-transfer coefficients to the highest possible multiple-quantum coherence in a system of coupled spins as the number of spin-$\frac{1}{2}$ nuclei increases from 1 to 5.

description of preparation and mixing processes. Using the composite rotation approach, the analysis of pulse sequences can be performed without reference to a specific spin system. Although the analysis has general applicability, here it will be limited to weakly coupled spin systems. The resulting composite rotation operator can be used in place of a detailed analysis of the behavior of a spin system at each step in a pulse sequence. In this section, the basic two-pulse chemical-shift-dependent and three-pulse chemical-shift-independent preparation sequences will be analyzed by this method.

As a simple example, consider the analysis of the symmetric sequence $\tau/2$—$(\pi)_x$—$\tau/2$. Using arguments similar to those used to derive Eq. [19], we write the operator describing this sequence as (14)

$$\exp(-i \sum \pi J_{mn}\tau 2 I_{mz}I_{nz}) \exp(-i \sum \pi I_{lx}) \quad [122]$$

Using this as a starting point, it is possible to derive simple composite rotation expressions that describe the three-pulse chemical-shift-independent preparation sequence discussed previously. After inclusion of the operators for the $\pi/2$ pulses, the unitary operator for the even-order multiple-quantum preparation sequence, $(\pi/2)_x$—$\tau/2$—$(\pi)_x$—$\tau/2$—$(\pi/2)_x$, is

$$U_x(\tau) = \exp\left(-i \sum \frac{\pi}{2} I_{lx}\right) \exp(-i \sum \pi J_{mn}\tau 2 I_{mz}I_{nz})$$

$$\times \exp(-i \sum \pi I_{lx}) \exp\left(-i \sum \frac{\pi}{2} I_{lx}\right) \quad [123]$$

After inserting the unit operator, $E = \exp[i \sum (\pi/2) I_{lx}] \exp[-i \sum (\pi/2) I_{lx}]$, and, realizing that $\exp(-i \sum 2\pi I_{lx})$ has no effect, we write this as

$$U_x(\tau) = \exp(-i \sum \pi J_{mn}\tau 2 I_{my}I_{ny}) \quad [124]$$

This preparation sequence can be generalized to include uniform phase shifting of all pulses, $(\pi/2)_\phi$—$\tau/2$—$(\pi)_\phi$—$\tau/2$—$(\pi/2)_\phi$. For this sequence, the operator is

$$U(\tau, \phi) = \exp(-i \sum \phi I_{lz}) \exp(-i \sum \pi J_{mn}\tau 2 I_{my}I_{ny}) \exp(i \sum \phi I_{lz}) \quad [125]$$

For $\phi = \pi/2$, this becomes

$$U_y(\tau) = \exp(-i \sum \pi J_{mn}\tau 2 I_{mx}I_{nx}) \quad [126]$$

The general form of Eqs. [124] and [126] is often referred to as a bilinear rotation operator. There is a formal similarity to the usual J-modulation operator,

$$\exp(-i \sum \pi J_{mn}\tau 2 I_{mz}I_{nz}) \quad [127]$$

4. PRINCIPLES OF MULTIPLE-QUANTUM SPECTROSCOPY

The composite rotation operator forms in Eqs. [124] and [126] for the preparation sequences can be used to calculate the response of a specific spin system starting from some initial condition. Thus one can start from equilibrium and determine the specifics of the coherence transfer effected by the preparation sequence without recourse to specific calculations for each pulse and delay in the sequence. To complete the analysis, we must consider the effect of the composite rotation operators in detail. The interaction of the linear angular momentum operators with the bilinear terms in the three composite rotation operators of Eqs. [124], [126], and [127] satisfy the following commutator, with cyclic permutations of the operator terms (14):

$$[I_{k\lambda}, 2I_{k\mu}I_{l\xi}] = i2I_{k\nu}I_{l\xi} \qquad [128]$$

where $\lambda, \mu, \nu = x, y, z$ and cyclic permutations, and $\xi = x, y$ or z.

The effect of these bilinear operators on the linear angular momentum operators has a simple geometric interpretation (14) as shown in Fig. 4-21. The composite rotation operator of Eq. [124] can be used to follow the excitation process of even-order multiple-quantum coherences. Starting from equilibrium, the density operator for the spin system at the end of the preparation sequence is

$$\sigma(\tau) = U_x(\tau)\sigma_0 U_x^+(\tau) \qquad [129]$$

where $\sigma_0 = a \sum_l I_{lz}$. As an illustration, consider a simple AX spin system for

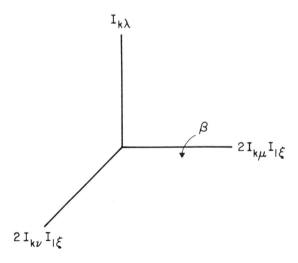

Fig. 4-21. Geometric interpretation of the effect of bilinear rotation operators on linear angular momentum operators. For J-coupling interaction, $\beta = \pi J_{kl} t$.

which Eq. [129] becomes

$$\sigma(\tau) = \exp(-i\pi J\tau 2 I_{1y}I_{2y})a(I_{1z}+I_{2z})\exp(i\pi J\tau 2 I_{1y}I_{2y})$$
$$= a[(I_{1z}+I_{2z})\cos \pi J\tau + 2(I_{1x}I_{2y}+I_{1y}I_{2x})\sin \pi J\tau] \quad [130]$$

Here double-quantum coherence is excited with optimum magnitude when $\tau = n/2J$, where n is an odd integer. This result has been easily derived using composite rotations and the simple vectorial relationships of Fig. 4-21 without the detailed machinations of Section III,A.

The composite rotation operator approach can be easily extended to systems of more than two spins. The exponential of a sum can be written as a product of exponential operators for each term if the terms commute. This is true for the bilinear operators considered here. The general composite rotation of the type given in Eqs. [124]–[127] can be considered as a cascade of bilinear rotations by all coupled pairs of spins in the system. With this approach it is possible to write down the effect of the excitation sequence on a three-spin system as given in Eq. [119].

The odd-order multiple-quantum preparation sequence $(\pi/2)_x$—$\tau/2$—$(\pi)_x$—$\tau/2$—$(\pi/2)_y$ can be written in composite operator form using a procedure similar to that used in the previous development for the even-order sequence (14). These preparation sequences can be combined into a formulation for a general multiple-quantum preparation sequence. One such general sequence is

$$\left(\frac{\pi}{2}\right)_\phi - \frac{\tau}{2} - (\pi)_\phi - \frac{\tau}{2} - \left(\frac{\pi}{2}\right)_{\phi+\eta}$$

To simplify the operator description, it is possible to insert a unit operator at any point in the sequence. If a unit operator is inserted before the last pulse in the operator description of this sequence, the unitary operator description, with the use of Eqs. [30] and [125], for this general sequence becomes

$$U(\tau, \phi, \eta) = \exp[-i\sum(\eta+\phi)I_{lz}]\exp(-i\sum \pi J_{mn}\tau 2 I_{my}I_{ny})$$
$$\times \exp\left(-i\sum_l \eta I_{ly}\right)\exp\left(+i\sum_l \phi I_{lz}\right) \quad [131]$$

Note that the phase shift η of the last pulse has the equivalent effect of shifting the phase of the transverse coherence and effecting a rotation about the y axis. Starting from equilibrium, the system would contain only longitudinal magnetization represented by a density operator, $\sigma_0 = a\sum_l I_{lz}$. The equation of motion for the spin system in terms of the density-operator formulation is

$$\sigma(\tau, \phi, \eta) = \exp\left(-i\sum_l \phi I_{lz}\right) \sigma(\tau, \eta) \exp\left(i\sum_l \phi I_{lz}\right) \quad [132]$$

where

$$\sigma(\tau, \eta) = U(\tau, \eta)\sigma_0 U^+(\tau, \eta) \quad [133]$$

and

$$U(\tau, \eta) = \exp(-i\eta I_z) \exp(-i\sum \pi J_{mn}\tau 2 I_{my}I_{ny}) \exp(-i\sum \eta I_{ly}) \quad [134]$$

For $\eta = 0°$, the even-order multiple-quantum preparation sequence of Eq. [124] is recovered; for $\eta = 90°$, the effect is as though the longitudinal magnetization is first rotated into the transverse plane, then further rotated by the bilinear interaction, and finally phase shifted. Starting from equilibrium, the sequence with $\eta = 90°$ will create odd orders of multiple-quantum coherence. It is interesting to note that equal amounts of odd and even orders of multiple-quantum coherence are created for $\eta = 45°$. The general preparation sequence described here can be used to create selectively odd or even orders of coherence or create equal amounts of both, based on the specific dependence on τ as represented by Eq. [131], independent of chemical shifts.

The two-pulse chemical-shift-dependent preparation sequence can also be formulated in terms of bilinear rotations. As will be shown, the two-pulse sequence has a formal similarity to the general three-pulse sequence as described by Eq. [134]. Consider a two-pulse preparation sequence of the form

$$(\pi/2)_{-x} — \tau — (\pi/2)_x$$

The phase of the first pulse is inverted in order to make the sequence formally similar to the refocused sequence of the previous analysis. The unitary operator for the present sequence is

$$U(\tau) = \exp\left(-i\sum \frac{\pi}{2} I_{lx}\right) \exp(-i\sum \pi J_{mn}\tau 2 I_{mz}I_{nz})$$

$$\times \exp(-i\sum \omega_l \tau I_{lz}) \exp\left(+i\sum \frac{\pi}{2} I_{lx}\right) \quad [135]$$

where ω_l is the precession frequency of spin l in the rotating frame of reference. A unit operator, $E = \exp[+i\sum (\pi/2) I_{lx}] \exp[-i\sum (\pi/2) I_{lx}]$, can be inserted between the chemical shift and J-coupling terms. This allows the use of rotation properties to simplify the above operator form to (14)

$$U(\tau) = \exp(-i\sum \pi J_{mn}\tau 2 I_{my}I_{ny}) \exp(+i\sum \omega_l \tau I_{ly}) \quad [136]$$

The formal similarity of the unitary operators of Eqs. [134] and [136] leads to an equivalent interpretation of their effectiveness in creating multiple-quantum coherences. Each will create both even and odd orders of coherence. The former operator allows direct control over the orders by choosing the phase η of the final pulse in the sequence, while the latter is dependent on the chemical shift of the particular spins in the system under study.

The two-pulse preparation sequence can be generalized to include phase-shifted pulses:

$$(\pi/2)_{\pi+\phi} - \tau - (\pi/2)_{\phi+\eta}$$

The unitary operator for this sequence is

$$U(\tau, \phi, \eta) = \exp(-i \sum \phi I_{lz}) U(\tau, \eta) \exp(i \sum \phi I_{lz}) \quad [137]$$

where the $U(\tau, \eta)$ operator can be simplified by a procedure outlined in the previous analysis. The resulting operator can be written

$$U(\tau, \eta) = \exp(-i \sum \eta I_{lz}) \exp(-i \sum \pi J_{mn} \tau 2 I_{my} I_{ny}) \exp[+i \sum (\omega_l \tau - \eta) I_{ly}]$$

[138]

From the first term, it can be seen that a choice of $\eta = 0°$ or $90°$ will determine the dependence of the amplitude of a particular order of coherence on the chemical shift, either $\cos(\omega_l \tau)$ or $\sin(\omega_l \tau)$. The form depends on whether even or odd orders of coherence are desired. It is apparent from this analysis that both the two- and the three-pulse sequences rely on the same fundamental processes for the preparation of multiple-quantum coherences. The particulars of the spin system and the desired result dictate which is more appropriate. Both sequences can be used for uniform excitation (57, 80), as will be discussed in a later section.

C. Evolution of Multiple-Quantum Coherences

Since multiple-quantum coherences are not directly observable, two-dimensional multiple-quantum spectroscopy relies on the mechanism of indirect detection of high-order coherences by transfer to observable single-quantum coherences. From the point of view of the operator formulation, there is nothing special to distinguish multiple-quantum from single-quantum coherences. The effect of pulses and periods of free precession on any order of coherence is treated in a like manner. The main distinction between simple single-quantum and multiple-quantum coherences is one of observability. In fact, free precession or evolution can be described by the same simple vectorial picture.

To describe evolution, a formal analogy can be drawn between the transverse angular momentum operators of Eqs. [3] and [4] for single-quantum coherence and multiple-quantum coherence operators expressed in terms of single-element operators. For example, the transverse components of a total N-spin, p-quantum coherence can be written (14)

$$\{N\text{-spin, }p\text{-QC}\}_x = \tfrac{1}{2}(I_1^+ I_2^+ \cdots I_N^+ + I_1^- I_2^- \cdots I_N^-) \quad [139]$$

$$\{N\text{-spin, }p\text{-QC}\}_y = \frac{1}{2i}(I_1^+ I_2^- \cdots I_N^- - I_1^- I_2^+ \cdots I_N^-) \quad [140]$$

Since this represents a total-spin multiple-quantum coherence, all N spins make a transition and the product-operator forms for these coherences contain a transverse component of angular momentum operator for every spin.

The evolution for a total-spin multiple-quantum coherence in an N-spin system is not affected by the coupling between the N spins. This is apparent upon noting that the coupling commutes with the product-operator form

$$[I_{j\delta}I_{k\varepsilon}, I_{jz}I_{kz}] = 0 \quad [141]$$

where δ and $\varepsilon = x, y$, or z. The evolution of total-spin multiple-quantum coherences is affected only by the chemical shift precession, according to the equation (14)

$$\exp(-i\mathcal{H}_0 t)\{N\text{-spin, }p\text{-QC}\}_x \exp(i\mathcal{H}_0 t)$$
$$= \{N\text{-spin, }p\text{-QC}\}_x \cos(\omega_{\text{eff}} t) + \{n\text{-spin, }p\text{-QC}\}_y \sin(\omega_{\text{eff}} t), \quad [142]$$

where $p = \Sigma_{j=1}^N \Delta m_j$, $\omega_{\text{eff}} = \Sigma_{j=1}^N \Delta m_j \omega_j$. An analogous expression holds for the evolution of the y component of a total-spin multiple-quantum coherence.

The form of Eq. [142] for the evolution of this coherence is analogous to that for a transverse component of single-quantum coherence. It is possible to illustrate the free precession of a multiple-quantum coherence as a vector rotating in a plane transverse to the static magnetic field. The effect of J coupling can be included in a straightforward manner as for single-quantum coherence. This is equivalent to describing the process of evolution in terms of single-transition operators. The forms of the operators in Eqs. [139] and [140] are the same as these of the appropriate single-transition operators for the transition considered here.

For partial-spin multiple-quantum coherences, $\{N, q\text{-spin}, p\text{-QC}\}$, an equation of the type of Eq. [142] would still be appropriate when considering only chemical shift evolution. In this case, an additional definition, $q = \Sigma_{j=1}^N |\Delta m_j|$, for an N-spin system is employed. A partial-spin multiple-quantum coherence involves transitions of only q spins in a system of N coupled spins ($q < N$). The product-operator representation of this type of coherence contains a transverse coherence operator for each of the q spins that make a transition. Passive spins for this type of coherence are represented by a z component of the angular momentum operator.

A partial-spin multiple-quantum coherence precession frequency is modulated by J coupling to passive spins. The effective coupling constant between the active spins and a passive spin m is (14)

$$J_{\text{eff}} = \sum_{\substack{k=1 \\ k \neq m}}^{N} \Delta m_k J_{km} \qquad [143]$$

The J evolution for coupling to passive spin m is

$$\exp\left(-i \sum_k \pi J_{km} \tau 2 I_{kz} I_{mz}\right) \{N, q\text{-spin}, p\text{-QC}\}_x \exp\left(i \sum_k \pi J_{km} \tau 2 I_{kz} I_{mz}\right)$$

$$= \{N, q\text{-spin}, p\text{-QC}\}_x \cos(\pi J_{\text{eff}} \tau)$$

$$+ 2 I_{mz} \{N, q\text{-spin}, p\text{-QC}\}_y \sin(\pi J_{\text{eff}} \tau) \qquad [144]$$

The complete evolution of a partial-spin multiple-quantum coherence is described by the chemical shift evolution of the type shown in Eq. [142] and the J-coupled evolution of Eq. [144] for coupling to each passive spin.

Using this formulation, the evolution of any order of coherence can be calculated in a straightforward manner. The evolution of any order of multiple-quantum coherence can be visualized as a rotation of a coherence "vector" about the main magnetic field much as magnetization "vectors" are visualized as rotating about the main field.

D. Single-Pulse Mixing and Coherence-Transfer Selection Rules

The indirect detection of multiple quantum coherence evolution is possible with two-dimensional Fourier transformation techniques. The energy-level pattern of a coupled spin system can be mapped with a knowledge of the available transitions, and information from multiple-quantum spectroscopy can complement that available from conventional single-quantum spectroscopy.

As shown in earlier discussion of the two- and three-spin systems, the coherence-transfer processes of preparation and mixing have a complicated yet straightforward dependence on the tip angle and phase of the rf pulses in the measuring sequence. It is sometimes possible to follow the coherence transfer with the geometric formulation of product operators. Up to this point, the product operator approach has been used for both the two- and three-spin systems. This allows a description in terms of more familiar vector rotations. However, the evolution of coherences is cumbersome when

4. PRINCIPLES OF MULTIPLE-QUANTUM SPECTROSCOPY

described in product-operator terms. Evolution is more easily followed in terms of single-element or single-transition operators. The latter give a geometric representation of the process, which is easier to visualize. This representation is also convenient because spectra appear as individual resonance lines for each coherence present during evolution and detection periods.

When considering coherence transfer between specific pairs of energy levels, it is most convenient to use the single-transition operators as described in Eqs. [26]–[28]. The example of coherence transfer with a single rf pulse illustrates the procedure. This also allow rules for coherence transfer (25) from multiple-quantum to observable single-quantum coherence to be developed which aid in spectral interpretation. Consider the transfer process from coherence between energy levels t and u to coherence between energy levels r and s. In a two-dimensional experiment, the (t,u) coherence evolves during t_1 in a manner described by Eqs. [47]–[49]. Assuming the coherence is initially along the x axis, we can describe the evolution of the (t, u) coherence at time t_1 by

$$\sigma_{tu}(t_1) = a(I_x^{tu} \cos \omega_{tu}t_1 + I_y^{tu} \sin \omega_{tu}t_1) \qquad [145]$$

At time t_1, an rf pulse effects a coherence transfer from the (t,u) to an observable (r,s) coherence. The pulse can have arbitrary tip angle and phase described by an operator of the form shown in Eq. [30]. The observable (r,s) magnetization originating from coherence (t,u) can be written, using Eq. [59]:

$$M_{rs,tu}(t_1, t_2) = \langle \mu_x^{rs} \sigma_{tu}(t_1, t_2) \rangle + i \langle \mu_y^{rs} \sigma_{tu}(t_1, t_2) \rangle$$
$$= \gamma\hbar[\langle I_x^{rs} R\sigma_{tu}(t_1)R^+ \rangle$$
$$+ i\langle I_y^{rs} R\sigma_{tu}(t_1)R^+ \rangle] \exp(i\omega_{rs}t_2) \qquad [146]$$

From Eq. [145] and evaluation of the trace operators, this becomes (27)

$$M_{rs,tu}(t_1, t_2) = \gamma\hbar a\{(R_{su}R_{tr}^+) \exp[i(\omega_{rs}t_2 + \omega_{tu}t_1)]$$
$$+ (R_{st}R_{ur}^+) \exp[i(\omega_{rs}t_2 - \omega_{tu}t_1)]\} \qquad [147]$$

The rotation-operator matrix-element product terms in parentheses represent complex coherence-transfer coefficients which describe the amplitude and phase of the transfer process. The frequency components of the exponential terms represent the resonance positions in the two-dimensional spectrum. In the presence of magnet inhomogeneities, the first term in Eq. [147] will not refocus any dephasing due to inhomogeneity, since it has an additive combination of frequency components. This is denoted as an antiecho term. The second term has a subtractive combination of frequency components and will refocus inhomogeneities; it is denoted as an echo term. The two-dimensional spectrum has two resonances, each with $F_2 = \omega_{rs}/2\pi$, positioned at $F_1 = \pm\omega_{tu}/2\pi$.

The amplitude and phase of the resonances are described by the value of the complex coherence-transfer coefficients. A pulse rotation operator $R(\beta, \phi)$ for a system of N weakly coupled spin-$\frac{1}{2}$ nuclei can be written in the form of Eq. [30]. A matrix element R_{st} of an operator of this form is then

$$R_{st} = \exp[-i(m_s - m_t)\phi]\left\langle s \left| \prod_{k=1}^{N} \exp(-i\beta I_{kx}) \right| t \right\rangle \qquad [148]$$

With I_{kx} expressed in terms of raising and lowering operators, this becomes (20, 21, 25)

$$R_{st} = (-i)^{\Delta_{st}}\left(\cos\frac{\beta}{2}\right)^{(N-\Delta_{st})}\left(\sin\frac{\beta}{2}\right)^{\Delta_{st}} \exp[-i(m_s - m_t)\phi] \qquad [149]$$

where Δ_{st} is the number of spins that flip in the transition from energy level t to energy level s (20, 25). The complex coherence-transfer coefficients (20, 25) are

$$R_{st}R_{ur}^+ = (i)^{(\Delta_{ur}-\Delta_{st})}\left(\cos\frac{\beta}{2}\right)^{(2N-\Delta_{st}-\Delta_{ur})}\left(\sin\frac{\beta}{2}\right)^{(\Delta_{st}+\Delta_{ur})}$$

$$\times \exp[-i(m_s - m_t + m_u - m_r)\phi] \qquad [150]$$

These coefficients are central to the description of coherence transfer. A system of spins occupying an energy level can be characterized by their Zeeman quantum number, that is, by specifying the orientation of each spin as either up or down with respect to the main magnetic field. The transfer coefficient describes the effect of a rotation operation on the system of spins. The factor Δ_{st} in Eq. [149] defines the number of spins changing their orientation as they are transferred from level t to s during the rf pulse. Each spin flip involves a factor of $i \sin(\beta/2)$. For a given rf-pulse phase ϕ, the phase of the resulting magnetization depends on the difference between the spin flip numbers {in Eq. [150], $(\Delta_{ur} - \Delta_{st})$} and the amplitude depends on the sum {in Eq. [150], $(\Delta_{ur} + \Delta_{st})$}. The transfer coefficients can be viewed as describing the detailed reorientation of spins during the application of an rf pulse.

The indirect detection of zero- and double-quantum coherence in an AX spin system was discussed in an earlier section. In Eqs. [69] and [74], it was shown that the transfer coefficients are of equal magnitude for $\beta = n\pi/2$, where n is an integer. Considering the energy-level diagram of Fig. 4-3, the general result using Eq. [147] for transfer of zero-quantum coherence between energy levels 2 and 3 or double-quantum coherence between levels

1 and 4 to an observable single-quantum coherence, say between levels 1 and 2, is described by the same transfer coefficients as Eqs. [69] and [74]. Thus a detailed description for transfer from a specific multiple-quantum coherence to an observable single-quantum coherence can be treated using a formulation as shown in Eq. [147].

The description of the observable magnetization given in Eq. [147] represents only a portion of the total magnetization transferred by the application of an rf pulse. It is possible that a coherence transfer by the rf pulse will result in a magnetization which exactly cancels another portion of the transferred magnetization, effectively making these coherence-transfer pathways not observable. If coherence is transferred to degenerate frequencies and transfer coefficients have opposite sign, then the magnetizations will cancel and the pathway will appear forbidden. Consider two transfer pathways: $(t,u) \to (r,s)$ and $(t,u) \to (p,q)$. These pathways will appear forbidden if (25)

$$\omega_{rs} = \omega_{pq} \quad [151]$$

$$R_{su}R_{tr}^+ = -R_{qu}R_{tp}^+ \quad [152]$$

and

$$R_{st}R_{ur}^+ = -R_{qt}R_{up}^+ \quad [153]$$

The frequencies may be identical because of symmetry considerations, or they may not be resolved due to magnetic inhomogeneities or unresolved couplings. Also, the magnetization may have opposite amplitude if the tip-angle dependence is identical and the phase factor is opposite (25).

These considerations lead to a number of coherence-transfer selection rules which are derived from the restrictions above (25). To see how these rules are developed, consider the situation where the observable magnetization from transitions (r,s) and (p,q) are part of the same spin multiplet. If these transitions differ only in the polarization of a spin which is passively involved in the multiple-quantum transition (t,u), no cancellation will occur even if $\omega_{rs} = \omega_{pq}$. However, if the two transitions (r,s) and (p,q) differ only in the polarization of a spin actively involved in the multiple-quantum transition (t,u), then cancellation occurs when the two multiplet lines from (r,s) and (p,q) coalesce. Coalescence can occur under low-resolution conditions where the coupling between the spin of the observed multiplet and active spin are not resolved. The following coherence-transfer selection rules are a consequence of these considerations (25).

1. A multiple-quantum coherence involving n spins can be transferred to the single-quantum transitions of a *passive* spin only if the couplings between the *passive* spin and all n *active* spins are resolved.

2. A multiple-quantum coherence involving n spins can be transferred to the single-quantum transitions of an *active* spin only if the couplings between all *active* spins can be resolved.

The last selection rule is a consequence of the second rule above.

3. A multiple-quantum coherence involving equivalent nuclei cannot be transferred to the single-quantum transitions of any of these equivalent nuclei because the coupling among equivalent nuclei is effectively zero.

To illustrate the meaning of these coherence-transfer selection rules, a diagrammatic representation of the double-quantum spectrum of both an AMX and an A_2X spin system is shown in Fig. 4-22. In both cases the J_{AM} coupling is set to zero. In the top spectrum for the AMX system, *direct connectivity* between the AX and MX spins results in double-quantum resonances at frequencies along F_1 which are the sum of the Larmor frequencies of each active spin in the double-quantum coherence. The other resonance, at Ω_X along F_2 and $\Omega_A + \Omega_M$ along F_1, represents the appearance of an AM double-quantum coherence detected through the passive X spin. This type of resonance allows one to distinguish a situation with separate AM and M'X systems in which M and M' are chemically equivalent from that in which there is an AMX with $J_{AM} = 0$ (25). The presence of the AM double-quantum coherence at the position of the passive X spin means that there is a *remote connectivity* between spin A and X along the linear chain, A—M—X. The bottom spectrum diagrammed in Fig. 4-22 represents the result for the double-quantum spectrum of an A_2X system. This result can be generated from the AMX spectrum by letting $\Omega_A = \Omega_M$, $J_{AX} = J_{MX}$, and holding $J_{AM} = 0$. In this case, the double-quantum coherence involving the magnetic equivalent A and M spins can be detected on the passive spin X. The appearance of this double-quantum coherence at the position of the passive spin indicates *magnetic equivalence*. In this case the double-quantum frequency is twice the Larmor frequency of the equivalent spin.

By determining the pattern of connectivities within a spin system, double-quantum spectroscopy can be used to characterize the system under study by a relatively simple procedure. A particularly striking example of this procedure (25) is shown in Fig. 4-23. This is the double-quantum spectrum of an $A_aM_mX_x$ spin system. The conventional one-dimensional spectrum consists of three multiplets centered at the positions of the three equivalent spins. As shown in the spectrum, the pattern of connectivities is consistent with $a, m, x = 2$ and $J_{AX} \approx 0$.

The tip-angle dependence of single-pulse mixing has been treated for the example of the AX spin system in Section III,C. For more complex systems, the explicit form of the transfer coefficients given in Eq. [147] can be used, with the selection rules above, to predict or interpret a multiple-

4. PRINCIPLES OF MULTIPLE-QUANTUM SPECTROSCOPY 331

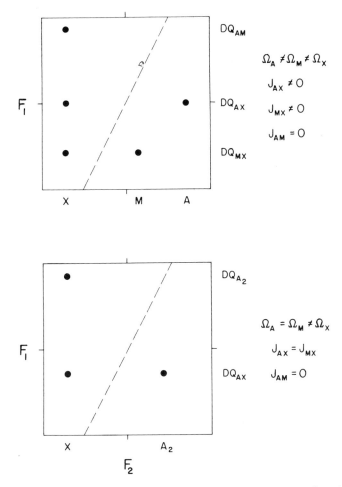

Fig. 4-22. Diagrammatic representation of the double-quantum spectra of an AMX and A_2X spin system. The top spectrum represents the result for an AMX spin system in the case where $\Omega_A \neq \Omega_M \neq \Omega_X$, $J_{AX} \neq 0$, $J_{MX} \neq 0$, and $J_{AM} = 0$. The bottom spectrum represents the result for the A_2X system in the case where $\Omega_A = \Omega_M \neq \Omega_X$, $J_{AX} = J_{MX}$, and $J_{AM} = 0$. [Based on Fig. 3 of Braunschweiler et al. (25).]

quantum spectrum. Examples of the tip-angle effects for a mixing pulse equal to $3\pi/4$ are shown in Figs. 4-14, 4-15, and 4-16. In particular, compare the zero-quantum spectra shown in Figs. 4-8 and 4-14. The same asymmetric pattern of resonances along F_1 and F_2 is apparent in both spectra, indicating that the tip-angle dependence follows the same trend for increasingly complex systems.

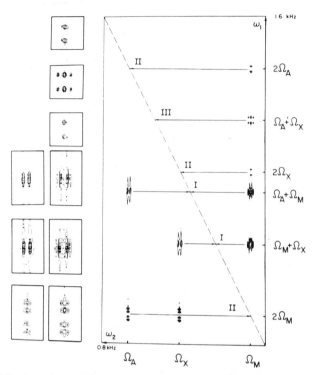

Fig. 4-23. Two-dimensional double-quantum spectrum of 3-aminopropanol ($DOCH_2CH_2CH_2ND_2$) in the absolute-value display mode. The labeled arrows indicate resonances for (I) direct connectivity, (II) magnetic equivalence, and (III) remote connectivity. The individual multiplets are reproduced on an expanded scale on the left. This molecule consists of an $A_aM_mX_x$ network of coupling with $a, m, x \geq 2$ and $J_{AX} \simeq 0$. [Reprinted with permission from L. Braunschweiler, G. Bodenhausen, and R. R. Ernst, *Mol. Phys.* **48**, 535 (1983). Copyright 1983 Taylor & Francis, Ltd.]

E. Uniform Excitation and Indirect Detection of Multiple-Quantum Coherence

In a complex spin system, the number of possible coherences and coherence transfer pathways can become quite large. The preparation of a particular coherence will be dependent on the J coupling among the spins in the coupled network and the details of the preparation sequence. As with all methods, the desired result dictates the details of the techniques used to obtain the result. As discussed in a previous section, the composite refocused sequence

$$\left(\frac{\pi}{2}\right)_\phi - \frac{\tau}{2} - (\pi)_\phi - \frac{\tau}{2} - \left(\frac{\pi}{2}\right)_{\phi+\eta}$$

provides an efficient preparation sequence with control over the coherence order excited and the phase of the resulting coherence. Also, order discrimination is easily accomplished with multistep phase cycling of rf phase ϕ. The general evolution of the excited coherences can be followed with the formulation of Section V,C. One is then left with the problems of controlling the J-dependent coherence excitation and choosing a mixing sequence to allow the indirect detection of the desired multiple-quantum coherence. In general, the single-quantum coherence observable after the mixing process will have a complicated amplitude and phase dependent on the details of the mixing sequence (25).

The simplest mixing sequence for indirect detection of multiple-quantum coherence consists of a single rf pulse. This mixing pulse transfers multiple-quantum to single-quantum coherences with an amplitude and phase dependent on the phase and tip angle of the pulse. In some situations, tip-angle effects can be used to advantage (25-27, 71), leading to a simplification of the result. However, for single-pulse mixing, it is usually necessary to examine two-dimensional spectra in magnitude rather than phase-sensitive form to avoid complications from phase variations. Also, the J dependence of the preparation and complicated amplitude dependence of single-pulse mixing will make spectral interpretation problematic. A uniform excitation and indirect detection scheme independent of J coupling and with predictable phases is advantageous and is possible with the symmetrized preparation and mixing sequences (57, 80, 95-97) which are next described.

The preparation of multiple-quantum coherence in scalar coupled systems relies on the formation of antiphase single-quantum coherences during the preparation period through J modulation. Thus coherence which initially starts out in-phase is prepared into antiphase, and it is then transferred into multiple-quantum coherence by a sequence such as the composite refocused sequence discussed in Section V,B. The coherence-transfer process necessary for indirect detection transfers multiple-quantum coherence into the same antiphase single-quantum coherences. One can extend the mixing sequence to include further J modulation so that the antiphase coherences will once again be in phase, thus gaining control over the phase of the observed magnetization.

The basic idea is to reverse the effect of the preparation sequence. The ideal experiment would provide an effective time-reversal during the mixing period (3), thus allowing the system to recover completely its original in-phase condition. In the general experiment, the system initially in equilibrium is prepared into the desired state by propagator $U(\tau)$. The coherences evolve for a time t_1 and then are mixed into the final observable coherences by propagator $V(\tau)$. Assume that the preparation propagator can be written as a product of individual operators representing pulses and delays in the preparation sequence

$$U(\tau) = ABC \cdots Z \qquad [154]$$

where each step is represented by a letter. In order to reverse the effect of the preparation sequence, the mixing sequence should ideally represent the inverse unitary transformation,

$$V(\tau) = U^+(\tau) = (ABC \cdots Z)^+ = Z^+ \cdots C^+ B^+ A^+ \qquad [155]$$

The result is the application of reflection symmetry to the order of elements in the pulse sequence. To provide complete time-reversal, each element must have its operator effectively time-reversed. If, for a particular element,

$$A(\tau_i) = \exp(-i\mathcal{H}\tau_i/\hbar) \qquad [156]$$

and if

$$\exp\left(-i \sum_m \phi I_{mz}\right) \mathcal{H} \exp\left(i \sum_m \phi I_{mz}\right) = -\mathcal{H} \qquad [157]$$

then it is possible effectively to reverse the time propagation for these elements by phase shifting the appropriate rf pulses (3). Unfortunately, the J-coupling interaction is bilinear in spin operators so that phase shifting cannot effect a time-reversal, as can be seen by considering the form of Eq. [124]. Following the prescription above for reflection symmetrization and phase shifting when appropriate, one can only accomplish an approximate time-reversal for systems with J coupling. However, quite useful results can be obtained with this modification to the basic sequences for multiple-quantum spectroscopy.

The composite refocusing sequence used for preparing the spin system with the desired coherence can be used to derive a symmetric mixing sequence (57, 80) as shown in Fig. 4-1e. The first pulse in the preparation sequence rotates equilibrium z magnetization into transverse coherence. For a completely symmetric preparation and mixing sequence, some transverse coherences at the end of the mixing sequence will be returned to the z axis and would not be observable. Hence the symmetrized mixing sequence does not contain the final $\pi/2$ pulse complementary to the first pulse in the preparation sequence.

The J dependence of the amplitude and phase of the observed signal is a product of the J dependence during preparation and mixing. The J dependence for the preparation of a p-quantum coherence in a system of N coupled, spin-$\frac{1}{2}$ nuclei can be written for a particular spin I_1 as (57)

$$f_1^{(p)}(\tau) = \prod_{i=2}^{P} \sin(\pi J_{1i}\tau) \prod_{j=p+1}^{N} \cos(\pi J_{1j}\tau) \qquad [158]$$

4. PRINCIPLES OF MULTIPLE-QUANTUM SPECTROSCOPY

This form assumes spin 1 is active for the p-quantum coherence. If spin 1 is passive, one cosine factor will be changed to a sine factor. After the evolution interval, the p-quantum coherence can be mixed to single-quantum coherence on the same spin or on another spin in the coupling network. For mixing to the same spin, the J dependence is the square of this factor and hence positive for any value of the preparation and mixing time τ, and the observed single-quantum coherences return to the in-phase condition. For mixing to another spin, the product of preparation and mixing functions will average to zero for a sum of various τ values. Thus it is possible to select only in-phase coherences by averaging over a series of τ values.

The resulting one- or two-dimensional spectra can be presented in the phase sensitive mode. Without the use of symmetrized sequences, the antiphase character of the multiplets could result in cancellation of overlapping multiplets and requires magnitude display in two-dimensional spectra with loss of resolution. For symmetrized preparation and mixing with τ averaging, the resulting multiple-quantum filtered one-dimensional spectrum will appear with positive absorption lineshapes (57). An example is shown in Fig. 4-24. Part (a) is the conventional spectrum of this mixture, and part (b) is the double-quantum filter spectrum with symmetric preparation and mixing. The amplitude of the AMX multiplet on the right of the double-quantum filtered spectrum is reduced by the averaging process for the symmetrized sequence. This results from a cancellation of magnetization derived from the various multiple coherence pathways available to the three-spin system when it is passed through a double-quantum filter. There are six possible double-quantum coherences available to this three-spin system. Each spin contributes actively in four of these coherences and

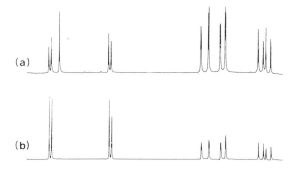

Fig. 4-24. Double-quantum filtered one-dimensional spectrum of a mixture of benzene (A_6 system), 2,3-dibromothiophene (AX system), and acrylonitrile (AMX system). (a) Conventional spectrum. (b) Spectrum after double-quantum filtering with the symmetrical preparation/mixing sequence shown in Fig. 4-1e. The preparation/mixing time τ has been varied in steps of 8 msec from $\tau = 8$ msec to $\tau = 736$ msec. [From Sørensen, Levitt, and Ernst (57).]

passively in two. As will be discussed next, the relative peak amplitude of a particular spin for the symmetric sequence will be positive if the spin is actively involved in the multiple-quantum coherence and negative if it is passively involved. Thus the double-quantum filtered spectrum for the AMX portion is reduced to one-third its amplitude relative to the AX portion of the same spectrum, since the AX spins both contribute actively to their single double-quantum coherence. As the size of a spin-coupling network increases, this type of cancellation will result in a reduction in the peak amplitude for a given order of coherence.

In a two-dimensional multiple-quantum spectrum obtained using a symmetric preparation and mixing sequence, cancellation would not occur because the multiple-quantum coherences would be resolved in the ω_1 dimension. With phase-sensitive two-dimensional acquisition and display, multiple-quantum coherences appear at the resonances of active or passive spins with opposite signs (80); therefore intensities will be positive for one and negative for the other. This can be illustrated using the example of double-quantum spectroscopy (80). Assume the initial $\pi/2$ pulse in the sequence creates a y component of magnetization on spin I_1.

1. Two-spin terms:

$$-I_{1y} \xrightarrow{J\tau} 2I_{1x}I_{2z} \xrightarrow{(\pi/2)_x} -2I_{1x}I_{2y} \xrightarrow{(2Q)} -(I_{1x}I_{2y} + I_{1y}I_{2x})$$

$$\xrightarrow{(\pi/2)_x} -I_{1x}I_{2z} \xrightarrow{J\tau} -\tfrac{1}{2}I_{1y} \qquad [159]$$

The arrows indicate evolution or coherence transfer, and terms have been neglected that do not contribute to the selected pathway. The (2Q) indicates double-quantum filtering. In this case, I_1 is actively involved in the double-quantum coherence.

2. Three-spin terms:

$$-I_{1y} \xrightarrow{J\tau} 4I_{1y}I_{2z}I_{3z} \xrightarrow{(\pi/2)_x} 4I_{1z}I_{2y}I_{3y} \xrightarrow{(2Q)} -2I_{1z}(I_{2x}I_{3x} - I_{2y}I_{3y})$$

$$\xrightarrow{(\pi/2)_x} -2I_{1y}I_{2z}I_{3z} \xrightarrow{J\tau} \tfrac{1}{2}I_{1y} \qquad [160]$$

For this three-spin term, I_1 is passively involved in the double-quantum coherence. The relative sign of observed I_1 magnetization at the end of the sequence is opposite the sign of the two-spin terms.

3. Four-spin terms:

$$-I_{1y} \xrightarrow{J\tau} -8I_{1x}I_{2z}I_{3z}I_{4z} \xrightarrow{(\pi/2)_x} 8I_{1x}I_{2y}I_{3y}I_{4y}$$

$$\xrightarrow{(2Q)} 4(I_{1x}I_{2y}I_{3y}I_{4y} + I_{1y}I_{2x}I_{3x}I_{4x}) \xrightarrow{(\pi/2)_x} 4I_{1x}I_{2z}I_{3z}I_{4z} \xrightarrow{J\tau} -\tfrac{1}{2}I_{1y} \qquad [161]$$

In this case the I_1 spin is actively involved in the double-quantum coherence and has the same sign as the two-spin term.

In general, filtering through n-spin terms results in a peak in the two-dimensional spectrum with a sign given by $(-1)^n$ for any order of coherence selected by the multiple-quantum filter. The two-dimensional multiple-quantum spectra will have resonances for active spins with one sign and passive spins with the opposite sign. For double-quantum filtered spectra, resonances from remotely connected or magnetically equivalent spins will appear with an opposite sign to that of resonances from directly connected spins. Several examples are shown in Rance *et al.* (80).

The process of averaging the J dependence over a variation in τ during preparation and mixing has limitations in practice. Relaxation can cause a loss of desired signal intensity, and averaging a finite number of τ values can result in incomplete suppression of undesired transfer pathways. The addition of a "purging pulse" or "z filter" (80) immediately before data acquisition can remove undesirable components. The suppression due to τ averaging can be improved by using a symmetric preparation and mixing sequence as shown in Fig. 4-1c (80). The effect of the preparation sequence is described by Eqs. [137] and [138]. Considering spin I_1, the functional form of the τ dependence for the symmetrized sequence applied to an N-spin system can be written as

$$\cos^2(\omega_1\tau)[f_1^{(p)}(\tau)]^2 \quad \text{for} \quad \eta = 0° \qquad [162]$$

and

$$-\sin^2(\omega_1\tau)[f_1^{(p)}(\tau)]^2 \quad \text{for} \quad \eta = 90° \qquad [163]$$

If a subtractive combination of the observed response is formed for $\eta = 0°$ and $\eta = 90°$, the τ dependence reduces to $[f_1^{(p)}(\tau)]^2$. The transfer from and to the same spin results in observed coherence independent of resonance offset ω_{1x}, as well as inhomogeneities. All other transfer pathways retain the ω dependence, and therefore the τ averaging process is more effective. This form of the symmetrized sequence is less sensitive by a factor of two than the spin-echo symmetrized sequence described previously. However, the improved suppression may be desirable in some situations at the expense of sensitivity.

VI. Summary

A review of recent literature shows the increasing interest and application of multiple-quantum nuclear magnetic resonance spectroscopy. Techniques and applications have been developed for both homonuclear and heteronuclear coupled spin systems. Multiple-quantum spectroscopy has been applied to a range of molecular systems from small simple molecules to large complex biomolecules such as proteins. In many respects, the development of multiple-quantum spectroscopy has matured nuclear magnetic resonance spectroscopy to its full potential. It is possible to observe either directly or indirectly a resonance for every transition between energy levels of a coupled spin system which is allowed by symmetry. Thus all information available from nuclear magnetic resonance spectroscopy is accessible to measurement. The resonance frequency as well as information on the dynamics of relaxation, exchange, and diffusion can be determined for any order of coherence possible in a coupled spin system. As with many aspects of nuclear magnetic resonance spectroscopy, the problem may be too much information. In this case, selective excitation and filtering procedures can be used to obtain only the desired information.

This chapter has been developed to introduce the principles of multiple-quantum spectroscopy. As such, many aspects have been given only cursory treatment or simply mentioned in passing. Perhaps the most significant topic worthy of further treatment is the study of relaxation, exchange, and diffusion. In this context, the reader is referred to the discussion in Bodenhausen (2) and Weitekamp (3). The primary emphasis of this chapter has been on high-resolution multiple-quantum spectroscopy of spin-$\frac{1}{2}$ nuclei in the liquid state. It is hoped that this choice of emphasis has allowed a simplication of presentation which has permitted the important principles of multiple-quantum spectroscopy to become apparent. The same general principles apply to systems in the solid state where the dipolar interaction among spins dominates (2, 3, 98).

Acknowledgments

Special thanks are extended to Dr. Ray Freeman and my fellow graduates with whom I spent many happy and productive hours as a student. Their tutoring and collaboration were invaluable to my understanding, and this work owes much to them. Specifically, I would like to acknowledge many helpful discussions with Dr. Ad Bax, Dr. Stewart Kempsell, Dr. Malcolm Levitt, and Dr. A. J. Shaka. I would like to thank Nelda Smith and Katherine Nash for their long, hard work typing the manuscript. Also, I would like to thank Debra Neill-Mareci for her support and encouragement and for providing the artwork. This work was supported by a grant from the National Institutes of Health (P41-RR-02278) and the Veterans Administration Medical Research Service.

References

1. A. Abragam, "The Principles of Nuclear Magnetism." Oxford Univ. Press, London and New York, 1961.
2. G. Bodenhausen, *Prog. NMR Spectrosc.* **14**, 137 (1981).
3. D. P. Weitekamp, *Adv. Magn. Reson.* **11**, 111 (1983).
4. J. Jeener, *Ampere Int. Summer Sch., Basko Polje, Yugosl.*, 1971 (unpublished).

5. W. P. Aue, E. Bartholdi, and R. R. Ernst, *J. Chem. Phys.* **64**, 2229 (1976).
6. R. Freeman, *Proc. R. Soc. London, Ser. A* **373**, 149 (1980).
7. A. Bax, "Two-Dimensional Nuclear Magnetic Resonance in Liquids." Delft Univ. Press (Reidel), Dordrecht, Netherlands, 1982.
8. A. Abragam, "The Principles of Nuclear Magnetism," p. 289, Eq. (69), p. 308, Eq. (119). Oxford Univ. Press, London and New York, 1961.
9. E. L. Hahn, *Phys. Rev.* **76**, 145 (1949).
10. H. Y. Carr and E. M. Purcell, *Phys. Rev.* **94**, 630 (1954).
11. E. B. McNeil, C. P. Slichter, and H. S. Gutowsky, *Phys. Rev.* **84**, 1245 (1951); E. L. Hahn and D. E. Maxwell, *Phys. Rev.* **84**, 1246 (1951); E. L. Hahn and D. E. Maxwell, *Phys. Rev.* **88**, 1070 (1952).
12. P. L. Corio, "Structure of High-Resolution NMR Spectra," pp. 188, 436. Academic Press, New York, 1966.
13. K. Gottfried, "Quantum Mechanics." Benjamin, New York, 1974.
14. O. W. Sørensen, G. W. Eich, M. H. Levitt, G. Bodenhausen, and R. R. Ernst, *Prog. Nucl. Magn. Reson. Spectrosc.* **16**, 103 (1983).
15. F. J. M. van de Ven and C. W. Hilbers, *J. Magn. Reson.* **54**, 512 (1983).
16. K. J. Packer and K. M. Wright, *Mol. Phys.* **50**, 797 (1983).
17. P. L. Corio, "Structure of High-Resolution NMR Spectra," p. 101. Academic Press, New York, 1966.
18. S. Vega and A. Pines, *J. Chem. Phys.* **66**, 5624 (1977); A. Wokaun and R. R. Ernst, *J. Chem. Phys.* **67**, 1752 (1977); S. Vega, *J. Chem. Phys.* **68**, 5518 (1978); M. Mehring, E. K. Wolff, and M. E. Stoll, *J. Magn. Reson.* **37**, 475 (1980); R. Kaiser, *J. Magn. Reson.* **40**, 439 (1980).
19. S. Meiboom and D. Gill, *Rev. Sci. Instrum.* **29**, 688 (1958).
20. A. Schaeublin, H. Hoehener, and R. R. Ernst, *J. Magn. Reson.* **13**, 196 (1974); W. P. Aue, E. Bartholdi, and R. R. Ernst, *J. Chem. Phys.* **64**, 2229 (1976).
21. G. Bodenhausen and R. Freeman, *J. Magn. Reson.* **36**, 221 (1979).
22. G. Bodenhausen, G. Wagner, M. Rance, O. W. Sørensen, K. Wüthrich, and R. R. Ernst, *J. Magn. Reson.* **59**, 542 (1984).
23. M. Mehring, E. K. Wolff, and M. E. Stoll, *J. Magn. Reson.* **37**, 475 (1980).
24. G. A. Morris and R. Freeman, *J. Am. Chem. Soc.* **101**, 760 (1979).
25. L. Braunschweiler, G. Bodenhausen, and R. R. Ernst, *Mol. Phys.* **48**, 535 (1983).
26. T. H. Mareci and R. Freeman, *J. Magn. Reson.* **48**, 158 (1982).
27. T. H. Mareci, D. Phil. Thesis, Oxford Univ., 1982.
28. L. Müller, *J. Magn. Reson.* **59**, 326 (1984).
29. A. Bax, R. Freeman, and S. P. Kempsell, *J. Am. Chem. Soc.* **102**, 4849 (1980).
30. A. Bax, R. Freeman, and S. P. Kempsell, *J. Magn. Reson.* **41**, 349 (1980).
31. A. Bax and R. Freeman, *J. Magn. Reson.* **41**, 507 (1980).
32. A. Bax, R. Freeman, and T. A. Frenkiel, *J. Am. Chem. Soc.* **103**, 2102 (1981).
33. A. Bax, R. Freeman, T. A. Frenkiel, and M. H. Levitt, *J. Magn. Reson.* **43**, 478 (1981).
34. D. L. Turner, *Mol. Phys.* **44**, 1051 (1981).
35. O. W. Sørensen, R. Freeman, T. Frenkiel, T. H. Mareci, and R. Schuck, *J. Magn. Reson.* **46**, 180 (1982).
36. D. L. Turner, *J. Magn. Reson.* **49**, 175 (1982).
37. A. D. Bax and T. H. Mareci, *J. Magn. Reson.* **53**, 360 (1983).
38. D. L. Turner, *J. Magn. Reson.* **53**, 259 (1983).
39. M. H. Levitt and R. R. Ernst, *Mol. Phys.* **50**, 1109 (1983).
40. L. Müller, *J. Am. Chem. Soc.* **101**, 4481 (1979).
41. R. Richarz, W. Ammann, and T. Wirthlin, *J. Magn. Reson.* **45**, 270 (1981).

42. A. Wokaun and R. R. Ernst, *Chem. Phys. Lett.* **52**, 407 (1977).
43. A. D. Bain, *J. Magn. Reson.* **56**, 418 (1984).
44. G. Bodenhausen, H. Kogler, and R. R. Ernst, *J. Magn. Reson.* **58**, 370 (1984).
45. A. Bax, P. G. deJong, A. F. Mehlkopf, and J. Smidt, *Chem. Phys. Lett.* **69**, 567 (1980).
46. J. R. Garbow, D. P. Weitekamp, and A. Pines, *J. Chem. Phys.* **79**, 5301 (1983).
47. C. J. R. Counsell, M. H. Levitt, and R. R. Ernst, *J. Magn. Reson.* **64**, 470 (1985).
48. S. Schäublin, A. Höhener, and R. R. Ernst, *J. Magn. Reson.* **13**, 196 (1974).
49. W. S. Warren, S. Sinton, D. P. Weitekamp, and A. Pines, *Phys. Rev. Lett.* **43**, 1791 (1979).
50. G. Bodenhausen and C. M. Dobson, *J. Magn. Reson.* **44**, 212 (1981).
51. D. M. Doddrell, D. T. Pegg, and M. R. Bendall, *J. Magn. Reson.* **48**, 323 (1982).
52. P. J. Hore, E. R. P. Zuiderweg, K. Nicolay, K. Dijkstra, and R. Kaptein, *J. Am. Chem. Soc.* **104**, 4286 (1982).
53. P. J. Hore, R. M. Scheek, A. Volbeda, and R. Kaptein, *J. Magn. Reson.* **50**, 328 (1982).
54. A. J. Shaka and R. Freeman, *J. Magn. Reson.* **51**, 169 (1983).
55. P. J. Hore, R.N. Scheek, and R. Kaptein, *J. Magn. Reson.* **52**, 339 (1983).
56. H. Bildsøe, S. Dønstrup, H. J. Jakobsen, and O. W. Sørensen, *J. Magn. Reson.* **53**, 154 (1983).
57. O. W. Sørensen, M. H. Levitt, and R. R. Ernst, *J. Magn. Reson.* **55**, 104 (1983).
58. M. H. Levitt and R. R. Ernst, *Chem. Phys. Lett.* **100**, 119 (1983).
59. O. W. Sørensen, U. B. Sørensen, and H. J. Jakobsen, *J. Magn. Reson.* **59**, 332 (1984).
60. J. M. Bulsing and D. M. Doddrell, *J. Magn. Reson.* **61**, 197 (1985).
61. U. B. Sørensen, H. J. Jakobsen, and O. W. Sørensen, *J. Magn. Reson.* **61**, 382 (1985).
62. U. Piantini, O. W. Sørensen, and R. R. Ernst, *J. Am. Chem. Soc.* **104**, 6800 (1982).
63. H. Kessler, H. Oschkinat, O. W. Sørensen, H. Kogler, and R. R. Ernst, *J. Magn. Reson.* **55**, 329 (1983).
64. M. Rance, O. W. Sørensen, G. Bodenhausen, G. Wagner, R. R. Ernst, and K. Wüthrich, *Biochem. Biophys. Res. Commun.* **117**, 479 (1983).
65. M. A. Thomas and A. Kumar, *J. Magn. Reson.* **61**, 540 (1985).
66. M. H. Levitt, C. Radloff, and R. R. Ernst, *Chem. Phys. Lett.* **114**, 435 (1985).
67. J. Boyd and C. Redfield, *J. Magn. Reson.* **68**, 67 (1986).
68. A. Minoretti, W. P. Aue, M. Reinhold, and R. R. Ernst, *J. Magn. Reson.* **40**, 175 (1980).
69. G. Pouzard, S. Sukumar, and L. D. Hall, *J. Am. Chem. Soc.* **103**, 4209 (1981).
70. D. P. Weitekamp, J. R. Garbow, and A. Pines, *J. Chem. Phys.* **77**, 2870 (1982).
71. T. H. Mareci and R. Freeman, *J. Magn. Reson.* **51**, 531 (1983).
72. P. H. Bolton, *J. Magn. Reson.* **52**, 326 (1983).
73. G. Wagner and E. R. P. Zuiderweg, *Biochem. Biophys. Res. Commun.* **113**, 854 (1983).
74. J. Boyd, C. M. Dobson, and C. Redfield, *J. Magn. Reson.* **55**, 170 (1983).
75. A. Bax, R. H. Griffey, and B. L. Hawkins, *J. Magn. Reson.* **55**, 301 (1983).
76. A. Bax, R. H. Griffey, and B. L. Hawkins, *J. Am. Chem. Soc.* **105**, 7188 (1985).
77. P. H. Bolton, *J. Magn. Reson.* **57**, 427 (1984).
78. D. H. Live, D. G. Davis, W. C. Agosta, and D. Cowburn, *J. Am. Chem. Soc.* **106**, 6104 (1984).
79. J. Boyd, C. M. Dobson, and C. Redfield, *J. Magn. Reson.* **62**, 543 (1985).
80. M. Rance, O. W. Sørensen, W. Leupin, H. Kogler, K. Wüthrich, and R. R. Ernst, *J. Magn. Reson.* **61**, 67 (1985).
81. J. Boyd, C. M. Dobson, and C. Redfield, *FEBS Lett.* **186**, 35 (1985).
82. F. J. M. van de Ven, C. A. G. Haasnoot, and C. W. Hilbers, *J. Magn. Reson.* **61**, 181 (1985).
83. J. Santoro, F. J. Bermejo, and M. Rico, *J. Magn. Reson.* **64**, 151 (1985).
84. G. Otting and K. Wüthrich, *J. Magn. Reson.* **66**, 359 (1986).
85. L. Müller, R. A. Schiksnis, and S. J. Opella, *J. Magn. Reson.* **66**, 379 (1986).
86. D. J. States, R. A. Habenkorn, and D. J. Ruben, *J. Magn. Reson.* **48**, 286 (1982).
87. D. Marion and K. Wüthrich, *Biochem. Biophys. Res. Commun.* **113**, 967 (1983).

88. J. Keeler and D. Neuhaus, *J. Magn. Reson.* **63**, 454 (1985).
89. A. Bax, R. Freeman, and G. A. Morris, *J. Magn. Reson.* **42**, 164 (1981).
90. R. Freeman, T. A. Frenkiel, and M. H. Levitt, *J. Magn. Reson.* **44**, 409 (1981).
91. E. L. Hahn, *Phys. Rev.* **80**, 580 (1950).
92. A. A. Maudsley, A. Wokaun, and R. R. Ernst, *Chem. Phys. Lett.* **55**, 9 (1978).
93. A. Bax, T. Mehlkopf, J. Smidt, and F. Freeman, *J. Magn. Reson.* **41**, 502 (1980).
94. W. K. Rhim, A. Pines, and J. S. Waugh, *Phys. Rev. B* **3**, 684 (1971).
95. W. S. Warren, D. P. Weitekamp, and A. Pines, *J. Chem. Phys.* **73**, 2084 (1980).
96. D. P. Weitekamp, J. R. Garbow, J. B. Murdoch, and A. Pines, *J. Am. Chem. Soc.* **103**, 3579 (1981).
97. D. P. Weitekamp, J. R. Garbow, and A. Pines, *J. Magn. Reson.* **46**, 529 (1982).
98. M. Munowitz and A. Pines, *Science* **233**, 525 (1986).

5

Applications of Two-Dimensional NMR Experiments in Liquids

GEORGE A. GRAY

VARIAN ASSOCIATES
PALO ALTO, CALIFORNIA 94303

I. Introduction	343
II. Instrumental Requirements	345
III. *J*-Resolved 2D NMR	351
A. General Nature of the *J*-Resolved Experiment	351
B. Homonuclear *J*-Resolved Spectra	352
C. Heteronuclear *J*-Resolved Spectra	358
D. Some Applications of *J*-Resolved 2D Spectroscopy	361
IV. Chemical Shift Correlation 2D NMR	364
A. Homonuclear Shift Correlation	364
B. Heteronuclear Shift Correlation	382
V. Shift Correlation by Relayed Coherence-Transfer 2D NMR	402
A. Homonuclear Relayed Coherence Transfer	402
B. Heteronuclear Relayed Coherence Transfer	405
VI. Magnetization-Transfer 2D NMR	411
A. Homonuclear Magnetization Transfer	411
B. Heteronuclear Magnetization Transfer	419
VII. Zero- and Multiple-Quantum 2D Methods	420
A. Zero-Quantum 2D Experiments	420
B. Double-Quantum 2D Experiments	422
C. Higher-Order Multiple-Quantum Methods	425
D. Indirect Detection of Rare Spins Using 2D NMR	426
References	429

I. Introduction

A major component in the wide expansion of NMR techniques in the recent past has been the class of two-dimensional NMR experiments. These experiments have opened an entirely new approach to problem solving, and, coupled with the introduction of new data systems which permit easy and fast processing and output, they have fundamentally changed the strategic thinking performed by a spectroscopist in experimental design. The aim of this chapter is to review the major 2D high-resolution NMR experiments and to point out their relative strengths and weaknesses, as well as to give representative examples of their use.

A one-dimensional NMR experiment can be viewed as some method of exciting an NMR signal (usually achieved by one or more pulses and delays, comprising what is termed a "pulse sequence"), followed by signal detection during a fixed time period. This time-domain signal is then Fourier-analyzed, producing a frequency-domain spectrum. A two-dimensional spectrum has two frequency axes, forming a plane in frequency space with each point described by an NMR intensity. Two frequency axes imply two time axes prior to the double Fourier analysis.

As implemented in NMR spectrometers, one of these axes is always the chemical shift axis for the observe nucleus, resulting from the normal Fourier analysis of the free-induction decay. The other axis is solely determined by the events in the pulse sequence. As initially described by Jeener (1), the general 2D NMR experiment is characterized by several time periods:

$$\text{Preparation} \ldots \text{Evolution}(t_1) \ldots \text{Mixing}(t_m) \ldots \text{Detection}(t_2)$$

The preparation period is necessary to bring the system to a known state; for example, this could include adequate time to bring the spins to an equilibrium magnetization followed by a 90° pulse to place the magnetization in the transverse plane. During the evolution period, the spins are allowed to evolve within a specific environment that may be different from the environment in the other time periods. The evolution time is usually incremented through a range of values, each experiment with a particular value resulting in a separate FID. The mixing period may be as short as the length of a pulse (appropriate for transverse magnetization or spin population changes within a spin system) or as long as several seconds when longitudinal magnetization is to be exchanged.

The result is a collection of FIDs, the number of which equals the number of different values of the length of the evolution period. Each of the FIDs is then transformed in the usual way, resulting in a collection of spectra (Fig. 5-1)—the familiar inversion-recovery T_1 experiment is an example of this stage of the process. A new collection of "FIDs" is assembled by taking values of spectral intensity found at a frequency f in each spectrum and arranging these in a data table. This is repeated for every f value in the spectrum, resulting in N "interferograms" where N is the number of points in the original spectrum. These are each then Fourier-transformed, resulting in N spectra. These N spectra, presented on a two-dimensional plot, are

5. APPLICATIONS OF 2D NMR EXPERIMENTS IN LIQUIDS

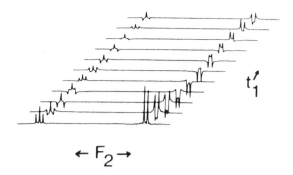

Fig. 5-1. Two-dimensional data after the first Fourier transformation. Peaks exhibit a phase or amplitude modulation or both as a function of the evolution time. The nature of this modulation depends on the particular pulse sequence used for the experiment.

now characterized by *two* frequency axes. One frequency (F_2) is always that of the normal observe spectrum, since it results from the Fourier transform of normal FIDs. The significance of the remaining (F_1) frequency is determined by the pulse sequence used.

The data-processing procedure is essentially common to all 2D NMR experiments. It usually involves first performing the normal 1D FT on all the free-induction decays, a collection typically ranging from 16 to 1024 separate FIDs. The data in each FID may be weighted by one or more functions prior to Fourier transformation. Although in principle the second FT may be done immediately, it is customary to transpose the resulting matrix of t_1 versus F_2 spectral data to a format of F_2 versus t_1 for convenience in using the normal instrument FT program. After weighting these data if desired, the new interferogram for each F_2 is subjected to a second transformation, to produce finally an F_1 versus F_2 two-dimensional array.

II. Instrumental Requirements

Spectrometers obtained commercially during the period 1978–1980 and after usually possess the necessary hardware and software to perform a 2D NMR experiment, at least for the major types of experiment. Several improvements have been made since then, particularly in the area of data systems.

The radio-frequency and analog parts of the spectrometers also have major demands if quality 2D NMR is to be performed. The rf transmitter must be controlled by the pulse programmer and capable of at least 0, 90, 180, and 270° phase shifts under pulse programmer control. In general, the shorter the value of the 90° pulse, the better—in order to minimize precession during the pulse resulting from off-resonance effects. Tradeoffs are usually made, however, in this property for increased sensitivity and reliability. In practice, 90° pulses for protons are typically 5 to 35 μsec for observe coils in normal 5-mm probes. These values are acceptable for most proton 2D experiments in liquids. When proton pulses are generated by a decoupler, increased power is usually used since the coils are often of greater diameter and length. Proton 90° pulses are important in heteronuclear shift correlation and other 2D experiments. Again, values of 20 to 100 μsec are typical, depending on the probe and spectrometer.

Just as in the case of the observe rf transmitter, the decoupler must also be under pulse programmer control, including gating on and off, modulation, power level, and rf phase, including, at a minimum, shifts of 0, 90, 180, and 270°. It is further desirable to be able to control the decoupler power during the actual execution of the pulse sequence for some special kinds of experiments in which only single proton types are to be irradiated.

The receiver must be capable of high dynamic range, independent of the size of the analog-to-digital converter, to produce highly linear response—particularly in the case of aqueous solutions for proton 2D experiments. Extended ADCs are also useful in these situations, since 90° pulses are usually required and solvent suppression techniques are often inadequate. The preamplifier, receiver, and digitizer must be of highest stability and reproducibility to minimize the contribution of F_1 noise.

The spectrometer must be accurately calibrated in terms of 90° pulse widths for the observe nucleus. If special decoupling modulations are used which require knowledge of the decoupling field strength, this should also be determined accurately, ideally on the sample of interest. If probe tuning characteristics change upon going from a calibration sample to the analytical sample, it is likely that the observe and/or decouple 90° pulse widths have changed. Normally it is simple to determine quickly the observe 90° pulse length by single-transient measurements at a variety of flip angles, either by finding the maximum signal corresponding to the 90° value, or better, by determining the passage through null signal at the 360° value. The decoupler 90° pulse measurement can be done in a variety of ways; these are described in Chapter 1.

The amount of sample needed for 2D NMR is determined by the sensitivity of the spectrometer and the time available for the experiment. The data for a COSY or 2D J spectrum can usually by collected in as little as a few minutes for a sample in which a normal proton spectrum can be taken in a few transients. Data processing can be done in as little as 30 sec, and plotting in a few minutes. As samples become more dilute or weaker nuclei are examined, times increase. As a general rule, the length of data accumulation is proportional to the desired resoution in F_1, and 64–1024 FIDs are usually taken. For minimal times, the number of transients per FID can be estimated by observing the number necessary to attain a signal-to-noise ratio of 5–10:1 on the peak or peaks of interest in a normal 1D experiment. Thus if four transients are found to give a S:N in this range and a resolution of 10 Hz is adequate to establish the nature of a peak in a spectral window of 1000 Hz, then at least 100 FIDs are required (the number equal to 1000/10), each having at least four transients, leading to a total of 400 transients. Because delay times are often about a second and FID acquisitions an order of magnitude less, the total experiment run time would be about 400 sec. Bookkeeping, pulse sequence, and instrumental delays add to this time, but it is obvious that time should not be a major obstacle in the use of 2D NMR.

A 2D spectrum is almost never obtained without first running a simple 1D spectrum. All the normal considerations of sample preparation and parameter choice apply equally to 2D experiments. Since many of the pulse sequences do employ extensive phase cycling to cancel peaks, it is very important to pay attention to relaxation, steady-state, and repetition time aspects. When the NMR signal is determined by the T_1 of the observed nucleus, this value should be estimated or an approximate value determined. Then the equilibration delay should be set to between T_1 and, for the most demanding experiments, $5T_1$. It may be disadvantageous to degas or nitrogen-purge samples, since this will normally extend proton T_1 values to seconds or tens of seconds without comparable increase in information content.

If the NMR signal is determined by polarization transfer, the relevant T_1 is that of the nucleus from which the polarization is obtained, normally the proton. The same considerations as described above apply for these experiments. Since the experiments have time periods in which proton magnetization precesses, anything which shortens proton T_2 values reduces the net transferred magnetization, reducing observe nucleus sensitivity and, in some cases, resolution in F_1. Therefore, paramagnetic impurities which have little effect on normal carbon-13 resonances can influence the resulting heteronuclear polarization experiment in a much more dramatic way. As molecular size increases and T_2 becomes shorter, the resulting 2D sensitivity and resolution are impaired. The extent of the effect is a function of the individual sample and of the molecular motions present in the sample.

Heteronuclear 2D experiments should be preceded by 1D spectra of both nuclei if shift correlations are being sought. This permits proper choice of rf frequency positions and spectral windows in both dimensions, and provides reference spectra to be used in interpreting the 2D data. Optimal equilibration delays can be estimated through a rapid examination of peak height as a function of repetition time in a series of experiments. A repetition period of 2 sec may give a doubling of peak height relative to a repetition period of 0.5 sec. However four times as many transients can be obtained in the same total time with the latter, resulting in the same net time-averaged signal-to-noise ratio. This does not mean that the two prospective experiments are equivalent in their performance. Since many 2D experiments require a minimum of 8, 16, 32, ..., or 256 transients, it may be better to use the faster recycle time, consistent with proper steady-state considerations. Those sequences involving cancellation of strong peaks may, however, require the longer delay to satisfy T_1 factors.

At higher magnetic fields the increase in hertz per ppm forces larger and larger data tables for equal chemical shift regions. In some experiments, resolution is paramount and often it is possible to attain higher resolution by carefully folding (aliasing) peaks. In F_2 this is done by proper placement of the rf transmitter and choice of spectral width. A 1D spectrum will reveal any bad overlaps, and readjustment of these parameters can usually place the folded peaks in regions of noise where they will cause no confusion. Folded peaks will be progressively attenuated the farther they are away from the spectral window. This is not true, however, in the F_1 dimension, where there is no equivalent of the electronic audio filter which is present in F_2. Hence, it is quite feasible to have extensive folding in F_1 without attenuation of the folded signals. For this reason the 1D equivalent of F_1 should be examined or estimated prior to setting the conditions which determine the maximum value of F_1.

Execution of the 2D experiment varies in difficulty depending on the capability and vintage of the spectrometer. Recent advances in software have made obtaining 2D data as simple as or simpler than obtaining 1D data. Powerful and flexible MACRO languages have been developed which permit the user to insert a sample, enter one command, and have the spectrometer obtain 1D data from which it chooses conditions for the 2D experiment, runs the experiment, processes the data, and plots the spectrum, all automatically!

There are several choices open to the operator in any given 2D experiment. The results of these choices, in addition to molecule-related factors, determine the sensitivity, resolution, and lineshape of the 2D data. Levitt, Bodenhausen, and Ernst (2) have reviewed the important factors controlling sensitivity in 2D NMR, independent of the type of experiment. They propose that only the signal envelope $S(t_1, t_2)$, the weighting functions in the two dimensions, and the required resolution in each dimension are relevant in determining the actual sensitivity of the 2D NMR experiment. First, they recommend that some estimate be made of the likely relaxation times and signal envelope function prior to the start of the experiment. Second, the repetition time in the experiment should always exceed the relevant T_1 (normally that of the observe nucleus, but that of the protons for polarization transfer experiments) to avoid saturation and to reduce problems of artifactual noise parallel to F_1. Third, a minimum number of FIDs should be taken. This number is determined by the required resolution in F_1: the linewidth in this dimension is given by the inverse of the maximum evolution time. Fourth, if data space is limited, the smallest possible sampling rate which does not lead to interolerable folding may be used; however, the sampling rate is irrelevant for sensitivity. Fifth, the maximum t_2 should be as large as possible. This is not as important if a properly matched analog filter is used. And sixth, prior to each Fourier transformation the time-domain data should be weighted by a function matched to the experimental data.

It is normal to apply analog filtering during t_2, so that a choice of $t_2(\max)$ should be made based on the minimum resolution $[1/t_2(\max)]$ in the 1D spectrum necessary to separate relevant peaks. Minimization of both $t_1(\max)$ and $t_2(\max)$ permits more transients per FID to be collected during the available experimental period. If cross-peak multiplet patterns are to be accurately reproduced, the relevant linespacing will determine $t_2(\min)$ and $t_1(\min)$.

Often sensitivity in the 2D experiment is determined by the purity of the spectrum, particularly the noise in traces parallel to F_1. This noise is actually NMR signal not appearing at its proper F_1 value. The noise source is usually non-steady state behavior or spectrometer instability. The factors influencing F_1 noise have been investigated by Mehlkopf et al. (3).

As the use and importance of 2D NMR has grown, the instrumental capabilities necessary to exploit it have also been enhanced. Pulse programmers have evolved from those suitable only for simple pulsed FT experiments to those having multiple phase shifts, indeterminate sequence length limit, dynamic branching, high-level language flexibility, and complete control of observe and decoupler channels, as well as many additional capabilities. As will be seen in the following sections, these features are crucial for the proper execution of the desired pulse sequences and therefore the ability to do the NMR experiment.

Once the data are collected, the next important technological requirement is a data system to process it and to output the results. In the last few years enormous strides have taken place which made 2D NMR spectroscopy as simple and as fast as 1D NMR has been for most users. Some of the characteristics of these modern systems are:

1. Large-capacity hard-disk storage. This is essential for the storage of large 2D data tables and system software. A single 2D experiment of intermediate size can easily occupy on the order of one megaword of disk space if raw data is stored separately from processed data, and the associated documentation of parameters is complete. Since users seldom wish to destroy data immediately before performing the next experiment, it is clear that optimal use of 2D NMR requires large storage capacity.

2. Large CPU memory. This promotes speed of processing, since large blocks of data can be handled without frequent disk accesses.

3. Array processor. This is ideally suited to the multiple repetitive operations characterizing 2D data processing. It can also be used in other calculations for general speed and responsiveness.

4. Multiple processors. These permit simultaneous data acquisition, calculation, operator input, and data output without slowing the system speed or limiting operator flexibility.

5. Large and informative display terminal. Usually in color, this permits the viewing of large data sets in FID, interferogram, or 2D FT form in spectral or contour format. The number, variety, and speed of these displays determine the range of possible operator response to the data, since in many cases results may be taken directly from the screen.

6. "Real-time 2D NMR." This capability enables the user to view the 2D data during the execution of the experiment, allowing early setup of output format, error checking at an early state, and possible stopping of the experiment as soon as the data are satisfactory, as is customarily done in 1D NMR.

7. Interactive weighting display. This involves "real-time" examination of the effect of FID and interferogram weighting for optimal performance.

8. High-speed output devices. High-speed plotters and/or printers are essential for good productivity. Although stacked plots are useful for looking at the "real" data, this is usually done on the system monitor. Output of this type is typically reserved for figures in published papers or for special purposes. Most often, data are plotted as contours of intensity. Modern systems can produce a high-quality contour plot in a few minutes. Often data are transferred to a work station for further processing and/or plotting, either over a high-speed parallel network or by physically transporting magnetic media such as floppy diskettes, hard disks, or magnetic tape.

With these common elements of experiment design, data processing, and data output, we can now focus directly on those aspects which make individual 2D NMR experiments unique, namely, their respective pulse sequences, spectral information content, and applications to solving chemical problems of structure and dynamics. These experiments fall into two major categories, designed to produce J-resolved or shift-correlation spectra.

III. J-Resolved 2D NMR

A. GENERAL NATURE OF THE J-RESOLVED EXPERIMENT

Many pulse sequences have been developed in one-dimensional NMR which exploit the properties of spin echoes. These all rely on some estimation of a J coupling, either homonuclear or heteronuclear. One of the favorable things about a 2D approach is that no estimation of J is necessary, since the process itself explores the entire range of J values. All that is necessary is to allow the spins to be affected by J interactions during the evolution period. The environment during the evolution period may include the interactions of spin coupling, the chemical shift—which governs the rate of precession (relative to the transmitter frequency as zero, that is, in the rotating frame)—and magnetic inhomogeneity, which causes identical nuclei to have different precession rates. Refocusing pulses, as seen before, can remove some, or all, of these interactions. Following the 90° "preparation" pulse, off-resonance spin vectors begin to dephase, or precess. Use of a strong nonselective 180° refocusing pulse on the observe nucleus midway through the evolution period reverses for all observe spins this dephasing arising from the off-resonance or chemical shift effect. Thus, for every value of t_1, each spin vector is returned to the same orientation it possessed after the first 90° pulse, that is, chemical shift effects are removed. The resultant 2D spectrum might be thought to be rather simple, as it is indeed for noncoupled nuclei—singlets are centered at coordinates ($F_1 = 0$, $F_2 = \delta_X$), that is, singlets are unaffected by the presence of the evolution period except for T_2-related reduction of intensities. As a function of the evolution period, their intensities are described by zero-frequency decays and hence $F_1 = 0$ for each singlet.

As the name implies, J-resolved 2D NMR is useful in extracting spin-coupling information. To understand this, recall the action of the refocusing pulse in the spin-echo pulse sequence for the X spins in an AX spin system, where A might be ^1H and X might be ^{13}C, as described in Chapter 1. The action of the 180°(A) pulse is to interchange the spin labels, allowing further divergence of the spin components. In the absence of an A pulse, the X components would always identically refocus at $\tau = t_1$, for no t_1-dependent modulation of the FID is generated, and only T_2 relaxation is probed. With the A pulse, however, *J-dependent* modulation is introduced and the phase separation of the X vectors varies as a function of J and of the time allowed for diverging. This information is coded into the phase of the X nucleus signals at the start of t_2, the data accumulation period. A full experiment, then, explores this dependence by repetition for a whole array of regularly spaced t_1 periods from zero to some selected maximum value of t_1.

The two major versions of this experiment are homonuclear and heteronuclear J-resolved 2D NMR. In the former, only one 180° pulse is necessary since the 180° pulse affects both nuclei in the spin system. For heteronuclear experiments, it is also customary to decouple the A nucleus from X during t_2 for sensitivity improvement, rather than having the AX coupling information present in both frequency dimensions. Equivalent information to that from the above "proton flip" experiment may be obtained in the heteronuclear case by simply turning on the decoupler half-way through the evolution period, in place of the 180°(A) pulse (Fig. 5-2). Since the spin component divergence is unsymmetrical with respect to the 180°(X) pulse, the divergence is not refocused and the J-coupling dependence is encoded into the phase of the decoupled X spin.

The heteronuclear J-resolved 2D experiment (4-6) produces completely separate information in the two domains, X nucleus chemical shift in F_2 and AX coupling in F_1. This is particularly valuable, for example, in ^{13}C-^1H cases, since complete coupling patterns can be obtained for each ^{13}C *without overlap of adjacent patterns.* Information can be extracted easily (Fig. 5-3) in contrast to direct observation of highly overlapped coupled 1D spectra. Absorption-mode presentation is possible, allowing full resolution of the heteronuclear coupling patterns (7, 8).

B. HOMONUCLEAR J-RESOLVED SPECTRA

The homonuclear J-resolved 2D spectrum differs from the heteronuclear in that AX coupling is active in *both* t_1 and t_2, and is thus reflected in the existence of spin multiplets in both dimensions. The 2D spectrum of a homonuclear AX system can be represented in contours of intensity (looking "top-down") (Fig. 5-4). Each spin pattern is present, although no one slice

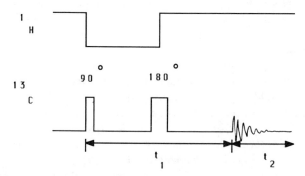

Fig. 5-2. Heteronuclear 2D J pulse sequence. Refocusing of J modulation is prevented by gating the decoupler. The phase of the detected decoupled signal is therefore affected by the presence of J coupling.

5. APPLICATIONS OF 2D NMR EXPERIMENTS IN LIQUIDS 353

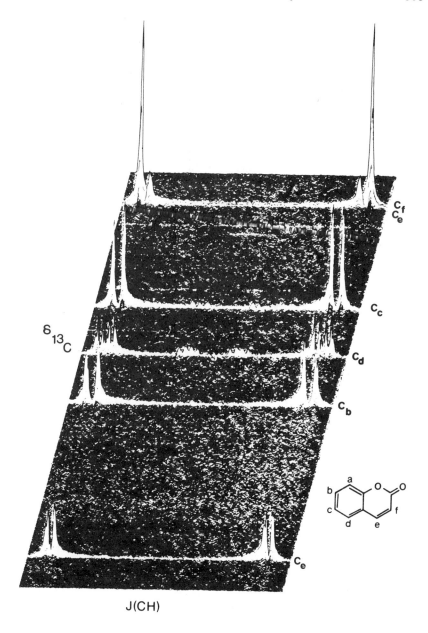

Fig. 5-3. Heteronuclear 2D J spectrum of coumarin. The large spacings arise from the one-bond carbon–hydrogen coupling, while the smaller splittings come from long-range C–H couplings.

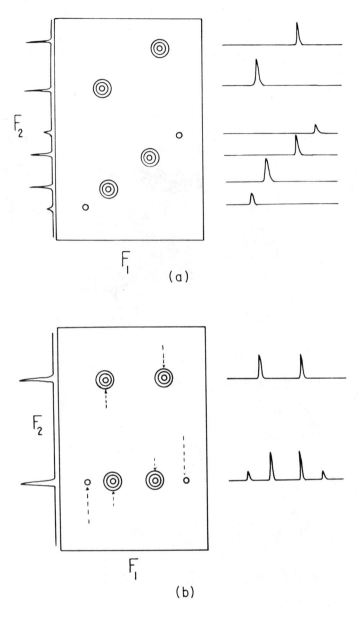

Fig. 5-4. Homonuclear 2D J data in contour format. The spectra are most conveniently presented in contours of intensity where each level represents a multiple of the minimum intensity plotted. (a) Multiplets appear tilted at a 45° angle. The doublet and the quartet are spread out in both dimensions, making measurements difficult. (b) Rotated 2D J data are obtained by processing the data such that multiplets belonging to one chemically shifted nucleus appear at one value of F_2.

perpendicular to F_2 contains a whole pattern (9). A computer process known as "tilting," "rotating," or "shearing" can be used to realign the spin patterns so that they become perpendicular to F_2, by tilting them by 45°. This permits display and plotting of "slices" of the 2D data—an F_1 spectrum at some specified F_2 value (10). These slices are the full coupled spin patterns shown in Fig. 5-4.

The spin-echo nature of the homonuclear J-resolved experiment produces narrow lines, and thus very highly resolved multiplets. In highly congested, but first-order, proton spectra, this technique can be of immense value. The value of this experiment for complicated organic molecules is evident from Fig. 5-5. Second-order effects do produce additional lines in the spectrum, which are not easily interpreted. These fall at positions which do not tilt onto an F_2 position which is a true chemical shift. They can complicate "projected" spectra where the F_1 domain is collapsed, leaving a chemical-shift-only spectrum, a procedure which is useful in giving single "one-line-per-nucleus" spectra (Fig. 5-4).

Homonuclear J-resolved 2D data are normally projected from a data presentation in absolute-value or magnitude form, since a "phase twist" (8) is present in phased data and overlapping lines may interfere with one another. The long tails from the absolute-value lineshape arise from the dispersion-mode component and can give very unsatisfactory lineshapes in projected data. Bax et al. (11) have overcome this problem by shaping the FIDs or interferograms by a mathematical function which forces sine-bell-like symmetry and generates what appears to be a full echo, hence the term "pseudo-echo processing." The transform of these data has no dispersion-mode characteristics and absolute-value spectra therefore can have absorption-like lineshapes as illustrated in Fig. 5-6. Of course, signal is degraded at either end of the pseudo-echo processed data, significantly reducing sensitivity. Blümich and Ziessow (12) have described a "skyline" projection technique which gives superior resolution since it can handle projections of phased data.

The problem of "phase twist" in homonuclear 2D J spectra has been addressed in two separate treatments. Shaka, Keeler, and Freeman (13) have described a computational method for correcting for the phase twist by factoring out the theoretical response and replacing it with an absorption mode representation. True absorption-mode homonuclear 2D J data have been obtained by Williamson (14) for two-spin systems using the same basic pulse sequence together with another measurement in which the pulse sequence is augmented by an additional 90° pulse just prior to acquisition. Data from both experiments are combined for every value of t_2 for each value of t_1, generating quadrature (complex) data points. In the second measurement, there is effective reversed precession during t_1, and co-addition of the two experiments converts phase modulation into amplitude

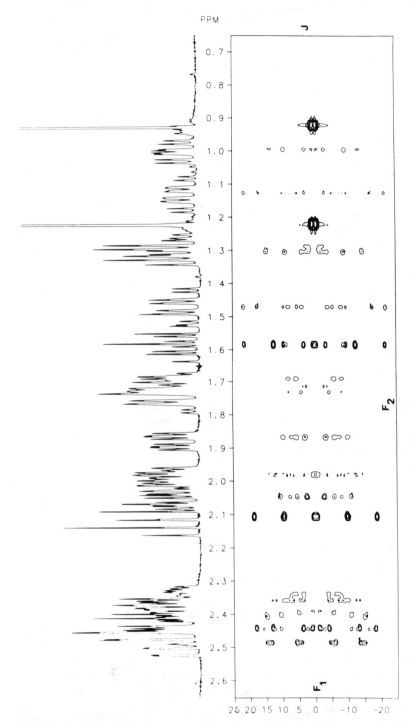

Fig. 5-5. The value of the homonuclear 2D J experiment for a complicated proton spectrum is demonstrated in this figure for a steroid. From the 2D contour plot, the individual proton shifts are obvious and couplings can be measured easily.

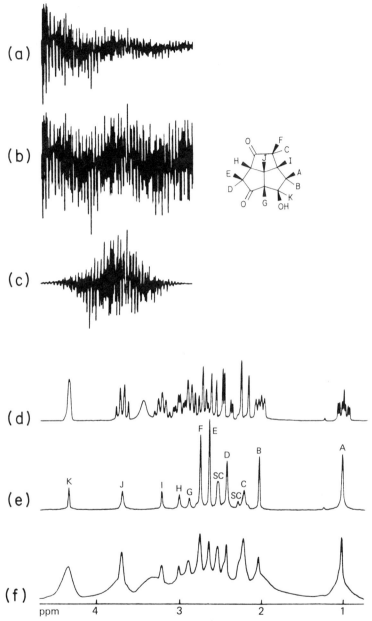

Fig. 5-6. The effects of data processing on absolute-value output: (a) normal FID without weighting; (b) effect of weighting the FID with an increasing exponential function; (c) effect of weighting the data in (b) with a Gaussian function, creating a "pseudo-echo"; (d) transform of (a); (e) projection of 2D J spectrum rotated by 45° with pseudo-echo weighting in t_2; (f) projection of the same spectrum without pseudo-echo weighting. [From Bax, Freeman, and Morris (11).]

357

modulation, similar to the way in which quadrature information is generated in shift correlation experiments, which will be described later. Pure absorption-mode spectra without phase twist can be produced, permitting higher resolution for complex molecules.

When the multiplet patterns are expected to have mirror symmetry about their midpoints, it is possible to apply a symmetrizing algorithm to the tilted data (15). This has the advantage of improving resolution by removing the effect of the absolute-value tails, which do not have mirror symmetry after the tilt.

Nagayama (16) has reported theoretical and experimental details on the effects of several types of homodecoupling schemes in homonuclear 2D J NMR. These experiments can be used to simplify 2D J spectra, assign resonances, and obtain pure-phase 2D-J resolved data. Two-dimensional spin-echo spectra and analysis have been obtained in oriented AB systems by Kumar and Khetrapal (17) and for oriented AB_2 systems by Turner (18). In the latter paper, a general method was described for computation of spectra for arbitrary refocusing flip angles as a guide to understanding resonance offset effects.

For macromolecules the normal $90°-\tau-180°-\tau-acquire$ pulse sequence produces considerable loss of sensitivity because of the irreversible loss of signal via short T_2 values. Macura and Brown (19) have proposed beginning data acquisition immediately following the 180° pulse.

Bodenhausen et al. (20) considered the homonuclear AB, ABX, and ABC spin systems. They documented the existence of negative peaks in the J spectra arising from the second-order spin system. Kumar (21) has analyzed 2D spin-echo spectra for AX, AB, A_2X, AMX, and ABX spin systems, using either a selective or a nonselective 180° pulse. The 2D spin-echo spectra were found to be symmetric for nonselective excitation of all coupled spins. Analytical expressions for frequencies and intensities were derived for heteronuclear spin systems, to be considered in the next section, as well as for homonuclear cases.

C. Heteronuclear J-Resolved Spectra

Heteronuclear 2D J spectra can have spurious resonances related to the main signals but appearing at different places in the spectrum. These artifacts arise from missetting of the flip angles from the true 90° or 180° values or from spatial inhomogeneities in the rf field. Bodenhausen, Freeman, and Turner (22) proposed the phase-cycling scheme "Exorcycle," described in Chapter 1, which cancels any magnetization that is unaffected by the 180° refocusing pulse as well as canceling xy magnetization which is created from any z magnetization present before the 180° pulse. The same workers also considered the effects of strong coupling on heteronuclear 2D J spectra (23, 24).

While Exorcycle does eliminate the nonideal aspects of the observe nucleus 180° pulse, in the "proton-flip" version of the heteronuclear 2D J experiment there is no compensation for errors in the proton 180° pulse. Errors of this type can also introduce artifacts (25). However, Freeman and Keeler (26) have shown that by using a composite proton 180° pulse (27), these errors are eliminated.

Polarization transfer can be used in heteronuclear J-resolved 2D NMR (28, 29) for signal enhancement, shorter equilibration periods, and elimination of nonprotonated resonances—all the same features as in INEPT or DEPT. In this experiment the refocusing period following polarization transfer becomes a variable time, rather than the $1/4J$, $1/2J$, and $3/4J$ values normally used for refocused INEPT or variable flip angle for DEPT. The polarization-transfer part of the sequence includes a $1/2J$ period, as usual, and thus the 2D sequence can be tailored to XH coupled pairs which have J values of a particular magnitude, such as certain long-range couplings for quaternary carbons whose normal ^{13}C T_1 values may be too long to permit ordinary heteronuclear 2D J experiments.

Bax and Freeman (30) have introduced an elegant method for generating specific heteronuclear 2D J data. Their pulse sequence is identical to that of the normal "proton-flip" except that a weak ($\gamma B_2/2\pi = 25$ Hz) proton 180° pulse is applied to *one* proton selectively in place of the strong nonselective 180° pulse normally used midway through the evolution period (Fig. 5-7). This pulse inverts the inner (long-range coupling) satellites of a proton while leaving the outer (one-bond coupling) satellites unaffected. Thus only long-range couplings to this *one* proton modulate the observe nucleus spin echo, and therefore there is only *one* such modulation for any observe nucleus resonance. Since only long-range couplings are involved in the F_1 dimension, good digitization is possible. The experiment probes the local environment of *one* proton only and thus is especially valuable in structural studies where extensive long-range XH couplings can produce heavily overlapped heteronuclear 2D J spectra, not to speak of the situation in normal coupled 1D spectra. The ability to identify which carbons are in the neighborhood of a specific proton will undoubtedly be of value in spectra analysis of congested spectra and unknown structures.

Bax (31) and Rutar (32, 33) have separately proposed a novel class of heteronuclear J-resolved 2D experiments which allow choice of either one-bond or long-range 2D J spectra. The preparation phase may be just establishment of nuclear Overhauser effect followed by a 90°(X) pulse or polarization transfer sequence such as DEPT. The X nucleus pulse sequence is as in the normal heteronuclear 2D J experiment. However, the proton

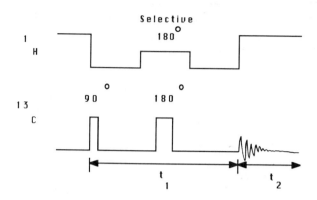

Fig. 5-7. Sequence for selective heteronuclear 2D *J* spectra. The selective low-power proton pulse inverts only over a narrow bandwidth, perhaps 20 Hz. The satellites for those protons directly bound to a carbon-13 are outside this bandwidth and therefore essentially unaffected by the inversion pulse. Couplings to these protons do not modulate the ^{13}C signal during the evolution period. Carbons a few bonds away from the inverted proton do have their signals modulated. Thus the 2D *J* spectrum shows couplings for only those carbons coupled by long-range interactions to the inverted proton.

180° pulse is replaced by the "bilinear" cluster $90°_x-1/2J-180°_x-1/2J-90°_{\pm x}$ where *J* is the value of the direct ^{1}H-X coupling constant, as described in Chapter 1, Section IV,B. As in the normal 2D *J* pulse sequence, *only protons inverted by this element* modulate the X-nucleus signal and therefore only these protons give rise to multiplets in the final 2D spectrum.

The fascinating property of this new sequence is that, with one choice of the final proton pulse phase *only* long-range X-H couplings appear, while choice of the opposite phase produces a 2D spectrum in which *only* direct one-bond X-H couplings are observed. In either case small F_1 ranges of 20-40 Hz can be used for good digitization, and extensive folding can be permitted in F_1 for one-bond couplings for good precision of measurement, and later corrected. A clear example of this is seen in Fig. 5-8, where the same number of data points in F_1 was used for each of the two experiments. For the case of $90°_x(H)$ as the last pulse, protons experiencing a large $^1J(CH)$ are inverted, while protons remote from, or only weakly coupled to, a ^{13}C, are restored to normal magnetization, and therefore there is a clean separation based on *J*. Reversing the phase of the last 90° proton pulse reverses the sense of polarization and the effect in the 2D experiment.

It is possible to obtain proton-proton coupling constants from 2D NMR spectra *indirectly*, that is, through observation of the resonance of another nucleus. Bodenhausen (34) proposed inserting simultaneous proton and observe nucleus 180° pulses midway through the evolution period in a magnetization-transfer 2D experiment devised by Maudsley and Ernst (35). This collapsed the proton chemical shift in F_1, leaving only J_{HH} and J_{XH} contributing to the J spectra in F_1. For proton spectra which are highly congested, this technique allows the dispersion of the X nucleus (e.g., ^{13}C or ^{31}P) to spread the H-H multiplets, as opposed to the situation where the spreading depends on the proton shifts as in ordinary homonuclear 2D J spectroscopy. Analogous approaches based on INEPT (36) or DEPT (37-39) which result in only H-H multiplets in F_1, suppressing the XH interaction, have also been described. These have considerable potential for extraction of stereochemical information for molecules for which multiplets in the normal proton spectrum overlap far too much for interpretation.

Rutar (40-42) has described a heteronuclear 2D J experiment in which the one-bond coupling of the observe nucleus to the proton is eliminated as above but both the long range XH and the H-H couplings modulate the detected signal. Thus, proton-proton couplings can be extracted, a result which may be impossible from the proton spectrum because of overlap or magnetic equivalence.

Keeler (43) has published a method that uses heteronuclear 2D J experiments for assignment of proton multiplicity in ^{13}C spectra. Since each F_1 trace represents the ^{13}C-^{1}H coupled multiplet, this type of information is, in principle, available after only a few t_1 values. Unfortunately, the truncation of a small t_1 data set leads to "sinc" wiggles in the F_1 slices, obscuring the desired information. Keeler proposes convolving those slices with the theoretical sinc function to extract the relevant NMR information. This was automated and produced low-resolution J spectra useful for multiplicity characterizations independent of the assumptions about relaxation, J, and pulse calibration which must be made in spectral editing techniques.

D. SOME APPLICATIONS OF J-RESOLVED 2D SPECTROSCOPY

Homonuclear 2D J experiments have been reported for the assignment of the proton spectrum of L-menthyldichlorophosphine (44). Second-order effects in the 2D J spectrum of 1,6-diazathianthrene were examined as a function of observe frequency (60, 100, 150, 200, 270, and 360 MHz) by Puig-Torres et al. (45). They found that the eight additional lines observed for the AB pair moved progressively outward from $F_1 = 0$ and diminished in intensity as the observe frequency increased. Ezell, Thummel, and Martin (46) used these second-order effects to identify strongly coupled members of three-spin systems in 2-(2'-pyridyl)-1,8-naphthyridine. Musmar et al. (47) used similar techniques in assignments for benzo[2,3]phenanthro[4,5-bcd]thiophene. Benn and Riemer (48) used homonuclear 2D J methods to identify and assign stereoisomers in which heterocouplings were present.

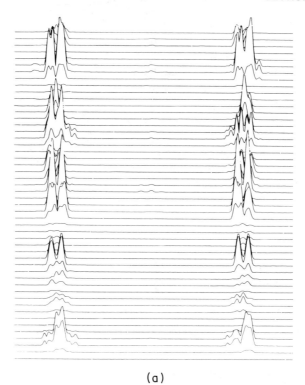

(a)

Fig. 5-8. Partial heteronuclear 2D J data for 2-chloronaphthalene: (a) normal heteronuclear 2D J data showing large one-bond J(CH) splittings as well as smaller splittings from two- and three-bond couplings; (b) long-range heteronuclear 2D J experiment for the same region. Although a finer digitization was used for F_2 in this experiment, the same number of values of the evolution period (128), the controlling factor in F_1 resolution, was used for each experiment. Since the maximum value of F_1 was only 50 Hz in this spectrum, compared to 250 Hz in (a), five times better resolution was available to characterize the long-range splittings.

Biopolymers have received considerable attention from NMR users, and 2D J studies applied to amino acids (49) and proteins have appeared (50, 51). Carbohydrates are particularly suitable for analysis because of the compressed chemical shift range and severe overlap. Coxon (52) determined the 2D J-resolved proton spectrum of hydroxyl-coupled alpha and beta-D-glucose, Hall et al. (53) looked at alpha, beta-D-xylopyranose, Dabrowski et al. (54) examined oligosaccharides isolated from human milk, Gagnaire et al. (55) studied capsular heteroglycans and cellulose triacetate, and Hall et al. (56) applied the technique to mono- and disaccharides. Yamada et al. (57) used 2D J techniques in glycosphingolipids, while glycopeptides from ovalbumin were characterized by Bruch and Bruch (58). Prostaglandin spectra have been simplified using 2D J methods (59), Lankhorst et al. (60)

(b)

Fig. 5-8 (cont.)

examined a trinucleoside disphosphate, and Gippert and Brown (61) have used J-resolved 2D spectra to sort out proton resonances corresponding to triad and tetrad configurations of polyvinyl alcohol.

Homonuclear J-resolved 2D NMR need not be always applied to protons. Colquhoun and McFarlane (62) used the experiment to simplify and assign the phosphorus spectrum of polyphosphorus compounds. Harris et al. (63) applied it to silicon-29 spectra of silicate solutions. Niedermeyer and Freeman (64) as well as Taravel and Vignon (65) have reported the use of 2D J NMR in multiply-enriched carbon-13 compounds.

Heteronuclear 2D J-resolved applications have been reported for 1-azathianthrene (66), all-*trans*-retinal (67), a polycyclic multifunctional nitrile (68), N,N'-diacetylchitobiose (69), and lupane, a demethylated triterpene (70).

Davis et al. (71) have modified the selective 2D J experiment described above by Bax and Freeman (30) by replacing the observe nucleus 90° pulse with an INEPT pulse sequence to obtain a greatly increased observe nucleus magnetization, particularly relevant for their application, the determination of nitrogen-15 couplings to protons.

IV. Chemical Shift Correlation 2D NMR

A. HOMONUCLEAR SHIFT CORRELATION

1. *COSY, SECSY, and Data Treatment*

Perhaps the most widely used 2D NMR experiments are those that relate chemical shifts of different nuclei, such that F_1 and F_2 represent chemical shift axes in both homonuclear and heteronuclear cases. Again, the environment during the evolution time dictates the interpretation of the new information. The simplest homonuclear experiment is that introduced by Jeener (1), also known in one form as COSY or correlation spectroscopy (72-74). The most widely used pulse sequence is rather simple, consisting of $90°$—t_1—$90°$ followed by acquisition. The resulting peaks in the 2D plot fall in several categories: axial, diagonal, and off-diagonal. The axial peaks lie along $F_1 = 0$ and result from longitudinal magnetization sampled by the last pulse. The diagonal peaks are due to magnetizations which remain associated with the same spins before and after the second $90°$ pulse; that is, the magnetization has the same frequency or is changed to another frequency within the same multiplet during t_1 and t_2. The off-diagonal peaks confirm the fact that magnetization present at one frequency in t_1 is transferred to another frequency after the second $90°$ "mixing" pulse. This is only possible if a nucleus at one of these frequencies shares the same spin system with a nucleus of the second frequency, that is, they are J-coupled. This last feature forms the basis of the widespread use of homonuclear shift-correlated 2D NMR, since it gives information equivalent to that from spin-decoupling. Furthermore, the spread of information in two dimensions makes the interpretation easier. Proper phase cycling (74) can remove the axial peaks, considerably simplifying the experiment.

A somewhat related correlation procedure, the SECSY (*s*pin *e*cho *c*orrelation *s*pectroscop*y*) method, reported by Nagayama, Wüthrich, and Ernst (75), uses the pulse sequence $90°$—$t_1/2$—$90°$—$t_1/2$—*acquire*. The F_1 in this case becomes a difference frequency axis, while the intensity which appears along the diagonal in COSY now resides along $F_1 = 0$. Cross peaks revealing J interactions between nuclei at two different values of F_2 are found above and below the $F_1 = 0$ line. The principal advantage of this method is that a smaller F_1 size is needed if only couplings between nuclei confined to a small portion of the chemical shift range are important, for example, only aromatics or only aliphatics. This choice then limits the maximum range of F_2 for the nuclei involved, and therefore, the size of the excursion from $F_1 = 0$ for a cross peak. Bain, Bornais, and Brownstein (76) have analyzed the general SECSY experiment for the AX, A_2X, AMX, A_2X_2, AA'XX', and A_2X_3 spin systems as a function of the flip angle of the second pulse. The primary motivation for the SECSY experiment is conservation of data space. This was relevant because in the earlier days of COSY the carrier had to be placed at one end of the chemical shift range with a corresponding doubling of spectral width, since quadrature detection in F_1 was not used. Bax, Freeman, and Morris (74) devised a phase-cycling scheme which allows F_1 quadrature detection, and subsequently the original Jeener, or COSY, pulse sequence has been used to an increasingly predominant extent.

Returning to the Jeener 2D spectrum, we note that it does not exhibit a simple phase pattern. On-diagonal peaks are in-phase, but cross peaks are mutually antiphase within each multiplet (Fig. 5-9). This property has important implications with regard to the procedure for data acquisition. Since the cross-peak components start off antiphase, poor digitization in F_1 (small number of t_1 values) can result in self-cancellation of peaks in the 2D spectrum. As the number of t_1 values increases, the characterization of the multiplet components gets increasingly better. Examination of the 2D data while the experiment proceeds may produce the observation that the cross peaks gain intensity faster toward the end of the data collection, as a result of the above factors. Care must be taken, then, not to terminate a homonuclear chemical shift correlation experiment too early or to weight the front part of the interferograms too much by data processing. Proper data processing can emphasize cross peaks at the expense of diagonal intensity by matching the natural time development of the off-diagonal magnetization. Since these start off antiphase, focus at $1/2J$, and then decay,

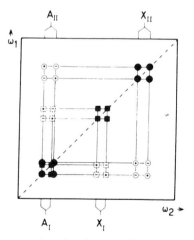

Fig. 5-9. Schematic representation of a phase-sensitive homonuclear Jeener spectrum for two two-spin systems. Filled symbols indicate dispersive lines, while open symbols are absorptive and alternate in sign. In absolute-value or magnitude mode, all signals are of the same sign.

a suitable weighting function would be symmetrical about $t_2 = 1/2J$, that is, a shifted sine-bell or pseudo-echo (77).

In order to avoid the phase twist and multiplet cancellation problems, COSY spectra in the early years of their use were displayed and plotted in absolute value (magnitude) mode. This has the undesirable feature that long tails arise from the dispersive part of the signal. If the data in t_1 and t_2 are processed with a weighting function such that they become symmetric about the midpoints of the respective time axes, presentation of the transformed data reveals an absorption-like lineshape, even though absolute-value mode is used. This reduces the likelihood of peaks being misassigned from accidental crossing of peak tails and also increases the feasibility of identifying cross peaks near the diagonal. Lorentzian-gaussian and/or convolution difference resolution enhancement functions have been used for this purpose.

Homonuclear shift correlation spectra can be rich in peaks to the point of possible confusion. In particular, errors in quadrature detection in F_2 (ghosts) and errors in F_1 pseudo-quadrature detection can produce small images across F_1 and $F_2 = 0$ (see Fig. 5-10). Errors in F_2 quadrature detection resulting from imbalance in the two phase detectors are normally eliminated by CYCLOPS four-transient phase cycling (78). In minimal COSY experiments, this phase cycling is not used; it can be added to these experiments, producing a minimum of 12 to 16 transients per FID. The user often chooses the minimum cycle for speed and therefore runs the risk of images in F_2. They are, however, always identifiable as present on the "other" diagonal, the "antidiagonal." The images are spaced equally from $F_2 = 0$ and do not have a corresponding intensity on the other side of the principal diagonal (Fig. 5-10a).

The F_1 images usually result from a failure to satisfy the requirements of the basic experiment. For example, in the four-transient version, the desired signals add coherently but their F_1 images add destructively, with as much positive as negative contribution. For an ideal system at equilibrium this cancellation is exact. If there is more magnetization generated in the first transient than in each of the three subsequent transients, there will be a small net image present along F_1 spaced at the same distance from $F_1 = 0$ as the genuine peak (see Fig. 5-10b). Again, these images will be on the antidiagonal and will be missing a partner on the other side of the principal diagonal. It is very important for the user to establish the existence of these images before forcing symmetry on the data matrix, since this process may put false intensity in the spectrum, depending on the algorithm used. One type of procedure is to multiply intensity at (F_1, F_2) by the intensity at (F_2, F_1), taking the square-root and replacing both previous values by this new value, for all values of (F_1, F_2) and (F_2, F_1). If, however, the value

(F_1, F_2) has intensity coming from a "tail" of a large peak, while (F_2, F_1) corresponds to an image, the resulting symmetrized data will have a false peak present in the spectrum. Other methods such as taking the minimum value of the pair will result in less error but will still have intensity corresponding to the small false signal.

Symmetrization, as introduced by Baumann *et al.* (79, 80), is useful for improving the clarity of a Jeener 2D spectrum, since it attenuates the impact of F_1 noise (resulting from spectrometer instabilities) and artifacts possibly present in slices in F_1 perpendicular to F_2. It also has the potential feature of about 40% improvement in sensitivity, since the signals are highly

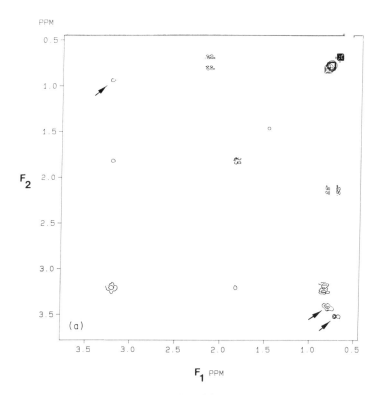

Fig. 5-10. Errors in COSY data collection. (a) Images across $F_2 = 0$, indicated by the arrows, occur from use of minimum four-transient phase cycling, omitting the usual four-transient phase cycle to correct quadrature imbalance. (b) Images across $F_1 = 0$ result from incomplete cancellation of the image peak in the four-transient double-quadrature experiment. Using steady-state pulses and more than four transients usually solves this problem. In either case, false peaks occur only on the "other" diagonal. Symmetrization may not always remove these so that data before and after symmetrization should be examined if "correlations" are found on the "other" diagonal. (*Figure continues.*)

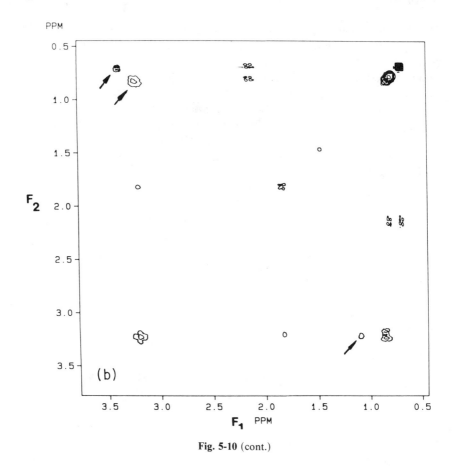

Fig. 5-10 (cont.)

correlated while the noise is, in principle, random. If data have been symmetrized without inspection beforehand, any intensity appearing along the antidiagonal should be viewed skeptically and interpreted accordingly.

2. *Modification of the Jeener Experiment; Use of Long-Range Couplings*

In a variation of the Jeener 2D experiment, the second pulse is set at a value less than 90° (73, 77). Magnetization is transferred predominantly to connected transitions, simplifying the cross-peak patterns of the 2D spectrum and increasing the intensity of cross peaks relative to diagonal peaks. A 45° pulse is commonly used here, but some cross-peak intensities can be selectively enhanced relative to the diagonal peaks by choosing other values such as 60°.

The size of the homonuclear J coupling does play a part in the size of the cross peak observed. Maximum transfer of magnetization between coupled spins A and B occurs when $\sin[\pi J(AX)t_1]\exp[-t_1/T_2(A)]$ is largest. On the other hand, the maximum detected signal from this transfer occurs when $\sin[\pi J(AX)t_2]\exp(-t_2/T_2(X))$ is at a maximum. If $J(AX)$ is very small, this may occur at very large values of t_1 and t_2, values perhaps incompatible with the data storage available. It is not necessary, however, to digitize the signals during all of t_2, or to begin t_1 values at zero (77). The addition of small delays just following both pulses in the Jeener pulse sequence, somewhat similar to SUPERCOSY, to be described below, permits postponing acquisition until the maximum sensitivity period. Presentation in the absolute-value mode eliminates any worries about frequency-dependent phase shifts arising from the delay in starting digitization. This matching of data collection to the size of relevant couplings makes the experiment extremely useful for establishing the existence of long-range couplings. If $\pi J(AX)$ is much smaller than T_2 of A and X, then the optimal centers of the pseudo-echo function should be at $t_1 = T_2(A)$ and $t_2 = T_2(X)$. Cross peaks may be observed for couplings too small to be resolved in normal 1D NMR.

Choice of the proper phase cycling can ameliorate the effects of poor magnetic-field homogeneity. When the phase cycling is selected to detect the coherence transfer echo (77), there is a refocusing of the detected signal in t_2, independent of magnetic-field homogeneity. Magnetization is always transferred within one molecule, so that macroscopic variation of magnetic field has no effect on the efficiency of the transfer. For the experiment designed for detection of very small couplings, pseudo-echo or sine-bell processing is therefore shaped relative to the actual T_2 rather than to T_2^*.

Batta and Liptak (81) exploited this capability to establish interglycosidic linkages in oligosaccharides via the four-bond $J(HCOCH)$ coupling (estimated to be lower than 0.2 Hz). Postpulse delays of 0.4 sec were needed here, making the technique useful only for protons with relatively long T_2, since, at a minimum, 0.8 sec elapses between the first pulse and data acquisition. Steffens et al. (82) have used the same techniques to establish proximity relationships and relative stereochemistries of substituents in natural and synthetic organic compounds. While proton NMR is routinely used for these objectives through analysis of coupling patterns, the Jeener experiment with postpulse delay was able to establish J couplings over four and five bonds. Figure 5-11 shows both types of COSY spectra for a natural product, while Fig. 5-12 demonstrates the utility in establishing identities of gem dimethyl groups. A further use was demonstrated by Wynants and Van Binst (83), who were concerned with assignment of aromatic residues in polypeptides. Nuclear Overhauser effects in these are very small, and

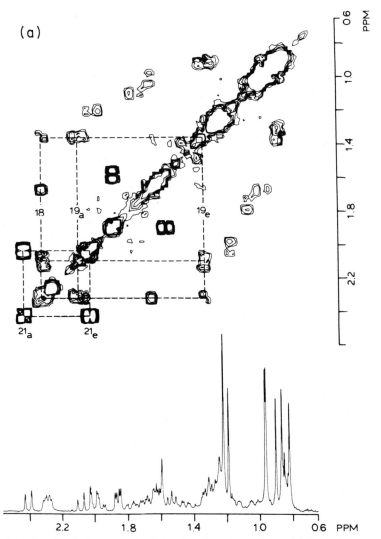

Fig. 5-11. Results of using postpulse delays in a Jeener experiment. (a) Normal spectrum. (b) Spectrum with 0.2-sec delay inserted after each pulse. Spectra are from the molecule:

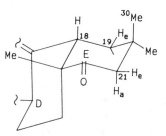

[Reprinted with permission from Steffens *et al.*, *J. Am. Chem. Soc.* **105**, 1669 (1983). Copyright 1983 American Chemical Society.]

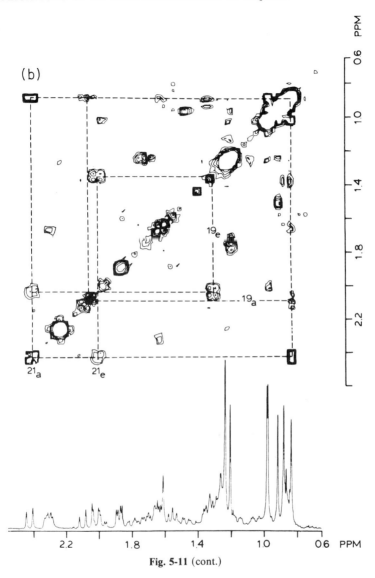

Fig. 5-11 (cont.)

assignments have generally been made by synthesis of deuterated peptides or CIDNP. The expected small four-bond coupling to beta protons and five-bond coupling to the alpha protons were used to identify the aromatic residues in polypeptides containing such amino acids as lysine, threonine, cystine, phenylalanine, and tryptophane (Fig. 5-13).

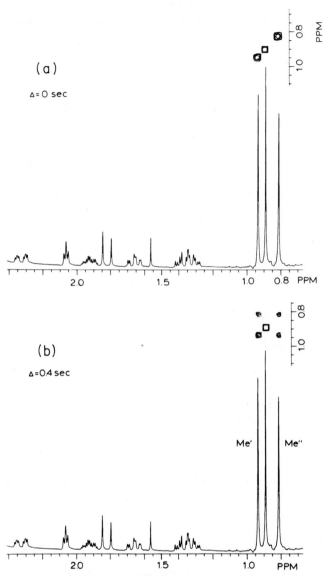

Fig. 5-12. Use of postpulse delays in Jeneer experiments. (a) Normal spectrum; (b) spectrum with 0.4-sec delay inserted after each pulse. Spectra are from the molecule:

[D. G. Lynn (private communication).]

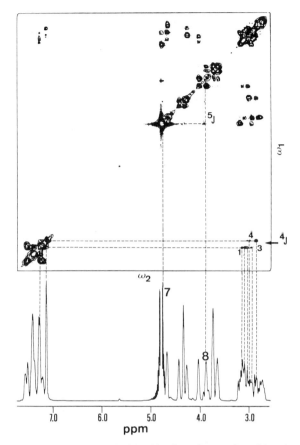

Fig. 5-13. Use of postpulse delays for identification of aromatic residues in polypeptides. The dotted lines show four-bond connectivities between C_2 and beta protons of aromatic residues and a five-bond coupling between alpha protons of two adjacent amino acids. [Reprinted with permission from Wynants and Van Binst, *Biopolymers*, **23**, 1799 (1984). Copyright 1984 John Wiley & Sons, Inc.]

Bax, Byrd, and Aszalos (84) have analyzed the dependence of cross-peak intensity on J in homonuclear shift correlation experiments and have found that proper weighting of the FID and interferogram data can improve the intensity of a cross peak by more than an order of magnitude for protons coupled to several other protons. This improvement is effected by matching the weighting function to the expected theoretical coherence-transfer echo. The function is specific for the selected J and nonmatched for other magnetization components. This "multiplet-selective" filter also reduces t_1 "noise," since it reduces the contribution of diagonal peaks. The method is also applicable in homonuclear double-quantum and relayed coherence-transfer experiments, as well as in heteronuclear relayed coherence-transfer studies.

While Jeener data have often been presented in absolute-value mode, it is now general practice to use phase-sensitive mode for higher resolution. This is illustrated in Fig. 5-14, where both types are given. Output devices must be able to plot data of only one sign, or make the data of opposite signs distinct by use of color or, as in the case here, by different appearance. The regions of interest here are off the diagonal so that dispersive components of diagonal intensity do not interfere.

Several procedures for producing phased 2D data have been developed. States, Haberkorn, and Ruben (85) described a phase-cycling and data-

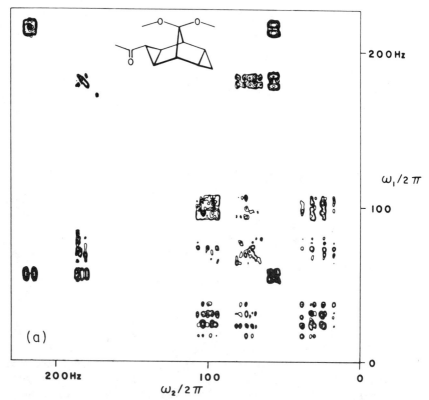

Fig. 5-14. A comparison of absolute-value and phased homonuclear correlation data. (a) Normal COSY spectrum in absolute-value mode. (b) Same sample with spectrum obtained by TOCSY sequence and displayed in phased mode. Not only have related coherence peaks appeared, but the normal COSY peaks are much better resolved in the phased spectrum. [From Braunschweiler and Ernst (86).]

5. APPLICATIONS OF 2D NMR EXPERIMENTS IN LIQUIDS 375

routing scheme and applied it in particular to exchange experiments. The method of time-proportional phase incrementation (TPPI) was described in Chapter 2. A third method is that of Braunschweiler and Ernst (86), who obtain phased COSY data by beginning with a single 90° pulse, which is followed by an evolution period to frequency-label the spins. Next, a rapid series of pulses is applied *during a mixing period*; this is followed by detection. The resultant data for a selection of mixing times are coadded to form the final data set as in Fig. 5-15. The pulse sequence during the mixing period is designed to factor out chemical shifts, leaving only the isotropic coupling, and can be as simple as repetitive 180° pulses, either of constant or of alternated phases. An additional feature of the technique is that all pairs of nuclei present in a spin system show cross peaks, even if they are not directly coupled, because of multiple-step relayed coherence transfer generated by the pulsing during the mixing period. Essentially, the 180° pulses propagate the magnetization throughout the individual spin

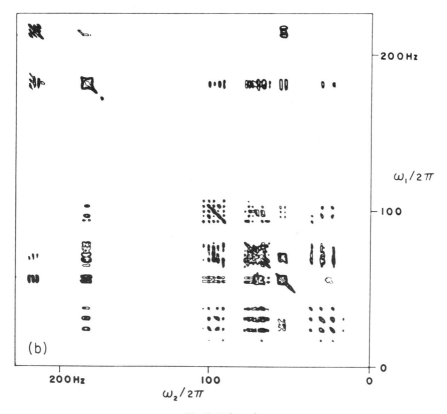

Fig. 5-14 (cont.)

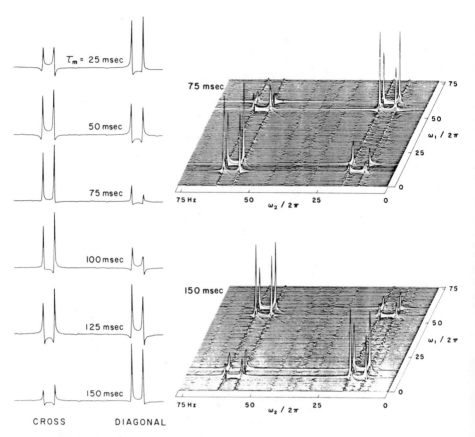

Fig. 5-15. Construction of a phased 2D TOCSY spectrum. Several 2D data sets are coadded, canceling the dispersive components. Six phased slices at one value of F_1 are given for various lengths of the mixing period. The 2D data sets for mixing times of 75 and 150 msec are given, along with the result of summing the six different complete 2D data sets. [From Braunschweiler and Ernst (86).]

systems. Of course, isolated spin systems would now show mutual cross peaks. Because of the total nature of this correlation, the technique is referred to as TOCSY (*to*tal *c*orrelation *s*pectroscop*y*). The price is time, of course, since the entire 2D experiment with full F_1 and F_2 resolution is performed for each value of the mixing time.

A variation of COSY called SUPERCOSY has been proposed by Kumar, Hosur, and Chandrasekhar (87). This pulse sequence replaces the last 90° pulse in the COSY sequence with the element $[-\tau-180°_x-\tau-90°_x-\tau-180°_x-\tau-]$ where τ is of the order of $1/3J(\text{HH})$. These workers demonstrated that enhanced cross peaks were produced, diagonal peaks

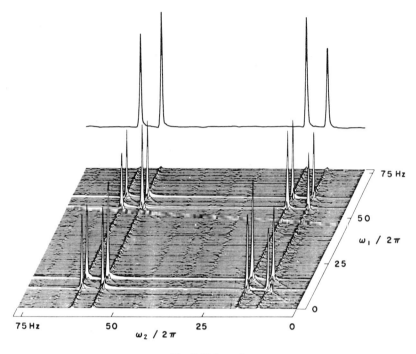

Fig. 5-15 (cont.)

were attenuated, t_1 noise was reduced, and selectivity for different ranges of coupling constants was possible.

Often, highly crowded spectra can benefit by reduction of multiplet structure, allowing recognition of chemically shifted protons in highly congested and overlapped regions. One example is that of homonuclear shift correlation of polymers of mixed tacticity, or, in a similar vein, large polypeptides or proteins. The need for resolving distinct protons becomes acute in these cases. Bax and Freeman (77) proposed a technique for collapsing multiplet structure in F_1 by using the modified pulse sequence shown in Fig. 5-16. They emphasized that a flip angle less than 90°, such as 45° or 60°, should be used for the last pulse to avoid mutual cancellation of antiphase cross peaks. A projection onto the F_1 axis can then give a proton "stick" spectrum as in Fig. 5-17. Second-order and strongly coupled systems can give anomalous intensities and false peaks, however.

Brown (88) revised the above method to permit comparable resolution in F_1 but without the large intensity variations, so that accurate measurements of high polymer tacticity could be made. This was achieved by scaling in F_1; that is, the apparent chemical shift of spin A in F_1 is scaled by the

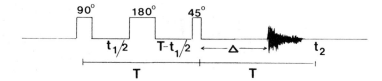

Fig. 5-16. Pulse sequence of Bax and Freeman (77) for decoupling multiplets in F_1; T and t represent fixed delays.

factor $(2-k)F_1(A)$ and its couplings are scaled by the factor $k \Sigma_i \alpha_{ia}^{tu} J_{ia}$, where i runs over all spins, k is defined in the pulse sequence diagram in Fig. 5-18, and $\alpha_{ia}^{tu} = \pm\frac{1}{2}$. The F_1 decoupling described in the preceding paragraph is simply the case with $k=0$. Sufficient resolution of cross peaks may be attained by choosing a value of k intermediate between 0 and 1. The case of $k=0.29$ is illustrated by the proton spectrum of polyvinyl alcohol in Figs. 5-19 and 5-20. The F_1 scaling technique is particularly suitable for polymers and high-molecular-weight compounds because the typically short T_2 values restrict the size of the delay used for F_1 decoupling

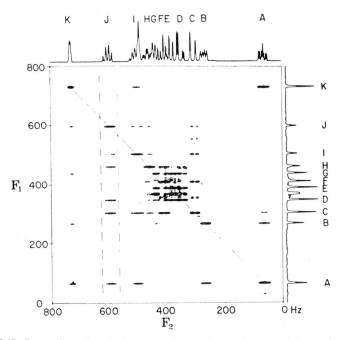

Fig. 5-17. Decoupling of F_1 in Jeener spectra. Collapse of the multiplet structure in F_1 produces less complex spectra and projections which are proton "stick" spectra. [From Bax and Freeman (77).]

5. APPLICATIONS OF 2D NMR EXPERIMENTS IN LIQUIDS

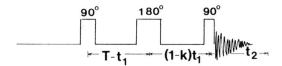

Fig. 5-18. Pulse sequence of Brown (88) for decoupling of multiplets in F_1; T is a fixed delay and k is a constant.

in the Bax and Freeman version. Applications to proteins of F_1 decoupling in COSY, RELAY, and SECSY correlation experiments were also treated in detail by Rance *et al.* (89).

Fine structure in two-dimensional homonuclear correlation experiments contains, in principle, coupling information not observable in normal 1D experiments. Oschkinat and Freeman (90) have published techniques which show how this fine structure may be extracted. Phased-mode presentation is used, following the method of States *et al.* (85), and two data tables are constructed, one with cross peaks in absorption and the other with diagonal peaks in absorption. Addition of traces from each at the same F_2 value leaves only one line of the $J(AX)$ multiplet, subtraction leaves the other line, and in both cases the lineshapes are unperturbed by overlap, permitting an estimate of $J(AX)$.

3. Applications of Homonuclear Correlation

Proton homonuclear shift correlation studies have been reported for 9-α-hydroxyestrone methyl ether (91), a gorgosterol (47), a polycyclic

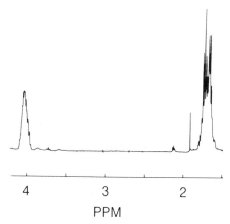

Fig. 5-19. Proton spectrum at 360 MHz of 10% poly(vinyl alcohol) in D_2O at 70°C. [From Brown (88).]

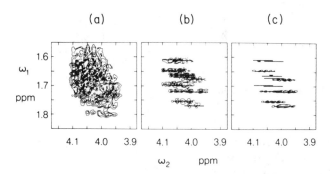

Fig. 5-20. Cross-peak region for the same solution as in Fig. 5-19. (a) Normal COSY. (b) With F_1 scaling of $k = 0.29$ and $\tau = 0.58$ sec. (c) With F_1 scaling of $k = 0$ and $\tau = 0.56$ sec. The bars indicate cross peaks between different triad and tetrad configurational sequences in the polymer. [From Brown (88).]

phenanthrothiophene (92), a substituted naphthyridine (46), and cynarin (93). Carbohydrates have been studied by Szilagyi (94), who analyzed the proton spectrum of cellobiononitrile octaacetate; Dabrowski and Hanfland (95), who examined a ceramide pentadecasaccharide; Prestegard et al. (96), who studied gangliotriaosylceamide; and Bernstein and Hall (97), who proposed using COSY and interunit NOE to sequence oligosaccharides—in this case, a model example, allyl-β-D-galactopyranosyl-1β-4-D-glucopyranoside.

Small proteins and polypeptides have received considerable attention, particularly trypsin inhibitor (98–100). Kessler et al. (101) examined cyclo(L-pro-L-pro-D-pro), while King and Wright (102) used COSY as a general method for assignment of C-2 and C-4 proton resonances in individual histidine residues in plastocyanin and myoglobin. A review of 2D techniques as applied to ^{1}H NMR of proteins has been published. It collects and discusses the many experimental techniques developed in the study of proteins by the groups of Ernst and Wüthrich (103).

Solvent suppression techniques are important for 2D NMR, particularly because 2D experiments usually require 90° pulses. When the solvent is protonated the resulting signal can easily overload the receiver or exceed the limit of the ADC employed. High-dynamic-range receivers and extended-range ADCs certainly help, but they are usually insufficient when operating at higher magnetic fields or when the solvent is H_2O. Some of the techniques employed to reduce solvent signals are discussed in Chapter 6, Section III.

Although the vast majority of homonuclear 2D shift correlations have been in proton spectroscopy, other studies have appeared with correlation methods applied to ^{31}P in cellular phosphates (104), ^{11}B in polyhedral boranes (105, 106), and ^{183}W in heteropolytungstates (Fig. 5-21) (107, 108). Other applications, such as establishing ^{13}C-^{13}C linkages in enriched metabolites, have been demonstrated in experiments using specifically labeled nutrients for biosynthesis (109–111).

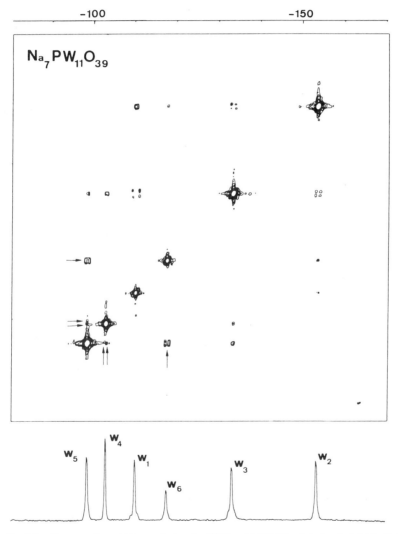

Fig. 5-21. Homonuclear shift correlation for ^{183}W at 16.67 MHz. Solution is 1 M in D_2O. [Reprinted with permission from Brevard et al., J. Am. Chem. Soc. **105**, 7059 (1983). Copyright 1983 American Chemical Society.]

B. Heteronuclear Shift Correlation

1. *The Heteronuclear Correlation Experiment*

The second major form of shift-correlation 2D NMR is the heteronuclear version. The experiment as proposed by Maudsley *et al.* (35, 112) can be visualized as generating proton magnetization, for example, by a 90° pulse, and letting the magnetization precess for a time t_1. The extent of precession in the rotating frame is proportional to the distance of the resonance from the proton transmitter frequency, and therefore the phase of the magnetization is dependent on the *proton shift*. Simultaneous 90° pulses on the proton and the X observe nucleus transfer magnetization to the scalar-coupled X nucleus—just as in the INEPT or DEPT experiment. The phase of the X nucleus magnetization is hence coded with the chemical shift of the coupled proton. The experiment is carried out for the full range of t_1 values to produce the 2D spectrum. Since the X nucleus shift is present in F_2, both chemical shifts for the XH pair are identified by extrapolation to the individual axes from the XH peak in the 2D data.

No net magnetization is transferred to the X nucleus, for the multiplet components are antiphase and a 2D experiment of this kind shows the XH couplings in both dimensions, in addition to the H and X chemical shifts. This sequence can be modified as shown in Fig. 5-22 by introducing a 180°(X) pulse midway during the t_1 period to refocus any effect of XH couplings on the protons in the t_1 period, and thus in F_1. Small delays are inserted prior to and after the simultaneous 90° pulses to prevent the cancellation of out-of-phase signals when decoupling is performed during t_2 (the delays allow antiphase components to refocus). Broadband decoupling of H during acquisition removes the XH coupling from the F_2 dimension (113, 114). Proper phase cycling can remove the axial ($F_1 = 0$) peaks which

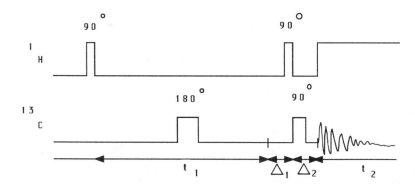

Fig. 5-22. Pulse sequence for heteronuclear chemical shift correlation.

have no intrinsic interest and also permits the proton transmitter, usually the decoupler, to be centered in the proton spectrum with pseudo-quadrature detection in F_1 (115). This allows finer digitization when data tables are restricted in size.

In the same manner as for homonuclear 2D shift correlation, the magnetization associated with any proton transition is modulated by couplings to other protons. This modulation, along with the proton shift, determines the sign of the proton polarization just prior to the second 90°(H) pulse, and hence this information is encoded into the resulting X-nucleus magnetization at the time of the 90°(X) "read" pulse. Therefore, all signals appearing in the final X/H heteronuclear 2D shift correlation spectrum will have fine structure corresponding to the homonuclear J-coupling pattern for the appropriate protons.

Since this experiment involves a polarization transfer, the repetition time is governed by the T_1 of the relevant protons. Signal enhancement is again determined by the ratio of the nuclear moments, a factor of 4 for carbon-13 and 10 for nitrogen-15. These two factors make this experiment highly sensitive and practical for small samples. Of course, nonprotonated X nuclei will not be observed if the refocusing delays used in the pulse sequence are set to be appropriate for one-bond couplings. If the delays are set for long-range couplings, these nonprotonated nuclei can give signals in the 2D spectrum.

2. *Application to ^{13}C and ^{15}N Spectroscopy*

The exploitation of heteronuclear chemical shift 2D spectroscopy has begun, but gives promise of a vastly greater scope of application to structural problems in the future. Ammann *et al.* (70) have illustrated how 2D methods can be used to assign protons and carbons in lupane, a C_{30} triterpene containing only carbon and hydrogen. Heteronuclear J-resolved 2D was used to assign the number of protons to each carbon, and heteronuclear chemical shift correlation allowed assignment of the associated proton shifts and many of the homonuclear H-H coupling constants. Ikura and Hikichi (116) used the same techniques on *d*-biotin. A mixture of allyl nickel complexes was also studied by Benn (117), and Morris and Hall (118) have examined a series of carbohydrates.

Rinaldi and Solomon (68) made extensive use of heteronuclear 2D chemical shift correlation in determining the relative configuration of a substituted tetracyclododecane. Waugh and Berlin (119) applied the technique to proton assignments in 2-anthrylalkanoic acids, while Joseph-Nathan *et al.* were able to resolve the controversy over the assignment of methyl resonances in angelic and tiglic acids (120) and to assign the carbon and proton spectra of cyperene (121). Other complex assignment problems were analyzed similarly by Gampe et al. (122) for cellobionitrile octaacetate and by Csuk, Müller, and Weidmann (123) for several alkylidene furanuronolactones. Heteronuclear 2D shift correlation NMR was also used by Haslinger et al. (124) in the spectral analysis of maleopimaric acid methyl ester, by Wernly and Lauterwein (67) on all-*trans*-retinal, by Martin et al. (125) on phenanthro(1,2-b)thiophene, and by Schmitt and Günther (126) on 9,11-bisdehydrobenzo(18)annulene.

Polypeptides have been a particular fruitful area for application of this technique. Kessler and Schuck (127) used a combination of homo- and heteronuclear shift correlation 2D NMR to make and correct assignments of carbon and proton resonances for cycloinopeptide A, a cyclic nonapeptide isolated from linseed oil. In a related study, Kessler et al. (128) studied the cyclic tripeptide cyclo(pro-pro-nitrobenzyl-gly), using C/H correlation to assign the carbon spectrum. Haslinger and Kalchhauser (129) applied the same approach to the cyclic tetrapeptide chlamydocin.

Nitrogen-15/proton heteronuclear shift correlation is emerging as a powerful tool in the assignments of nitrogen-15 resonances. Gray (130), Kessler et al. (131), and Hawkes et al. (132) have all provided examples of the utility of this experiment, particularly in view of the narrow range of peptide nitrogen shifts and the sensitivity of these shifts to sequence and conformational factors. Two-dimensional methods allow nonsubjective judgments to be made regarding assignments. For example, in the pentapeptide CBZ-gly-pro-leu-ala-pro (130), the known regularities in carbon shifts were used to assign the alpha carbons (which were well separated in shift). A C/H correlation experiment (Fig. 5-23) allowed direct assignment of the connected alpha protons. The vicinally coupled NH protons were then unambiguously assigned by a homonuclear shift correlation experiment (Fig. 5-24). Finally, an N/H correlation 2D experiment provided the direct assignment of the amide nitrogens, four of which occurred within a 10-ppm shift range (Fig. 5-25). Extension of these techniques to larger polypeptides and proteins is direct and facile. Westler, Ortiz-Polo, and Markley (133) studied a small serine proteinase inhibitor at a 12-mM concentration using C/H chemical shift correlation at 50.3 MHz. The observed methyl correlations are illustrated in Fig. 5-26. As complexity of data increases, methods such as these will become critical for extracting spectral information. Chan and Markley (134, 135) obtained C/H correlation data for another protein, ferrodoxin, while Kojiro and Markley (136) studied the blue copper protein, plastocyanin.

Coffin, Limm, and Cowburn (137) have reported a C/H correlation study of bovine high-density lipoprotein. Polynucleotides have also been examined. Lankhorst et al. (138) reported C/H correlation spectra for several

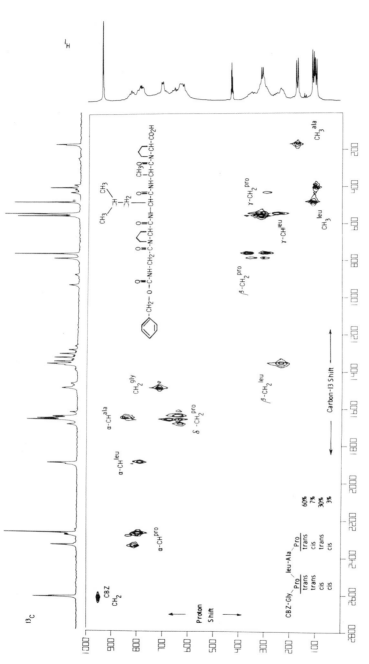

Fig. 5-23. Carbon-hydrogen chemical shift correlation for 0.2 M CBZ-gly-pro-leu-ala-proline in DMSO-d_6. Correlation peaks indicate a chemical bond between the protons and the carbon at shifts projected from the two axes. One dimensional spectra are included for convenience. The additive nature and the spread of carbon shifts permit assignment of the carbons, from which the proton assignments may be made. [Reprinted with permission from Gray, *Org. Magn. Reson.* **21**, 111 (1983). Copyright 1983 John Wiley & Sons, Ltd.]

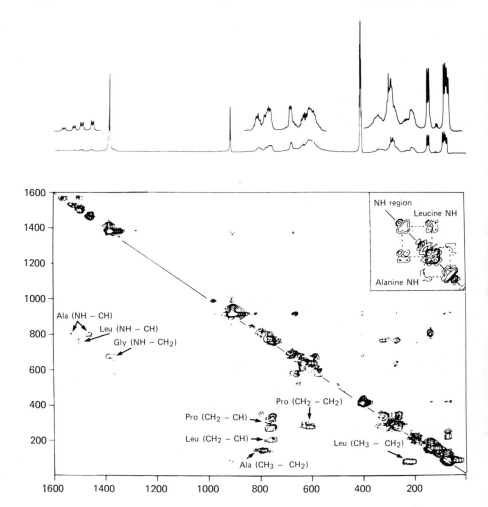

Fig. 5-24. From the assignments of the alpha protons and this homonuclear shift correlation spectrum (lower left triangle), direct assignment of the NH protons is possible. The upper left triangle (and expansion) have chemical exchange/NOESY data. [Reprinted with permission from Gray, *Org. Magn. Reson.* **21**, 111 (1983). Copyright 1983 John Wiley & Sons, Ltd.]

oligoribonucleotides. Beloeil *et al.* (139) combined both homonuclear and heteronuclear shift correlation techniques to perform a total assignment of the proton and carbon spectra of the ionophorus antibiotic X.14547 A, a molecule for which normal 1D techniques were insufficient. Proton-carbon correlations also aided in the assignments in tricyclo(7.3.1.0[2,7])tridecane (140).

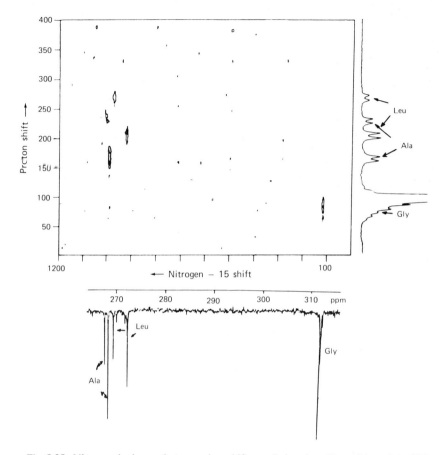

Fig. 5-25. Nitrogen–hydrogen heteronuclear shift correlation data. From this and the NH proton assignments in Fig. 5-24, the assignments of the nitrogen-15 resonances follow directly. [Reprinted with permission from Gray, *Org. Magn. Reson.* **21**, 111 (1983). Copyright 1983 John Wiley & Sons, Ltd.]

3. Correlation by Long-Range Coupling

As valuable as the above C/H correlations have been, they only illustrate one type of "connectivity," that of bonded nuclei. The user predetermines this by the choice of delays within the pulse sequence, a choice which is based on the value of the coupling between the protons and the nucleus to be observed, normally 125–170 Hz for protonated carbons. Nonprotonated as well as protonated carbons can have couplings to nearby protons, usually ranging from 1 to 10 Hz. If the selection of pulse sequence delays is based on these couplings, long-range heteronuclear correlation spectra may be

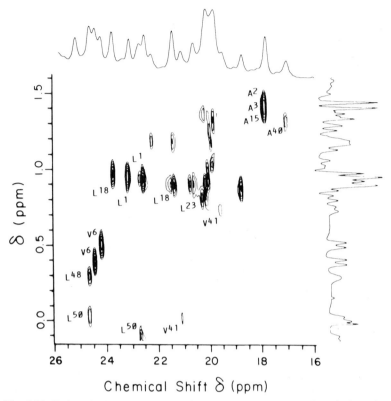

Fig. 5-26. Carbon–hydrogen chemical shift correlation for a 0.012 M solution of serine proteinase inhibitor. [From Westler et al. (133).]

obtained. Delays within the pulse sequence are usually about 15–20 times longer than for one-bond couplings, so that T_2 relaxation during the sequence can be an important cause of loss of sensitivity. Another cause is the spread in long-range J values for carbons coupled to many different types of protons. For these reasons it is common that not all possible correlations are observed. Nevertheless, the experiment can be particularly valuable in categorizing the local environment of a carbon.

A revealing example of this was published by Banerji, Saha, and Shoolery (141) in the determination of the structure of a reaction product of 2-methylindole and mesityl oxide. The proposed structures shown in Fig. 5-27 were both consistent with a variety of 1D and 2D data. However, carbon "C" shows long-long correlations to both pairs of methyl groups, indicating that only structure **1** is possible since carbon "C" would be expected to have similar three-bond CH couplings to all the methyl protons.

5. APPLICATIONS OF 2D NMR EXPERIMENTS IN LIQUIDS

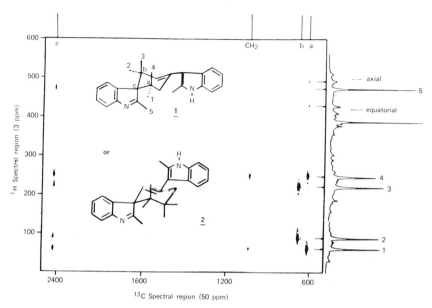

Fig. 5-27. Long-range carbon–hydrogen shift correlation. Nonprotonated carbons can exhibit very valuable correlations as in this reaction product for which two choices for a structure were possible. The presence of srong correlations to all four methyl carbons is consistent with only one of the structures. [Reprinted with permission from Banerji, Saha, and Shoolery, *Indian J. Chem., Sect. B* **22**, 903 (1983).]

Another area of promise for long-range correlation techniques is in the study of polypeptides. Wynants *et al.* (142) used these to correlate each alpha proton to the two flanking carbonyls belonging to the same and to the next amino acid, thus permitting sequencing independent of the existence of interunit homonuclear NOE, an interaction frequently proposed for use in confirming sequential arrangement (see Fig. 5-28). Detailed information on other long-range interactions is available from this experiment, as illustrated in Figure 5-29.

Kessler *et al.* (143) have tried to reduce the loss of signal due to proton T_2 relaxation during the evolution period by developing their COLOC pulse sequence (*co*rrelation spectroscopy via *lo*ng-range *c*oupling, Fig. 5-30). In a way similar to the F_1 decoupling performed in proton homonuclear correlation (77), the evolution time is "folded" into a fixed time period, here the Δ_1 delay. Figure 5-31 shows the application of COLOC to cyclo(-pro-phe-D-trp-thr-gly-) at 0.2 M in DMSO. The reduced length of the pulse sequence and the decoupling of the H–H splittings normally obtained in the heteronuclear shift correlation spectrum both improve the observed sensitivity.

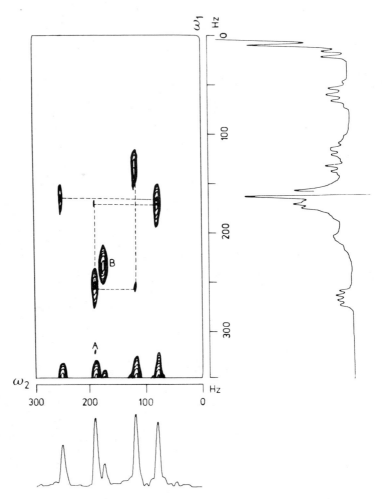

Fig. 5-28. Long-range carbon–hydrogen shift correlation for 0.16 M CH$_3$CO-Thr-Phe-Thr-Ser-NH$_2$ in DMSO-d6 at 30°C. The vertical axis corresponds to the alpha region of the proton spectrum. The horizontal axis shows the carbonyl carbon portion. A and B are folded peaks. Two of the carbons exhibit correlations to two amino acid alpha protons, while three different alpha protons each correlate with two different carbonyl carbons, thus permitting linkage assignments to be made. [From Wynants *et al.* (142).]

A somewhat different approach was taken by Bauer, Freeman, and Wimperis (144), who proposed the pulse sequence in Fig. 5-32. The first cluster of pulses acts as a 90° excitation pulse only for protons remote from the observe nucleus, but acts as a 180° pulse for directly bonded protons (145). The placement of the evolution time within a fixed constant time effectively decouples the proton–proton couplings in F_1. The pair of 90° pulses midway through the sequence transfers remote proton chemical shift and long-range CH coupling-encoded polarization to the observe nucleus, which then precesses from an initial antiphase doublet state, focusing after a time $1/J$. Remote protons are inverted by the second cluster of pulses, but directly bonded protons are unaffected. The second observe-nucleus 180° pulse then decouples any directly bonded protons, but also changes the spin states of the carbon magnetizations for remote-coupled spins as for a pair of carbon and proton 180° pulses in the normal spin-echo sequence. This 180° pulse also refocuses the effect of chemical shifts.

The net result of the last cluster of pulses is that magnetization transferred from protons via one-bond couplings, initially antiphase, is returned to an antiphase condition and is destroyed by the onset of broadband decoupling. This then eliminates one-bond couplings from the spectrum, leaving only doublets arising from long-range CH couplings. The two-dimensional spectrum has axes F_2 corresponding to the observe nucleus chemical shift and F_1 corresponding to the proton chemical shift. Each carbon having long-range couplings to protons may have one or more doublets appearing at one or more F_1 values corresponding to the shifts of those protons to which the carbon is coupled. Similarly, a given F_1 value may have several doublets in F_2, indicating the various carbons coupled to this individual proton.

This information was available from the selective heteronuclear 2D J experiment previously proposed by Bax and Freeman (30) where a soft pulse centered on only one proton was used to explore long-range couplings of that proton to nearby carbons. The advantage of this newer shift correlation experiment is that all possible protons are explored at once without the need for making prior choices. However, the longer sequence duration leads to greater losses from T_2 and from rf inhomogeneities than in the selective 2D J version. For this reason the latter should be used for those cases involving only one or a few protons of interest.

4. *Applications to Other Heteronuclei*

Aside from the carbon and nitrogen examples discussed above, there are several instances in which other nuclei have been related to protons by shift correlation 2D NMR. Bleich *et al.* (146) used a simplified version of the experiment to correlate protons and deuterium in small molecules. The example of methyl ethyl ketone is shown in Fig. 5-33. Boron-11 proton correlation spectra have been reported by Finster, Hutton, and Grimes (147) for 2,4-$C_2B_5H_7$ (Fig. 5-34) and by Venable, Hutton, and Grimes (106) for boron hydrides, carboranes, metallaboranes, and metallacarboranes. An illustration for a complex of pentamethylcyclopentadiene and a

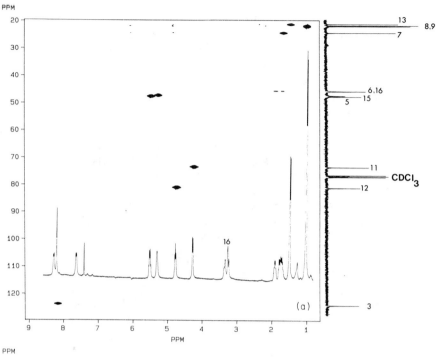
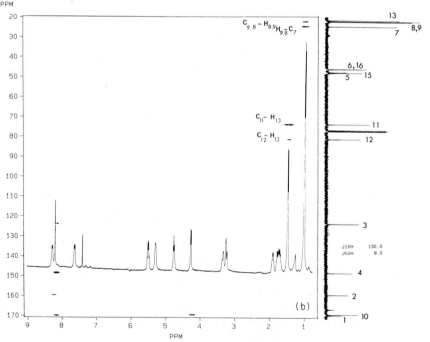

5. APPLICATIONS OF 2D NMR EXPERIMENTS IN LIQUIDS

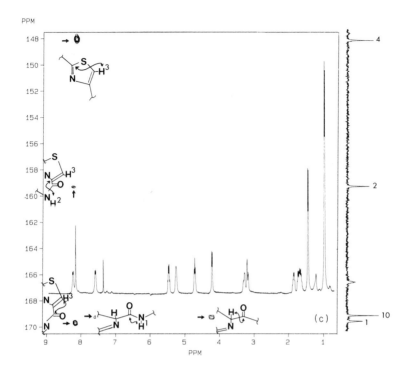

Fig. 5-29. Carbon–hydrogen shift correlation spectra for a solution of 80 mg of ulithiacyclamide in CDCl$_3$; hydrogen at 100 MHz.

(a) One-bond shift correlation. (b) Long-range correlation using a pulse sequence optimized for $J = 8$ Hz. (c) Expansion of (b) for the 148–170 ppm region showing detailed assignments.

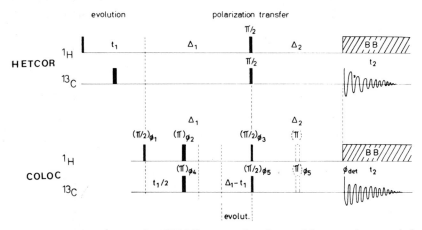

Fig. 5-30. Pulse sequence for COLOC compared to the usual heteronuclear correlation sequence. [From Kessler *et al.* (143).]

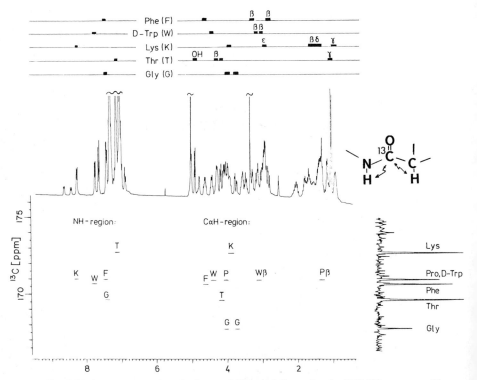

Fig. 5-31. Long-range carbon–hydrogen shift correlation using the COLOC sequence. The horizontal shift axis is for hydrogen. Each carbonyl at the right shows correlations that permit sequencing of the polypeptide *trans*-cyclo(-Pro-Phe-D-Trp-Lys-Thr-Gly-) in DMSO. The proton assignments are given above the figure. [From Kessler *et al.* (143).]

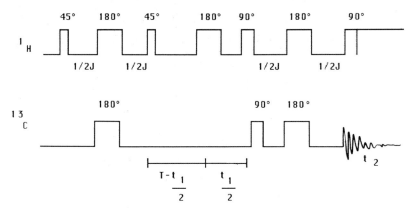

Fig. 5-32. Pulse sequence used to measure and assign long-range C-H couplings; J is the one-bond coupling constant and T is a fixed delay. The first part of the sequence excites only protons remote from carbon-13 and minimizes one-bond correlations. The last part of the sequence refocuses only carbon-13 vectors for remote protons. [From Bauer, Freeman, and Wimperis (144).]

cobalcarborane is given in Fig. 5-35. Burum (148) analyzed the experiment using density-matrix methods to identify the optimum delay times within the sequence for observing quadrupolar nuclei, confirming the results obtained by Pegg *et al.* (149), and applied the results to correlation of boron-11 with hydrogen in decaborane.

Fluorine-19/proton shift correlation 2D NMR was used by Gerig (150) to help understand the proton spectra of fluorinated polymers. The shift

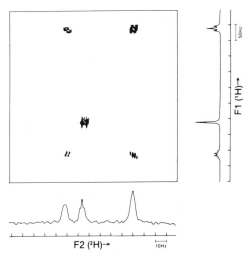

Fig. 5-33. Hydrogen–deuterium chemical shift correlation for methyl ethyl ketone. [From Bleich *et al.* (146).]

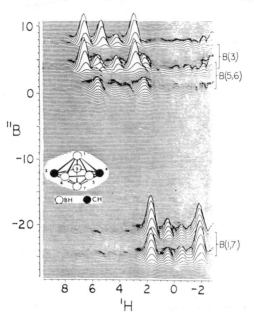

Fig. 5-34. Hydrogen–boron chemical shift correlation for 2,4-$C_2B_5H_7$. [Reprinted with permission from Finster et al., J. Am. Chem. Soc. **102**, 400 (1980). Copyright 1980 American Chemical Society.]

correlation spectrum for poly(p-fluorostyrene) is given in Fig. 5-36. Projection of the data onto the proton shift axis produces a subspectrum for only those protons coupled to the fluorine. Contrast this to the normal proton spectrum given in Fig. 5-37. Detail such as this can allow interpretation of proton chemical shifts that are normally unresolved.

Bolton and Bodenhausen (151, 152) and Bolton (153, 154) have used phosphorus/hydrogen 2D shift correlation techniques to obtain H–H couplings in cellular phosphates. The phosphorus serves to pick out proton subspectra containing conformationally relevant couplings. Pardi et al. (155) have obtained $^1H/^1H$ and $^{31}P/^1H$ 2D chemical shift correlation spectra from the backbones of oligonucleotides. The latter spectra contain F_1 multiplets for each ^{31}P chemical shift value. These correspond to the ^{31}P-coupled protons on the two sugars linked together by the phosphate, thus directly establishing the linkage between the sugars.

Lai et al. (156) relied on shift correlation of phosphorus with hydrogen to assign protons on the deoxyribose sugar rings in the oligonucleotide d(ApGpCpT) by chemically labeling the phosphodiesters and by observing the correlations of this added phosphorus to the H-5′ and H-3′ protons nearby. Bolton (157) has reviewed the general P/H shift correlation method.

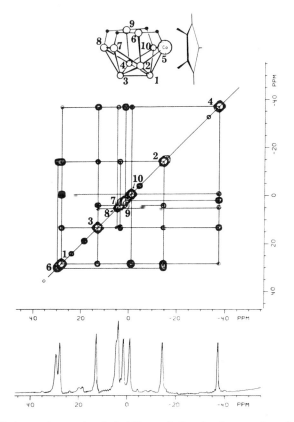

Fig. 5-35. Hydrogen-boron chemical shift correlation for the complex of pentamethylcyclopentadiene and cobalcarborane. [Reprinted with permission from Venable et al., J. Am. Chem. Soc. **106**, 29 (1984). Copyright 1984 American Chemical Society.]

Heteronuclear chemical shift correlation techniques can be used to infer spin–lattice relaxation times of the protons attached to the observe nucleus (158). This is accomplished by saturation of the protons and observe nucleus followed by a variable time t (saturation recovery) during which saturation of the observe nucleus is continued by repeated pulsing and the attached proton remagnetizes. The process is followed by the normal H/X 2D shift correlation experiment. The dependence of the 2D peak intensities on t is then used to extract ^{1}H T_1 values by exponential analysis.

5. Modifications of the Correlation Experiment

Most of the above heteronuclear shift correlation techniques produce spectra which have H-H spin multiplets in the F_1 dimension. These can be

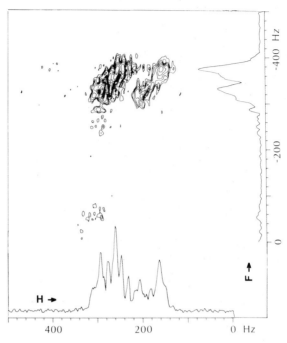

Fig. 5-36. Hydrogen-fluorine chemical shift correlation for poly(p-fluorostyrene). [Reprinted with permission from Gerig, *Macromolecules* **16**, 1797 (1983). Copyright 1983 American Chemical Society.]

invaluable in certain situations where the pattern is obscured in the 1D spectrum. At other times sensitivity considerations would argue for collapsing these multiplets, thereby gaining at least a factor of two in intensity. In other cases, highly congested 2D spectra could have overlapping H/C correlations. Bax (159) has developed a method for collapsing these multiplets which essentially replaces the single 180°(X) pulse at the midpoint of the evolution time with the element $90°_x(H)—1/2J—180°_x(H)—180°_x(X)—1/2J—90°_{-x}(H)$ (Fig. 5-38). This is another application of the bilinear cluster proposed by Garbow, Weitekamp, and Pines (160) to invert selectively protons remote from the observe nucleus, but to leave unaffected directly bonded protons. Thus, all modulation of the finally detected X signal from remote protons as a function of the evolution time is refocused, and their effect is removed from the detected 2D shift correlation peaks. The resultant 2D shift correlation spectra are characterized by single peaks for X–H bonds. Slices in F_1 show proton singlets at the appropriate chemical shift. Of course, a projection onto the F_1 axis would give a proton "stick" spectrum (see Fig. 5-39 for a comparison of these two methods.)

5. APPLICATIONS OF 2D NMR EXPERIMENTS IN LIQUIDS

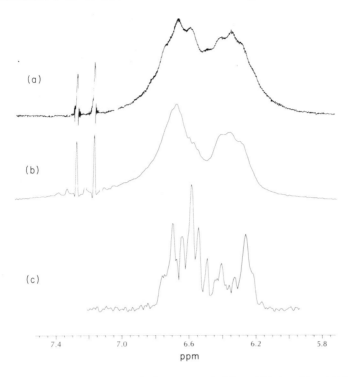

Fig. 5-37. Proton spectrum of poly(p-fluorostyrene): (a) at 100 MHz; (b) at 300 MHz; (c) from the projected H/F shift correlation data. [Reprinted with permission from Gerig, *Macromolecules* **16**, 1797 (1983). Copyright 1983 American Chemical Society.]

Rutar (161) and Wilde and Bolton (162) have subsequently reported related pulse sequences for suppression of H-H splittings in heteronuclear 2D shift correlation NMR. Rutar (161) illustrated the case for nonequivalent protons within a methylene group where there are two peaks, each split by the geminal H-H coupling. He also demonstrated the results for misset delays in the pulse sequence arising from a spread in J values. When J differs by more than 10–15%, artifacts appear flanking the multiplet-decoupled peak. These artifacts are, in fact, the heteronuclear 2D J spectrum for that particular X nucleus. Interestingly, this multiplet—split by the large one-bond X-H and smaller long-range H-H couplings—is not centered about the shift-correlation peak, but is displaced to high frequency by about one-half the one-bond XH coupling. These unwanted peaks can complicated spectral interpretation and decrease sensitivity. Wilde and Bolton (162) showed that, fortunately, these artifacts are canceled by proper phase cycling, leaving the desired singlets.

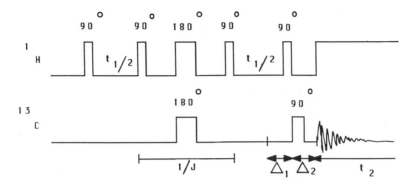

Fig. 5-38. Proton sequence of Bax (159) for heteronuclear shift correlation with decoupled multiplets; J is the one-bond coupling constant.

The polarization transfer step within the heteronuclear 2D shift correlation pulse sequence can be effected by a DEPT sequence, as described in Chapter 1 (163). The selectivity for CH, CH_2, and CH_3, for example, within the DEPT sequence, is made possible by the choice of flip angle in the final proton pulse. Linear combinations of different 2D data sets for different values of flip angle can produce CH-, CH_2-, and CH_3-only C/H correlation

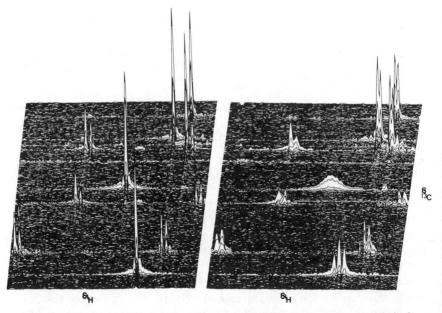

Fig. 5-39. A comparison for menthol of normal HETCOR (heteronuclear correlation) on the right and multiplet-decoupled HETCOR on the left.

spectra, thus simplifying crowded spectra. Levitt et al. (164) showed spectra (Fig. 5-40) for $CH + CH_3$ and for CH_2 which can be extracted from phase-sensitive data with the aid of a 135° proton pulse. Nakashima, John, and McClung (165) reported selective heteronuclear 2D DEPT C/H spectra obtained by using linear combinations of 2D data. Of course, a simple J-modulated spin-echo 1D spectrum or a series of DEPT spectra can usually suffice to establish, with high resolution, the number of protons attached to each carbon.

The DEPT analog of the multiplet-decoupled X/H correlation experiment has been used on 9-α-fluorocortisol by Wong, Rutar, and Wang (166), who observed the H-F and C-F couplings in the 2D spectrum. Since specific spin states are linked, they were able to relate the signs of the H-F and C-F couplings: these have the same sign if the doublets have the same "tilt" in both dimensions.

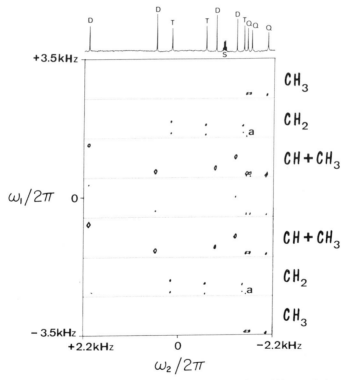

Fig. 5-40. Multiplet-separated carbon-hydrogen heteronuclear shift correlation data for menthol. The central region and peaks marked by "a" are artifacts resulting from phase inhomogeneities and spread in J values. [Reprinted with permission from Levitt, Sørensen, and Ernst, Chem. Phys. Lett. **94**, 540 (1983). Copyright 1983 North-Holland Physics Publishing.]

All of these techniques rely on a single coherence transfer to establish shift correlation. This usually involves X-nucleus detection, with its concomitant lower sensitivity relative to that of the proton. Bodenhausen and Ruben (167) have proposed a *two-step* process whereby magnetization is transferred from H to X, then evolves for a time as X magnetization, and finally is *reconverted* by a reverse magnetization transfer to the protons and detected with high sensitivity. Proper phase cycling ensures that only these transferred signals time-average coherently. The method requires X-nucleus decoupling to obtain highest sensitivity. The disadvantages of the technique are the large number of pulses required (10), exposing the spectral quality to pulse imperfections, and the T_2 attenuation of magnetization throughout the several delays. The same indirect technique was used to obtain the ^{199}Hg chemical shift in organomercurial phosphates by Vidnsek, Roberts, and Bodenhausen (168).

V. Shift Correlation by Relayed Coherence-Transfer 2D NMR

Any technique capable of relating at least three nuclei would permit establishment of *connectivity* patterns. These would be invaluable in establishing structural assignments. For example, if in a propyl group the methyl carbon has already been assigned, ^{1}H/^{13}C shift correlation permits assignment of the methyl protons. Any technique that permits assignment of the adjacent methylene protons could then permit subsequent assignment of the carbon to which they are attached via normal one-bond heteronuclear shift correlation. The key is the connection between the protons and how to achieve it. Once this is made, all carbons and protons can, in principle, be assigned and a molecular framework established. These capabilities have been realized by Eich, Bodenhausen, and Ernst (169) and Bolton and Bodenhausen (170), who reported, respectively, homonuclear and heteronuclear 2D techniques for relaying coherence between spins in a coupled network. These methods produce correlations not only with bonded nuclei in the heteronuclear case, but also, in extended versions, with next-nearest and more remote neighbors. Combined with normal ^{1}H/X shift correlation 2D NMR, this permits the desired connectivities to be made. Another approach to producing remote relay peaks is the TOCSY experiment of Braunschweiler and Ernst (86), described in Section IV,A.

A. Homonuclear Relayed Coherence Transfer

In the homonuclear case, the second 90° pulse of a homonuclear 2D shift correlation $90°_A$—t_1—$90°_B$ pulse sequence is replaced with a $90°_C$—τ—$180°$—τ—$90°_D$ sequence (Fig. 5-41). (The letters A to D simply label the position of each 90° pulse in the sequence.) The $90°_C$ pulse may

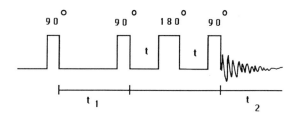

Fig. 5-41. Pulse sequence for RELAY. The periods t permit propagation of coherence to remote protons.

be thought of as "mixing" coherences or magnetizations obtained from the very first $90°_A$ pulse within the spin network. These coherences then evolve, with the 180° pulse serving to refocus multiplets. Finally, the $90°_D$ pulse monitors the extent of distribution of these magnetizations among the various nuclei in the spin system. Incrementation of t_1 leads to the normal 2D experiment. Any new correlations which appear, relative to the normal $90°-t_1-90°$ experiment, indicate relayed coherence, or neighboring *noncoupled* spins.

Wagner (171) has applied homonuclear RELAY to proteins, relying on the NH proton correlating with the beta proton of the amino acid in cases of alpha-proton degeneracy. He reports the use of a $90°-t_1-90°-t_1-180°-t_1-90°-t_2-180°-t_2-90°$ multiple-relay experiment which gives NH correlations all the way to gamma protons.

Figure 5-42 illustrates the differences between hononuclear COSY and RELAY for the case of polypeptides (171). The extra NH to beta-proton

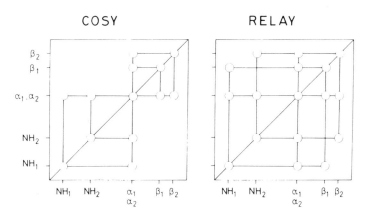

Fig. 5-42. Schematic representations of the correlations obtained from COSY and RELAY experiments. [From Wagner (171).]

correlations further establish the linkages within the individual amino acids. These were documented by Wagner (171) for 20 mM BPTI (basic pancreatic trypsin inhibitor) in water, as shown in Fig. 5-43. Both relayed intensities and direct J connectivities appear in the spectrum, but undesired axial and nuclear Overhauser-related transfer peaks can be eliminated by proper four-step phase cycling. The relayed peaks have positive absorption lineshapes, while diagonal and direct J cross peaks have mixed phases.

King and Wright (102) applied homonuclear RELAY to macromolecules, in particular copper(I) plastocyanin (molecular weight 10,600). As indicated in Fig. 5-44, new peaks appear in the RELAY data which are not present in the COSY spectrum. These come from the AA'BB'C spin systems of phenylalanine residues. The solid line connectivity is given for the C/BB' coupling.

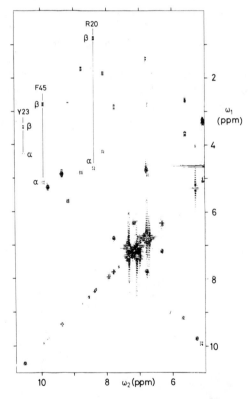

Fig. 5-43. Relayed coherence transfer data for 20 mM BPTI in D$_2$O at 360 MHz. Direct and relayed connectivities from both alpha and beta protons to the NH protons are given for three residues. [From Wagner (171).]

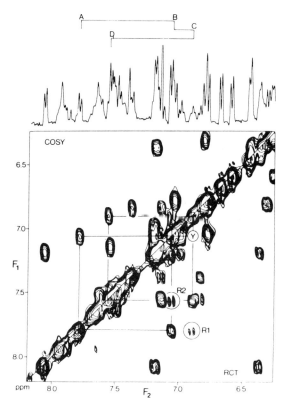

Fig. 5-44. Combined COSY and RELAY coherence transfer for carbonmonoxyleghemoglobin c_1. Two relayed peaks are identified. [From G. King and P. E. Wright, *J. Magn. Reson.* **54**, 328 (1983).]

B. Heteronuclear Relayed Coherence Transfer

The heteronuclear case uses the same basis sequence as the homonuclear case except for the additional 180° and 90° observe nucleus pulses as shown in Fig. 5-45. In this way the information concerning proton–proton couplings is transferred to the X observe nucleus. Thus one carbon, for example, could have two or three peaks in the RELAY 2D spectrum, one coming from its bonded proton and the other one or two from protons two bonds away. Since in most cases only vicinal proton–proton couplings are resolved, only protons on the neighboring atom provide relayed intensity, simplifying spectral analysis. Bolton (172) reported heteronuclear RELAY data for pentanol using a 16-transient phase cycle. A clear example of the value of the technique is illustrated in the HETCOR and RELAY data for a cyclic polypeptide in Fig. 5-46 (173).

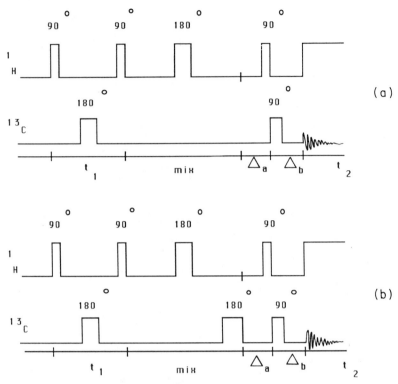

Fig. 5-45. Pulse sequences of (a) Bolton (172) and (b) Bax (174) for heteronuclear relayed coherence transfer. "Mix" is a delay for propagation of proton coherence to a neighboring proton, and Δ_a and Δ_b are the normal delays found in HETCOR for attainment of coherent spin vector phase before and after the polarization transfer.

One main virtue of the RELAY technique is that, as opposed to multiple-quantum 2D connectivity experiments such as INADEQUATE, the relay experiment obtains signals from molecules in which there is only *one* carbon-13. Apart from factors such as proton T_2 and relayed coherence-transfer efficiencies, the RELAY experiment should have a substantial sensitivity advantage. It should be noted, however, that practical experience indicates that the heteronuclear RELAY 2D experiment is significantly less sensitive than the ordinary ^{13}C/^{1}H chemical shift correlation experiment. Bax (174) has reported an improvement in the heteronuclear RELAY 2D experiment where the number of transients per cycle is reduced to four and the sensitivity is increased by as much as 40%. This can allow older spectrometers, with limited phase-cycling capability, to be used in RELAY 2D NMR techniques. An example obtained with this improved sequence is given in Fig. 5-47 for 2-acetonaphthalene (174).

RELAY can also be used in a form of H—X—H rather than the customary H—H—X. Applying this to diethylphosphite as a test case, Delsuc *et al.* (175) observed cross peaks between the methylene and phosphite protons, indicating a mutual coupling to the phosphorus; no cross peaks were noted when the phosphorus pulses were omitted. Similarly, Neuhaus *et al.* (176) saw cross peaks between noncoupled protons, each of which was coupled to the phosphorus in triethyl phosphate. This method can serve to match neighboring units within an oligonucleotide once one proton is assigned.

Heteronuclear RELAY has also been applied by Bigler, Ammann, and Richarz (177) to triglucoses, where they were able to assign all carbon resonances as well as prove the carbon connectivities in compounds for which the proton spectra are exceedingly complex and difficult to assign or interpret.

One of the problems associated with heteronuclear RELAY is that one-bond correlations contribute appreciable intensity to the final data, causing possible confusion and potential overlap. Kogler *et al.* (178) have attempted to solve this problem by the use of so-called "*J* filtering." Normal heteronuclear RELAY and their modified version are shown in Fig. 5-48. The low-pass *J*-filtering stage converts the proton magnetization created by the first 90°(H) pulse into unobservable heteronuclear two-spin coherence, with progressively better stages of conversion as additional carbon-13 pulses are applied. Remote protons are essentially unaffected since the interpulse delays are so short. The neighbor proton magnetization is reduced to a value of 2%, while only 13% of the remote proton magnetization is lost when three-stage filtering is used, with interpulse delays of 3.79, 2.87, and 2.30 msec. Thus, at the beginning of the proton evolution time the only surviving proton magnetization is that belonging to protons not bonded to carbon-13. A normal heteronuclear RELAY experiment then follows (Fig. 5-48). Comparative data for both experiments are presented in Fig. 5-49. *J* filtering is effective for one-bond *J*'s from 125 to 225 Hz with suppressions of up to 99%. Care must be taken, however, in situations where proton T_2 values are short, as is typical of large molecules, since approximately 8 msec is added to the duration of the pulse sequence.

Bolton has extended multiple-quantum techniques to the RELAY experiment (179). This overcomes the problem encountered when neighbor and remote protons are degenerate or nearly so. Similar information has been obtained by Sørensen and Ernst (180) using "pseudo-multiple-quantum" techniques. After an initial 90° pulse on the protons, another 90° proton pulse halfway through the evolution time generates sum and difference frequencies for the coupled protons (these may later be sorted out by proper phase cycling into pseudo zero- and double-quantum data). The resulting magnetizations are then transferred by DEPT cycle to the directly connected

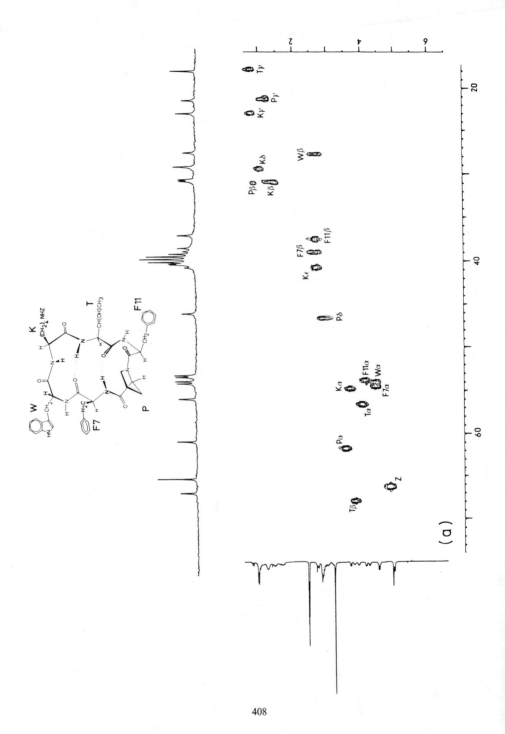

408

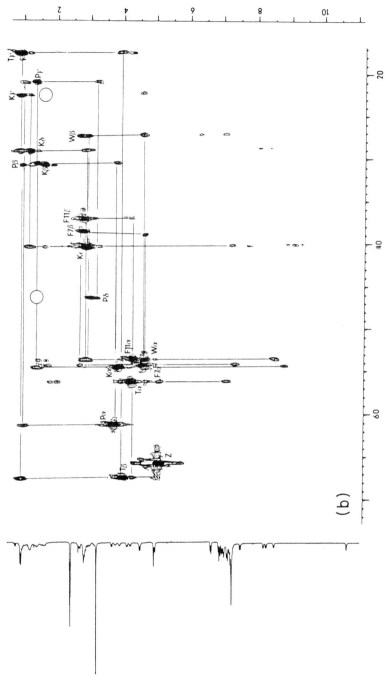

Fig. 5-46. Heteronuclear correlation for a cyclic polypeptide. (a) Normal spectrum. (b) Relayed spectrum. [Reprinted with permission from Kessler et al., J. Am. Chem. Soc. 105, 6944 (1983). Copyright 1983 American Chemical Society.]

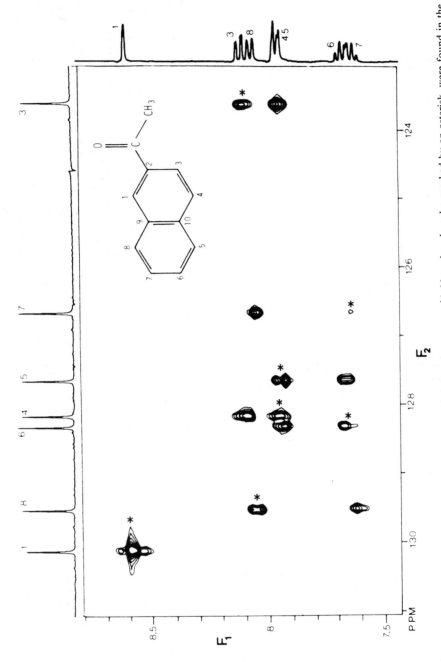

Fig. 5-47. Relayed coherence spectrum for 2-acetonaphthalene in acetone-$d6$. Nonrelayed peaks, marked by an asterisk, were found in the normal HETCOR spectrum. [From Bax (174).]

5. APPLICATIONS OF 2D NMR EXPERIMENTS IN LIQUIDS

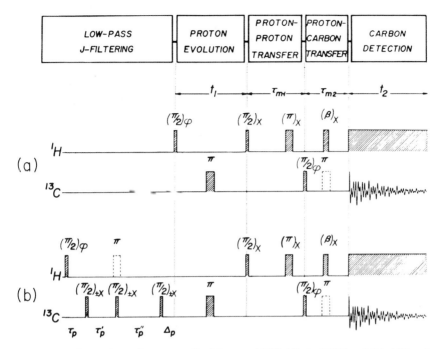

Fig. 5-48. Comparison of sequences for (a) normal RELAY via DEPT and (b) J-filtered relayed coherence transfer. [From Kogler *et al.* (178).]

carbons. No fixed delay to rephase homonuclear couplings is necessary. Signals for both neighbor and remote protons are present for each carbon, as in the basic RELAY experiment.

VI. Magnetization-Transfer 2D NMR

Experiments which can probe intramolecular interactions, proximity of nuclei, and chemical exchange have wide utility in studies of both structure and dynamics. The 2D experiments which have been described above include *transverse* (x, y) magnetization transfer, while those below are experiments wherein *longitudinal* (z) magnetization is transferred. The nuclear Overhauser effect is one example of population redistribution via an incoherent process. Other types include saturation transfer experiments where the mechanism is chemical exchange.

A. HOMONUCLEAR MAGNETIZATION TRANSFER

Lineshape analysis at or near coalescence, as well as selective saturation or inversion with subsequent measurements of propagation of magnetization

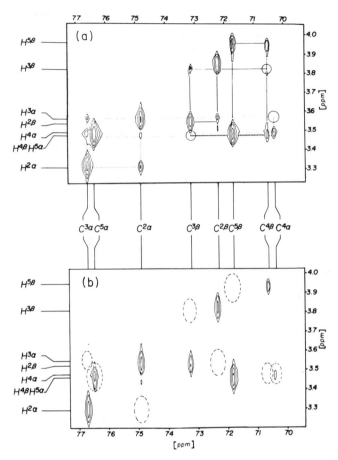

Fig. 5-49. Partial heteronuclear relayed coherence transfer spectra of glucose with alpha and beta anomers. (a) Normal RELAY spectrum. (b) With neighbor peak suppression using a series of three filtering pulse sequences. Some relayed peaks are missing because of mismatch in the mixing interval. [From Kogler *et al.* (178).]

throughout the spin system, has been used for studying exchange. The first technique can be very difficult for broad coalescence peaks and requires certain temperatures to be established. The latter is very useful for a limited number of lines but becomes time-consuming for more than a few lines and very difficult for closely spaced lines.

A general method requiring no particular special conditions would have a great deal of attractiveness for studies of chemical and biochemical dynamics. A class of 2D experiments fulfilling this goal was proposed by Jeener *et al.* (181) and Meier and Ernst (182) and elaborated by Macura and Ernst (183). The basic pulse sequence is $90°—t_1—90°—t_{mix}—90°$. The second 90° pulse can be thought of as restoring to the z axis transverse magnetization created by the first 90° pulse. The component and direction of the resulting z-axis magnetization are functions of the precession of the spins during t_1. Some components will be positive, some negative, and some zero, and the components oscillate as a function of t_1. The resultant z magnetizations will be in a nonequilibrium state after t_1, and spin–lattice relaxation processes will attempt restoration of equilibrium. Since relaxation of a proton is primarily dipolar and via other protons, this nonequilibrium magnetization will then be redistributed by mutual spin flip to other protons. The final 90° pulse monitors the extent of magnetization transfer. The 2D experiment samples all degrees of magnetization transfer as it increments t_1.

The 2D shift-correlation spectrum is characterized by the usual diagonal peaks coming from magnetization remaining at the same frequency in t_1 and t_2. The same phase cycling used in homonuclear shift correlation can remove the axial peaks. The off-diagonal intensities arise from longitudinal magnetization which has been transferred from one type of spin to another. This might be from true chemical exchange occurring during the mixing time, or it could be simply from mutual dipolar relaxation. This sequence has been used to indicate NOEs in polypeptides and proteins (184, 185). It is effective in macromolecules, where there is much slower molecular tumbling and more favorable NOE. Changing t_{mix} allows examination of the *rates* of magnetization transfer, and thus proximity of protons from NOE or the kinetics of chemical exchange processes (186).

Misleading cross peaks may occur in the 2D data for this experiment. These arise via normal homonuclear correlation and are generated from the last two pulses in the sequence (187). This can make interpretation unreliable or cause unnecessary repetition of the experiment as a function of t_{mix}. Macura, Wüthrich, and Ernst (188) have proposed a technique which takes advantage of the difference in behavior between magnetization transfer, which is incoherent in nature, and the interfering zero- and multiple-quantum processes, which are coherent. The last two oscillate in time during t_{mix}, while the NOE or chemical exchange cross peaks grow and decay exponentially. Since the latter decay away slowly, a variation of the mix period through a range of values during the 2D experiment will force the oscillatory coherences to appear at positions displaced from their normal positions along the F_1 direction, destroying the symmetry usually associated with the homonuclear correlation spectrum. Symmetrization about the diagonal will retain only symmetric data, thus eliminating J-correlation cross peaks. Choice of proper phase cycling (187) and use of a homospoil pulse in the mix period are also useful techniques to reduce undesirable zero- and multiple-quantum coherence transfers.

In some situations it may be desirable to keep a fixed mixing time. Macura, Wüthrich, and Ernst (189) have modified the basic $90°$—t_1—$90°$—t_{mix}—$90°$ sequence by inserting a $180°$ pulse in the t_{mix} period. Its position is a function of t_1 and occurs at a time $(t_{mix} + kt_1)/2$. As t_1 increases, the $180°$ pulse moves from half-way through the mixing period to some time later. As a result, zero- and multiple-quantum J-coherence-transfer peaks are shifted in F_1, and symmetrization eliminates them efficiently. A more detailed analysis of selection of time intervals for efficient filtering of zero-quantum contributions has been made by Rance et al. (190).

Bodenhausen et al. (191) have examined in depth the exchange 2D experiment to assess errors by missetting of $90°$ pulse values and/or rf field inhomogeneities leading to a spread in effective flip angle. New signals appear in the 2D spectrum which arise from dipolar order and are produced by product operator terms of type $2I_{kz}I_{lz}$. These "zz" signals can be differentiated from the genuine exchange signals arising from I_{kz} terms in the mixing period by exploiting their different behavior when a $180°$ pulse is placed in the mixing period: "zz" signals appear as antiphase multiplets and give essentially the equivalent of a double-quantum filtered COSY spectrum. The key to the differentiation is that the bilinear $2I_{kz}I_{lz}$ terms are invariant under the action of a $180°$ pulse, while the I_{kz} terms arising from the magnetization transfer in either chemical exchange or NOE are affected. Thus, by subtracting data sets accumulated with and without a $180°$ pulse at the beginning of the mixing period, the "zz" signals can be canceled. Alternatively, a spectrum with only "zz" signals may be obtained by adding the two data sets. This can be accomplished either by accumulating two separate data sets or by modifying the pulse sequence to automatically insert the $180°$ pulse for half of the transients with appropriate summing to achieve the desired effect. If "zz" signals are desired, the second and third normally $90°$ pulses should be set to $45°$. Suppression of "zz" signals is further enhanced by using composite pulses for the second and third pulses.

Some of the more interesting problems of chemical exchange arise in biochemical areas, particularly proton exchange in water. Schwart and Cutnell (192) have used the 2D-NOE approach to study NH exchange with solvent protons in water. Here the large solvent resonance makes the normal "hard-pulse" version inapplicable because of the 10^5 dynamic range between solute and solvent protons. These workers replaced the final monitoring $90°$ pulse with a long soft pulse which produces a null in excitation at the water resonance. Because the last pulse suppresses the water signal, cross peaks corresponding to magnetization transfer between NH do not appear. Of course, this means that symmetrization techniques will fail if used here. Since the cross peaks for *intermolecular* chemical exchange cannot arise from J correlations, the lack of symmetrization is not critical as in NOE

2D studies where spin-coupled nuclei may also participate in cross relaxation.

As molecular systems become more complex and as the chemical shift difference between mutually relaxed or exchanging spins gets small, the presence of the large *diagonal* peaks in the 2D plot becomes more of a problem. Bodenhausen and Ernst (193) have developed difference 2D experiments which do cancel the diagonal ridge, leaving the cross peaks unaffected. The 2D experiment is first performed in the normal manner (A), then with zero mixing time (B), and finally in one of two possible ways: (C) all nuclei are saturated by a train of 90° pulses followed by a mix period and then (B); or (D) all nuclei are inverted by a 180° pulse followed by a mix period and then (B). The saturation-recovery difference technique requires combination of signals as $A - B + C$, which gives a vanishing diagonal. The inversion-recovery difference technique combines the data as $A - \frac{1}{2}B + \frac{1}{2}D$, again nulling the diagonal. These techniques could be combined into one pulse sequence to produce the diagonal-less spectrum directly, given a sufficiently complex pulse programmer, since normally eight-transient phase cycling and incremented mix times are also necessary to eliminate J cross peaks (187).

Another combination experiment is a combined COSY-NOESY pulse sequence in which the normal homonuclear shift correlation acquisition is followed by a third 90° pulse and a separate acquisition (194, 195). In this procedure, the first acquisition period serves as the mixing period. This places a minimum on the length of the mixing period and could rule out the use of the above techniques to discriminate against J-correlation peaks. If the mixing period is short enough, however, an additional delay can be placed prior to the last pulse and used as above to discriminate against J cross peaks in the NOESY portion of the data. The technique offers an economy of time, since the full NOESY data set can be acquired with only a small percentage of extra time.

The need to maximize efficiency in determination of exchange rates has led Bodenhausen and Ernst to invent the "accordion" approach (196, 197). The basic pulse sequence is presented in Fig. 5-50. The mixing time is equal to kt_1 and is incremented regularly along with t_1. The constant k ranges from 10 to 100. Information on rates is present in the *lineshapes* of the cross peaks. The phase-sensitive F_1 lineshape of a cross peak can be subjected to direct line-shape analysis as a superposition of Lorentzian lines, each producing a rate constant determining its width. Alternately, the lineshape can be subjected to a reverse Fourier transformation for the same purpose. Another technique is to take sums and differences of two F_1 slices at the two chemical shifts for exchanging spins. These linear combinations produce pure Lorentzians. The linewidths at half height of the signals in the sum

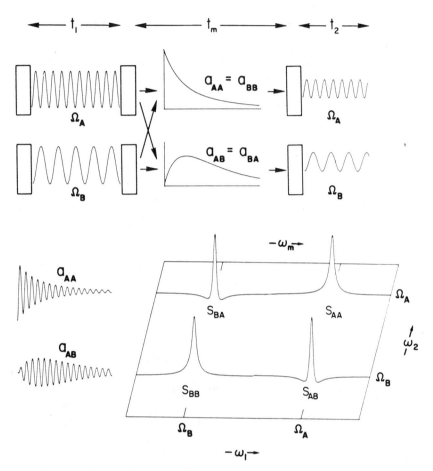

Fig. 5-50. The "accordion" pulse sequence (top) with lineshape functions appropriate for two-site exchange. The mixing time is not constant but is incremented in proportion to the evolution time. The individual magnetizations for each site are indicated by oscillatory signals, while longitudinal magnetization is modified in the mixing period by the mixing functions, which show magnetization staying on one site or being transferred, either A to B or B to A. Interferograms at frequencies for A and B give composite lineshapes where the cross-peak shape is determined by the exchange rate. [From Bodenhausen and Ernst (196).]

are direct measures of the relaxation rate. The corresponding difference linewidths can be used to calculate directly the chemical exchange rate constants. This last method can be readily applied in most spectrometers by using the normal add–subtract software.

"Accordion" methods work well in heteronuclear systems but have problems in homonuclear J-coupled cases, due to zero-quantum effects, as indicated above. Bremer, Mendz, and Moore (198) have developed a technique they SKEWSY, or skewed exchange spectroscopy, which overcomes the zero-quantum interfering signals and allows direct calculations of rates of internuclear magnetization transfer. They precede the normal three-pulse sequence by a preparation period in which a 180° pulse inverts the magnetization and the spins are allowed to precess and exchange for a fixed time, usually of the same order as the mixing period prior to detection. The mixing period may be randomized to eliminate zero-quantum coherence as above. The resulting 2D data are processed in the same way, but the exchange cross peaks are asymmetrical about the diagonal. No symmetrization is permitted, since this difference in intensity is the object of the analysis. No lineshape analysis is required for the method, only measurement of crosspeak intensities. Mathematical analysis involving simple matrix manipulations produces a rate-constant matrix for the magnetization transfer process. Errors may be minimized by repeating the experiment with a variety of preparation exchange periods during which the mixing time is held constant; alternately, an accordion-like approach could be used, incrementing the preparation time as a function of t_1, with subsequent deconvolution of cross sections in F_1.

Finally, Wagner (199) has published a method which combines J-related coherence transfer with incoherent magnetization transfer in a sequence which then has the feature of permitting the transfer of NOE cross peaks from a crowded region to an uncrowded part of the 2D spectrum, that is, 2D relayed coherence-transfer-NOE spectroscopy. This can be accomplished in either of two ways, the first involving magnetization transfer followed by coherence transfer, and the second, the reverse order. The sequences can be thought of as modifying the basic 90°—t_1—90°—mix—90°—t_2 pulse train by inserting a 90°—τ—180°—τ— segment either prior to the mixing period (the former) or after the mixing period (the latter). Figure 5-51 shows relayed NOESY data for 20 mM BPTI in H_2O. Peaks labeled $R_{N,J}$ show magnetization first propagated via dipolar interactions and then transferred via J coupling to another spin. Peaks labeled N_J and N_N refer to correlations to neighboring spins via only coherent or incoherent processes. All N_N peaks are strongly suppressed relative to normal NOESY data. The value of τ may be chosen to favor specific ranges of coupling constants, further enhancing the ability to target specific residues in a protein. Macura, Kumar, and Brown (200) obtained similar data but were able to achieve superior solvent suppression by using a homonuclear relayed *double-quantum* 2D experiment.

Nonexchangeable proton resonances in d(C-G-C-G-A-A-T-T-C-G-C-G) in D_2O have been examined using exchange 2D spectra by Hare *et al.* (201), leading to unambiguous assignment of all but two of the protons

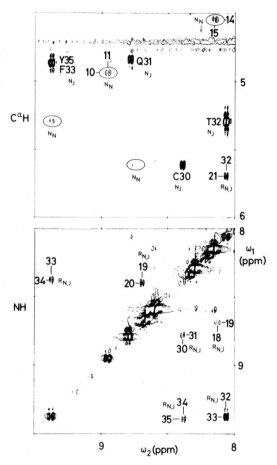

Fig. 5-51. Relayed NOESY data for BPTI in H_2O at 36°. The upper section shows partial data illustrating remote (R) and neighboring (N) connectivities for NH and alpha protons. All neighboring magnetization transfer correlations (N_N, weak peaks enclosed in circles) are strongly suppressed, while connectivities progressing via J coupling (N_J) and ($R_{N,J}$) are clear. [From Wagner (199).]

in the molecule and to the finding that the helix is right-handed and similar to beta-DNA. Van de Ven, de Bruin, and Hilbers (202) applied the same experiment to ribosomal protein E-L30, using mixing times of 100–150 msec, and were able to infer the existence of one or more hydrophobic domains, at least two helical regions, and the presence of an antiparallel beta sheet involving two threonines. Wu, Westler, and Markley (203) directed the technique to the assignment of imidazolium NH resonances in the spectrum of turkey ovomuccoid protein. The use of exchange 2D experiments allowed confirmation of isomer interconversion in a pentapeptide, which exists in several isomeric forms because of the presence of two prolines (130). Here, NH protons from alanine and leucine appear in the form of major (60%) and minor (30%) isomers, and the remainder have nonresolved shifts. The 2D NOE spectrum in Fig. 5-24 revealed cross peaks between major and minor leucine NH protons, as well as cross peaks for the alanine major and minor isomers. Willem *et al.* (204) used this method to make rotamer assignments and a motional study of tetra-*o*-tolylcyclopentadienone.

Two-dimensional ^{31}P magnetization transfer techniques have been applied to the study of metabolism in rats by Garlick and Turner (205), and by Balaban, Kantor, and Ferretti (206), in particular to the enzyme-catalyzed exchange of the terminal phosphate in ATP with creatine phosphate interconversion. Boyde *et al.* (207) compared this latter 2D approach to one-dimensional saturation-transfer techniques, particularly in regard to *in vivo* analysis.

B. Heteronuclear Magnetization Transfer

Carbon-13 chemical exchange networks have been explored by Huang, Macura, and Ernst (208) using the techniques described above. Complications from J correlations are absent from these natural-abundance spectra. Normal $90°$—t_1—$90°$—t_{mix}—$90°$ sequences were used, in the presence of proton decoupling, which was applied to provide NOE, sensitivity, and simplicity. Refocused INEPT was used in place of the first 90° pulse to prepare the initial magnetization, in order to gain additional sensitivity. This extension suggests future flexibility and selectivity, since the INEPT portion of the sequence can be tailored for specific J magnitudes (long-range or direct), thus allowing very precise control over the site from which magnetization can evolve. Huang *et al.* (208) applied 2D strategies to the classic exchange problems of ring puckering in decalin, bond shift in bullvalene, and solvation-shell exchange in aluminum complexes. The longer relaxation times of carbon-13 can actually be put to advantage in these studies, since they permit longer mixing times for slower exchange processes. The other major advantage of the 2D technique is that it can be performed on a very slowly exchanging system whose lines are still narrow.

Rinaldi (209) has described a heteronuclear version of the 2D magnetization transfer experiment. The $90°$—t_1—$90°$—t_{mix} portion of the normal sequence is retained, with an observe nucleus 180° pulse midway through t_1 to decouple the X and H spins. Cross relaxation between X and H occurs during the mixing period; this is followed by a $90°(X)$ pulse, and the magnetization is sampled under full proton decoupling. The signal is modulated according to the amount of cross relaxation occurring during the mixing period. Cross peaks in the 2D spectrum reflect the existence of this cross relaxation, which is typically dipolar in nature. Note that J coupling is not necessary between the correlated nuclei. Existence of cross peaks in stereochemically rigid systems, particularly for quaternary carbons, can permit local neighborhoods to be mapped via the presence of dipolar relaxation. Rinaldi pointed out the value of such an experiment for probing metal ion binding sites in complexes with biological or organic ligands.

Yu and Levy (210, 211) have also reported two-dimensional heteronuclear NOE experiments. They applied the same pulse sequence to monitor the dipolar interactions between quaternary carbons and nearby protons in camphor and fluoranthene, and between phosphorus and protons in ATP.

VII. Zero- and Multiple-Quantum 2D Methods

A. Zero-Quantum 2D Experiments

The above correlation methods rely on single-quantum coherence or magnetization transfer. Bolton (212) has modified the heteronuclear shift correlation experiment to detect only zero-quantum coherence transfer from protons to the observe nucleus. The pulse sequence $90°(H)$—τ—$180°(H)180°(X)$—τ—$90°(H)90°(X)$—t_1—$90°(H)$ has phase cycling so that the X signals are modulated only by the zero-quantum coherence as a function of t_1. Proton–proton couplings are present in the resultant 2D spectra, and slices perpendicular to F_2 can give proton spectra of those protons coupled to X. Linewidths are predicted to be only 60% of those obtained in the conventional heteronuclear 2D shift correlation experiment, and this is useful for extracting the proton–proton fine structure. The frequency of this modulation is the difference between the frequency offset of the proton from the proton transmitter and the frequency offset of the observe nucleus from the observe transmitter. The advantages of zero-quantum spectroscopy are the weaker dependence of F_1 multiplets on field inhomogeneity, and the absence of the heteronuclear X–H coupling in the F_1 domain when there is only one proton coupled to the observe nucleus. Experimental $^{31}P/^1H$ zero-quantum correlation 2D data show well-resolved H–H coupling patterns for cytidine 3′-phosphate and phosphothreonine (Fig. 5-52).

Müller (213) has proposed a homonuclear shift correlation technique which establishes connectivities. The standard $90°$—τ—$180°$—τ—$90°$ pulse sequence normally used to excite double-quantum coherence does not efficiently excite zero-quantum coherence in a weakly coupled spin system. The $90°$—τ—$90°$ sequence does excite zero-quantum coherence, but the amplitudes of the transitions are dependent on both J and chemical shift.

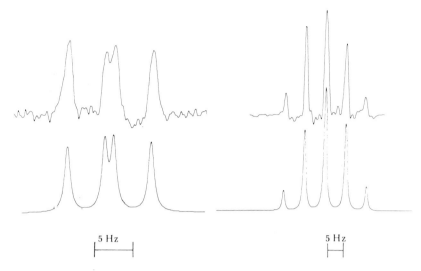

Fig. 5-52. Heteronuclear zero-quantum correlations for (top left) cytidine 3′-phosphate and (top right) phosphothreonine. The bottom simulations are of phosphorus-decoupled proton spectra. [From P. H. Bolton, *J. Magn. Reson.* **52**, 326 (1983).]

Selection of the value of τ requires a prior knowledge of the NMR spectrum. Müller proposed a 90°—τ—180°—τ—45° sequence to excite zero-quantum coherence independent of chemical shift. Proper phase cycling suppresses uncoupled signals, and the zero-quantum coherence is converted back into observable magnetization after the evolution time by a final pulse of either 45° or 90°. Use of the 90° value generates spectra similar to 2D INADEQUATE, while use of a 45° value produces spectra similar to SECSY.

Bolton (214) has also reported zero-quantum homonuclear correlation work where he was able to reduce substantially the normally large $F_1 = 0$ signals arising from longitudinal magnetization following the last pulse in the 90°—τ—90°—t_1—90° pulse sequence. No phase cycle is capable of distinguishing between the contributions from this longitudinal magnetization and the zero-quantum coherence when the final pulse is 90°. Bolton recognized that flip angles of 45° and 135° for the final pulse give zero-quantum signals having inversion symmetry about $F_1 = 0$, but that longitudinal magnetization-induced signals have constant phase. Therefore subtracting data accumulated with both phase settings cancels the latter while retaining the former, a procedure Bolton refers to as "flip-angle filtering." Two separate 2D data sets may be accumulated and subsequently subtracted, or the pulse sequence may be executed with the final pulse flip angle changed for half the transients, giving one data set. Zero-quantum coherence may be obtained more reliably by use of a variable mixing time.

B. Double-Quantum 2D Experiments

Bax, Freeman, and Kempsell (215, 216) have defined a technique relying on selectively detecting double-quantum coherence in order to establish direct carbon–carbon connectivity. This experiment suppresses resonances from isolated ^{13}C nuclei while retaining the signals from molecules with two neighboring labels (the 1D version of this technique was named INADEQUATE.) The basic pulse sequence of $90°—\tau—180°—\tau—90°—90°$ uses the first two 90° pulses to generate double-quantum coherence. This evolves for a short time and is reconverted into single-quantum magnetization by the last 90° pulse. Only double-quantum related signals survive the 32-transient phase cycling in the experiment. Suppression of up to a factor of 1000 are realized. The experiment is very demanding on sensitivity and spectrometer stability, since the signals are from pairs of carbon-13 nuclei, representing 0.01% in natural abundance, and are further split by CC coupling. Still, the information is direct and can unequivocally link carbon atoms. The two-dimensional version of this technique treats the period between the first two 90° pulses as the evolution period. Slices in the resultant F_1 dimension at the carbon chemical shifts in F_2 give J spectra with inhomogeneities reduced just as in the analogous J-resolved homonuclear and heteronuclear experiments described earlier. Phase-sensitive plots give very highly resolved multiplets assignable to a precision of ±0.03 Hz or better.

Bax, Freeman, and Frenkiel (217) proposed using the same pulse sequence but treating the period between the last two 90° pulses as the evolution period. The double-quantum coherence precesses during this period with a frequency characteristic of the *sum* of the chemical shifts of coupled CC pairs. The F_1 axis becomes a double-quantum frequency axis in the 2D data with both carbons carrying the modulation in t_1 and thus having intensity at the same F_1 value. This coincidence of peaks at a common value of F_1 is direct proof of the existence of a coupling between the carbons. The size of the coupling can readily differentiate between bonded carbons and long-range coupled carbons; the period between the first two 90° pulses is normally set to produce optimum double-quantum coherence for bonded carbons.

Data from this experiment are most conveniently plotted as a F_1/F_2 contour intensity plot. The F_1 signals are comprised of two doublets symmetrically disposed about a line of slope 2 in frequency units. This is of great help in assigning peaks in noisy spectra, since the constraints are several and the F_2 values are already available from the 1D spectrum. A powerful property of this general type of experiment is the ability to determine the carbon–carbon connectivity pattern in the molecular backbone by beginning at some F_2 chemical shift value, finding any intensities along F_1 at this F_2 value, and pairing up these intensities with any corresponding intensities sharing the same F_1 value. This establishes all carbon connectivities of the carbon which resonates at this value of F_2. Extending the process establishes further connectivities. An example of this is given in Fig. 5-53 for the organic molecule cedrol (218).

Bax et al. (219) improved the method by introducing a technique permitting quadrature detection in both dimensions. This involved taking half the data in the normal manner but inserting a novel 45_z° composite pulse in the evolution period for the other half of the data. This induces a 90° phase shift in the double-quantum coherence, permitting quadrature detection. Mareci and Freeman (220) achieved quadrature detection in F_1 by the simpler method of increasing the last pulse in the INADEQUATE sequence to 135°, which results in selective detection of the coherence transfer echo. Antiecho responses are greatly attenuated and the spectral width in F_1 can be reduced. The inherent low sensitivity of this experiment has been improved by including polarization-transfer techniques such as INEPT prior to the INADEQUATE part of the sequence (221).

Turner has proposed a somewhat different approach to detecting carbon–carbon connectivities (222-224). Instead of the above pulse sequence,

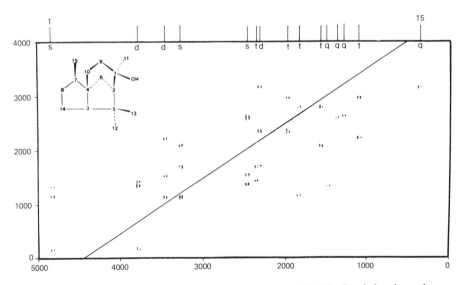

Fig. 5-53. Carbon–carbon connectivity plot for cedrol at 75 MHz. Bonded carbons share common values of double-quantum coherence (verrtical axis). The scale is in hertz. Chemical shift positions and protonation type are given at the top. [Reprinted with permission from Shoolery, *J. Nat. Prod.* **47**, 226 (1984).]

Turner proposes a Hahn echo method involving a $90°—\tau—180°—\tau—90°—90°—t_1/2—90°—t_1/2—t_2$ pulse sequence. The resulting spectrum is similar in format to the homonuclear shift correlation spectrum with a principal diagonal about which correlated (bonded) carbons are symmetrically displaced. There are no diagonal peaks, however—these are suppressed by phase cycling. Turner has analyzed sensitivity considerations (223, 224), comparing the various 2D methods of obtaining carbon–carbon connectivity information, but several of his conclusions have been challenged by Bax and Mareci (225).

Robinson and Turner (226) have reported carbon–carbon connectivities using a 2D method for monensin sodium salt. Other studies include those on a trimer of biacetyl (227), a biosynthetically ^{13}C-labeled riboflavin (228), and an organochromium complex (229). Buchbauer et al. (140) used 2D INADEQUATE in assigning the proton and carbon signals in tricyclo(7,3,1,0[2,7])tridecane, while Berger and Zeller (230) used it to complete the CC coupling matrix in azulene.

Levitt and Ernst (231) have applied the methods of composite (or sandwich) pulses (27, 232–235) to the INADEQUATE experiment. They found this modification to be essential for optimum double-quantum excitation, principally to correct for off-resonance effects.

The 2D version of INADEQUATE can be applied to proton NMR. This is complicated by the presence of three or more coupled spins as opposed to the simpler AX or AB case for ^{13}C in natural abundance. Mareci and Freeman (236) showed that the 2D spectrum can be simplified by using a value between 90° and 180° for the flip angle of the last pulse. As in INADEQUATE, the members of a pair of coupled protons have intensity at the same double-quantum frequency, symmetrically displaced about a diagonal of slope 2. There are no strong diagonal peaks as there are in COSY spectra to complicate proton–proton correlations for protons of similar shift.

Boyd et al. (237) applied the technique to proteins, with particular emphasis on the characterization of tyrosine, tryptophan, and phenylalanine residues. Kessler et al. (128) used spin filtering to obtain a homonuclear 2D J spectrum in which the singlets were suppressed. Intensity distortions and difficulties associated with variation of excitation efficiency and J were avoided by using the symmetric excitation/detection cycle of Sørensen, Levitt, and Ernst (238).

Two-dimensional deuterium double-quantum spectra were obtained by Ramachandran et al. (239). These spectra were of solute molecules in an ordered phase and exhibited quadrupolar splittings along F_2 and both homo- and heteronuclear interactions along F_1.

C. HIGHER-ORDER MULTIPLE-QUANTUM METHODS

Multiple-quantum filtering is the use of multiple-quantum coherence selection for the simplification of correlation data (240). Proper selection of the *order* of the multiple quantum coherence (MQC) present between the evolution and detection periods can eliminate singlets and higher-order multiplets if desired. The pulse sequence $90°—t_1—90°—90°—t_2$ has phase cycling programmed for the desired order of MQC present between the last two 90° pulses. Singlets cannot produce MQC and thus are canceled. Diagonal peaks are also greatly attenuated. Higher-order filtering can allow selection of certain spin systems with other overlapping resonances canceling. This can be of great value in macromolecular spectra.

Rance *et al.* (241) applied double-quantum filtering to a phased homonuclear chemical shift correlation 2D experiment. This allowed elimination of the strong water singlet and the dispersive character of the diagonal peaks, permitting identification of cross peaks lying immediately adjacent to the diagonal. An illustration of this experiment is given in Fig. 5-54 where the strong singlet from the benzene solvent is greatly reduced in the DQF spectrum, leaving the multiplets coming from the spin system in the 2,3-dibromothiophene and acrylonitrile.

Spin filtering has also been separately reported by Shaka and Freeman (242). When this was applied as four-quantum filtering, all peaks with

CONVENTIONAL COSY

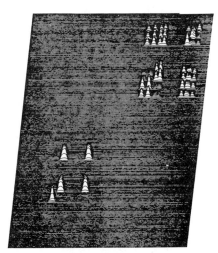
DOUBLE-QUANTUM FILTERED COSY

Fig. 5-54. Comparison of normal COSY (left) and double-quantum-filtered COSY (right).

homonuclear 2D shift-correlated spectrum involving two- and three-spin systems were eliminated. Phase shifters capable of intermediate values were necessary for this version of spin filtering. One-dimensional spectra using multiple-quantum filtering were reported in both versions as well, the same pulse sequences with fixed time being used. The resultant 1D spectra have lower-order spin patterns eliminated.

D. Indirect Detection of Rare Spins Using 2D NMR

While polarization-transfer techniques in both 1D and 2D NMR have been used commonly for sensitivity enhancement, it has usually involved rare-spin detection—for reasons of convenience and absence of a large background signal from the abundant spins not coupled to the rare spin. Normal heteronuclear chemical shift correlation experiments are of this type, even though the normal objective is shift correlation, not signal detection. Detection of rare spins indirectly, via the abundant spin, can

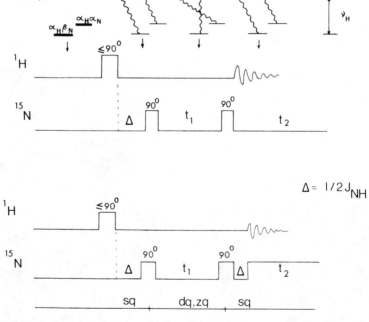

Fig. 5-55. Indirect detection via multiple quantum coherence transfer. Double- and single-quantum coherences are established by the first nitrogen-15 pulse. These evolve during t_1 and are sampled by the monitoring 90° pulse. Addition of a $1/2J$ delay prior to detection allows decoupling of the ^{15}N for sensitivity enhancement. [From Bax, Griffey, and Hawkins (245).]

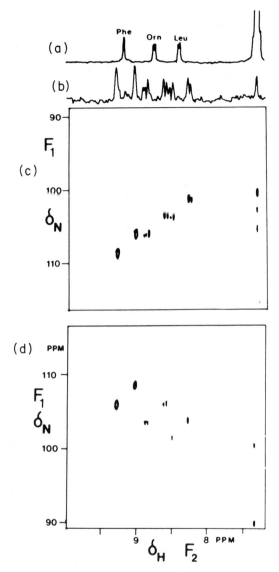

Fig. 5-56. Detection of natural abundance nitrogen-15 and correlation to attached protons using zero- and double-quantum heteronuclear shift correlation. (a) Normal proton spectrum of 18 mM gramacidin S in DMSO. (b) Projection of 2D spectrum showing only protons coupled to the natural-abundance ^{15}N. (c) The 2D plot calculated from the double-quantum signals. (d) The 2D plot calculated from the zero-quantum signals. Both (c) and (d) are derived from the same data set. Acquisition time is three hours. [Reprinted with permission from Bax, Griffey, and Hawkins, *J. Am. Chem. Soc.* **105**, 7188 (1983). Copyright 1983 American Chemical Society.]

bring large sensitivity gains, particularly when the abundant spin has a large magnetogyric ratio. This experiment is much more demanding since the large background signal must be canceled to reveal signal from protons coupled to the rare spin.

Müller (243) made early proposals for this type of indirect detection wherein he developed pulse sequences which created higher-order multiple-quantum coherences which were then observed in the abundant spin after conversion to single-quantum signals.

Bax et al. (224, 245) and Griffey et al. (246) have developed a technique for obtaining heteronuclear shift correlations via zero-quantum and double-quantum coherence in which protons are detected in order to provide higher sensitivity. The first part of the sequence (Fig. 5-55) converts all longitudinal proton magnetization coupled to an X spin into zero- and double-quantum coherences. These are converted back into transverse magnetization following an evolution period. Fewer pulses are involved than in the double INEPT technique of Bodenhausen and Ruben (167), and proper phase cycling can allow use of the technique in protonated solvents. The high sensitivity of the experiment allowed ^{15}N-^{1}H shift correlations to be obtained on 18 mM solutions in 5-mm tubes in three hours (Fig. 5-56). The value of this experiment can be seen in Fig. 5-57, from which Foxall and Zens (247) assigned individual ^{15}N chemical shifts in 1 mM tRNA(lysine) using uniformly enriched [^{15}N]uridine bases. Similar N-H shift correlations have been reported by Redfield (248), Roy et al. (249), and Live et al. (250).

Fig. 5-57. Nitrogen-15-proton multiple quantum shift correlation for tRNA having all uridines and uridine derivatives uniformly enriched in ^{15}N. Solvent is 10% D_2O–90% H_2O; concentration is 1 mM; 5-mm sample tube; total time, 17.5 h. [D. Foxall and A. Zens (private communication).]

References

1. J. Jeener, *Ampere Int. Summer Sch.*, Basko Polje, Yugosl., 1971 (unpublished).
2. M. H. Levitt, G. Bodenhausen, and R. R. Ernst, *J. Magn. Reson.* **58**, 462 (1984).
3. A. F. Mehlkopf, D. Korbee, T. A. Tiggleman, and R. Freeman, *J. Magn. Reson.* **58**, 315 (1984).
4. L. Müller, A. Kumar, and R. R. Ernst, *J. Chem. Phys.* **63**, 5490 (1975).
5. G. Bodenhausen, R. Freeman, and D. L. Turner, *J. Chem. Phys.* **65**, 839 (1976).
6. G. Bodenhausen, R. Freeman, R. Niedermeyer, and D. L. Turner, *J. Magn. Reson.* **24**, 291 (1976).
7. P. Bachmann, W. P. Aue, L. Müller, and R. R. Ernst, *J. Magn. Reson.* **28**, 29 (1977).
8. G. Bodenhausen, R. Freeman, R. Niedermeyer, and D. L. Turner, *J. Magn. Reson.* **26**, 133 (1977).
9. W. P. Aue, J. Karhan, and R. R. Ernst, *J. Chem. Phys.* **64**, 4226 (1976).
10. K. Nagayama, P. Bachmann, K. Wüthrich, and R. R. Ernst, *J. Magn. Reson.* **31**, 133 (1978).
11. A. Bax, R. Freeman, and G. A. Morris, *J. Magn. Reson.* **43**, 333 (1981).
12. B. Blümich and D. Ziessow, *J. Magn. Reson.* **49**, 151 (1982).
13. A. J. Shaka, J. Keeler, and R. Freeman, *J. Magn. Reson.* **56**, 294 (1984).
14. M. P. Williamson, *J. Magn. Reson.* **55**, 471 (1983).
15. J. D. Mersh and J. K. M. Sanders, *J. Magn. Reson.* **50**, 171 (1982).
16. K. Nagayama, *J. Chem. Phys.* **71**, 4404 (1979).
17. A. Kumar and C. L. Khetrapal, *J. Magn. Reson.* **30**, 137 (1978).
18. D. L. Turner, *J. Magn. Reson.* **46**, 213 (1982).
19. S. Macura and L. R. Brown, *J. Magn. Reson.* **53**, 529 (1983).
20. G. Bodenhausen, R. Freeman, G. A. Morris, and D. L. Turner, *J. Magn. Reson.* **31**, 75 (1978).
21. A. Kumar, *J. Magn. Reson.* **30**, 227 (1978).
22. G. Bodenhausen, R. Freeman, and D. L. Turner, *J. Magn. Reson.* **27**, 511 (1977).
23. R. Freeman, G. A. Morris, and D. L. Turner, *J. Magn. Reson.* **26**, 373 (1977).
24. G. Bodenhausen, R. Freeman, G. A. Morris, and D. L. Turner, *J. Magn. Reson.* **28**, 17 (1977).
25. G. Bodenhausen and D. L. Turner, *J. Magn. Reson.* **41**, 200 (1980).
26. R. Freeman and J. Keeler, *J. Magn. Reson.* **43**, 484 (1981).
27. M. H. Levitt and R. Freeman, *J. Magn. Reson.* **33**, 473 (1979).
28. D. M. Thomas, M. R. Bendall, D. T. Pegg, D. M. Doddrell, and J. Field, *J. Magn. Reson.* **42**, 298 (1981).
29. V. Rutar and T. C. Wong, *J. Magn. Reson.* **53**, 495 (1983).
30. A. Bax and R. Freeman, *J. Am. Chem. Soc.* **104**, 1099 (1982).
31. A. Bax, *J. Magn. Reson.* **52**, 330 (1983).
32. V. Rutar, *J. Am. Chem. Soc.* **105**, 4095 (1983).
33. V. Rutar, *J. Magn. Reson.* **56**, 87 (1984).
34. G. Bodenhausen, *J. Magn. Reson.* **39**, 175 (1980).
35. A. A. Maudsley and R. R. Ernst, *Chem. Phys. Lett.* **50**, 368 (1977).
36. G. A. Morris, *J. Magn. Reson.* **44**, 277 (1981).
37. D. T. Pegg and M. R. Bendall, *J. Magn. Reson.* **55**, 114 (1983).
38. V. Rutar, *J. Magn. Reson.* **56**, 413 (1984).
39. V. Rutar and T. C. Wong, *J. Magn. Reson.* **60**, 333 (1984).
40. V. Rutar, *J. Phys. Chem.* **87**, 1669 (1983).
41. V. Rutar, *J. Am. Chem. Soc.* **105**, 4496 (1983).
42. V. Rutar, *J. Phys. Chem.* **87**, 5280 (1983).

43. J. Keeler, *J. Magn. Reson.* **56**, 463 (1984).
44. M. Feigel, G. Hagele, A. Hinke, and G. Tossing, *Z. Naturforsch.* B **37**, 1661 (1982).
45. S. Puig-Torres, R. T. Gampe, G. E. Martin, M. R. Wilcott, and K. Smith, *J. Heterocycl. Chem.* **20**, 253 (1983).
46. E. L. Ezell, R. P. Thummel, and G. E. Martin, *J. Heterocycl. Chem.* **21**, 817 (1984).
47. M. J. Musmar, A. J. Weinheimer, G. E. Martin, and R. E. Hurd, *J. Org. Chem.* **48**, 3580 (1983).
48. R. Benn and W. Riemer, *Z. Naturforsch.* B **36**, 488 (1981).
49. K. Nagayama, K. Wüthrich, P. Bachmann, and R. R. Ernst, *Naturwissenschaften* **64**, 581 (1977).
50. K. Nagayama, *Adv. Biophys.* **14**, 139 (1981).
51. C. Grathwohl and K. Wüthrich, in "Perspectives in Peptide Chemistry" (A. Eberle, R. Geiger, and T. Wieland, eds.), p. 249. Karger, Basel, 1981.
52. B. Coxon, *Anal. Chem.* **55**, 2361 (1983).
53. L. D. Hall, G. A. Morris, and S. Sukumar, *Carbohydr. Res.* **76**, C7 (1979).
54. J. Dabrowski, H. Egge, and U. Dabrowski, *Carbohydr. Res.* **114**, 1 (1983).
55. D. Y. Gagnaire, F. R. Taravel, and M. R. Vignon, *Macromolecules* **15**, 126 (1982).
56. L. D. Hall, S. Sukumar, and G. R. Sullivan, *J.C.S. Chem. Commun.* p. 292 (1979).
57. A. Yamada, J. Dabrowski, P. Hanfland, and H. Egge, *Biochim. Biophys. Acta* **618**, 473 (1980).
58. R. C. Bruch and M. D. Bruch, *J. Biol. Chem.* **257**, 3409 (1982).
59. G. Kotovych, G. H. M. Aarts, and T. T. Nakashima, *Can. J. Chem.* **59**, 1449 (1981).
60. P. P. Lankhorst, C. M. Groeneveld, G. Wille, J. H. van Bloom, and C. Altona, *Rec. Trav. Chim. Pays-Bas* **101**, 253 (1982).
61. G. P. Gippert and L. R. Brown, *Polym. Bull.* **11**, 585 (1984).
62. I. J. Colquhoun and W. McFarlane, *J.C.S. Chem. Commun.* p. 484 (1982).
63. R. K. Harris, M. J. O'Connor, E. H. Curzon, and O. W. Howarth, *J. Magn. Reson.* **57**, 115 (1984).
64. R. Niedermeyer and R. Freeman, *J. Magn. Reson.* **30**, 617 (1978).
65. F. R. Taravel and M. R. Vignon, *Nouv. J. Chim.* **6**, 37 (1982).
66. S. Puig-Torres, G. E. Martin, J. J. Ford, M. R. Willcott, and K. Smith, *J. Heterocycl. Chem.* **19**, 1441 (1982).
67. J. Wernly and J. Lauterwein, *Helv. Chim. Acta* **66**, 1576 (1983).
68. P. L. Rinaldi and R. G. Solomon, *J. Org. Chem.* **48**, 3182 (1983).
69. R. C. Bruch, M. D. Bruch, J. H. Noggle, and H. B. White, *Biochem. Biophys. Res. Commun.* **123**, 555 (1984).
70. W. Ammann, R. Richarz, T. Wirthlin, and D. Wendisch, *Org. Magn. Reson.* **20**, 260 (1982).
71. D. G. Davis, W. C. Agosta, and D. Cowburn, *J. Am. Chem. Soc.* **105**, 6189 (1983).
72. W. P. Aue, E. Bartholdi, and R. R. Ernst, *J. Chem. Phys.* **64**, 2229 (1976).
73. K. Nagayama, A. Kumar, K. Wüthrich, and R. R. Ernst, *J. Magn. Reson.* **40**, 321 (1980).
74. A. Bax, R. Freeman, and G. A. Morris, *J. Magn. Reson.* **42**, 164 (1981).
75. K. Nagayama, K. Wüthrich, and R. R. Ernst, *Biochem. Biophys. Res. Commun.* **90**, 305 (1979).
76. A. Bain, J. Bornais, and S. Brownstein, *Can. J. Chem.* **59**, 723 (1981).
77. A. Bax and R. Freeman, *J. Magn. Reson.* **44**, 542 (1981).
78. D. I. Hoult and R. E. Richards, *Proc. R. Soc. London, Ser. A* **344**, 311 (1975).
79. R. Baumann, A. Kumar, R. R. Ernst, and K. Wüthrich, *J. Magn. Reson.* **44**, 76 (1981).
80. R. Baumann, G. Wider, R. R. Ernst, and K. Wüthrich, *J. Magn. Reson.* **44**, 402 (1981).
81. G. Batta and A. Liptak, *J. Am. Chem. Soc.* **106**, 248 (1984).
82. J. C. Steffens, J. L. Roark, D. G. Lynn, and J. L. Riopel, *J. Am. Chem. Soc.* **105**, 1669 (1983).

83. C. Wynants and G. Van Binst, *Biopolymers* **23**, 1799 (1984).
84. A. Bax, R. A. Byrd, and A. Aszalos, *J. Am. Chem. Soc.* **106**, 7632 (1984).
85. D. J. States, R. A. Haberkorn, and D. J. Ruben, *J. Magn. Reson.* **48**, 286 (1982).
86. L. Braunschweiler and R. R. Ernst, *J. Magn. Reson.* **53**, 521 (1983).
87. A. Kumar, R. V. Hosur, and K. Chandrasekhar, *J. Magn. Reson.* **60**, 143 (1984).
88. L. R. Brown, *J. Magn. Reson.* **57**, 513 (1984).
89. M. Rance, G. Wagner, O. W. Sørensen, K. Wüthrich, and R. R. Ernst, *J. Magn. Reson.* **59**, 250 (1984).
90. H. Oschkinat and R. Freeman, *J. Magn. Reson.* **60**, 164 (1984).
91. D. Leibfritz, E. Haupt, M. Feigel, W. E. Hull, and W.-D. Weber, *Liebigs Ann. Chem.*, p. 1971 (1982).
92. M. J. Musmar, R. T. Gampe, G. E. Martin, W. J. Layton, S. L. Smith, R. D. Thompson, M. Iwao, M. L. Lee, and R. Castle, *J. Heterocycl. Chem.* **21**, 225 (1984).
93. I. Horman, R. Dadoud, and W. Ammann, *J. Agric. Food Chem.* **32**, 538 (1984).
94. L. Szilagyi, *Carbohydr. Res.* **118**, 269 (1983).
95. J. Dabrowski and P. Hanfland, *FEBS Lett.* **142**, 138 (1982).
96. J. H. Prestegard, T. A. W. Koerner, P. C. Demou, and R. K. Yu, *J. Am. Chem. Soc.* **104**, 4993 (1982).
97. M. H. Bernstein and L. D. Hall, *J. Am. Chem. Soc.* **104**, 5553 (1982).
98. K. Nagayama and K. Wüthrich, *Eur. J. Biochem.* **114**, 365 (1981).
99. G. Wagner and K. Wüthrich, *J. Mol. Biol.* **155**, 347 (1982).
100. A. S. Arseniev, G. Wider, F. J. Joubert, and K. Wüthrich, *J. Mol. Biol.* **159**, 323 (1982).
101. H. Kessler, W. Bermel, A. Friedrich, G. Krack, and W. E. Hull, *J. Am. Chem. Soc.* **104**, 6297 (1982).
102. G. King and P. E. Wright, *Biochem. Biophys. Res. Commun.* **106**, 559 (1982).
103. G. Wider, S. Macura, A. Kumar, R. R. Ernst, and K. Wüthrich, *J. Magn. Reson.* **56**, 207 (1984).
104. J. M. Van Divender and W. C. Hutton, *J. Magn. Reson.* **48**, 272 (1982).
105. T. L. Venable, W. C. Hutton, and R. N. Grimes, *J. Am. Chem. Soc.* **104**, 4716 (1982).
106. T. L. Venable, W. C. Hutton, and R. N. Grimes, *J. Am. Chem. Soc.* **106**, 29 (1984).
107. C. Brevard, R. Schimpf, G. Tourne, and C. M. Tourne, *J. Am. Chem. Soc.* **105**, 7059 (1983).
108. P. J. Domaille, *J. Am. Chem. Soc.* **106**, 7677 (1984).
109. G. Gray, unpublished work.
110. M. Ubukata, J. Uzawa, and K. Isono, *J. Am. Chem. Soc.* **106**, 2213 (1984).
111. D. J. Ashworth and I. J. Mettler, *Biochemistry* **23**, 2252 (1984).
112. A. A. Maudsley, A. Wokaun, and R. R. Ernst, *J. Magn. Reson.* **28**, 303 (1977).
113. G. Bodenhausen and R. Freeman, *J. Magn. Reson.* **28**, 471 (1977).
114. G. Bodenhausen and R. Freeman, *J. Am. Chem. Soc.* **100**, 320 (1978).
115. A. Bax and G. A. Morris, *J. Magn. Reson.* **42**, 501 (1981).
116. M. Ikura and K. Hikichi, *Org. Magn. Reson.* **20**, 266 (1982).
117. R. Benn, *Z. Naturforsch. B* **37**, 1054 (1982).
118. G. A. Morris and L. D. Hall, *J. Am. Chem. Soc.* **103**, 4703 (1981).
119. K. M. Waugh and K. D. Berlin, *J. Org. Chem.* **49**, 873 (1984).
120. P. Joseph-Nathan, J. R. Wesener, and H. Günther, *Org. Magn. Reson.* **22**, 190 (1984).
121. P. Joseph-Nathan, E. Martinez, R. L. Santillan, J. R. Wesener, and H. Günther, *Org. Magn. Reson.* **22**, 308 (1984).
122. R. T. Gampe, M. Alam, A. J. Weinheimer, G. E. Martin, J. A. Matson, M. R. Willcott, R. R. Inners, and R. E. Hurd, *J. Am. Chem. Soc.* **106**, 1823 (1984).
123. R. Csuk, N. Müller, and H. Weidmann, *Monatsh. Chem.* **115**, 93 (1984).
124. E. Haslinger, H. Kalchhauser, and W. Robien, *Monatsh. Chem.* **113**, 805 (1982).

125. G. E. Martin, S. L. Smith, W. J. Layton, M. R. Wilcott, M. Iwao, M. L. Lee, and R. N. Castle, *J. Heterocycl. Chem.* **20**, 1367 (1983).
126. P. Schmitt and H. Günther, *Angew. Chem., Int. Ed. Engl.* **22**, 499 (1983).
127. H. Kessler and R. Schuck, in "Peptides 1982" (K. Blaha and P. Malon, eds), p. 797. de Gruyter, Berlin, 1983.
128. H. Kessler, H. Oschkinat, O. W. Sørensen, H. Kogler, and R. R. Ernst, *J. Magn. Reson.* **55**, 329 (1983).
129. E. Haslinger and H. Kalchhauser, *Tetrahedron Lett.* **24**, 2553 (1983).
130. G. A. Gray, *Org. Magn. Reson.* **21**, 111 (1983).
131. H. Kessler, W. Hehlein, and R. Schuck, *J. Am. Chem. Soc.* **104**, 4534 (1982).
132. G. E. Hawkes, L. Y. Lian, and E. W. Randall, *J. Magn. Reson.* **56**, 539 (1984).
133. W. M. Westler, G. Ortiz-Polo, and J. L. Markley, *J. Magn. Reson.* **58**, 354 (1984).
134. T.-M. Chan and J. L. Markley, *J. Am. Chem. Soc.* **104**, 4010 (1982).
135. T.-M. Chan and J. L. Markley, *Biochemistry* **22**, 5996 (1983).
136. C. L. Kojiro and J. L. Markley, *FEBS Lett.* **162**, 52 (1983).
137. S. Coffin, M. Limm, and D. Cowburn, *J. Magn. Reson.* **59**, 268 (1984).
138. P. P. Lankhorst, C. Erkelens, C. A. G. Haasnoot, and C. Altona, *Nucleic Acids Res.* **111**, 7215 (1983).
139. J. C. Beloeil, M. A. Delsuc, J. Y. Lallemand, G. Dauphin, and G. Jeminet, *J. Org. Chem.* **49**, 1797 (1984).
140. G. Buchbauer, A. Fischlmayr, E. Haslinger, W. Robien, H. Vollenkle, and C. Wassmann, *Monatsh. Chem.* **115**, 739 (1984).
141. J. Banerji, R. Saha, and J. N. Shoolery, *Indian J. Chem., Sect. B* **22**, 903 (1983).
142. C. Wynants, K. Hallenga, G. Van Binst, A. Michel, and J. Zanen, *J. Magn. Reson.* **57**, 93 (1984).
143. H. Kessler, C. Griesinger, J. Zarbock, and H. R. Loosli, *J. Magn. Reson.* **57**, 331 (1984).
144. C. Bauer, R. Freeman, and S. Wimperis, *J. Magn. Reson.* **58**, 526 (1984).
145. S. Wimperis and R. Freeman, *J. Magn. Reson.* **58**, 348 (1984).
146. H. Bleich, S. Gould, P. Pitner, and J. Wilde, *J. Magn. Reson.* **56**, 515 (1984).
147. D. C. Finster, W. C. Hutton, and R. N. Grimes, *J. Am. Chem. Soc.* **102**, 400 (1980).
148. D. P. Burum, *J. Magn. Reson.* **59**, 430 (1984).
149. D. T. Pegg, D. M. Doddrell, W. M. Brooks, and M. R. Bendall, *J. Magn. Reson.* **44**, 32 (1981).
150. J. T. Gerig, *Macromolecules* **16**, 1797 (1983).
151. P. H. Bolton and G. Bodenhausen, *J. Magn. Reson.* **43**, 339 (1982).
152. P. H. Bolton and G. Bodenhausen, *Chem. Phys. Lett.* **89**, 139 (1982).
153. P. H. Bolton, *J. Magn. Reson.* **45**, 239 (1981).
154. P. H. Bolton, *J. Magn. Reson.* **46**, 91 (1982).
155. A. Pardi, R. Walker, H. Rappoport, G. Wider, and K. Wüthrich, *J. Am. Chem. Soc.* **105**, 1652 (1983).
156. K. Lai, D. Shah, E. DeRose, and D. G. Gorenstein, *Biochem. Biophys. Res. Commun.* **121**, 1021 (1984).
157. P. H. Bolton, in "Biomolecular Stereodynamics" (R. H. Sarma, ed.), pp. 437-453. Academic Press, New York, 1981.
158. A. G. Avent and R. Freeman, *J. Magn. Reson.* **39**, 169 (1980).
159. A. Bax, *J. Magn. Reson.* **53**, 517 (1983).
160. J. R. Garbow, D. P. Weitekamp, and A. Pines, *Chem. Phys. Lett.* **93**, 504 (1982).
161. V. Rutar, *J. Magn. Reson.* **58**, 306 (1984).
162. J. A. Wilde and P. H. Bolton, *J. Magn. Reson.* **59**, 343 (1984).
163. M. R. Bendall and D. T. Pegg, *J. Magn. Reson.* **53**, 144 (1983).

164. M. H. Levitt, O. W. Sørensen, and R. R. Ernst, *Chem. Phys. Lett.* **94**, 540 (1983).
165. T. T. Nakashima, B. K. John, and R. E. D. McClung, *J. Magn. Reson.* **57**, 149 (1984).
166. T. C. Wong, V. Rutar, and J.-S. Wang, *J. Am. Chem. Soc.* **106**, 7046 (1984).
167. G. Bodenhausen and D. J. Ruben, *Chem. Phys. Lett.* **69**, 185 (1980).
168. D. A. Vidnsek, M. F. Roberts, and G. Bodenhausen, *J. Am. Chem. Soc.* **104**, 5452 (1982).
169. G. Eich, G. Bodenhausen, and R. R. Ernst, *J. Am. Chem. Soc.* **104**, 3731 (1982).
170. P. H. Bolton and G. Bodenhausen, *J. Magn. Reson.* **46**, 306 (1982).
171. G. Wagner, *J. Magn. Reson.* **55**, 151 (1983).
172. P. H. Bolton, *J. Magn. Reson.* **48**, 336 (1982).
173. H. Kessler, M. Bernd, H. Kogler, J. Zarbock, O. W. Sørensen, G. Bodenhausen, and R.R. Ernst, *J. Am. Chem. Soc.* **105**, 6944 (1983).
174. A. Bax, *J. Magn. Reson.* **53**, 149 (1983).
175. M. A. Delsuc, E. Guittet, N. Trotin, and J. Y. Lallemand, *J. Magn. Reson.* **56**, 163 (1984).
176. D. Neuhaus, G. Wider, G. Wagner, and K. Wüthrich, *J. Magn. Reson.* **57**, 164 (1984).
177. P. Bigler, W. Ammann, and R. Richarz, *Org. Magn. Reson.* **22**, 109 (1984).
178. H. Kogler, O. W. Sørensen, G. Bodenhausen, and R. R. Ernst, *J. Magn. Reson.* **55**, 157 (1983).
179. P. H. Bolton, *J. Magn. Reson.* **54**, 333 (1983).
180. O. W. Sørensen and R. R. Ernst, *J. Magn. Reson.* **55**, 338 (1983).
181. J. Jeener, B. H. Meier, P. Bachmann, and R. R. Ernst, *J. Chem. Phys.* **71**, 4546 (1979).
182. B. H. Meier and R. R. Ernst, *J. Am. Chem. Soc.* **101**, 6441 (1979).
183. S. Macura and R. R. Ernst, *Mol. Phys.* **41**, 95 (1980).
184. A. Kumar, R. R. Ernst, and K. Wüthrich, *Biochem. Biophys. Res. Commun.* **95**, 1 (1980).
185. C. Bosch, A. Kumar, R. Baumann, R. R. Ernst, and K. Wüthrich, *J. Magn. Reson.* **42**, 159 (1981).
186. A. Kumar, G. Wagner, R. R. Ernst, and K. Wüthrich, *J. Am. Chem. Soc.* **103**, 3654 (1981).
187. S. Macura, Y. Huang, D. Suter, and R. R. Ernst, *J. Magn. Reson.* **43**, 259 (1981).
188. S. Macura, K. Wüthrich, and R. R. Ernst, *J. Magn. Reson.* **46**, 269 (1982).
189. S. Macura, K. Wüthrich, and R. R. Ernst, *J. Magn. Reson.* **47**, 351 (1982).
190. M. Rance, G. Bodenhausen, G. Wagner, K. Wüthrich, and R. R. Ernst, *J. Magn. Reson.* **62**, 497 (1985).
191. G. Bodenhausen, G. Wagner, M. Rance, O. W. Sørensen, K. Wüthrich, and R. R. Ernst, *J. Magn. Reson.* **59**, 542 (1984).
192. A. L. Schwartz and J. D. Cutnell, *J. Magn. Reson.* **53**, 398 (1983).
193. G. Bodenhausen and R. R. Ernst, *Mol. Phys.* **47**, 319 (1982).
194. C. A. Haasnoot, F. J. M. Van de Ven, and C. W. Hilbers, *J. Magn. Reson.* **56**, 343 (1984).
195. A. Z. Gurevich, I. L. Barsukov, A. S. Arseniev, and V. F. Bystrov, *J. Magn. Reson.* **56**, 471 (1984).
196. G. Bodenhausen and R. R. Ernst, *J. Magn. Reson.* **45**, 367 (1981).
197. G. Bodenhausen and R. R. Ernst, *J. Am. Chem. Soc.* **104**, 1304 (1982).
198. J. Bremer, G. L. Mendz, and W. J. Moore, *J. Am. Chem. Soc.* **106**, 4691 (1984).
199. G. Wagner, *J. Magn. Reson.* **57**, 497 (1984).
200. S. Macura, N. G. Kumar, and L. R. Brown, *J. Magn. Reson.* **60**, 99 (1984).
201. D. R. Hare, D. E. Wemmer, S.-H. Chou, G. Drobny, and B. R. Reid, *J. Mol. Biol.* **171**, 319 (1983).
202. F. J. M. van de Ven, S. H. de Bruin, and C. W. Hilbers, *FEBS Lett.* **169**, 107 (1984).
203. X. Wu, W. M. Westler, and J. L. Markley, *J. Magn. Reson.* **59**, 524 (1984).
204. R. Willem, A. Jans, C. Hoogzand, M. Gielen, G. Van Binst, and H. Pepermans, *J. Am. Chem. Soc.* **107**, 28 (1985).
205. P. B. Garlick and C. J. Turner, *J. Magn. Reson.* **51**, 536 (1983).

206. R. S. Balaban, H. L. Kantor, and J. L. Ferretti, *J. Biol. Chem.* **258**, 12787 (1983).
207. J. Boyd, K. M. Brindle, I. D. Campbell, and G. K. Radda, *J. Magn. Reson.* **60**, 149 (1984).
208. Y. Huang, S. Macura, and R. R. Ernst, *J. Am. Chem. Soc.* **103**, 5327 (1981).
209. P. L. Rinaldi, *J. Am. Chem. Soc.* **105**, 5167 (1983).
210. C. Yu and G. C. Levy, *J. Am. Chem. Soc.* **105**, 6994 (1983).
211. C. Yu and G. C. Levy, *J. Am. Chem. Soc.* **106**, 6533 (1984).
212. P. H. Bolton, *J. Magn. Reson.* **57**, 427 (1984).
213. L. Müller, *J. Magn. Reson.* **59**, 325 (1984).
214. P. H. Bolton, *J. Magn. Reson.* **60**, 342 (1984).
215. A. Bax, R. Freeman, and S. P. Kempsell, *J. Magn. Reson.* **41**, 349 (1980).
216. A. Bax, R. Freeman, and S. P. Kempsell, *J. Am. Chem. Soc.* **102**, 4849 (1980).
217. A. Bax, R. Freeman, and T. A. Frenkiel, *J. Am. Chem. Soc.* **103**, 2102 (1981).
218. J. N. Shoolery, *J. Nat. Prod.* **47**, 226 (1984).
219. A. Bax, R. Freeman, T. A. Frenkiel, and M. H. Levitt, *J. Magn. Reson.* **43**, 478 (1981).
220. T. H. Mareci and R. Freeman, *J. Magn. Reson.* **48**, 158 (1982).
221. O. Sørensen, R. Freeman, T. Frenkiel, T. Mareci, and R. Schuck, *J. Magn. Reson.* **46**, 180 (1982).
222. D. L. Turner, *Mol. Phys.* **44**, 1051 (1981).
223. D. L. Turner, *J. Magn. Reson.* **49**, 175 (1982).
224. D. L. Turner, *J. Magn. Reson.* **53**, 259 (1983).
225. A. Bax and T. H. Mareci, *J. Magn. Reson.* **53**, 360 (1983).
226. J. A. Robinson and D. L. Turner, *J. Chem. Soc.* 150 (1982).
227. J. Hudec and D. L. Turner, *J.C.S. Perkin II* 951 (1982).
228. P. J. Keller, Q. L. Van, A. Bacher, J. F. Kozlowski, and H. G. Floss, *J. Am. Chem. Soc.* **105**, 2505 (1983).
229. R. Benn, *Angew. Chem.* **94**, 633 (1982).
230. S. Berger and K.-P. Zeller, *J. Org. Chem.* **49**, 3725 (1984).
231. M. H. Levitt and R. R. Ernst, *Mol. Phys.* **50**, 1109 (1983).
232. M. H. Levitt, *J. Magn. Reson.* **48**, 234 (1982).
233. M. H. Levitt, *J. Magn. Reson.* **50**, 95 (1982).
234. M. H. Levitt and R. Freeman, *J. Magn. Reson.* **43**, 65 (1981).
235. R. Freeman, S. P. Kempsell, and M. Levitt, *J. Magn. Reson.* **38**, 453 (1980).
236. T. H. Mareci and R. Freeman, *J. Magn. Reson.* **51**, 531 (1983).
237. J. Boyd, C. M. Dobson, and C. Redfield, *J. Magn. Reson.* **55**, 170 (1983).
238. O. W. Sørensen, M. H. Levitt, and R. R. Ernst, *J. Magn. Reson.* **55**, 104 (1983).
239. R. Ramachandran, A. C. Kunwar, H. S. Gutowsky, and E. Oldfield, *J. Magn. Reson.* **60**, 352 (1984).
240. U. Piantini, O. W. Sørensen, and R. R. Ernst, *J. Am. Chem. Soc.* **104**, 6800 (1982).
241. M. Rance, O. W. Sørensen, G. Bodenhausen, G. Wagner, R. R. Ernst, and K. Wüthrich, *Biochem. Biophys. Res. Commun.* **117**, 479 (1983).
242. A. J. Shaka and R. Freeman, *J. Magn. Reson.* **51**, 169 (1983).
243. L. Müller, *J. Am. Chem. Soc.* **101**, 4481 (1979).
244. A. Bax, R. Griffey, and B. L. Hawkins, *J. Am. Chem. Soc.* **105**, 7188 (1983).
245. A. Bax, R. Griffey, and B. L. Hawkins, *J. Magn. Reson.* **55**, 301 (1983).
246. R. Griffey, C. D. Poulter, A. Bax, B. L. Hawkins, Z. Yamaizumi, and S. Nishimura, *Proc. Natl. Acad. Sci. USA* **80**, 5895 (1983).
247. D. Foxall and A. Zens, unpublished data.
248. A. G. Redfield, *Chem. Phys. Lett.* **96**, 537 (1983).
249. S. Roy, M. Z. Papastavros, V. Sanchez, and A. G. Redfield, *Biochemistry* **23**, 4395 (1984).
250. D. H. Live, D. G. Davis, W. C. Agosta, and D. Cowburn, *J. Am. Chem. Soc.* **106**, 6104 (1984).

6

Applications of 2D NMR to Biological Systems

J. H. PRESTEGARD
CHEMISTRY DEPARTMENT
YALE UNIVERSITY
NEW HAVEN, CONNECTICUT 06511

I. Introduction	435
II. The Basic 2D NMR Experiments for Spectral Assignment and Structure Elucidation	437
A. J-Resolved Spectroscopy	438
B. Scalar-Coupling-Correlated Spectroscopy	442
C. Cross-Relaxation-Correlated Spectroscopy	446
III. Advanced Methods for Biological Samples	449
A. Multiple-Quantum Coherence	449
B. Relayed Coherence	455
C. Choice of Processing Methods	457
D. Heteronuclear Experiments	461
E. Two-Dimensional Spectra in H_2O	464
IV. Structural Analysis of Biomolecules	470
A. Relationship of Cross Relaxation to Structure	470
B. Application to Proteins	475
C. Application to Polynucleotides	479
D. Application to Oligosaccharides	482
References	485

I. Introduction

The utility of NMR in the analysis of structural properties of biological macromolecules has been recognized for the past two decades. The reasons for its utility are straightforward: it is one of the few structural methods applicable to biomolecules in a medium approaching a physiological environment, it offers an extraordinary range of sensitivities to kinetic processes, and it often allows focus on properties of discrete functional sites in otherwise unmanageably complex systems. One-dimensional NMR experiments for extracting structural information had evolved during this period and had, in fact, been applied in a variety of studies involving moderate-sized biological molecules (1-4). The basic strategies for structural analysis of biomolecules are well laid out in these early works. (1) Using unique spectral properties or selective modification, assign as many resonances to specific residues as possible. (2) Using through-bond spin–spin couplings or through-space magnetization transfer, extend these assignments to adjacent residues. (3) Through a combination of effects on scalar coupling constants, nuclear Overhauser effects, and spin relaxation, investigate structural characteristics of assigned resonances. (4) Use these characteristics to test models derived from other structural studies such as x-ray crystallography.

Despite these well-defined strategies, progress toward structural elucidation using one-dimensional NMR experiments was relatively slow. The reasons lie more in the overwhelming quantity of data required than in a fundamental flaw in the strategies. Consider the fact that a protein of molecular mass 10,000 daltons or 90 residues would have on the order of 500 protons, most with distinct chemical shifts. With scalar coupling these would give rise to more than a thousand spectral resonances. Superimposed on this enormous amount of data is the fact that not all resonances are well resolved. Consider the case of nucleic acids, where there are only four commonly occurring, chemically distinct, residues. Degeneracy of resonances in cases like this is a severe limitation to analysis using one-dimensional methods.

The potential that 2D NMR methodology offered in overcoming limitations of resolution and the systematic presentation of large quantities of data was actually recognized rather quickly. Shortly after the first applications of 2D NMR to organic molecules by Ernst, Freeman, and others, Wagner, Kumar, and Wüthrich published the first in their series of classic works on bovine pancreatic trypsin inhibitor (BPTI) (5). The strategy which eventually led to the complete assignment of proton resonances and the verification of crystal structure, interestingly enough, remained much the same as developed for 1D NMR: improve resolution as much as possible, assign a few key resonances, extend these assignments to adjacent residues by a combination of experiments relying on scalar coupling and spin-relaxation effects, extract structural constraints from experiments based on the distance dependence of spin relaxation, and compare observed distances with those found in predicted structures. Progress was, however, enormously faster. The experiments used in these steps included *J*-resolved spectroscopy, scalar-coupling-correlated spectroscopy (SECSY, spin-echo-correlated spectroscopy; COSY, coupling-correlated spectroscopy), and cross-relaxation-correlated spectroscopy (NOESY, nuclear Overhauser effect spectroscopy).

Our first objective will be to achieve a better understanding of some of these basic experiments. The experiments have been introduced in previous chapters of this volume; however, we will review some qualitative aspects here and also make some practical suggestions relating to choice of experimental parameters.

Following the pedagogical intent of this volume, we will present some rather simple examples in this first section, using a single residue representative of those found in complex carbohydrates, 1-O-methyl-β-D-galactose. Aside from being a ubiquitous compound, it has the advantage of having a moderately complex proton spectrum lying in a very small segment of the normal chemical shift range. This small shift range facilitates the rapid collection and processing of 2D data sets. Experimental details are given in the text or figure captions so that duplication of the results could be pursued as an exercise.

Experimental procedures have, of course, advanced a great deal in the years following 1981 and the introduction of the basic 2D experiments. Some of the new experimental procedures are very important to the successful application of 2D methods to biological macromolecules. We will discuss these advances as they apply to biomolecules in the second segment of this chapter. Translation of data obtained by both basic and advanced techniques into specific assignments and structural determinations will be reserved for a third section dealing with applications to specific proteins, nucleic acids, and carbohydrates.

II. The Basic 2D NMR Experiments for Spectral Assignment and Structure Elucidation

Many of the basic 2D NMR experiments can be understood in terms of the one-dimensional experiments from which the strategy for structural analysis was derived. While this is not a rigorous approach, it serves to point out some of the important limitations as applied to biomolecules. We will use this approach to discuss three basic two-dimensional experiments used in the early evaluation of protein structure: J-resolved spectroscopy, COSY, and NOESY. Because these experiments remain important tools today, we will also take the opportunity to discuss optimal parameter selection for these experiments as they apply to current studies of macromolecules.

A. J-Resolved Spectroscopy

Although J-resolved experiments have become less important in macromolecule studies as methods for extracting coupling constants from COSY spectra have been devised, they are by far the easiest 2D NMR experiments to understand. They are also the least time-consuming of the 2D NMR experiments to perform. The J-resolved experiments are based on the J modulation of spin echoes, described in Chapter 1, Section II,B, that occurs in the 90°—τ—180°—τ—*aquire* pulse sequence commonly used for the determination of transverse relaxation times, T_2 (6). Spin-echo experiments were developed to separate the heterogeneous contributions to relaxation of transverse magnetization (such as magnetic field inhomogeneities) from homogeneous contributions which relate more directly to molecular motions. Contributions to the spin Hamiltonian linearly dependent on I_z cause density matrix elements related by total spin inversion to precess in opposite directions during the time period τ. Under the influence of a 180° rf pulse these elements for collections of spin-$\frac{1}{2}$ nuclei interchange values, and refocus to their original values at the end of the period 2τ. An ability to refocus heterogeneous contributions at the time of observation leads to improved resolution of J-resolved spectra. This resolution improvement is not often realized in protein spectra because it is relatively rare that field inhomogeneities dominate decay of transverse magnetization and hence line widths.

In larger molecules, the value of J-resolved spectra stems more from spectral simplification associated with an ability to separate offsets of resonances due to scalar coupling from offsets of resonance resulting from chemical shift. Chemical shift offsets depend on I_z just as do effects caused by field heterogeneities, and the amplitude of the refocused magnetization at the time of observation is not modulated by chemical shift. Scalar coupling, however, depends on an $\mathbf{I} \cdot \mathbf{I}$ operator instead of an I_z operator. In first-order spectra of collections of nonequivalent nuclei, $\mathbf{I} \cdot \mathbf{I}$ contributions can be approximated by $I_{z1}I_{z2}$. Elements related by total spin inversion evolve identically with respect to J and appear to have precessed with constant velocity throughout the period 2τ. Hence, the amplitude of the echo at the time of observation is modulated only by spin–spin coupling constants. J modulation has been used extensively to edit complex spectra from biomolecules in one-dimensional experiments. In the work of Dobson *et al.*, for example, choice of a delay between 90° and 180° pulses of appropriate length to invert the lines of doublets, along with subtraction from a normal spectrum, has been used to retain selectively all doublets in a complex spectrum (7).

This same type of simplification is realized in the two-dimensional experiment based on J modulation, but there is no necessity to know in advance the exact delay that inverts certain multiplet components. The delay between pulses is instead systematically incremented to form a t_1 time domain. For each point in the t_1 time domain, data are acquired beginning at the peak of the echo and extending onward in time. Magnetization components in this t_2 time domain are modulated by both chemical shift and spin coupling, much as in a normal one-dimensional spectrum. After Fourier transformation in t_2, each peak is found to be phase modulated in the t_1 time domain (the echo time) by coupling constants only. The spin couplings are normally removed from the F_2 dimension after transformation in the t_1 dimension by shifting each row of the data set by an amount corresponding to its increment in the F_1 frequency domain. Projection of the data set onto the F_2 axis has the effect of producing a completely homonuclear-decoupled one-dimensional spectrum. Columns taken at the peak of each resonance display the multiplet structures for that resonance.

Figure 6-1 shows a 2D J-resolved spectrum and a 1D spectrum of 1-O-methyl-β-D-galactose. The structure of this molecule is shown in Fig. 6-2. This molecule is small enough to show some improvement in resolution in the J dimension. For example, the broad doublet at 3.86 ppm is shown to be a doublet of doublets. It is assigned to H4, which, because of its equatorial configuration, is expected to have small couplings to H5 and H3. The H5 multiplet at 3.63 ppm, which experiences extensive coupling to H4, H6, and H6', and lies in a crowded region, can be resolved to the point where all couplings are observable. This is a task difficult to achieve in the one-dimensional spectrum (Fig. 6-1b).

Second-order effects tend to produce distortions of 2D J-resolved spectra both in terms of column linearity and numbers of 2D peaks. This is illustrated by the H5, H6, and H6' resonances in Fig. 6-1 (3.63–3.78 ppm). These effects for amino acid spectra have been analyzed extensively by Wider *et al.* (8). While there is some potential for residue identification based on the patterns that result, the effects are most often an inconvenience. In particular, broadening of lines in projections and, in some cases, production of low-amplitude spurious peaks complicate analysis.

Selection of parameters for a J-resolved experiment is fairly straightforward. Data storage and data processing times are minimal, so a reasonable data set size can be retained in the t_2 dimension. By looking at a single FID for the sugar we see that our homogeneity-limited T_2^* is about 0.25 sec. An acquisition time of approximately two times T_2^*, or 0.5 sec, is adequate to retain inherent resolution in the F_2 dimension. With a quadrature sweep width of ±512 Hz and associated digitizing rate of 1024 Hz, this

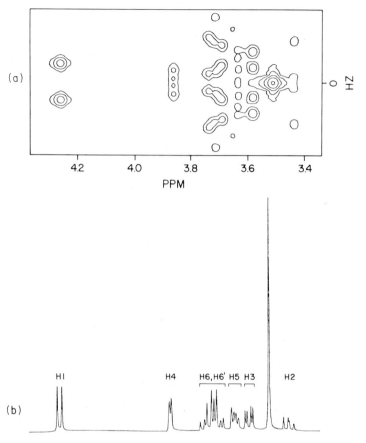

Fig. 6-1. (a) J-Resolved and (b) 1D spectra of 1-O-methyl-β-D-galactose at 500 MHz. The sample is approximately 20 mM in D_2O in a 5-mm NMR tube run at 303 K. Both 1D and 2D sets were acquired in quadrature using 512 complex points with the transmitter just downfield of the H1 resonance. Only the upfield halves of the data sets are displayed. For the J-resolved spectrum, each of the 32 t_1 files was the result of 32 acquisitions with a repetition rate of approximately 2 sec. Processing was accomplished with zero filling and sine-bell weighting in both dimensions. The spectrum was tilted to remove J contributions from the F_2 dimension and symmetrized.

corresponds to a t_2 data set size of approximately 1 K, one half from each of the quadrature channels. The sweep width in the F_1 dimension needs to be only slightly larger than the expected width of the largest multiplet. In our sugar, this multiplet probably corresponds to that for the proton in position 5, which is coupled to three other protons. If all couplings were 9 Hz, a maximum multiplet width of 27 Hz would be expected. A slightly

6. APPLICATIONS OF 2D NMR TO BIOLOGICAL SYSTEMS

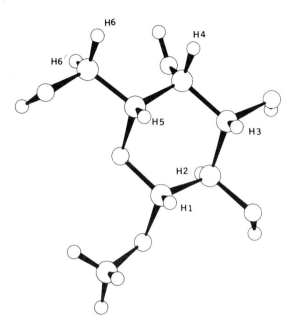

Fig. 6-2. Structure of 1-O-methyl-β-D-galactose.

larger F_1 spectral width of ± 16 Hz is chosen, since a ratio of F_1 and F_2 dimensions equal to a power of 2 is convenient for subsequent tilting operations. We expect the inherent T_2 to be somewhat longer than our observed T_2 for a molecule of this type, perhaps half a second. To achieve the resolution in the F_1 dimension associated with this long T_2, we need to acquire a sufficient number of F_1 time domain points to achieve an acquisition time of two times T_2, or one second. The digitizing rate dictated by our sweep width of ± 16 Hz would suggest about 32 points. Since experimental accuracy is not terribly dependent on recovery of longitudinal magnetization, recycling times can be trimmed to the order of T_1, or two seconds. Further details are given in the legend of Fig. 6-1.

For typical protein samples, an F_2 spectral width five times larger must often be used; but because of more rapid FID decays, data set size in the F_2 dimension may not need to be increased by more than two times that used for our sugar. Ranges of multiplet splittings would not be more than two times larger and inherent T_2 values would be shorter, leaving 32 t_1 points, with about one-half the t_1 increment specified above as a good choice. The total acquisition time and storage space are thus very modest, and J-resolved spectra are likely to continue to find application even in macromolecule studies.

B. Scalar-Coupling-Correlated Spectroscopy

Two-dimensional experiments designed to elucidate scalar connectivities are analogous to spin decoupling experiments in the sense that both provide an aid to spectral assignment through an ability to connect resonances from vicinal protons in a linear molecular framework. Examples of application to many types of biomolecules exist, but there is a particularly good example in the successive spin decoupling experiments used to extend resonance assignments from the well-resolved anomeric proton resonances of a sugar residue (H1 in Fig. 6-2) around the ring to resonances from other protons (9, 10). These same general procedures are extensively used in applications of 2D spin-coupling-correlated experiments to more complex oligosaccharides, proteins and nucleic acids, but the 2D experiments avoid many problems associated with nearly degenerate resonances. Specific examples will be discussed later in this chapter. For now, we wish to improve our understanding of the basic experiments.

The origin of effects establishing connectivities in decoupling and 2D scalar connectivity experiments is actually quite distinct. In the case of spin decoupling, energy levels are altered in the presence of an applied magnetic field to eliminate splittings due to scalar couplings. In the case of 2D experiments, coherence intensities associated with one multiplet line are transferred to a coherence assigned to a line in a multiplet of a proton scalar coupled to the proton of the first multiplet. The latter is more analogous to early one-dimensional double-resonance experiments with weak rf fields in which one line of a multiplet is saturated and the amplitude perturbation on scalar-coupled multiplet components is observed (11), except that here transverse, rather than z, magnetizations are involved. Aside from the inherent improvement in efficiency, the transfer of coherence can be done without the distortions associated with incomplete spin decoupling and the presence of an rf field during acquisition. Two-dimensional experiments of the scalar-coupling type, such as COSY, SECSY, and MQCO, thus all share advantages of both efficiency and spectral clarity.

The simplest experiments are two-pulse experiments, usually executed with a pair of 90° pulses separated by an incrementable delay. The first pulse creates transverse magnetization, or one-quantum coherence. Components for each line in the spectrum then precess at their characteristic frequencies. Transfer of magnetization to another resonance at the time of the second pulse depends on both the relative and the absolute phases of the donating multiplet components. Thus, as the delay between pulses is incremented, the magnetization of the receiving resonance is modulated by the precessional frequency of the donating resonances. In the COSY experiment, data are acquired immediately after the second 90° pulse. Transformation of each FID results in a spectrum resembling a one-dimensional experiment except that amplitudes of resonances vary with t_1 as a function of both their own characteristic frequencies and the frequencies of resonances to which they are coupled. Transformation in the t_1 dimension produces a plot such as that depicted in Fig. 6-3. The data are again for 1-O-methyl-β-D-galactose, the molecule depicted in Fig. 6-2. The strong diagonal represents magnetization modulated in both t_1 and t_2 periods at the same characteristic frequency and yields a reasonable approximation of the 1D spectrum. The off-diagonal peaks arise from scalar connectivities. Construction of a pair of horizontal and vertical lines from a cross peak to two points on the diagonal serves to identify scalar-connected resonances. With successive constraints of this type, one can begin at the anomeric resonance at 4.27 ppm, easily assigned by its single splitting of 8 Hz and its far downfield position, and identify in turn H2, H3, H4, and H5 resonances.

Optimization of this basic experiment has consumed a good deal of research, and rightly so. This is a time-consuming experiment, not so much

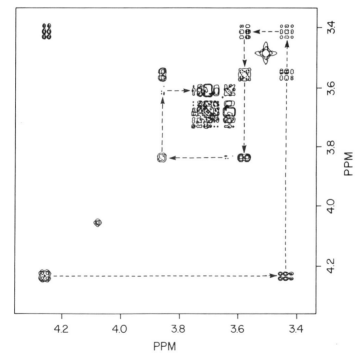

Fig. 6-3. COSY spectrum of 1-O-methyl-β-D-galactose at 490 MHz. The sample conditions are as given in Fig. 6-1. Sweep widths in both directions are restricted to ±250 Hz. The spectrum was collected in quadrature with 512 complex points in t_2 and 128 points in t_1, each incremented by 0.002 sec. Each t_1 point was the result of 16 scans with a repetition rate of 2 sec and four dummy scans prior to the beginning of acquisition. Processing employed a sine-bell function in both dimensions, zero filling to 256 in t_1, and symmetrization about the diagonal. The dashed lines illustrate the assignment scheme.

because of the lack of inherent sensitivity, but because of the large number of t_1 points that must be collected. First, transfer of magnetization does depend on spin–spin coupling and occurs with optimum efficiency when multiplet components are out of phase. For a doublet, this optimum occurs after a t_1 period of length one-half the reciprocal of the J splitting. Quite aside from resolution considerations, data must therefore be acquired over a moderately long t_1 period of the order of 0.1 sec where doublets with couplings as small as 5 Hz may be of interest. (Note that the H4–H5 cross peak of Fig. 6-3 is weak because the maximum t_1 was less than $1/2J_{45}$.) If we were dealing with a protein spectrum in which connectivities occur over the entire 10-ppm range, and if we did not have quadrature in the F_1 dimension, as was the case in many early applications, we would need a large number of t_1 files. At 500 MHz we would need a digitizing rate of 10 kHz, and, over 0.1 sec, 1024 t_1 files. Given that each file would need sensitivity and resolution approaching that of a 1D experiment, it is easy to see that this could exhaust the data storage capacity of the spectrometer and the patience of any spectroscopist.

Quadrature is fairly easily added to the basic two-pulse scheme. One approach can be understood by realizing that the last 90° pulse is selecting magnetization of only one phase for transfer. If a second sequence is acquired with a 90° phase shift of the last pulse and added or subtracted, quadrature in the F_1 dimension is achieved (12). In employing quadrature, one actually has the choice of mixing positive frequencies in t_1 with positive or negative frequencies in t_2. The usual convention is the latter, selection of coherence-transfer echoes, and it has some sensitivity advantages where heterogeneous broadening contributes to line widths (13). A second approach employs time-proportional phase incrementation, in which phases are incremented for each new t_1 point to make all frequencies appear to precess in the positive direction (14). Thus, with slightly more complex sequences and the placement of the transmitter in the center of the spectrum, the requirements for sweep width and for the number of t_1 points are reduced by a factor of two.

In a few cases it is possible to reduce further the t_1 domain size by altering the definition of t_1 and t_2 periods. In the SECSY version of this coupling correlated experiment, the t_1 period is actually split so that half occurs before, and half occurs after, the final 90° pulse. In other words, acquisition is delayed. By appropriate phase cycling, magnetization from coherences which evolve at positive frequencies before the pulse can be transferred to those which evolve at negative frequencies after the pulse, yielding effective modulation at the differences of frequencies of the coupled resonances. Moreover, the transferred magnetization precesses for only half the total t_1 period, and the F_1 axis becomes the difference in resonance frequencies divided by two. The SECSY experiment is actually one of the first scalar-connectivity experiments employed with macromolecules (15). Time saving because of spectral width reduction and the need for a smaller number of t_1 points is obvious. This is, however, achieved at a cost. The first half (the most intense part) of the FID is not collected, thus reducing signal to noise in each t_1 file. In protein studies where resonances at one extreme of the spectrum are often coupled to resonances at the other, little is saved by having F_1 representing differences in frequencies, and sweep widths can only be reduced by a factor of two. The associated time savings are usually not enough to compensate for the sensitivity loss, and use of the COSY format is favored.

Phase cycling employed in either experiment is actually more complex than required for quadrature detection. One normally wants to eliminate DC levels and direct contribution of magnetization that recovers during t_1. A sequence that accomplishes this is presented in Table 6-1. The necessity of averaging quadrature receiver and pulse imperfections adds another level of complexity. One approach is to understand the dependence of the desired signal on pulse phase and to combine phases in every way consistent with coherent averaging of that signal.

Recycling times for the above experiments depend on T_1 relaxation times but in a rather indirect way. It is transverse magnetization with which we

TABLE 6-1

Phase Cycling Employed in the 2D COSY Experiment of Fig. 6-3

Acquisition number	Pulse 1 (90°)	Pulse 2 (90°)	Receiver
1	x	x	x
2	x	−x	x
3	x	y	−x
4	x	−y	−x
5	y	y	y
6	y	−y	y
7	y	−x	−y
8	y	x	−y
9	−x	y	x
10	−x	−y	x
11	−x	−x	−x
12	−x	x	−x
13	−y	−x	y
14	−y	x	y
15	−y	−y	−y
16	−y	y	−y

are dealing. In the interest of economy of time, recycling times are held short: about one T_1 for acquisition plus delay. The final 90° pulse can also be replaced with a 45° pulse. This tends to emphasize peaks arising from coherence transfer between directly connected transitions and reduces the complexity of both diagonal and cross peaks.

For the data presented in Fig. 6-3, we have reduced the sweep width in F_2 to 500 Hz by allowing the water resonance to fold back into a region between the anomeric resonance and the H4 resonance (4.08 ppm). This allows good F_2 resolution with a size of 1 K, or 512 complex points, and saves some data storage and transform time. The F_1 width is selected as ±250 Hz corresponding to an increment of 0.002 sec. To allow moderate resolution in the F_1 dimension, 128 files were acquired for a t_1 acquisition time of 0.256 sec. A recycling time of 2 sec allowed 70% recovery and a 90° final pulse was used. Note that this is a relatively short experiment in the example presented only because of the limited spectral range represented by sugar resonances.

C. Cross-Relaxation-Correlated Spectroscopy

The most commonly applied form of cross-relaxation-correlated spectroscopy is the NOESY or nuclear Overhauser effect spectroscopy experiment. Overhauser effects as originally introduced referred to the effect on resonance intensity of one spin when a second spin (actually an electron) was saturated (16). The two-dimensional experiment referred to above establishes correlations in a manner more analogous to the effects seen on selective inversion of resonances in one-dimensional studies (17), and cross-relaxation-correlated spectroscopy is probably a more descriptive term.

The primary experiment is executed by the application of three 90° pulses. The first two are separated by an incremented delay, t_1, and serve to modulate the z component of magnetization of each resonance in accord with its precessional frequency. Imagine the special case in which the first pulse establishes transverse magnetization and the delay, t_1, is just long enough for one complete rotation of the magnetization of a primary resonance. The second 90° pulse, if it is of the same phase as the first, will then selectively invert this resonance.

The second 90° pulse is followed by a mixing delay, a fraction of the longitudinal relaxation time in length. During this period, cross relaxation occurs. The z magnetization of inverted resonances is transferred to resonances coupled by through-space dipolar interactions. The process is quite analogous to that occurring in selective inversion-recovery experiments (17). It is important to realize that the cross-relaxation processes on which these experiments rely can be extremely complex. Some appreciation for this can be gained by reference to works such as those by Bothner-By and Noggle (18). Macura and Ernst have presented a theoretical framework for analysis of relaxation effects in two-dimensional experiments (19). It is ordinarily only in the limit of short mixing times that the amplitudes of cross-relaxation peaks are simply interpretable in terms of relaxation rates and, perhaps, internuclear distances for the pair of protons involved. Fortunately, distance dependence is very steep (r^{-6}), translating large errors in cross-relaxation measurements into small errors in distance. Also, rather qualitative restrictions on distances are often all that is needed for application to macromolecular structure. Analysis of data in quantitative terms will be discussed more fully in the applications presented later in this chapter.

After the mixing period, a third 90° pulse is applied. This transfers z magnetization, which is modulated during t_1 by the precession of a primary resonance, to the xy plane. During t_2, this xy magnetization will be modulated either by the same precession frequency or by a different frequency, depending on whether or not the z intensity was the result of transfer of magnetization to a secondary resonance via cross relaxation. The z magnetizations sampled in this way are observed to precess during t_2, and the t_2 and t_1 time domains are transformed in the usual way. The presentation is identical to the COSY format. In fact, procedures have been presented for the integration of the two experiments into a single acquisition (20, 21), and it is a common practice to integrate displays so that one side of the diagonal presents coupling-correlated data and the other side presents cross-relaxation-correlated data (22). Constructions to establish through-space connectivities are identical to those shown in Fig. 6-3.

Cross-relaxation 2D experiments are rare examples of experiments which are actually easier to perform on a large molecule than on a small molecule. For a pair of proximate nuclei, magnetization can be transferred either by zero-quantum or double-quantum processes. For small molecules the double-quantum pathway dominates. The result has at best half the efficiency of the total T_1 relaxation process. Thus, magnetization of the primary resonance returns to equilibrium before very much transfer can occur. For large molecules, the zero-quantum process, which can be many more times efficient than T_1 relaxation, dominates, and transfers are efficient.

Thus, we will not present a 2D cross-relaxation-correlated set for methyl galactose but will refer the reader to the set presented later in this chapter to illustrate oligosaccharide structural analysis (Fig. 6-19).

Selection of parameters for acquisition in the relaxation-correlated experiments is even more critical than that for the coupling-correlated experiment. Magnetization transferred between resonances of interest can be smaller and recovery between acquisitions is probably more important. Both make the cross-relaxation experiment more time-consuming. The increment for t_1 is, of course, dictated by the necessary sweep width. Experiments are run in quadrature using a phase-cycling scheme similar to that described for the COSY experiment. The number of t_1 points needed is not affected by considerations of transfer efficiency but only by considerations of resolution. Moreover, cross-peak components tend to be in phase and do not cancel when unresolved as they do in COSY experiments. Hence, it is often possible to truncate data set sizes in t_1 substantially while retaining a size in t_2 sufficient to give adequate resolution in one dimension. A large size in t_2 may be limited by data storage capacity or required processing time, but it does not make a significant contribution to acquisition time. For the data presented in Fig. 6-19, 1 K complex points were used for an F_2 sweep width of ± 1600 Hz, but the number of t_1 points was reduced to 192 while holding the F_1 sweep width at ± 1600 Hz by using an increment in t_1 of 0.0003125 sec.

Selection of a mixing delay can be quite critical. At early times transferred magnetization grows in a near-linear fashion so longer delays mean larger cross peaks. At longer times, intensities become difficult to interpret and eventually there result peaks from secondary transfers. Delays are thus chosen as a compromise between ones that are long enough to provide reasonably large perturbations of z magnetization of correlated resonances and those that are short enough to give interpretable results. If any quantitative interpretation is to be attempted, it is usually wise to perform experiments at two or more delays so that comparison of intensities of cross peaks in the experiments can be utilized to assess the validity of the short mixing time assumptions. Typical delays for a small protein are 0.1 and 0.3 sec (22). For more qualitative applications, where a single mixing time might be used, 150–170 msec gives a good signal-to-noise ratio for cross peaks with minimal interference from auto peaks. Most protons of the glycolipid in Fig. 6-19 have relaxation times of 1–2 sec at 30°C in DMSO. A mixing time of 0.125 sec was used along with a recycling time of 1.2 sec. This recycling time is a little short for good quantitative applications and should be lengthened where quantitative interpretations are anticipated.

One other point about mixing delays: they are sometimes randomly changed in length by approximately 10% during the course of acquisition in order to eliminate scalar-coupling-correlated cross peaks. That coupling-correlated peaks could arise is clear in that the first two pulses constitute exactly a COSY experiment. Since the transfer of scalar correlations is modulated in time, and information transferred via z magnetization decays monotonically, these small variations allow for cancellation of the former effects with little effect on the latter. Other options for selection of cross-relaxation peaks over coupling-correlated peaks have been proposed (23, 24), but were not used in the illustrations presented here.

Processing also requires some special consideration if data from cross-relaxation experiments are to be quantitatively interpreted. Technically, volumes of peaks in the 2D presentation should be used to quantitate peaks, and weighting functions which severely distort peak shapes, such as those used for resolution enhancement, should be avoided. In cases where it can be assumed that all peaks have the same width, these restrictions can be relaxed.

III. Advanced Methods for Biological Samples

Unlike the simple examples presented above, two-dimensional spectra of biological macromolecules are very susceptible to limitations imposed by inherently broad resonances, by spectral overlap due to the large number of resonances present, and by the presence of intense resonances from the solvent or the macromolecule itself. These problems are often compounded by the current method of two-dimensional data display: the contour plot, which often is unable to depict true resolution, particularly when resonances of interest ride on the sides of large resonances.

Several aids to spectral resolution have evolved to cope with these problems. Some of them involve modifications of the ways in which spectra are acquired. Some simply involve choice of appropriate processing and display parameters. We will first talk about some methods of data acquisition; then we will discuss aids to processing; then we will address the special problem of eliminating the water resonance in protonated media.

A. Multiple-Quantum Coherence

One-dimensional experiments, and the two-dimensional experiments discussed in the previous section, rely on the precessional properties of one-quantum coherence. In the language of time-dependent perturbation theory, these coherences are associated with $\Delta m_I = 1$ selection rules and give rise to the transverse magnetization that most spectrometers detect. The large numbers of $\Delta m_I = 1$ transitions, the precessional frequencies, and inherent relaxation properties dictate limits for resolution in experiments based on one-quantum coherence.

Multiple-quantum coherence experiments, as the name implies, involve precessional and relaxation properties of phenomena arising from transitions between levels with $\Delta m_I > 1$. The frequencies, relaxation times, and number of transitions are different and in many cases allow improved resolution (25, 26). Because only $\Delta m_I = \pm 1$ transitions can be directly excited or detected with standard spectrometer configurations, multiple-quantum coherences are generated and detected indirectly. A typical experimental procedure is outlined in Fig. 6-4. The usual generation of this type of

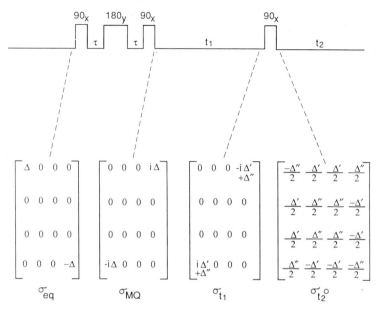

Fig. 6-4. Pulse sequence and density-matrix interpretation for a multiple-quantum experiment.

coherence is achieved by two or more pulses and is closely related to the INADEQUATE experiment introduced by Bax et al. for selective detection of ^{13}C-^{13}C coupled pairs in ^{13}C spectroscopy (27). After the coherence is produced, a t_1 time period is incremented to form the multiple-quantum evolution domain. The evolution period is followed by a single mixing pulse which samples multiple-quantum coherence and transforms it back into single-quantum coherence where it can be detected during period t_2.

Understanding these experiments requires a density-matrix or product-operator formalism. Treatments are presented in the book by Bax (28) and the book by Ernst et al. (29). Detailed discussions have also been presented in Chapters 2, 3, and 4 of this book. We will discuss the experiment here only to the point necessary to illustrate potential improvements in spectral presentation and to make reasonable decisions about acquisition parameters.

For a simple two-spin case ($I = \frac{1}{2}$) the pulse sequence can be followed by considering elements of a rotating frame, spin-deviation density matrix σ, as presented in the lower half of Fig. 6-4. Prior to excitation we assume the system to be at equilibrium so that only diagonal elements are populated. Only the deviation Δ from uniform population distribution is indicated.

The first 90° pulse of the excitation sequence can be considered to populate $\Delta m_I = \pm 1$ elements such as σ_{12} and σ_{21} from the corresponding diagonal elements, σ_{11} and σ_{22}. These elements give rise to transverse magnetization and are modulated by chemical shift, δ_A, and scalar coupling, $\pm J/2$, during τ. The 180° pulse serves to eliminate the effects of chemical shift modulation at the time of the second 90° pulse, a fact that helps produce uniform excitation across the spectral range. The second 90° pulse then populates double quantum elements, σ_{14} and σ_{41}, in proportion to differences in single quantum coherences (i.e., $\sigma_{12} - \sigma_{43}$). The delay period τ is obviously important since, at $\tau = 0$, populations of σ_{12} and σ_{43} are equal and no excitation would result. As discussed in Chapter 4, excitation, in the limit of long relaxation times, is optimized at $\tau = 1/4J$ for this simple two-spin case. In practice, shorter delays are used, for higher-order multiplets and when magnetization is lost through spin relaxation. Once double-quantum elements (2Q) are populated, they evolve during t_1 in a manner dependent on scalar coupling and chemical shifts of both spins. For this two-spin case, evolution is at the sum of the chemical shifts of spins A and B and does not show the effects of A-B scalar coupling.

The final 90° pulse in the sequence of Fig. 6-4 is responsible for returning double-quantum coherence to a single-quantum coherence, which can be detected. Multiple-quantum coherence cannot be detected directly, so this final pulse is absolutely essential. Phase and flip angle for this pulse are also important considerations. First, it is desirable to eliminate one-quantum coherences produced by this final pulse as well as one-quantum coherence produced by earlier pulses due to pulse imperfection, spin relaxation during τ, and an inability to choose an optimum τ for all multiplets. Desired orders of coherence can be selected by appropriate phase cycling as discussed in Chapters 2 and 4. We will use a scheme that selects for orders two, six, etc. (30, 31). Second, phase and duration of the final pulse are important in providing quadrature detection. For example, the multiple-quantum experiment as described in Fig. 6-4 produces amplitude modulation of detected magnetization as a function of t_1 and hence no quadrature detection. There are several ways of introducing quadrature detection. One is to use a 135° rather than a 90° pulse (32), another is to repeat the acquisition in such a way as to produce a 90° rotation of the multiple quantum coherence before detection (33), and perhaps the most efficient is to use time-proportional phase incrementation (14, 24). Quadrature detection in the t_1 domain through the superposition of two sets with a 90° phase change in the multiple-quantum coherence is relatively easy to understand. If we consider our first accumulation to have sampled the real component of 2Q magnetization, we need to phase shift the coherence by 90° and transfer it into the imaginary channel with the last pulse. Since pulse rotations have twice their one-quantum velocities in a two-quantum domain, rotation is accomplished with a composite 45°_z rotation using a composite pulse suggested by Bax *et al.* (34), and a 90° shift of the final pulse assures storage of the rotated 2Q coherence in a one-quantum coherence of appropriate quadrature relationship.

As seen in Fig. 6-5, the presentation of a transformed 2Q experiment is distinctly different from a COSY or NOESY spectrum in that cross peaks showing connectivities of resonances A and B have the same t_1 modulation frequency $\nu_A + \nu_B$. Hence the connectivity construct is a horizontal line. The data presented are again for 1-O-methyl-β-D-galactose and are plotted with the same f_2 width as the previous COSY and J-resolved spectra of this molecule (Figs. 6-1 to 6-3). It is useful in examining these plots to draw a diagonal at $F_1 = 2F_2$. Connected cross peaks then occur on horizontal lines at equal displacements from this diagonal. This is clear from the connectivity in Fig. 6-5 of a peak at +0.40 ppm for H1 with a peak at −0.43 ppm for H2.

It is useful to pair peaks in this way because it is often necessary to distinguish 2Q cross peaks showing simple pairwise coupling of spins from cross peaks resulting from more complex coupling. For example, in a linear AMX spin system with no direct A-to-X coupling, two-quantum coherences involving A and X can be transferred to the M resonance but not to A or X resonances. Thus a cross peak appears at a frequency ν_m in F_2 and $\nu_A + \nu_X$ in F_1. Cross peaks generated in this way have no partners symmetrically displaced across the diagonal and can easily be identified (35). Recently, methods to give such peaks special phase properties have also been introduced (36).

An example of a $\nu_A + \nu_X$ correlation peak appears in Fig. 6-5 for the H2 proton ($F_2 = -0.43$ ppm). Direct connectivities appear at F_1 values of −0.01 ppm (H1) and −0.68 ppm (H3). The indirect connectivity (circled in the figure) appears at 0.12 ppm, which is equal to the H1 shift (0.40) plus the H3 shift (−0.28). This type of connectivity has been used to advantage in spectra of proteins. Here, amide resonances can be connected with β resonances of amino acids through the appearance of a $\nu_A + \nu_X$ peak at the position of the α-proton resonance (37).

Since the multiple-quantum experiment provides similar scalar connectivity information to that from the COSY experiment, it is appropriate to discuss its relative advantages and disadvantages. The reasons for considering a multiple-quantum experiment are basically ones of spectral simplification. The power of the method was perhaps best illustrated in the work of Warren *et al.* on solid samples showing dipolar coupling (30). A first-order, one-quantum spectrum of a molecule having just six protons but lacking symmetry would have 192 lines. But as you observe higher and

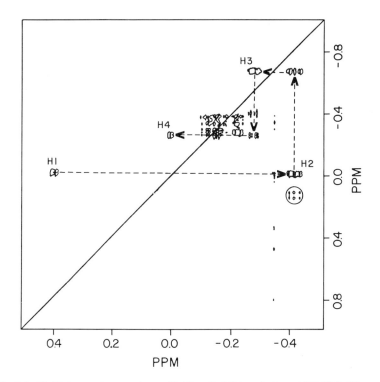

Fig. 6-5. Multiple-quantum spectrum of 1-O-methyl-β-D-galactose at 490 MHz. The sample conditions are as given in Fig. 6-1. Sweep width in F_2 is ± 250 Hz with 256 complex points. Sweep width in F_1 is ± 500 Hz with 128 points (increment 0.001 sec). Each t_1 point was the result of 16 scans at a cycle time of 2 sec. Processing employed a sine-bell function in both dimensions, zero filling to 256 in t_1. The set was processed as a phase-sensitive spectrum. Both positive and negative contours are plotted. Shifts are referenced to the carrier frequency in both dimensions. The F_2 dimension may be converted to a TMS scale by adding 3.85 ppm to the scale given. Direct connectivities are shown as dotted lines. An example of a remote connectivity is circled. A phase-cycling scheme is depicted below. The labels ϕ_1, ϕ_2, ϕ_4, and ϕ_5 correspond to pulses as diagrammed in Fig. 6-4: ϕ_3 is a 22.5° pulse, inserted just before ϕ_4, which is used to produce real and imaginary sets for quadrature processing in F_1; ϕ_{rec} is the receiver phase. The phase list is shown only for the real set. The imaginary set has ϕ_3 incremented by 180° and ϕ_5 incremented by 90°. Quadrature phase cycling in F_2 can be implemented by incrementing all phases in 90° steps and repeating the sequence.

Acquisition number	ϕ_1	ϕ_2	ϕ_3	ϕ_4	ϕ_5	ϕ_{rec}
1	x	y	x	y	x	x
2	y	$-x$	y	$-x$	x	$-x$
3	$-x$	$-y$	$-x$	$-y$	x	x
4	$-y$	x	$-y$	x	x	$-x$
5	x	$-y$	x	y	x	x
6	y	x	y	$-x$	x	$-x$
7	$-x$	y	$-x$	$-y$	x	x
8	$-y$	$-x$	$-y$	x	x	$-x$

higher orders of coherence, the number of possible resonances decreases until at $\Delta m_I = 6$ only one line is observed. The rationale in 2D spectra is similar—reduction in numbers of resonances. The lack of A–B scalar modulation in the t_1 time domain of an A+B transition reduces the number of peaks in a connectivity group. Also, one-quantum coherences can be eliminated from a 2Q spectrum. In a COSY spectrum it is often very difficult to observe connectivity resonances near the diagonal, because of the intensity of this region. Note that in the 2Q spectrum presented in Fig. 6-5, there are few peaks on the diagonal, or more appropriately no pronounced line at $F_1 = F_2$ or $F_1 = 2F_2$. Moreover, singlets, such as that from the O-methyl at −0.36 ppm, have been largely removed. These resonances have been eliminated by appropriate phase cycling.

One of the disadvantages of the MQC experiment is the large F_1 frequency domain required. Even with quadrature detection, the required t_1 domain for equal resolution is about twice the size of a COSY experiment ($\nu_A + \nu_B$). Some economies can be realized if coupling is within a limited region of the spectrum. Also, since excitation of cross peaks is independent of t_1, t_1 can be truncated to any extent that resolution can be sacrificed. Although multiplet components are antiphase in F_2, direct connectivity multiplets for simple spin systems are in phase in F_1, and cancellation is minimal in poorly resolved spectra.

It is sometimes convenient to retain the COSY presentation format used in so many other experiments (24). It is possible to retain this format and still utilize the ability of a MQC experiment to selectively retain only connectivity peaks. Rance et al. have presented a means of doing this in protein spectra based on a multiple-quantum filter (38). The sequence is a straightforward modification of that presented in Fig. 6-5. The 180° pulse is eliminated, the 2τ period becomes the t_1 period, and the t_1 period in the MQC experiment is fixed at a short value. It is much like a COSY experiment, except instead of observing after the second 90° pulse, one eliminates the transverse magnetization by systematic phase cycling, and coupling correlated information is saved in the 2Q coherence that this pulse excites. This information is read by the third 90° pulse in the same sense as that described for the MQC experiment.

The ideas behind multiple-quantum-coherence selectivity can, of course, be extended. Some extensions to identification of spin-system type are found in the work of Piantini et al. (39). In the assignment of spectra of macromolecules it is often useful to know what the spin system is. For example, alanine, dissolved in D_2O to eliminate the amide proton, is unique as an AX_3 spin system. Identifications can sometimes be made on the basis of selection rules for the appearance of diagonal and cross peaks when filters of different orders of coherence are used. Although extensive application of these methods has not as yet been made, they are mentioned here to indicate possible directions for future development.

A related point that will provide some useful applications is the ability to avoid the necessity of following connectivities through each and every resonance in a scalar-coupled sequence. This is often not possible because of a high level of degeneracy in certain regions, or the obscuring of certain regions by large solvent resonances.

For example, the fact that the position of three-quantum resonances for a linear AXY system depends on the shift of all three nuclei could allow connection of A and Y and identification of the shift of X without observing the X resonance. Again, these experiments will not be described in further detail, but relevant references are given (35, 39).

B. RELAYED COHERENCE

It is possible to establish remote connectivities by means other than multiple-quantum coherence. Such procedures can be viewed as successive one-quantum coherence transfers (40, 41). The sequence usually begins in a manner similar to a COSY sequence. A 90° pulse excites transverse magnetization and it is allowed to evolve during an incremented time period t_1. At the end of this period, magnetization is transferred to a second resonance by a second 90° pulse. Rather than sampling magnetization at this point, however, a second transfer is initiated by a third pulse. The delay between the second and third pulses can be adjusted to $1/2J$ to optimize transfer for typical coupling constants, and a 180° pulse can be placed in the middle of the delay to remove effects of chemical shift on transfer efficiency. There are some limitations and disadvantages in the experiment. Both single- and double-transfer cross peaks survive in the sequence as described, so it is necessary to acquire a normal COSY experiment for comparison. Recently, however, methods for minimizing single-transfer cross peaks have been devised (42). Intensities for cross peaks established on the basis of small couplings are also very low because of potential losses at both stages of transfer.

Typical experimental parameters are given in the caption of Fig. 6-6, along with a sequence in which delays can be set to minimize single-transfer peaks and which has quadrature in F_1. The figure shows a spectrum of 1-O-methyl-β-D-galactose. Note that, in addition to a connectivity between H1 and H2 (4.27 to 3.44), a relayed connectivity from H1 to H3 is seen at 3.58 ppm.

These experiments prove valuable for the analysis of protein spectra. Amide proton to α proton connectivities in COSY spectra are often difficult to see because α resonances are degenerate or lie near the H_2O resonance. Direct observation of relay peaks to β resonances bypasses these problems

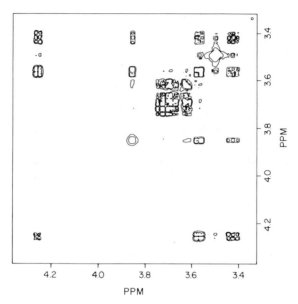

Fig. 6-6. Relayed coupling-correlated experiment on 1-O-methyl-β-D-galactose at 490 MHz. The experiment is the result of 128 t_1 points incremented by 0.001 sec for a sweep width of 500 Hz. Sixteen scans were taken for each file at a repetition rate of 2 sec. The F_2 dimension is ±500 Hz in 512 complex points. The carrier is at the left-hand edge, and only the upfield half of the spectrum is plotted. To produce quadrature in t_1 as well as to minimize normal COSY peaks, the following sequence can be used with phase cycling as shown: $90°-t_1-90°-\tau-180°-\tau-90°-t_2$.

Acquisition number	ϕ_1 (90°)	ϕ_2 (90°)	ϕ_3 (180°)	ϕ_4 (90°)	ϕ_{rec}
1	x	x	x	x	x
2	x	x	x	−x	x
3	x	−x	−x	x	x
4	x	−x	−x	−x	x
5	x	y	y	y	−x
6	x	y	y	−y	−x
7	x	−y	−y	−y	−x
8	x	−y	−y	y	−x

and makes identification of certain amino acid types more straightforward (43). Coherence transfers have been extended over even longer ranges recently, making observation of nearly all resonances of an amino acid spin system possible through transfer to the amide resonance (44).

It is important to realize that RELAY experiments can be used in heteronuclear as well as homonuclear experiments. In nucleic acids, for example, following a pattern of scalar couplings from one ribose ring to another would necessitate utilizing a ^1H-^{31}P coupling. Relayed coherence from both neighboring and remote protons in nucleotides has been observed (45, 46).

C. CHOICE OF PROCESSING METHODS

NMR spectroscopists are usually quite familiar with the process of sensitivity or resolution improvement through the choice of appropriate convolution functions. For optimum sensitivity, in one-dimensional spectra, multiplication of the time-domain signal by an exponential of the same decay constant as the FID itself is the common choice. This maintains the Lorentzian line shape and introduces line broadening by a factor of two.

Signal-to-noise improvement in a 2D spectrum, particularly when a macromolecule is involved, can be a far more complex process. In part, this arises because in practical terms we seek an ability to distinguish various resonances in a two-dimensional contour map. Resonances can be obscured in such a presentation, not only by random phase noise, but by tails from other larger resonances. They can also be obscured by vertical stripes called t_1 noise emanating from strong, narrow resonances. Removal of these obstructions is often more important than optimization of statistical signal-to-noise ratios.

Lineshape is a particularly important consideration. In a 2D plot, a Lorentzian line yields contours that are star-shaped, as illustrated in Fig. 6-7a. Lineshapes generated by multiplication by other functions such as sine bell and Lorentzian-to-Gaussian transformation are far better, as illustrated in Fig. 6-7b (28, 47). The Lorentzian-to-Gaussian transformation is usually applied by removing the normal exponential decay through multiplication by an exponential with decay constant set equal to the negative of the natural T_2 (LB = $1/\pi T_2$) followed by multiplication by a Gaussian function, $\exp(-t^2/T_g^2)$, in which T_g is chosen to be approximately equal to the time at which the last recognizable portion of an FID from macromolecular resonances disappears. One must also choose T_g small enough that the modified FID is near zero at the end of the time domain. If it is not, truncation effects produce oscillations near the base of resonances that can have deleterious effects on contour plots.

A steeper drop to a baseline can be achieved by use of functions that attenuate the beginning of the FID relative to its midpoint. The sine-bell function is an extreme case in which data actually start and end at zero. These functions normally introduce negative components near the base of peaks, which can be deleterious when phase-sensitive spectra are to be presented or integrals are to be used.

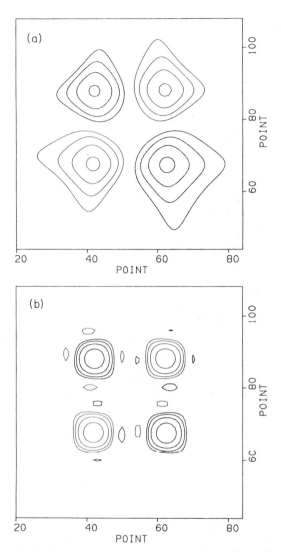

Fig. 6-7. Illustration of the effects of two common convolution functions: (a) exponential multiplication and (b) an unshifted sine-bell. The plots are from the cross peak of a simulated COSY experiment processed as a phase-sensitive spectrum. Both positive and negative contours are plotted.

All convolution functions have the ability to trade signal-to-noise ratio (S/N) for resolution. The choice is, therefore, always a compromise, and several trials may be necessary to find an optimum. Increased S/N can be achieved with a sine-bell function by means of a shift in phase. At 90° phase shift, a cosine function is generated. More S/N in the Lorentzian to Gaussian transformation is achieved by using smaller values of T_g.

The wings generated by natural lineshapes are often aggravated by the display of power or magnitude spectra in 2D plots. This is done to avoid the extremely complex phase properties of lines generated in 2D spectra. The double dispersive lines generated for diagonal peaks of a COSY spectrum are particularly apparent in the simulated spectrum on the left of Fig. 6-8, where a set of diagonal peaks is shown in the lower part and a set of cross peaks in the upper part. Rather than attempting to phase correct these peaks, the spectroscopist often generates a magnitude spectrum by taking the root-mean-square combination of real and imaginary sets. The problem with using magnitude spectra can be understood if one views a spectrum with Lorentzian lines adjusted for pure absorption in the real channel. The process of calculating a magnitude spectrum then adds a dispersion component from the imaginary channel to what would have been

Fig. 6-8. Simulation of normal COSY and multiple-quantum-filtered COSY of a two-spin system. The stacked plot on the left shows auto peaks (bottom) and cross peaks (top) from the normal COSY processed as a phase-sensitive spectrum. The stacked plot on the right shows auto peaks (top) and cross peaks (bottom) from the MQ-COSY processed by the method of States *et al.* (48). Both have been weighted by a cosine function in the t_1 and t_2 dimensions. The pulse sequence and phase cycling for the real part of the multiple-quantum-filtered set are as follows: $90°—t_1—90°—\tau—90°—t_2$.

Acquisition number	ϕ_1	ϕ_2	ϕ_3	ϕ_{rec}
1	y	x	x	x
2	y	x	y	-y
3	y	x	-x	-x
4	y	x	-y	y

The cycling for the imaginary set differs in that ϕ_1 is advanced by 90°. Quadrature phase cycling in F_2 may be implemented by incrementing all phases in 90° steps and repeating the above.

a perfect Lorentzian line. This dispersion component has maximum amplitude displaced by $(1/\sqrt{3})\Delta\nu_{1/2}$ where $\Delta\nu_{1/2}$ is the width at half height of the absorption peak. This greatly broadens the base of peaks unless efforts at resolution enhancement are made through the use of convolution functions.

There is a more straightforward means of dealing with the problems leading to the use of magnitude spectra than compensating with convolution functions. The procedure involves collection of spectra in modes that allow phasing of resonances for pure absorbance. Procedures have been developed for multiple-quantum spectra, multiple-quantum-filtered COSY spectra, and NOESY spectra, among others (38, 48). Most methods are based on a procedure outlined by States et al., in which spectra sampling real and imaginary components of t_1 evolution are stored separately rather than being co-added during acquisition. Experiments are selected in which t_1 evolution leads to amplitude modulation of t_2 signals in the real or imaginary channels. After the transformation in t_2, residual signals in the two sets can be phased to be pure absorption in real and imaginary channels, respectively, using one of the first t_1 files as a model. The dispersion component is then zeroed and the sets combined before transformation in t_1. The result is illustrated in the simulation on the right side of Fig. 6-8 for a multiple-quantum filtered COSY spectrum. Diagonal peaks are in the upper part and cross peaks are in the lower part. The procedure does retain a Lorentzian lineshape in both dimensions. Resolution can be further improved with the convolution functions mentioned above. While the procedures sound time-consuming, it is important to realize that better resolution and contour definition lead to an improvement in spectral quality that often more than compensates for the extra acquisition and processing time. A method for obtaining pure absorption spectra based on time-proportional phase incrementation has also been introduced. This gives similar results, perhaps with less processing effort (14, 49).

Phase-sensitive spectra offer some other advantages. One is that contour-plotting routines which are arranged to plot only positive peaks will yield only half the number of peaks in many phase-sensitive spectra because of an antiphase relationship of multiplet components. This often reduces the level of spectral complexity. The very characteristic pattern of positive and negative peaks could also prove very valuable in pattern recognition algorithms (50).

In addition to distorted line shapes, a factor important in obscuring wanted spectral information is the t_1 noise that appears as vertical lines of noise associated with intense narrow peaks in the t_2 dimension. For an example, see Fig. 6-5, where a prominent vertical stripe is seen at the frequency of the O-methyl resonance (-0.35 ppm). These lines arise from a wide array of instabilities in spectral acquisition. Imagine what would happen as the result of any factor that could lead to unreproducible peak amplitudes as t_1 was incremented (receiver gain, pulse width, rf phase, T_1 relaxation, spin echoes, etc.). These pseudo-random amplitude variations, when transformed in t_1, appear as noise (51).

Aside from extraordinary attention to purity of rf and precision of spectrometer performance, there are a few things which can be done to eliminate these undesirable characteristics. One is spectral symmetrization. COSY, RELAY, and short-mixing-time NOESY spectra, in particular, have peaks related by inversion across the diagonal. It is clear that having cross peaks on both sides is somewhat redundant and that averaging of diametrically opposed points could improve signal to noise.

A simple averaging procedure, while improving signal to noise, produces only a modest reduction in t_1 noise. There is a second procedure often employed in magnitude spectra in which each pair of points about the diagonal is replaced by the lower amplitude member of the pair (52). While this procedure does little for signal to noise in real cross peaks, it is clear that ridges of t_1 noise will be almost entirely eliminated since the vertical lines have no counterpart on reflection about the diagonal axis. Spectra such as those presented in Figs. 6-3 and 6-6 have been processed in this way. Note the improvement in comparison to Fig. 6-5. Despite the advantages symmetrization offers in eliminating t_1 noise, it should be used with caution. In cases where spectra have marginal digital resolution, it is quite possible to lose signal on symmetrizing spectra (53). In such situations the most intense points representing cross peaks may not correspond exactly to the ones related by reflection across the diagram. Replacing both symmetry-related points with the lower-amplitude member of the pair may entirely eliminate a cross peak. The problem is particularly severe in phase-sensitive spectra where sign changes occur in the middle of cross peaks.

D. HETERONUCLEAR EXPERIMENTS

Resolution and assignment of resonances in protein spectra go hand in hand as major problems in the spectroscopy of macromolecules. In several instances it has been possible to take advantage of the improved chemical shift dispersion and the ease of resonance assignment for one nucleus type to overcome assignment problems encountered when observing spectra of a second nuclear type (54). Two-dimensional methods for chemical shift correlation of two different nuclei have been presented in previous chapters.

A direct application of the $\delta^{13}C$-δ^1H shift correlation sequence to proteins has been made by Chan and Markley (55). This sequence is quite analogous to the homonuclear scalar-coupling-correlated sequences except that the pair of 90° pulses is split between the two frequency domains. The first 90° pulse excites transverse magnetization of the proton domain, and an incremented delay, t_1, allows evolution of proton magnetization. Simultaneous 90° pulses to ^{13}C and ^1H domains mix magnetization of coupled pairs, and ^{13}C magnetization is detected and transformed as a function of t_1 and t_2. The chemical shift of carbon is shown in the F_2 dimension and the chemical shift of protons coupled to those carbons is shown in the F_1 dimension. Because scalar coupling is the source of the correlation, only the 1 percent of the carbon sites bearing a ^{13}C actually gives rise to the spectrum.

Peaks from the experiment as discussed show heteronuclear multiplet structure with typical couplings of 100–200 Hz. The coupling makes the proton projection look rather abnormal, and it is often desirable to remove this coupling. A 180° pulse to carbon at the midpoint of t_1 serves to decouple protons from carbon in the t_1 dimension. A small delay of the order of $1/2J$ is left at the end of t_1 to ensure the antiphase relationship of multiplet components necessary for transfer. Decoupling of protons during acquisition also collapses multiplets in the t_2 dimension. Again a small delay is used to ensure that collapsed multiplet components, which are excited with equal and opposite phase, evolve to the point where they will not cancel upon decoupling.

The advantage of this experiment in the investigation of macromolecules can be severalfold. At natural abundance, ^{13}C shifts often aid in the assignment of proton shifts. The absence of resonances not coupled to carbons—that from water, for example—also adds to spectral simplification. Enrichment in ^{13}C at specific sites can also lead to the direct detection and assignment of resonances from directly bonded protons. An example of data acquired in this way is presented in Fig. 6-9.

Carbon-13 is not the only rare nucleus of interest. Nitrogen-15 observation has always been a topic of discussion among those interested in NMR of biological molecules. In both proteins and nucleic acids, nitrogen participates in hydrogen-bonding interactions important for the maintenance of secondary structure. Both chemical shifts and coupling constants of ^{15}N sites should respond to these interactions and provide valuable structural information.

Direct observation of ^{15}N is, however, difficult because of both the low abundance (0.3%) and the low receptivity of this nucleus. Recently, methods have been presented by Bax *et al.* that allow detection indirectly via protons (56). The methods are based on two-dimensional double-quantum NMR, and pulse sequences are derived from those discussed in Section III,A of this chapter and in Chapter 4 of this book. Excitation can be achieved with 90° pulses, first to the proton domain and then to the ^{15}N domain, and

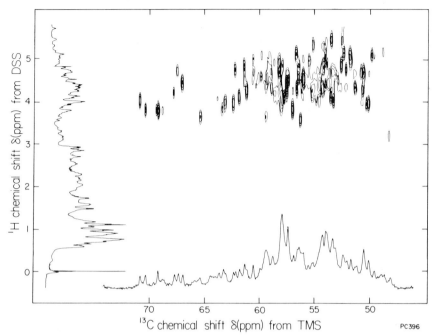

Fig. 6-9. Heteronuclear (^{1}H, ^{13}C) 2D contour plot for an oxidized *Anabaena variabilis* ferredoxin II sample. The region shown is for only a portion of the ^{1}H-^{13}C two-dimensional domain, the very large spectral regions requiring collection in separate experiments for high- and low-field regions. The sample was uniformly enriched to 20% in ^{13}C and run at 5.5 mM in 2.5 ml. At spectrometer frequencies of 200 MHz for ^{1}H and 50.3 MHz for ^{13}C, the data required 50 hr of accumulation for 128 sets of FIDs of 1024 data points each. One-dimensional proton and ^{13}C spectra are shown along the vertical and horizontal axes. The region selected shows predominantly alpha-proton–alpha-carbon connectivities. [Reprinted with permission from Chan and Markley, *Biochemistry* **22**, 5996 (1983). Copyright 1983 American Chemical Society.]

detection can be accomplished with a ^{15}N pulse after the double-quantum evolution. Magnetization is then detected in the proton domain. Since magnetization starts with protons rather than ^{15}N, signal intensity is improved by a factor of the ratio of gyromagnetic ratios. Since magnetization is detected as proton magnetization, an additional sensitivity enhancement of $(\gamma_H/\gamma_N)^2$ may be achieved. Thus, sensitivity, aside from losses due to decay of magnetization and the incomplete transfer normally present in multiple-quantum experiments, approaches that of a simple proton experiment. Experiments on natural abundance samples in the 100-mM range have been performed successfully (57).

The use of a sequence that places the last proton pulse at a time removed from observation is important in these experiments. For both proteins and nucleic acids, ^{15}N sites of interest are observed through exchangeable protons. This means that samples must be prepared in protonated water, and observation has all the complications of a water-suppression spectrum. Allowing a time interval before the observation period and after the last proton pulse helps greatly in avoiding saturation of receiver electronics.

Figure 6-10 presents data accumulated on an oxytocin sample. This cyclic nonapeptide (Fig. 6-11) illustrates the potential of the technique. The F_2 dimension, plotted as the vertical axis in this figure, is a normal proton spectrum except that the only resonances seen are ^{15}N-coupled, producing a series of doublets with 90-Hz ^{1}H–^{15}N coupling. The F_1 dimension in Fig. 6-10a is a straightforward heteronuclear double-quantum spectrum and displays the difference of ^{1}H and ^{15}N shifts. The experiment shown has been simplified by inserting a 180° ^{1}H refocusing pulse in the proton domain to remove proton couplings from the F_1 axis. It can be simplified further by appropriate tilting of the spectrum to remove the proton chemical shift from F_1. This has been done in Fig. 6-10b. The experiment shows excellent resolution. Since the proton spectrum has been previously assigned, the experiment also illustrates the ease with which correlated ^{15}N resonances can be assigned.

E. Two-Dimensional Spectra in H_2O

Among the special problems that two-dimensional NMR studies of biological macromolecules present, the necessity of operating in fully protonated media is perhaps the most severe. The necessity arises not simply from a zealous desire to approach physiological conditions but also from the importance of data that results from observation of exchangeable protons. As we will see later in the applications section, amide protons in proteins and imino protons in nucleic acids each provide essential information on the structure and dynamics of these molecules. The problem of protonated water as solvent is severe because of the enormous difference in signals produced by the water resonance (approximately 100 M) and the signals from the solute (typically in the millimolar range). This presents a dynamic range requirement on the order of 10^5. A 12-bit analog-to-digital converter found in most spectrometers can handle integers from 0 to 4095 and, even if a 16-bit ADC is employed, nonlinearity of other elements in the receiver section often limits accurate reproduction of the incoming signals.

Methods for dealing with high-dynamic-range signals have, of course, been developed for applications in one-dimensional studies. These have been reviewed recently by Hore (58). The methods range in complexity from something as simple as selective saturation of the water resonance

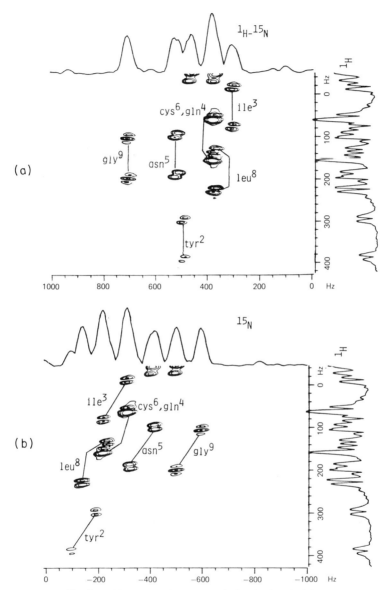

Fig. 6-10. The ^{15}N-^{1}H shift-correlated spectrum obtained using double-quantum spectroscopy at a proton frequency of 150 MHz. (a) Plot with the difference in ^{15}N and ^{1}H shifts along the horizontal axis. (b) Plot tilted to show only ^{15}N shifts along the horizontal axis. The sample is oxytocin in approximately 3 ml 95% H_2O/5% D_2O at 70 mM, pH 3. The ^{15}N sites are all at natural abundance. The data set required 64 t_1 files of 512 data points, each file incremented by 0.0005 sec in t_1. The set required seven hours of accumulation. [D. L. Live (unpublished results).]

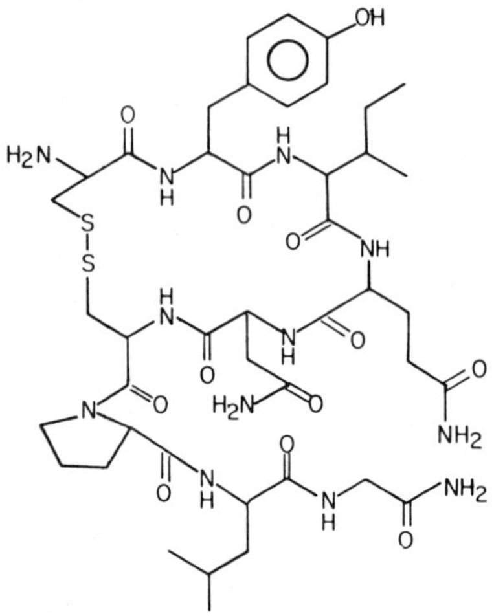

Fig. 6-11. Structure of oxytocin.

prior to observation, to rather complex multipulse sequences designed to excite a broad range of frequencies on either side of water, omitting water itself. The earliest in the latter class is, of course, the 21412 sequence suggested by Redfield and co-workers in 1975 (59). Deciding which method is optimum for a particular two-dimensional sequence and for a particular spectrometer is not an easy task.

With most spectrometers, reduction of the water signal by a factor of a few hundred is possible without unusual efforts. Reduction by a factor of 1000 or more requires consideration of the origin of the residual signal, which could come from B_1 inhomogeneity, from pulse imperfections, or from poor line shape due to B_0 inhomogeneity. Each case may respond best to a different choice of water suppression method. Effects of B_1 inhomogeneities may be eliminated by sequences designed to compensate for B_1 errors. Effects originating in B_0 inhomogeneities may respond best to pulse sequences with broad nulls near water (60). The point is that it may be necessary to experiment with a number of sequences on any particular spectrometer.

With respect to implementing these sequences in two-dimensional experiments, there are additional problems. First, there are multiple pulses to deal with. Transverse magnetization emanating from the first pulse in a COSY experiment can be just as devastating as the magnetization emanating from the second. Second, cross peaks in two-dimensional experiments owe their intensities to transfers that occur in two or more pulses. Selective excitation sequences do not generate particularly uniform excitation across a spectrum.

While loss of half the intensity in a single-pulse experiment may be acceptable, loss of half at each pulse in a multiple-pulse experiment may not be acceptable. Third, there is the problem of t_1 noise. Variations in residuals from intense peaks can make major contributions to this noise, and stability of the residual may be more important than its amplitude.

Wüthrich and co-workers reviewed the possibilities for water elimination relatively early in their studies and concluded that one of the simplest procedures worked best, simple presaturation of water using a minimal decoupler level applied at the water frequency (61). This irradiation may be applied at various times during a multipulse sequence, including cw irradiation for a period long enough to allow saturation before the beginning of the sequence, continuation of this irradiation through the t_1 period, and continuation of saturation using a time-shared homonuclear decoupling sequence during acquisition itself. Use of homodecoupling sequences was found to be undesirable because of noise generated, probably because of random variations of residuals during t_1. Saturation during t_1 was useful but generated some unwanted Bloch–Siegert shifts in the F_1 dimension that can complicate interpretation of spectra.

The COSY spectrum with simple preirradiation suffers from recovery of the water resonance as the t_1 period is lengthened. One cannot make the t_1 period arbitrarily short because of resolution consideration for the F_1 dimension, and because establishment of cross peaks requires enough t_1 evolution time to dephase multiplet components.

Other experiments, NOESY and relayed COSY, do not suffer as much from water recovery during t_1 because both have a period after t_1 and just before t_2 acquisition, during which water saturation can be applied a second time. Presaturation is more effective in these situations. As an illustration of the quality of spectra that can be obtained, we present in Fig. 6-12 a RELAY spectrum on a relatively large protein, cytochrome c (12.5 daltons) (62). Only the amide region is shown in the F_2 dimension. The F_1 dimension has two groups of peaks, direct amide-α proton connectivities at 5.2–3.2 ppm and relayed proton connectivities at 3.2–0.8 ppm. Note the lack of any interference from water at the right edge (1.3 ppm from water) and along the line $F_1 = 4.7$ where wings from the water resonance might be expected.

Presaturation cannot be effectively used on all molecular systems. First, there are cases in which protons giving rise to the resonances of interest exchange at moderate rates with the protons of water. If this exchange rate is large in comparison to the spin–lattice relaxation rate of the proton to be observed, the resonance of the latter will saturate along with water and it will not be observable. Amide protons involved in the secondary structure of proteins seldom exchange at rapid rates. Even solvent-exposed amide

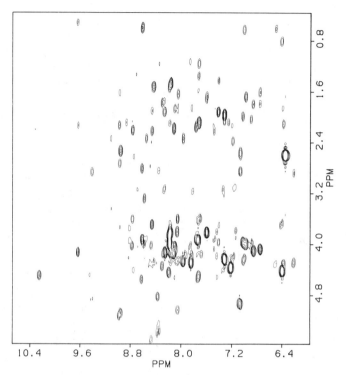

Fig. 6-12. Presaturation RELAY spectrum of cytochrome c. The sample is 12 mM in 10% deuterated, pH 5.7, phosphate buffer at 313 K. The data are the result of 128 acquisitions in each of 512 (1 K complex) t_1 files. Each acquisition was the result of a pulse sequence as described in the legend to Fig. 6-6, except that quadrature in t_1 was not employed and presaturation was applied for about 1 sec prior to the start of the pulse sequence and continued through the transfer period of $2\tau = 56$ msec. Data are weighted with sine-bell functions to 480 and 700 points in t_1 and t_2 respectively and zero-filled to 1 K × 2 K before transformation.

protons, when the pH is near 3, exchange at a low rate, so presaturation can be employed in most protein systems. In the case of imino protons in nucleic acids or in the case of small peptides at higher pH, exchange is faster and choices other than presaturation are often sought. Second, saturation cannot be carried out very selectively. Resonances from macromolecular protons lying near and under water are saturated along with water. Direct observation of these protons is not a primary concern because they would be extremely difficult to see against the water resonance anyway. However, with macromolecules at high field, very efficient equilibration of magnetization can occur among proximate protons through mutual spin flips (spin diffusion). This can transfer saturation throughout large regions of some molecules. Again, this is not a severe problem for proteins because there are relatively few protein resonances near water. For nucleic acids, a good fraction of the ribose or deoxyribose protons are near or under water and loss of resonance intensity throughout the spectrum can occur (63).

Selective pulses offer a viable alternative to presaturation, but there are several factors to consider. First, some selective pulses require that the transmitter be placed at one extreme of the spectrum. This may require very large F_1 and F_2 sweep widths, placing major demands on both acquisition time and data storage. The $45°$-τ-$45°$ sequence used by Moore and Kime, for example, falls in this class (64). One sets the transmitter several thousand hertz away from the water resonance and adjusts the frequency so that the water resonance executes a $180°$ rotation during τ. The second $45°$ pulse then returns the water magnetization to the z axis. Resonances near the transmitter rotate little during τ and experience something near a $90°$ pulse. If resonances of interest all lie in one region—imino protons, for example— this may be an ideal choice. The "jump-return" of Plateau and Gueron (65) in which the transmitter is placed on the water resonance frequency, on the other hand, allows smaller sweep widths and optimum excitation on both sides of water. The sequence can be described as x-τ-$(-x)$. Water that does not precess in the rotating frame of the transmitter is tipped to the y axis by the first pulse and returned to the z axis by the second pulse.

A second consideration is what to do about magnetization emanating from the earlier pulses in the sequence. Basus (66) has made effective use of water inversion pulses at the beginning of a sequence, coupled with a choice of a delay before the first pulse of the sequence that coincides with the null point in an inversion recovery curve. Water relaxation is usually much slower than relaxation of protons in a macromolecule. The first pulse therefore excites macromolecule resonances but not water. Intermediate pulses such as those that occur in NOESY experiments usually occur close in time to the first pulse and produce little water signal. The final pulse is then chosen to be a selective composite, such as the jump-return of Plateau and Gueron.

A third consideration related to the use of selective pulses stems from the necessity of carrying out extensive phase cycling in most 2D experiments. This sounds like a rather trivial problem but it is not. Some selective pulses—the jump-return, for example—require a set of pulses $(x, -x)$ closely matched in both phase and amplitude. When it is difficult to achieve this match with hardware, it can be done with small separate adjustments in the length of the two pulses or the addition of a third correction pulse of appropriate phase. Settings of corrections are often made more precise by attenuating the normal rf output of the spectrometer by a factor of two to four. The separate adjustment of pulse lengths or the addition of correction pulses, however, makes phase cycling very difficult. A choice of an experimental sequence that requires minimal cycling can prove advantageous.

As an example of a strategy that combines both consideration of the effect of early pulses in multiple-pulse 2D experiments and the minimization of required phase cycling of pulses near the end of the sequence, consider the application of a 2D multiple-quantum experiment to a water sample. The normal multiple quantum excitation, $90°_x$—τ—$180°_y$—τ—$90°_x$, has the effect of a 180° pulse on all singlet resonances in the spectrum, including water. This, in theory, leaves no x, y magnetization after the exciting pulse. Inspection of density matrix equations shows that the $90°_x$—τ—$180°_y$—τ—$90°_{-x}$ sequence is equally effective for double-quantum excitation. It has the net effect of a 0° pulse on all singlets, including water, and offers the additional benefit that it does not refocus transverse magnetization from prior pulses. After a delay, t_1, allowing precession of double-quantum coherence, transfer to the one-quantum domain can be accomplished using a pulse sequence which does not excite water, and the magnetization can be observed. Since the phase of the magnetization that has evolved through the double-quantum domain is affected by all pulses, it can be manipulated using only the excitation sequence. This is further removed in time from the point of observation, and phase cycling can be done with less concern for pulse imperfections. The phase of the last water-elimination pulse can be left fixed, with lengths and trim pulses tailored to one set of phases. An example of a spectrum acquired in this way is presented in Fig. 6-13, along with the phase-cycling procedure used (67).

IV. Structural Analysis of Biomolecules

The previous sections have focused on the mechanics of acquiring two-dimensional data sets for biomolecules. Of broader interest is the interpretation of those data in terms of structural properties. This has been accomplished for several classes of molecule, including peptides, nucleic acids, and carbohydrates. We will discuss applications to each of these in turn. We will first, however, digress to discuss the primary source of geometry information: cross-relaxation rates.

A. Relationship of Cross Relaxation to Structure

It is well recognized that dipole–dipole interactions provide a source of spin relaxation that is strongly dependent on the distance of internuclear separation: $1/r^6$. In proton NMR, where these interactions dominate spin relaxation, interpretation of various spin relaxation experiments in terms of specific distance constraints is possible. One-dimensional NOE experi-

6. APPLICATIONS OF 2D NMR TO BIOLOGICAL SYSTEMS

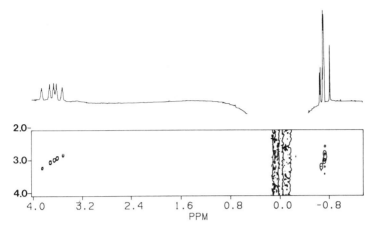

Fig. 6-13. Multiple-quantum water-suppression spectrum of 3.4 mM hexaglycine in 81% H_2O, 19% ^2H_2O, pH 1.9. The data were accumulated in 128 t_1 files at an F_1 sweep width of 2000 Hz. Each file was the result of 192 acquisitions into 1024 complex points at a sweep width of 2451 Hz and with a repetition rate of approximately 2.5 sec. The set was not acquired with quadrature in t_1 but could be by using the following pulse sequence, collecting two sets of data with ϕ_3 incremented by 180° and ϕ_5 and ϕ_6 incremented by 90° in the second set. Pulse sequence: 90°—τ—180°—τ—22.5°, 90°—t_1—90°—τ'—90°—t_2. Note that except for small τ', peak intensities are a complex function of both resonance frequencies and τ'.

Acquisition number	ϕ_1	ϕ_2	ϕ_3	ϕ_4	ϕ_5	ϕ_6	ϕ_{rec}
1	x	y	y	−x	x	−x	x
2	y	−x	−x	−y	x	−x	−x
3	−x	−y	−y	x	x	−x	x
4	−y	x	x	y	x	−x	−x
5	x	−y	y	−x	x	−x	x
6	y	x	−x	−y	x	−x	−x
7	−x	y	−y	x	x	−x	x
8	−y	−x	x	y	x	−x	−x

Quadrature phase cycling in F_2 may be added by incrementing all phases in 90° steps and repeating the above.

ments have, for years, provided a means of looking selectively at the cross-relaxation interactions between specific pairs of nuclei and extracting internuclear distances (68). This is not always a straightforward process. In the case of n distinguishable spins uncorrelated in their motion, the time variation in magnetization associated with each of the spins is best represented as a vector equation.

$$\begin{bmatrix} \dot{m}_1 \\ \dot{m}_2 \\ \vdots \\ \dot{m}_n \end{bmatrix} = - \begin{bmatrix} R_{11} & R_{12} & \cdots & R_{1n} \\ R_{21} & R_{22} & & \\ \vdots & & & \\ R_{n1} & & & \end{bmatrix} \times \begin{bmatrix} m_1 - m_1^0 \\ m_2 - m_2^0 \\ \vdots \\ m_n - m_n^0 \end{bmatrix} \quad [1]$$

In this equation, the R_{ij} terms are elements of a relaxation matrix describing the dependence of the rate of change of various magnetization components, $\dot{m}_i$, on the displacement of other components from their equilibrium values, $m_j - m_j^0$. For the experiments described here, we restrict ourselves to components of longitudinal magnetization. The elements of the R matrix are then given by

$$R_{ii} = \sum_j (W_0^{ij} + 2W_1^{ij} + W_2^{ii}) + R_{i1}$$
$$R_{ij} = W_2^{ij} - W_0^{ij} \quad [2]$$

where W_0, W_1, and W_2 are zero-, one-, and two-quantum transition probabilities for the pair of spins under consideration, and R_{i1} is a random field contribution to relaxation of the i spin. The transition probabilities are in turn given by

$$W_0^{ij} = C_{ij} J_{ij}(\omega_i - \omega_j)$$
$$W_1^{ij} = (3/2) C_{ij} J_{ij}(\omega_i) \quad [3]$$
$$W_2^{ij} = 6 C_{ij} J_{ij}(\omega_i + \omega_j)$$

where

$$C_{ij} = (1/10) \gamma^4 \hbar^2 r_{ij}^{-6} \quad [4]$$

and

$$J_{ij}(\omega) = \tau_c / [1 + (\omega \tau_c)^2] \quad [5]$$

In principle, different motions can modulate each interaction and each motion may be characterized by more than one correlation time, τ_c. However, we have neglected these complications by assuming that the same isotropic motion is common to all interactions.

Even with this assumption, it is clear that the time evolution of magnetization can be a very complex process depending on the deviation of all magnetization components from their equilibrium values and depending on all internuclear distances, r_{ij}. It is possible to approach this problem directly by solving the above set of linear coupled differential equations (18, 69). This would, however, require observations on a nearly complete set of resonances. For many macromolecules this is not possible because of spectral overlap.

The above equation can also be greatly simplified if only a single site has its magnetization m_k displaced from equilibrium and only the early stages of time evolution of magnetization are observed. Under these conditions, all $m_j - m_j^0 \neq m_k - m_k^0$ are set equal to zero and the set of linear differential equations represented by the above matrix equation greatly simplifies. Since $m_i - m_i^0$ remains close to zero at times near zero, its rate of change can be approximated as a simple linear function of r_{ik}^{-6}. If ratios of measurements involving known distances and unknown distances are examined, all constants and spectral density functions cancel leaving a simple expression relating distances to rates of change in magnetization:

$$(r_{ij}/r_{ik})^{-6} = (\dot{m}_k/\dot{m}_j) \qquad [6]$$

The cross peaks that appear in two-dimensional cross-relaxation or NOE spectra are an analog of one-dimensional NOE enhancements that can be used to extract specific distance information in macromolecules. Although selective perturbation of a resonance does not explicitly occur in these experiments, the origin of magnetization for each cross peak is made distinct by sinusoidal modulation of the perturbation at the resonance frequency of the donating spin, and the destination of the magnetization is made distinct by precession during the observation period at the frequency of the receiving spin. Macura and Ernst (19) have presented an analysis of cross-peak intensities in the limit where relaxation can be approximated using equations for the evolution of magnetization for pairs of coupled spins. After integration, intensities of cross peaks at any mixing time τ_m are given as follows:

$$I(\tau_m) = C\{\exp(-R_L\tau_m)[1 - \exp(-R_C\tau_m)]\} \qquad [7]$$

where R_L and R_C are the diagonal lattice leakage constant and the off-diagonal cross-relaxation constant, respectively. As above, R_C depends on the magnetic properties of the nuclei, the spectral density function of the interaction, and the inverse sixth power of the internuclear distance r_{ij}. The constant C also depends on the spectral density functions but reduces to $-1/4$ of the total two-spin magnetization in the limit of motional correlation times much longer than ω_0^{-1}. For short τ_m, the above equation can be linearized so that intensity $I(\tau_m)$ increases in simple proportion to τ_m, R_C, and the constant C.

If spectral density functions and magnetic properties can be assumed equal for all pairs, consideration of ratios of cross-peak intensities allows elimination of τ_m and all constants other than r_{ij}. In analogy to one-dimensional NOE experiments, comparison of intensities of cross peaks for nuclear pairs at known distances to intensities of cross peaks for nuclear pairs at unknown distances then allows calculation of distances as follows:

$$I_1(\tau_m)/I_2(\tau_m) = (r_{i2}/r_{i1})^6 \qquad [8]$$

The analysis can be extended to groups of equivalent nuclei by scaling cross-peak intensities by the number of equivalent receiving and the number of equivalent donating nuclei.

Some words of caution are in order when applying the above equation. The intensities are most appropriately taken from volume integrals. In two-dimensional spectra, where multiplicities of resonances spread cross peaks over both dimensions, intensities taken from cross sections through cross peaks can lead to rather large errors. Weighting functions should also be used with caution. When lines have different natural line widths, use of functions which truncate the early part of the FID, such as the sine-bell function, distorts relative intensities of cross peaks.

It is also important to remember the underlying assumptions in the derivation of the above equations. These include the assumption of weak scalar coupling (first-order spectra) and the use of short mixing times, τ_m. Normally, experiments are conducted with at least two τ_m values, 100 msec and 200 msec, and intensities of cross peaks are compared to validate the assumption of operation in a region where transfers are nearly linear in time.

The effects of having strong coupling are most serious when they produce cross peaks to spins that are at rather large distances from a spin under observation but show strong coupling to the observed spin or to a spin which is spatially proximate to the observed spin. When they arise from strong coupling, the intensities of these anomalous peaks scale approximately as J/δ and are usually not a problem for J/δ ratios less than 0.2.

Although the above cautionary notes seem very restrictive, the $1/r^6$ dependence on distance can hide a multitude of sins. When reference and sample cross peaks are of comparable magnitude, an error of a factor of two in relative intensity reduces to an error of less than 10% in relative distance. As an example, glycines in proteins frequently show strong NOE cross peaks from amide protons to both alpha protons, despite the fact that one is typically at a distance of 2.5 Å and the other is at a distance of 3.0 Å. This is because there is very efficient dipolar relaxation coupling between the two geminal alpha protons, and at typical choices of τ_m (150 msec) the assumption of linear time dependence of magnetization transfer is rather severely violated. The intensity of the primary cross peak for a 10-kDa protein will be in error by about a factor of two but the distance calculated will be in error by only a few tenths of an angstrom (70). This relative insensitivity to error makes the use of NOE cross-peak data possible in circumstances where simplifying assumptions are not well justified and where data quality may be poor. It is also possible to use the data by making

B. APPLICATION TO PROTEINS

Use of two-dimensional cross-relaxation data for structure determination of macromolecules is perhaps most advanced in the case of peptides and proteins. Normally the primary structure, the amino acid sequence, of the protein is known, and NMR data are used in secondary and tertiary structure determinations. Before this can be done, however, resonances must be assigned to specific residues in the sequence. A sequential assignment strategy, based on the known peptide sequence, has been outlined by Wüthrich and co-workers, and effectively applied to a number of proteins in the 5- to 10-kDa range (71-73). This includes the classic work on bovine pancreatic trypsin inhibitor.

We will illustrate the procedure with data on a protein of 8900 daltons, acyl carrier protein (ACP). The size of this protein is near the current limit of applicability of sequential assignments. It is, however, soluble to a concentration of more than 10 mM, and it has a high percentage of α-helical structure, allowing extraction of many sequential assignments from the well-resolved amide-amide connectivities alone. In applying the strategy, resonances, particularly alpha-proton and amide-proton backbone resonances, are first assigned to amino acid type. Scalar-coupling-correlated experiments of high sensitivity are necessary because a very high fraction of all possible connectivities must be observed to make the strategy successful. High concentrations and long acquisitions are necessary because connectivities stemming from groups with complex multiplet patterns or small vicinal couplings are often weak. Data for D_2O solutions as well as for H_2O may be required to achieve the necessary quality of results. Connectivity patterns allow grouping of resonances and classification as to spin system type (53). In a spectrum for a D_2O solution where amide protons are missing, these might be AX_3, AMX, AX, etc. Certain of these classifications correlate in a one-to-one fashion with amino acid type, for example, AX_3 with alanine and AX with glycine. In cases where spin system type does not allow a unique correlation, supplementary information from chemical shifts and NOE experiments often allows assignment. In aromatic amino acids such as tyrosine and phenylalanine, the alpha- and beta-proton portions of the spectrum show no scalar coupling to the protons in the aromatic region and the resonances in these portions appear as AMX spin systems indistinguishable from one another and indistinguishable from amino acids such as serine and aspartic acid. There is, however, a high probability of observing cross-relaxation interactions from protons at positions 2 and 6 in the aromatic rings with beta protons on the side chain. The unique shift in the aromatic region along with observation of this cross-relaxation connectivity would allow assignment of the entire spin system to an amino acid type.

Once alpha and beta protons are assigned to particular amino acids and connectivities to their respective amide protons are established from COSY or RELAY spectra in H_2O, assignment to particular places in the protein sequence can proceed. A small number of amino acids will occur only once in a protein of the size of ACP, and their resonances will immediately be assigned to a position in the sequence. Tyrosine 71, near the carboxy terminus of ACP, is a good example. If connectivities from the assigned amide or alpha proton of such a residue to the alpha proton or amide proton of the amino acid at position 70 or 72 could be observed, resonances grouped with these protons would be assigned. Unfortunately, scalar coupling from backbone protons on one residue to an adjacent residue is not observed in COSY spectra. Such a strategy must, therefore, rely on through-space connectivities observed in cross-relaxation-correlated spectra. Cross relaxation between amide protons on one amino acid and alpha, beta, or amide protons on an adjacent residue is conformation-dependent. Analysis of x-ray data on 16 proteins of well-defined structure shows, however, that one or more of the amide–alpha, amide–amide, or amide–beta distances, symbolized respectively by $d_{N\alpha}$, d_{NN}, and $d_{N\beta}$, has a very high probability of being within a range that will produce an observable NOE connectivity (74, 75). Figure 6-14 shows these distances in an extended segment of a polypeptide chain. Observation of such a connectivity identifies a pair of resonances as belonging to two amino acids at adjacent positions. By repeating the procedure, one could in principle begin at a unique site and sequentially assign groups of resonances throughout the molecule.

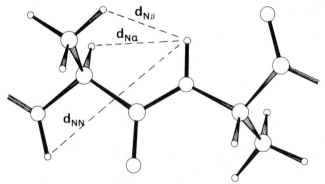

Fig. 6-14. Alanyl–alanine segment of an extended polypeptide chain showing probable cross-relaxation interactions.

6. APPLICATIONS OF 2D NMR TO BIOLOGICAL SYSTEMS

Figure 6-15 shows a down-field section of a 2D NOESY spectrum of ACP (76). This region shows only amide–amide connectivities, but it is clear that even with this limited set of information, rather long sequences of assignments can be made. The solid lines show a sequence of eight amide resonances, including that of the unique tyrosine residue, which have been assigned using the above strategy. Also shown is a sequence of five amide resonances including that of a unique arginine (R6). Such sequential connectivities can be corroborated and extended by examining regions showing alpha–amide connectivities. When connectivities are carried from the amide of one amino acid to an alpha proton of the next, a connectivity from an alpha proton to an amide in the same residue is best made in a COSY plot before proceeding to the next residue. Combined COSY-NOESY plots are useful in illustrating such sequential assignment pathways.

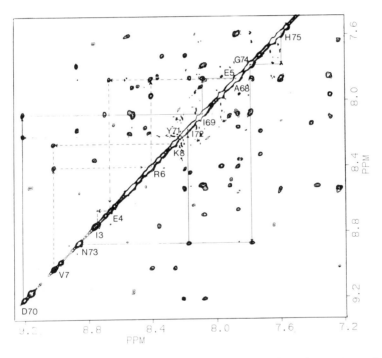

Fig. 6-15. Plot of the amide–amide section of a 2D NOE pure absorption spectrum of acyl carrier protein (ACP). The sample is 11 mM in 80% H_2O, 20% ^2H_2O, at 30°C, pH 6.1. The 490-MHz spectrum was acquired using a mixing time, τ_m, of 180 msec and 451 t_1 files of 2 K complex points each. Total acquisition time was 45 hr. The sequence for ACP is as follows:

STIEERVKKI$_{10}$-IGEQLGVKQE$_{20}$-EVTNNASFVE$_{30}$-DLGADSLOTV$_{40}$-ELVMALEEEF$_{50}$-DTEIPDEEAE$_{60}$-KITTVQAAID$_{70}$-YINGHQA.

In practice, the sequential assignment chain is broken periodically by the failure to observe a connectivity or by the degeneracy of an observed set of connectivities. A new starting point must then be identified. It is not always necessary to have unique amino acids as starting points. If alpha and amide protons are assigned as to amino acid type and three can be linked in sequence, there is only one chance in a hundred that the triad will occur more than once in a protein the size of ACP. Thus sequential assignment strategies are highly successful.

The next phase in structural analysis involves the assignment of various residues to secondary structural types: β-sheet, α-helix, or random coil. The intensities of NOESY cross peaks are of course very sensitive to internuclear distance. The values of interproton distances along the backbone, d_{NN}, $d_{N\alpha}$, and $d_{N\beta}$ as defined previously, vary significantly as secondary conformation is changed. In a β-sheet the distance $d_{N\alpha}$ is shortest, at less than 2.5 Å, with other distances greater than 3.5 Å. With short mixing times and low signal to noise, the $d_{N\alpha}$ peaks are likely to be the only observable connectivities. In an α-helix, both $d_{N\alpha}$ and d_{NN}, 3.4 and 2.8 Å, respectively, fall in a range which should show connectivities. While the cross peaks are a little harder to observe, the appearance of the d_{NN} peaks with intensities equal to or stronger than the $d_{N\alpha}$ peaks is a clear indication of an α-helical structure. Peaks for $d_{N\beta}$ connecting a residue to its neighbor on the amino terminus side (i to $i-1$) and remote $d_{N\alpha}$ peaks connecting a residue with one three units removed toward the amino terminus (i to $i-3$) also occur rather characteristically in an α-helical segment. These connectivity patterns can be combined with chemical shift tendencies such as that for β-sheet amides to occur further down-field, to reach rather definite secondary structure assignments. The connectivities in Fig. 6-15 are characteristic of α-helical structures.

The third step in structural determination involves analysis of the folding of the entire molecule into a three-dimensional tertiary structure. Perhaps the simplest approach would be to look for cross-relaxation effects that connect residues in different segments of secondary structure. This can, in favorable circumstances, produce a unique pattern of protein folding that can then be refined by a variety of procedures.

In principle, all observed cross-relaxation peaks could be converted to interproton distances and this set of distances used to determine a structure. These distances need not be extraordinarily accurate. There is an intriguing statement made in an early article by Havel, Crippen, and Kuntz (77) on the distance–geometry approach: "... complete specification of interresidue contacts and noncontacts at a 10 Å cutoff is sufficient to determine a structure with a precision comparable to most protein x-ray crystallographic studies." This has led to a distance–geometry approach to solving protein structures from NMR data (78). Here, matrices of upper and lower bounds for internuclear distances can be defined based on the simple presence or absence of cross-relaxation peaks along with other molecular bonding data, and a search for a set of distances lying between these bounds is instituted. The approach has been refined and successfully applied to a number of small proteins (79). More quantitative approaches integrating NMR data with molecular dynamics or molecular mechanics refinements of structure have also begun to appear (80, 81).

C. Application to Polynucleotides

One might anticipate that sequential assignment strategies and methods for tertiary structure determination would be applicable to other biopolymers. This has in fact proven to be the case. Ribonucleic acids consisting of linear sequences of adenosine (A), guanosine (G), uridine (U), and cytidine (C) monophosphates, as well as deoxyribonucleic acids, in which thymidine (T) replaces uridine, have been studied by two-dimensional NMR methods. When complementary sequences of four or more bases exist in these polymers there is a strong tendency for two chains to join in double-helical structures having interchain Watson–Crick base pairs (A–U or A–T, and G–C). Identification of regions showing such structures has been important in tertiary structure determination of tRNAs (82, 83). Several different forms of helical structure are also known to exist and are believed important in controlling fundamental steps in transcription. 2D NMR can contribute to the identification and characterization of these structures (84).

When bases are involved in hydrogen-bonded base pairs, such as an A–T pair, the imino protons involved in the hydrogen bond exchange slowly with protons in an aqueous solvent. These protons, which are depicted as small circles bonded by dashed lines to the bases of a single strand of a double helix in Fig. 6-16, give rise to discrete resonances in the 12–14 ppm region of the proton spectrum. If the resonances can be sequentially linked and identified as to base pair type, helical regions in a polymer of known sequence can be identified and properties such as thermal stability studied. Double helical structures normally place imino protons on sequential base pairs above or below one another at a distance of approximately 3 Å. At this distance nuclear Overhauser effects (NOEs) are useful in establishing sequential connectivities (85, 86). While one-dimensional studies have proven adequate in most cases, two-dimensional methods may become important as molecules become larger and spectra more complex.

Assignment of imino protons as to base-pair type can be done to a certain extent on the basis of chemical shift differences (82). NOE measurements between the imino proton and H2 of the adenine base are also useful in

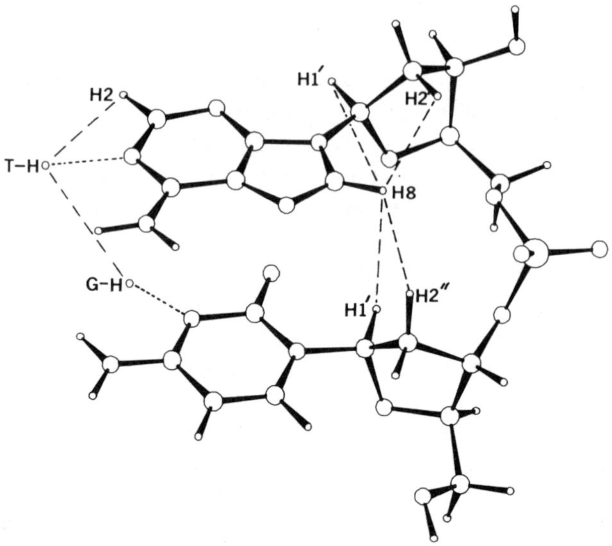

Fig. 6-16. Sequential nucleotides in DNA-B helix showing probable cross-relaxation interactions.

making these assignments. In principle, connection of a sufficient number of bases in a sequence could lead to unique placement in a known sequence. However, the smaller number of residue types in nucleic acids as compared to proteins makes assignment a little more difficult. It is frequently possible, however, to identify the imino protons of the terminal base pairs through their higher propensity for exchange with the solvent as temperature is raised, and to begin a sequential strategy from this point.

Two-dimensional methods have been relatively more important in the attempts to make use of ribose or deoxyribose protons and their cross-relaxation interactions with base protons for structural assignment (87–91). The three helical forms—A, the most common RNA structure, B, the most common DNA structure, and Z, the less common left-handed form—differ significantly in interproton distances. For example, in DNA having a B helix, the distance between H6 of the pyrimidine rings and H2′ of the attached deoxyribose rings is 1.8 Å and the distance between the H8 in a purine and the H2′ of the attached deoxyribose ring is 2.1 Å. In DNA having an A helix the corresponding distances change to 3.7 Å and 3.9 Å (92, 93). These changes are clearly within a range of distances easily differentiated on the basis of cross-relaxation effects.

Sequential assignment of ribose and deoxyribose protons has therefore become more and more important in recent years. Initial attempts at sequential assignments made use of heteronuclear scalar connectivities through the phosphate backbone (94). Fortunately, protons at positions 6 and 8, which are easily assigned to base type on the basis of multiplicity, chemical shift, proton exchange, and NOE effects, show cross-relaxation interactions not only with sugar protons in the same nucleotide but with sugar and base protons on the nucleotide to the 5' side of the sequence (see Fig. 6-16). Sequential assignment strategies based on homonuclear 2D NOE experiments have, therefore, also been developed (87–91). The most useful in B-type helices have been from H6 or H8 to H1' and H2" sugar protons. This is illustrated in Fig. 6-17 for a region showing aromatic proton to H1' connectivities. Examining a horizontal line at the chemical shift of an H1' proton assigned to a nucleotide near the terminus of the helix such as guanosine-2, one finds both an intraresidue connectivity to the guanosine H8 resonance and a connectivity to a cytidine H6 resonance. This second

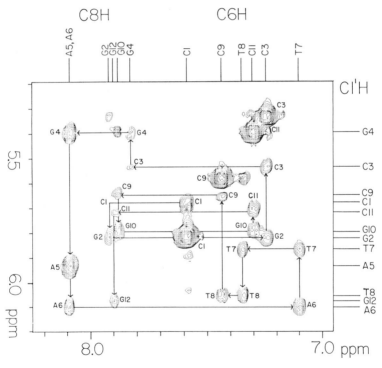

Fig. 6-17. Plot of the H6, H8 versus H1' region of a 2D NOE spectrum of a dodecanucleotide double helix, d(CGCGAATTCGCT). The sample is approximately 4 mM in D$_2$O at pH 7.0. The 500-MHz spectrum was acquired using a mixing time (τ_m) of 300 msec and a recycle time of 2.2 sec. It is the result of 512 t_1 files of 1 K complex points over a sweep width of 4400 Hz. The set was zero-filled once in t_1 and apodized with an unshifted skewed sine-bell function. [From D. R. Hare (88).]

connectivity is assigned to cytidine 3. Examination of a vertical line at the chemical shift of the cytidine-3 H6 resonance shows a second cross peak identifying the H1' resonance of this cytidine. Examination of a horizontal line at the chemical shift of the cytidine-3 H1' proton now shows a second cross peak to a guanosine H8 proton. This is assigned to guanosine-4. Following this procedure, a sequential series of base and sugar H1' protons can be assigned. In cases where degeneracies occur, similar connectivities involving the H2" protons can be used. COSY spectra can be used to extend assignments of H1' or H2" resonances to most of the other protons in a given sugar ring.

D. APPLICATION TO OLIGOSACCHARIDES

Carbohydrates also offer ample opportunity for the application of 2D NMR methods for structural analysis (95-97). Carbohydrates as a class are somewhat lower in molecular weight than proteins or polynucleotides, and may therefore seem to present a less formidable problem in structural analysis. The variety of linkage modes and the potential for branched as well as linear polymers, however, more than compensate for their smaller size. The variety of structures possible with a relatively small number of residues is perhaps the basis of the use of these molecules as receptors for a variety of biological agents active at the surface of cells.

Carbohydrates also present some special spectroscopic problems. The functional similarity of sugar residues leads to the concentration of resonances in a small spectral region, 3-5 ppm. Resolution problems therefore require operation at high fields and the use of a variety of resolution-enhancing methods including the relay and phase-sensitive techniques discussed in previous sections. Cross-relaxation experiments for members of this class in the molecular weight range 1000-3000 also suffer from very low efficiency because of the near-cancellation of effects from zero-quantum and two-quantum transitions when $\omega\tau_c$ approaches 1. Recent work using rotating-frame cross-relaxation experiments may provide a solution to these later difficulties (98, 99). In other cases, tendencies toward aggregation may actually push $\omega\tau_c$ into a range where large cross-relaxation effects are observed.

The example given below is drawn from studies on an octasaccharide glycolipid isolated from the spermatozoa of a freshwater bivalve (100). Residue composition for such a molecule is often known from conventional chemical analysis, and the major contribution of 2D NMR is in sequence determination and the characterization of linkages between residues. As a first step toward this determination, resonances are grouped into scalar-coupled sets corresponding to single residues, using a combination of COSY and RELAY experiments. Correlation of chemical shifts of several resonances with model compounds along with estimation of scalar coupling constants allows assignment to specific residue types. Since we have used monosaccharide spectra to illustrate acquisition techniques earlier in this chapter, we will not reproduce specific illustrations of scalar-coupling correlation experiments here. Let us emphasize, however, that assignment of a fairly large number of resonances in each ring is important since characterization of linkages is predicated on an ability to distinguish interresidue from intraresidue cross-relaxation interactions.

The basis for identification of linkages is similar to that in the sequential assignment of residues in nucleotides and proteins. For accumulations with short mixing times, cross-peak intensities in 2D NOE experiments are proportional to $1/r^6$ where r is the distance between proton pairs. This means that one sees significant connectivities only for interproton distances less than 3 Å. In Fig. 6-18, two glucose rings joined in a β1-4 linkage are shown. Let us focus on the anomeric proton, H1. Resonances for anomeric protons lie in a well-resolved region of the spectrum from 4 to 5 ppm. It is clear that this proton lies within 3 Å of several other protons. Two of these are in the same ring, H3 and H5 at 2.5 Å. Given the relatively fixed geometry of the glucose ring, we should always expect to see connectivities to these protons. Similar expectations arise for other sugars. For most β-linked sugar residues, one sees axial 1-3 and 1-5 connectivities at 2.5 Å. In β-mannosides one expects to see an additional equatorial 1-2 connectivity at 2.5 Å. For an α-linked sugar residue, one sees only an equatorial 1-2 connectivity at 2.5 Å. In other words, except for β-mannose, one expects to see one or two intraring connectivities. Since resonances arising from protons on the same ring have been grouped using COSY experiments, cross peaks arising from intraring interactions are readily identified.

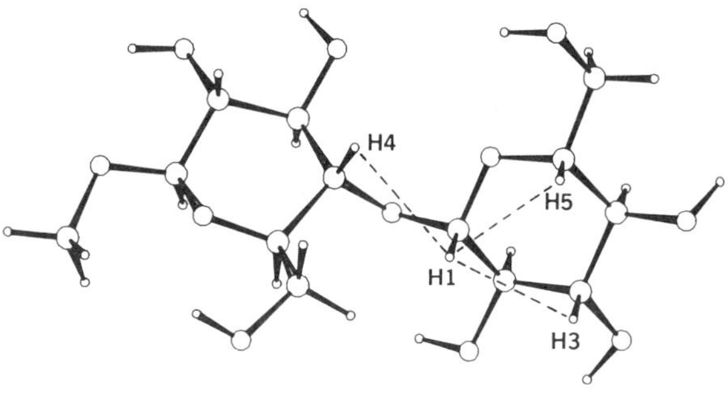

Fig. 6-18. Glucose-(β1-4)glucose segment of an oligosaccharide showing probable cross-relaxation interactions.

A 2D NOE set is presented in Fig. 6-19. Inspection of vertical lines extending upward from each of the anomeric resonances in the 4–5 ppm region shows three or more connectivities instead of the expected one or two. The origin of the extra connectivities can be seen in Fig. 6-18. In the geometry depicted, the *trans*-glycosidic H4 proton is also at approximately 2.5 Å. One would not at first expect the occurrence of the extra connectivities to be general, but consideration of the energetics of glycosidic rotational conformation by Lemieux and co-workers suggests that torsional angles within a small range are preferred (101) and that in all of these cases the anomeric proton is expected to be within 3 Å of a proton on the linkage site. By eliminating NOE cross peaks assignable to intraring interactions and choosing the largest remaining cross peak as a route to the trans-glycosidic proton, it should be possible to sequence and identify linkage sites for a variety of carbohydrates.

The vertical line shown in the figure is at the anomeric resonance of an *N*-acetylglucosamine residue (residue E). The H3 and H5 connectivities are easily seen and assigned. The largest additional connectivity (A2) lies at a chemical shift corresponding to the H2 proton on an α-mannose residue.

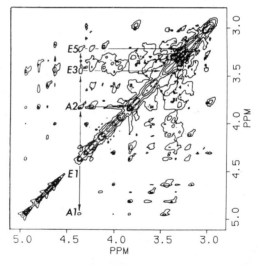

Fig. 6-19. Plot of the sugar region of a 2D NOE pure absorption spectrum of an octasaccharide glycolipid, 2 mM in d_6-DMSO/(3%)D$_2$O, GlcA(β1-4) (GalNAc(α1-3))Fuc(α1-4)GlcNAc(β1-2)Man(α1-3)(Xyl(β1-2))Man(β1-4)Glc(β1-1)Cer. The 490-MHz spectrum was collected as 192 t_1 files of 1K complex points each for a sweep width of 3205 Hz. The mixing time τ_m was 125 msec, the recycling time 1.2 sec, and the total acquisition time 10 hr. [Reprinted with permission from J. N. Scarsdale *et al.*, *Carbohydrate Research*. Copyright 1986 Elsevier Science Publishers.]

This identifies the linkage as an N-acetylglucosamine (β1-2)mannose linkage. Following this procedure for the other anomeric protons allows unambiguous characterization of five of the eight linkages in this octasaccharide glycolipid.

In cases where very high quality data can be obtained, it is possible to see more than one inter-ring cross peak. This is true for the N-acetylglucosamine residue in Fig. 6-19, where a weak connectivity to the mannose H1 proton (A1) is observed. The observation of multiple cross peaks offers the possibility of greatly restricting the number of glycosidic rotational isomers allowed. There has been some success in determining three-dimensional structures on the basis of such observations, but caution must be exercised since motional averaging can lead to cross-relaxation peak intensities which cannot be interpreted correctly on the basis of a single conformer.

We hope that the above examples present a convincing case for the potential of 2D NMR methods in structural analysis of macromolecules. The examples presented cover the three most common areas of biochemical application: protein, nucleic acid, and carbohydrate structure. It takes little imagination to greatly extend the list of applications, and, with the introduction of automated methods of structure analysis, the list will no doubt expand.

References

1. O. Jardetzky and G. C. K. Roberts, "NMR in Molecular Biology." Academic Press, New York, 1981.
2. F. M. Poulsen, J. C. Hoch, and C. M. Dobson, *Biochemistry* **19**, 2597 (1980).
3. A. G. Redfield, S. Roy, V. Sanches, J. Trapp, and N. Figuroa, "2nd Biomolecular Sterodynamics Conference" (R. Sarma, ed.), pp. 195-208. Academic Press, New York, 1981.
4. K. Wüthrich and G. Wagner, *J. Mol. Biol.* **130**, 1 (1979).
5. G. Wagner, A. Kumar, and K. Wüthrich, *Eur. J. Biochem.* **89**, 367 (1981).
6. E. L. Hahn and D. E. Maxwell, *Phys. Rev.* **88**, 1070 (1952).
7. I. D. Campbell, C. M. Dobson, and R. J. P. Williams, *Proc. R. Soc. London, Ser. A* **345**, 23 (1975).
8. G. Wider, R. Baumann, K. Nagayama, R. R. Ernst, and K. Wüthrich, *J. Magn. Reson.* **42**, 73 (1981).
9. L. D. Hall, *Adv. Carbohydr. Chem. Biochem.* **29**, 11 (1974).
10. J. F. G. Vliegenthart, L. Dorland, and H. van Halbeek, *Adv. Carbohydr. Chem. Biochem.* **41**, 209 (1983).
11. R. Freeman, *J. Chem. Phys.* **53**, 457 (1970).
12. A. Bax, R. Freeman, and G. A. Morris, *J. Magn. Reson.* **42**, 164 (1981).
13. A. A. Maudsley, A. Wokaun, and R. R. Ernst, *Chem. Phys. Lett.* **69**, 567 (1979).
14. D. Marion and K. Wüthrich, *Biochem. Biophys. Res. Commun.* **113**, 967 (1983).
15. K. Nagayama and K. Wüthrich, *Eur. J. Biochem.* **114**, 365 (1981).

16. A. W. Overhauser, *Phys. Rev.* **92**, 411 (1953).
17. A. G. Redfield and R. K. Gupta, *Cold Spring Harbor Symp. Quant Biol.* **36**, 405 (1972).
18. A. A. Bothner-By and J. H. Noggle, *J. Am. Chem. Soc.* **101**, 5152 (1979).
19. S. Macura and R. R. Ernst, *Mol. Phys.* **41**, 95 (1980).
20. A. Z. Gurevich, I. L. Barsukov, A. S. Arseniev, and V. F. Bystrov, *J. Magn. Reson.* **56**, 471 (1984).
21. C. A. G. Haasnoot, F. J. M. van de Ven, and C. W. Hilbers, *J. Magn. Reson.* **56**, 343 (1984).
22. G. Wider, K. H. Lee, and K. Wüthrich, *J. Mol. Biol.* **155**, 367 (1982).
23. A. Kumar, G. Wagner, R. R. Ernst, and K. Wüthrich, *J. Am. Chem. Soc.* **103**, 3654 (1981).
24. G. Bodenhausen, H. Kogler, and R. R. Ernst, *J. Magn. Reson.* **58**, 370 (1984).
25. G. Bodenhausen, *Prog. NMR Spectrosc.* **14**, 137 (1981).
26. D. Weitekamp, *Adv. Magn. Reson.* **11**, 111 (1983).
27. A. Bax, R. Freeman, and S. P. Kempsell, *J. Am. Chem. Soc.* **102**, 4849 (1980).
28. A. Bax, "Two-Dimensional NMR in Liquids." Reidel, Dordrecht, Netherlands, 1982.
29. R. R. Ernst, G. Bodenhausen, and A. Wokaun, "Principles of NMR in One and Two Dimensions." Oxford Univ. Press, London, 1987.
30. W. S. Warren, S. Sinton, D. P. Weitekamp, and A. Pines, *Phys. Rev. Lett.* **43**, 1791 (1979).
31. A. Wokaun and R. R. Ernst, *Chem. Phys. Lett.* **52**, 407 (1977).
32. T. H. Mareci and R. Freeman, *J. Magn. Reson.* **48**, 158 (1982).
33. V. W. Miner, P. M. Tyrell, and J. H. Prestegard, *J. Magn. Reson.* **55**, 438 (1983).
34. A. Bax, R. Freeman, T. Frenkiel, and M. H. Levitt, *J. Magn. Reson.* **43**, 478 (1981).
35. L. Braunschweiler, G. Bodenhausen, and R. R. Ernst, *Mol. Phys.* **48**, 535 (1983).
36. M. Rance, O. W. Sørensen, W. Leupin, H. Kogler, K. Wüthrich, and R. R. Ernst, *J. Magn. Reson.* **61**, 67 (1985).
37. G. Wagner and E. R. P. Zuiderweg, *Biochem. Biophys. Res. Commun.* **113**, 854 (1983).
38. M. Rance, O. W. Sørensen, G. Bodenhausen, G. Wagner, R. R. Ernst, and K. Wüthrich, *Biochem. Biophys. Res. Commun.* **117**, 479 (1983).
39. U. Piantini, O. W. Sørensen, and R. R. Ernst, *J. Am. Chem. Soc.* **104**, 6800 (1982).
40. P. H. Bolton and G. Bodenhausen, *Chem. Phys. Lett.* **89**, 139 (1982).
41. G. Eich, G. Bodenhausen, and R. R. Ernst, *J. Am. Chem. Soc.* **104**, 3731 (1982).
42. A. Bax and G. Drobny, *J. Magn. Reson.* **61**, 306 (1985).
43. G. Wagner, *J. Magn. Reson.* **55**, 151 (1983).
44. D. K. Sukumaran, G. M. Clore, A. Preuss, J. Zarbock, and A. M. Gronenborn, *Biochemistry* **26**, 332 (1987).
45. P. H. Bolton, in "Biological Magnetic Resonance" (L. J. Berliner and J. Reuben, eds.), Vol. 6, p. 1. Plenum, New York, 1984.
46. H. Kessler, M. Bernd, H. Kogler, T. Zarbock, O. W. Sørensen, G. Bodenhausen, and R. R. Ernst, *J. Am. Chem. Soc.* **105**, 6944 (1983).
47. G. Wider, S. Macura, A. Kumar, R. R. Ernst, and K. Wüthrich, *J. Magn. Reson.* **56**, 207 (1984).
48. D. J. States, R. A. Haberkorn, and D. J. Ruben, *J. Magn. Reson.* **48**, 286 (1982).
49. J. Keeler and D. Neuhaus, *J. Magn. Reson.* **63**, 454 (1985).
50. P. Pfandler, G. Bodenhausen, B. U. Meier, and R. R. Ernst, *Anal. Chem.* **57**, 2510 (1985).
51. A. F. Mehlkopf, D. Korbee, T. A. Tiggelman, and R. Freeman, *J. Magn. Reson.* **58**, 315 (1984).
52. R. Baumann, A. Kumar, R. R. Ernst, and K. Wüthrich, *J. Magn. Reson.* **44**, 76 (1981).
53. D. Neuhaus, G. Wagner, M. Vasak, J. H. R. Kagi, and K. Wüthrich, *Eur. J. Biochem.* **151**, 257 (1985).
54. A. Bax and G. A. Morris, *J. Magn. Reson.* **42**, 501 (1981).

55. T.-M. Chan and J. L. Markley, *Biochemistry* **22**, 5996 (1983).
56. A. Bax, R. H. Griffey, and B. L. Hawkins, *J. Magn. Reson.* **55**, 301 (1983).
57. D. H. Live, D. G. Davis, W. C. Agosta, and D. Cowburn, *J. Am. Chem. Soc.* **106**, 6104 (1984).
58. P. J. Hore, *J. Magn. Reson.* **55**, 283 (1983).
59. A. G. Redfield, S. D. Kunz, and E. K. Ralph, *J. Magn. Reson.* **19**, 114 (1975).
60. P. J. Hore, *J. Magn. Reson.* **56**, 535 (1984).
61. G. Wider, R. V. Hosur, and K. Wüthrich, *J. Magn. Reson.* **52**, 130 (1983).
62. A. J. Wand and S. W. Englander, *Biochemistry* **24**, 5290 (1985).
63. M. J. Kime and P. B. Moore, *Biochemistry* **23**, 1688 (1984).
64. M. J. Kime and P. B. Moore, *FEBS Lett.* **13**, 199 (1983).
65. P. Plateau and M. Gueron, *J. Am. Chem. Soc.* **104**, 7310 (1982).
66. V. J. Basus, *J. Magn. Reson.* **60**, 138 (1984).
67. J. H. Prestegard and J. N. Scarsdale, *J. Magn. Reson.* **62**, 136 (1985).
68. J. H. Noggle and R. E. Schirmer, "The Nuclear Overhauser Effect." Academic Press, New York, 1971.
69. E. T. Olejniczak, F. M. Poulsen, and C. M. Dobson, *J. Magn. Reson.* **59**, 518 (1984).
70. L. E. Kay, T. A. Holak, B. A. Johnson, I. M. Armitage, and J. H. Prestegard, *J. Am. Chem. Soc.* **108**, 4242 (1986).
71. W. Braun, G. Wider, K. H. Lee, and K. Wüthrich, *J. Mol. Biol.* **169**, 921 (1983).
72. K. Wüthrich, G. Wider, G. Wagner, and W. Braun, *J. Mol. Biol.* **155**, 311 (1982).
73. K. Wüthrich, *Biopolymers* **22**, 131 (1983).
74. M. Billeter, W. Braun, and K. Wüthrich, *J. Mol. Biol.* **155**, 321 (1982).
75. K. Wüthrich, M. Billeter, and W. Braun, *J. Mol. Biol.* **180**, 715 (1984).
76. T. A. Holak and J. H. Prestegard, *Biochemistry* **25**, 5766 (1986).
77. T. F. Havel, G. M. Crippen, and I. D. Kuntz, *Biopolymers* **18**, 73 (1979).
78. L. R. Brown, W. Brown, A. Kumar, and K. Wüthrich, *Biophys. J.* **37**, 319 (1982).
79. T. F. Havel and K. Wüthrich, *J. Mol. Biol.* **182**, 281 (1985).
80. R. Kaptein, E. R. P. Zuiderweg, R. M. Scheek, R. Boelens, and W. F. van Gunsteren, *J. Mol. Biol.* **182**, 179 (1985).
81. G. M. Clore, A. M. Gronenborn, A. T. Brunger, and M. Karplus, *J. Mol. Biol.* **186**, 435 (1985).
82. D. R. Kearns and R. G. Shulman, *Acc. Chem. Res.* **7**, 33 (1974).
83. S. Roy and A. G. Redfield, *Nucleic Acids Res.* **9**, 7073 (1981).
84. J. Feigon, A. H.-J. Wang, G. A. van der Marel, J. H. Van Boom, and A. Rich, *Nucleic Acids Res.* **12**, 1243 (1984).
85. S. H. Chou, D. R. Hare, D. E. Wemmer, and B. R. Reid, *Biochemistry* **22**, 3037 (1983).
86. D. J. Patel, S. A. Kozalowski, L. A. Marky, C. Broka, J. A. Rice, K. Itakura, and K. T. Breslauer, *Biochemistry* **21**, 428 (1982).
87. R. M. Scheek, N. Russo, R. Boelens, and R. Kaptein, *J. Am. Chem. Soc.* **105**, 2914 (1983).
88. D. R. Hare, D. E. Wemmer, S.-H. Chou, G. Drobny, and B. R. Reid, *J. Mol. Biol.* **171**, 319 (1983).
89. D. Frechet, D. M. Cheng, L.-S. Kan, and P. O. P. Ts'o, *Biochemistry* **22**, 5194 (1983).
90. J. Feigon, W. Leupin, W. A. Denny, and D. R. Kearns, *Biochemistry* **22**, 5943 (1983).
91. M. A. Weiss, D. J. Patel, R. T. Sauer, and M. Karplus, *Proc. Natl. Acad. Sci. USA* **81**, 130 (1984).
92. H. P. Westerink, G. A. van der Marel, J. H. Van Boom, and C. A. G. Haasnoot, *Nucleic Acids Res.* **12**, 4324 (1984).
93. D. J. Patel, S. A. Kozlowski, A. Nordheim, and A. Rich, *Proc. Natl. Acad. Sci. USA* **79**, 1413 (1982).

94. A. Pardi, R. Walker, H. Rapoport, G. Wider, and K. Wüthrich, *J. Am. Chem. Soc.* **105**, 1652 (1983).
95. J. Dabrowski and P. Hanfland, *FEBS Lett.* **142**, 138 (1982).
96. J. H. Prestegard, T. A. W. Koerner, P. C. Demou, and R. K. Yu, *J. Am. Chem. Soc.* **104**, 4993 (1982).
97. S. W. Homans, R. A. Dwek, D. L. Fernandes, and T. W. Rademacher, *Proc. Natl. Acad. Sci. USA* **82**, 6286 (1984).
98. A. A. Bothner-By, R. L. Stephens, and J. Lee, *J. Am. Chem. Soc.* **106**, 811 (1984).
99. A. Bax and D. G. Davis, *J. Magn. Reson.* **63**, 207 (1985).
100. J. N. Scarsdale, S. Ando, T. Hori, R. K. Yu, and J. H. Prestegard, *Carbohydr. Res.* **155**, 45 (1986).
101. H. Thorgerson, R. V. Lemieux, K. Bock, and B. Meyer, *Can. J. Chem.* **60**, 44 (1982).

7

Multiple-Resonance and Two-Dimensional NMR Techniques in Analysis of Fluorocarbon Compounds and Polymers

DERICK W. OVENALL AND
RAYMOND C. FERGUSON*

E. I. DU PONT DE NEMOURS & CO.
CENTRAL RESEARCH AND DEVELOPMENT DEPARTMENT
EXPERIMENTAL STATION
WILMINGTON, DELAWARE 19898

I. Introduction	489
II. Resolution Enhancement and Proton Decoupling	491
III. Two-Dimensional NMR	494
IV. Experimental Details	504
V. Summary	505
References	505

I. Introduction

High-resolution ^{19}F, ^{1}H, and ^{13}C NMR are complementary techniques for analysis for fluorocarbon compounds. Each has advantages that, along with the nature of the sample, determine the strategy of analysis. The chemical shift dispersions are in the order $^{19}F \geq {}^{13}C \gg {}^{1}H$, while the sensitivities run $^{1}H \approx {}^{19}F \gg {}^{13}C$. For these reasons, ^{19}F NMR is the method of choice (along with ^{1}H for compounds containing protons).

The frequency dispersion of ^{19}F is much larger than that of ^{13}C and requires correspondingly greater bandwidths and larger data arrays for FT NMR (Table 7-1). A major limitation on use of ^{13}C for fluorocarbon

---

* Present address: CONDUX, Inc., 228 Unami Trail, Newark, Delaware 19711.

TABLE 7-1

Characteristics and Experiment Requirements for FT NMR of ^{1}H, ^{13}C, and ^{19}F[a]

Nucleus	Receptivity	Shift range (ppm)	Frequency range (kHz)	Data table for 1 Hz resolution
^{1}H	1	10	4	8K
^{13}C	10^{-3}	300	30	64K
^{19}F	0.8	300	113	256K

[a] For resonance in 9.4-tesla magnetic field.

compounds is the decoupling requirement. Broadband ^{19}F decoupling from ^{13}C is feasible at low frequencies (1), but more efficient multiple-pulse techniques are needed at high frequencies (2). Simultaneous ^{19}F and ^{1}H decoupling is even more difficult (3, 4), and is not a standard option on commercial spectrometers. We will therefore discuss only ^{19}F and ^{1}H NMR.

Fluorine coupling constants can be large (Table 7-2) and multiplet patterns are often complex and poorly resolved. A number of techniques, particularly some of the two-dimensional techniques described elsewhere in this book, are useful in ^{19}F studies. Some of these and their uses are:

1. Resolution enhancement—improved multiplet resolution.
2. ^{1}H Decoupling—improved resolution; ^{1}H versus ^{19}F couplings.
3. Selective decoupling—identification of coupled groups; improved resolution.
4. 2D COSY—identification of coupled groups, connectivity.

These techniques are useful singly and in combinations. Modern Fourier transform spectrometers permit effective implementation and a variety of modifications suitable for special situations.

TABLE 7-2

Typical ^{19}F NMR Parameters

Chemical shift range	Coupling constants in aliphatic compounds	
300 ppm	$^2J(F, F)$	200–800 Hz
	$^3J(F, F)$	≤1 Hz
	$^4J(F, F)$	1–20 Hz
	$^2J(H, F)$	30–60 Hz
	$^3J(H, F)$	~10 Hz
	$^4J(H, F)$	<1 Hz

II. Resolution Enhancement and Proton Decoupling

The sensitivity of modern high-field, high-resolution FT NMR spectrometers is high, and solutions at 5% w/v concentration or less will give spectra with good signal-to-noise ratios (S/N) with a few pulses. High S/N free induction decays can be obtained with continued accumulation. Resolution enhancement increases noise, but tradeoffs between resolution and sensitivity can be advantageous (5). Figures 7-1 and 7-2 are examples in which the multiplets were at best poorly resolved without resolution enhancement.

Broadband proton decoupling provides immediate benefits in pattern simplification and resolution, when there are detectable proton couplings. Running both coupled and decoupled spectra is frequently helpful, because comparison of the spectra can quickly establish which ^{19}F resonances are proton-coupled.

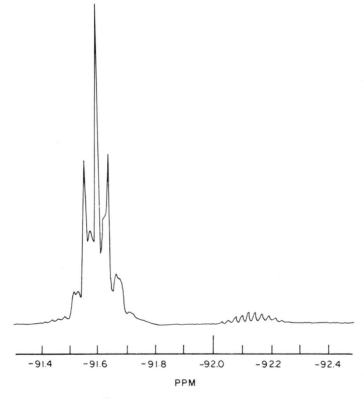

Fig. 7-1. The 376.5 MHz ^{19}F NMR spectrum of poly(vinylidene fluoride), 5% solution in N,N-dimethylacetamide, 25°C: $-CF_2-CH_2-CF_2-CH_2-CF_2-$ region, proton-coupled.

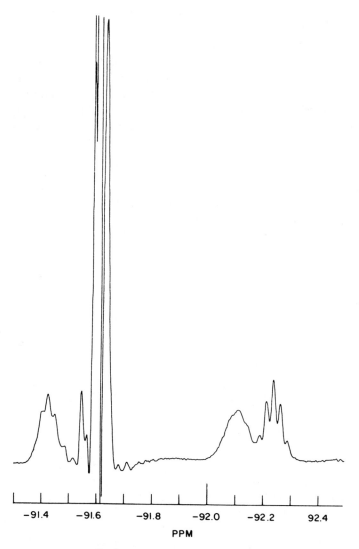

Fig. 7-2. The 376.5 MHz ^{19}F{^{1}H} NMR spectrum of poly(vinylidene fluoride). Same region and conditions as for Fig. 7-1, with broadband proton decoupling.

For example, resolution of the ^{19}F spectrum of poly(vinylidene fluoride), $(CH_2-CF_2-)_n$, is improved by broadband proton decoupling (and resolution enhancement). The spectrum has a number of CF_2 resonances from differing arrangements of monomer units in six-monomer sequences, as well as weak resonances due to nonchain structures (6–8).

7. ANALYSIS OF FLUOROCARBON COMPOUNDS AND POLYMERS

Figures 7-1 and 7-2 show the region of the spectrum due to the central CF_2 group in $-CF_2-CH_2-CF_2-CH_2-CF_2-$ sequences. The multiplet patterns of the coupled spectrum (Fig. 7-1) are not clear and are distorted by overlap. In the $^{19}F\{^1H\}$ spectrum (Fig. 7-2), the pattern at -92.15 ppm is resolved into a clean quintet at -92.25 ppm and an unresolved pattern at -92.1 ppm which is actually two overlapped quintets. Similarly, quintets are resolved at -91.4 and -91.6 ppm. The latter patterns are distorted by overlap with other multiplets associated with different eleven-carbon sequences having the same central five-carbon sequence (7). The quintet multiplicity is due to the $^4J(F, F) = 9$ Hz coupling to the fluorines in the two CF_2 groups beta to the central CF_2.

In contrast, the triplet multiplicity observed in the -94 to -95 ppm region (Fig. 7-3) identifies $-CH_2-CH_2-CF_2-CH_2-CF_2-$ sequences.

Combining selective ^{19}F and broadband ^1H decoupling helps identify spin-coupled groups. For example, irradiating the -92.25 ppm quintet collapsed the -94 to -95 ppm triplets (Fig. 7-4), thereby establishing that the central CF_2 groups in the two types of sequence are beta to each other. We will demonstrate later that two-dimensional techniques are more efficient for establishing these connectivity relationships.

Proton spectra of poly(vinylidene fluoride) provide little information about the main chain, but help identify other structures. For example, in Fig. 7-5 the very weak triplet $[^2J(H, F) = 54$ Hz$]$ at 6.13 ppm is assignable to a $-CH_2F$ group, and the triplet $[^3J(H, F) = 19$ Hz$]$ at 1.74 ppm to a $X-CH_2-CF_2-Y$ group, where X contains no proton or fluorine on the carbon alpha to the CH_2 (no resolvable splitting). Locating the fluorine resonances of these groups presents more of a challenge, because they are weak and possibly obscured by stronger resonances.

A differential-mode selective ^1H decoupling method achieves the desired objective. A weak unmodulated ^1H decoupling signal is applied at the center of the triplet and the ^{19}F FID is collected for one quadrature phase cycle. The decoupling signal is then shifted off resonance and a second cycle is collected and subtracted from the first. This process is repeated until adequate S/N is obtained in the difference spectrum.

With this technique, resonances unaffected by the selective ^1H decoupling cancel and give little or no response. Those which are narrowed by decoupling appear as broad peaks subtracted from sharper peaks as shown in Figs. 7-6 and 7-7. The results locate the weak CF_2H and $X-CH_2-CF_2-Y$ resonances at -116 and -108 ppm, respectively.

Similarly, a differential decoupling experiment in which a proton-coupled cycle is subtracted from a broadband proton-decoupled cycle is the most effective way to identify all fluorines coupled to protons in a single experiment.

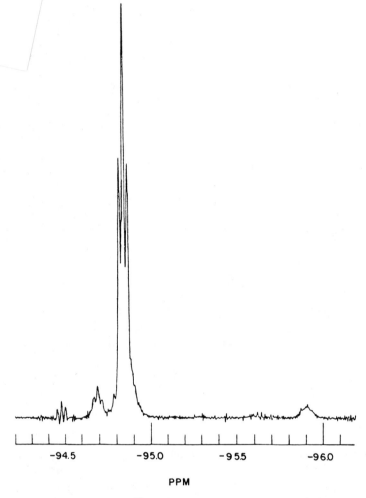

Fig. 7-3. The 376.5 MHz ^{19}F NMR spectrum of poly(vinylidene fluoride), $-CH_2-CH_2-CF_2-CH_2-CF_2-$ region. Same conditions as for Fig. 7-2.

III. Two-Dimensional NMR

Two-dimensional NMR experiments consist of collecting a series of FIDs from progressively timed spin-system interactions, generated in a particular way, which are transformed to produce a two (frequency)-dimensional plot of particular features of the spin system. These experiments generally have three distinct time periods. In the first, or preparation, period, the spins are allowed to relax, and then one or more rf pulses are applied to place the

7. ANALYSIS OF FLUOROCARBON COMPOUNDS AND POLYMERS   495

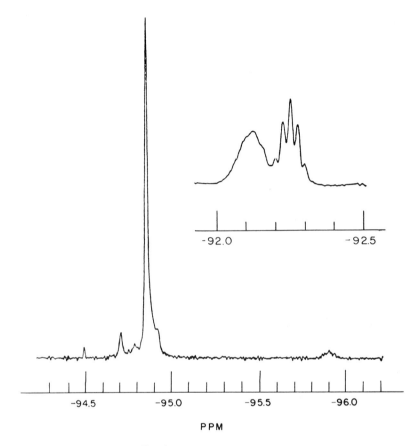

Fig. 7-4. The 376.5 MHz ^{19}F {^{1}H} NMR spectrum of poly(vinylidene fluoride), with selective ^{19}F decoupling at the −92.25 ppm quintet (inset). Same region as in Fig. 7-3.

spins in a specific state. During the second period, lasting a time t_1, the spin system evolves to a new state, which is affected by the chemical shifts, spin coupling constants, and relaxation times. The final period is for data collection, for a time t_2, as for a normal FID.

The experiment is repeated with equal increments to t_1 chosen to cover the desired frequency width. The array of FIDs that are functions of t_1 and t_2 is then transformed to give the two-dimensional response with frequency axes, f_1 and f_2. These are frequently displayed as intensity contour plots.

The hardware and software for such experiments are described elsewhere, and are available as standard options on commercial spectrometers, so that they will not be discussed further here.

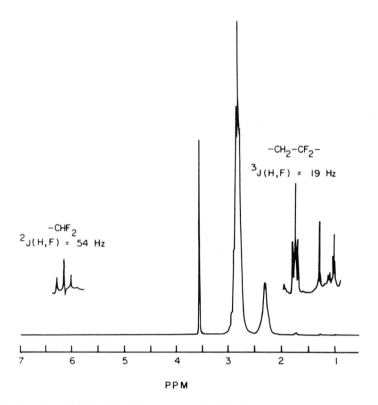

Fig. 7-5. The 400 MHz ^{1}H NMR spectrum of KYNAR 301 (Pennwalt) poly(vinylidene fluoride), 5% solution in dioxane-d_8. The two insets have been expanded in scale vertically by a factor of 32.

One of the more commonly used two-dimensional experiments is homonuclear chemical shift correlation spectroscopy or COSY. It has been used extensively in proton NMR, and applications to fluorine have been reported (9). The COSY experiment is shown diagramatically in Fig. 7-8. The response is presented as an intensity contour plot in which the spectrum appears along the diagonal and the off-diagonal responses indicate spin coupling between resonances at the corresponding f_1 and f_2 frequencies.

The utility of COSY is illustrated by assignment of the CF_2 region of the ^{19}F spectrum of perfluorooctanoate (Fig. 7-9). The assignments of the CF_3 (−81 ppm, not shown) and CF_2 groups 1 and 6 were easily made on the basis of chemical shift–structure correlations. Even with resolution enhancement, however, the multiplet structures of the CF_2 resonances were too complex and poorly resolved to permit further assignments.

7. ANALYSIS OF FLUOROCARBON COMPOUNDS AND POLYMERS 497

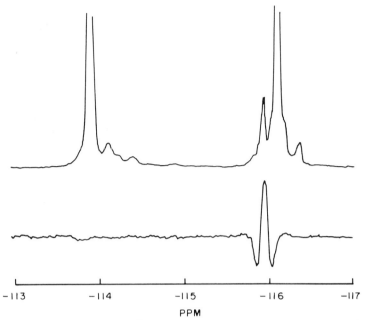

Fig. 7-6. Differential 376.5 MHz ^{19}F NMR spectrum of KYNAR 301, 5% solution in dioxane-d_8, with selective ^{1}H decoupling at 6.13 ppm. Upper, normal spectrum; lower, differential spectrum.

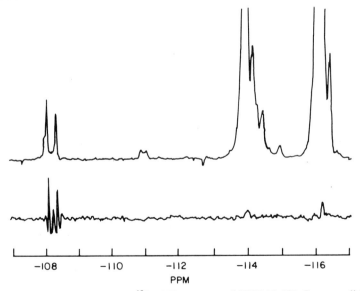

Fig. 7-7. Differential 376.5 MHz ^{19}F NMR spectrum of KYNAR 301. Same conditions as Figure 7-6, except with selective ^{1}H decoupling at 1.74 ppm. Upper, normal spectrum; lower, differential spectrum. The ^{1}H coupled resonances are at approximately −108 ppm.

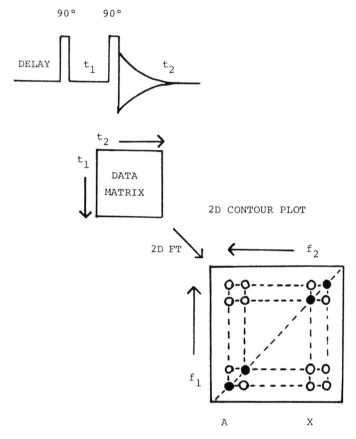

Fig. 7-8. Schematic description of homonuclear chemical shift correlation (COSY) spectroscopy.

The COSY spectrum (Fig. 7-9) permits the remaining assignments to be made unequivocally. Recalling that three-bond F–F couplings in saturated fluorocarbon compounds are less than 1 Hz, while four-bond couplings are 10 Hz or more (Table 7-2), experimental conditions were selected to emphasize the larger couplings. Starting with the identification of groups 1 and 6, the other CF_2 resonances are readily identified by the connectivities indicated by the COSY diagram. A very weak off-diagonal response between groups 2 and 3 indicates that the three-bond coupling, although small, is not zero.

In fluorocarbon compounds containing an asymmetric center, the fluorine resonances of CF_2 groups often appear as AB patterns, which can be readily

7. ANALYSIS OF FLUOROCARBON COMPOUNDS AND POLYMERS

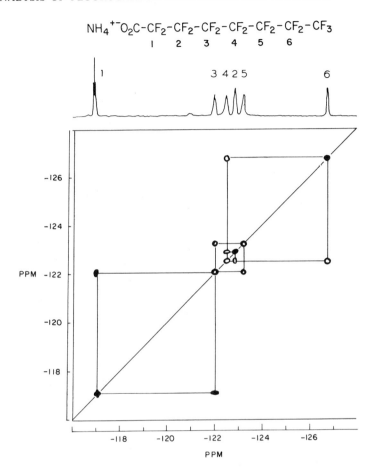

Fig. 7-9. The 376.5 MHz ^{19}F 2D COSY spectrum of ammonium perfluorooctanoate in acetone-d_6, CF_2 region.

identified by the COSY experiment. Figure 7-10 shows a COSY spectrum of the CF_2 resonances of perfluoromethylcyclohexane (10). In each of the three different kinds of CF_2 groups, the axial fluorine resonances occur at higher frequencies than the equatorial fluorine resonances, and show large geminal spin–spin couplings. The three pairs of AB patterns can be clearly identified from the off-diagonal peaks in the COSY spectrum.

A particularly useful feature of COSY is the ability to detect spin couplings between sets of nuclei even though the spin–spin splitting is not resolvable. To illustrate this, we return to the ^{19}F {^{1}H} spectrum of poly(vinylidene fluoride), in which several of the resonances have no resolvable

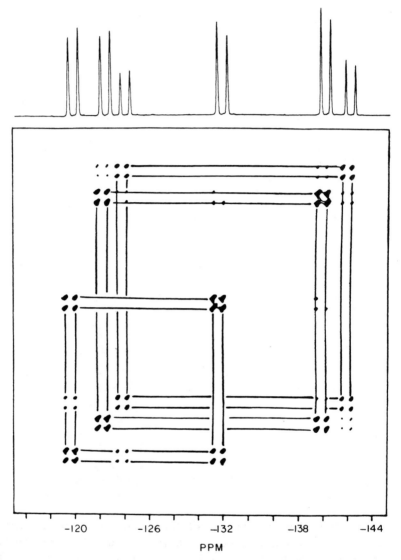

Fig. 7-10. The 376.5 MHz ^{19}F 2D COSY spectrum of perfluoro(methyl cyclohexane), in acetone-d_6/C_6F_6, CF_2 region. Each ring CF_2 produces an AB quartet; two pairs are equivalent, giving the 2:2:1 intensity ratios.

fine structure. Figure 7-11 shows the full spectrum and the COSY response, which establishes connectivities for the resonances of one of the major monomer sequence arrangements (a single tail-on enchainment in an otherwise all head-to-tail sequence).

7. ANALYSIS OF FLUOROCARBON COMPOUNDS AND POLYMERS

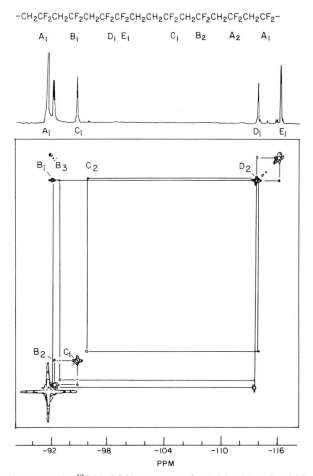

Fig. 7-11. The 376.5 MHz ^{19}F 2D COSY spectrum of poly(vinylidene fluoride). Broadband proton-decoupled. B_3, C_2, and D_2 are due to less probable 11-carbon sequences.

Figure 7-12 displays an expansion of the regions of the spectrum shown in Figs. 7-1 to 7-3, confirming the connectivity established by the selective ^{19}F decoupling experiment (Fig. 7-4), but displaying several other connectivities in the single experiment.

COSY was also valuable in analysis of the microstructure of poly(vinyl fluoride), $(-CH_2-CHF-)_n$ (9). The microstructure and spectra of this polymer reflect both regioisomerism (head-on versus tail-on additions) and stereoisomerism. The ^{19}F{^{1}H} spectrum and corresponding COSY spectrum of isoregic poly(vinyl fluoride) (Figs. 7-13 and 7-14) permit assignment of stereosequences.

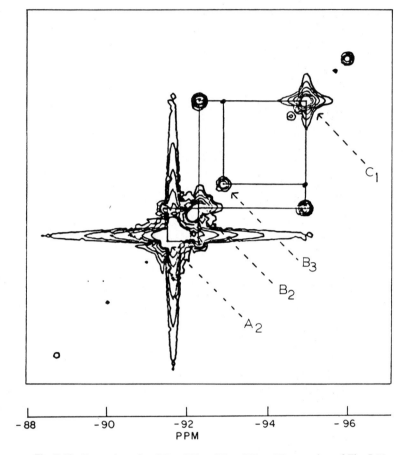

Fig. 7-12. Expansion of $-CF_2-CH_2-CF_2-CH_2-CF_2-$ region of Fig. 7-11.

The $^4J(F, F)$ couplings were detected even though small relative to the linewidths of the resonances. The three regions of the spectrum were assigned to the mm, mr, and rr triads as indicated in Fig. 7-13 (9). Racemic (r) or syndiotactic diads are pairs of adjacent asymmetric centers having opposite optical configuration (dl). Meso- (m) or isotactic diads have the same optical configurations (dd or ll). Triads, tetrads, pentads, etc. are denoted by a succession of diads.

The connectivities established by the COSY experiment lead to unambiguous assignments of the pentad resonances by means of tables or diagrams of connected sequences. For instance, the rrrr pentad must always be in an rrrrr or an rrrrm sequence. Thus, the rrrr resonance can be coupled

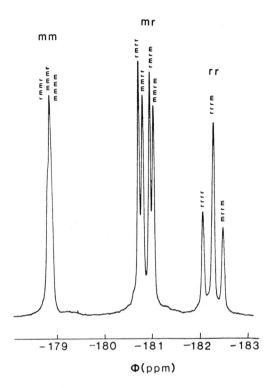

Fig. 7-13. The 188 MHz ^{19}F spectrum of isoregic poly(vinyl fluoride). [Reprinted from Bruch *et al.* (9), with permission of the authors and the American Chemical Society.]

only to the rrrm resonance through the $^4J(F, F)$ coupling of the type

$$\begin{array}{c}
\text{H} \quad\quad \text{F} \quad\quad \text{H} \quad\quad \text{F} \quad\quad \text{H} \quad\quad \text{H}\\
| \quad\quad\quad | \quad\quad\quad | \quad\quad\quad | \quad\quad\quad | \quad\quad\quad |\\
\text{C}-\text{CH}_2-\text{C}-\text{CH}_2-\text{C}-\text{CH}_2-\text{C}-\text{CH}_2-\text{C}-\text{CH}_2-\text{C}\\
| \quad\quad\quad | \quad\quad\quad | \quad\quad\quad | \quad\quad\quad | \quad\quad\quad |\\
\text{F} \quad\quad \text{H} \quad\quad \text{F} \quad\quad \text{H} \quad\quad \text{F} \quad\quad \text{F}\\
\text{r} \quad\quad\quad \text{r} \quad\quad\quad \lfloor\text{r}\rfloor \quad\quad \text{r} \quad\quad\quad \text{m}\\
\quad\quad\quad\quad\quad\quad ^4J(F, F)
\end{array}$$

Note that the rrrr resonance can interact with an rrmm or rrmr resonance only through a negligibly small $^6J(F, F)$ coupling. On the other hand, rrrm may also occur in rrrmm and rrrmr sequences, and thus the rrrm resonance can also be coupled to mmrr and rmrr resonances. These relationships and similar arguments lead to the pentad assignments made in the figures.

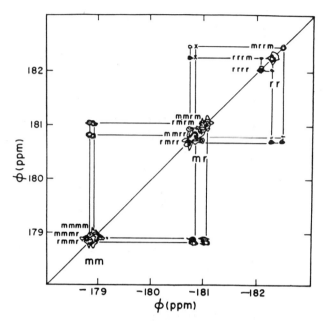

Fig. 7-14. 188 MHz ^{19}F 2D J-correlated spectrum of isoregic poly(vinyl fluoride). [Reprinted from Bruch et al. (9), with permission of the authors and the American Chemical Society.]

IV. Experimental Details

We will not attempt to define hardware and software requirements in any detail. The experimental details following describe how the experiments were done in our laboratory. The approaches are not necessarily appropriate or optimum for other spectrometer systems.

Fluorine-19 NMR spectra were taken with a Bruker WM-400 spectrometer operating at 376.5 MHz with a 5-mm probe. For selective fluorine decoupling, radio-frequency power was generated by quadrupling the output of a frequency synthesizer set at 94.1 MHz. This was gated with homonuclear decoupling pulses from the spectrometer, amplified to about 20 milliwatts, and applied to the probe through a directional coupler.

COSY experiments were run with data matrices of 256K points. This experiment uses the same spectral width in both dimensions. Parameters were chosen to give the same digital resolution in both dimensions. This allows the appearance of the contour plots to be improved and artifacts to be removed by symmetrization (11), a process in which intensities of points symmetrically disposed about the diagonal are compared, and the lower of the two inserted into both locations.

The number of t_2 data points was 1024. With quadrature detection this gives 512 points in the t_2 direction after transformation. There were 256 t_1 values run; these were zero-filled to 512 before transformation, to yield a two-dimensional spectrum with 512 points along each frequency axis and a total size of 256K.

Aliasing of peaks in the t_1 direction was avoided by appropriate phase cycling of the pulses. Figure 7-9 was obtained with a spectral width of 5000 Hz (13.3 ppm) in both dimensions, giving a digital resolution of about 10 Hz/point. Although this is coarse by one-dimensional standards, it is adequate for this problem. Sixteen acquisitions were collected for each t_1 value and a relaxation delay of 0.5 sec was used, giving a total data collection time of about one hour.

V. Summary

We have described a number of multiple-resonance and two-dimensional techniques applied to ^{19}F NMR of fluorocarbon compounds and polymers. The examples are by no means exhaustive, and, in fact, are a selection of conceptually and experimentally simple experiments. Other combinations of selective or broadband decoupling, difference spectra, and two-dimensional experiments can be devised for specific needs.

Techniques which work for other nuclei, particularly protons, can generally be applied to fluorine. The major experimental difficulties are due to the large frequency dispersion of the ^{19}F spectra, possible cross talk between ^{19}F and ^{1}H channels, and the need for combinations of frequencies and pulse sequences which may not be standard.

References

1. D. W. Ovenall and J. Chang, *J. Magn. Reson.* **25**, 361 (1977).
2. V. Sklenar and Z. Starcuk, *Org. Magn. Reson.* **72**, 662 (1977).
3. F. C. Schilling, *J. Magn. Reson.* **47**, 61 (1982).
4. A. E. Tonelli, F. C. Schilling, and R. E. Cais, *Macromolecules* **14**, 560 (1981).
5. A. G. Ferrige and J. C. Linden, *J. Magn. Reson.* **31**, 337 (1978).
6. R. E. Cais and J. E. Kometani, *Macromolecules* **17**, 1887 (1984).
7. R. E. Cais and J. A. Sloane, *Polymer* **24**, 179 (1983).
8. R. C. Ferguson and D. W. Ovenall, *Polym. Prepr., Am. Chem. Soc. Div. Polym. Chem.* **25**(1), 340 (1984).
9. M. D. Bruch, F. A. Bovey, and R. E. Cais, *Macromolecules* **17**, 2547 (1984).
10. R. E. Hurd, in "Relaxation Times," Vol. 2, No. 4. General Electric Co., Medical Systems Group, Fremont, California, 1981.
11. R. Baumann, G. Wider, R. R. Ernst, and K. Wüthrich, *J. Magn. Reson.* **44**, 402 (1981).

8

Recent Developments in Pulsed NMR Methods

WALLACE S. BREY

DEPARTMENT OF CHEMISTRY
UNIVERSITY OF FLORIDA
GAINESVILLE, FLORIDA 32611

I. RELAY and Observation by Hydrogen Detection	508
A. Homonuclear Relayed Coherence Transfer (H—H—H)	508
B. RELAY by the Path H—H—X	511
C. RELAY by the Path H—C—C	513
D. RELAY Originating on a Heteronucleus	514
E. Observation of Hydrogen Signals from Heteronuclear RELAY and Other Sequences	515
II. Homonuclear Correlation	520
A. Delayed COSY	520
B. Z-Filtered COSY	521
C. Heteronuclear-Filtered COSY	522
D. Multiple-Quantum-Filtered COSY	523
E. Difference Methods	524
F. Analyses and Applications	526
III. Isotropic Mixing Experiments	528
IV. Applications of Zero-Quantum Coherence	531
A. Homonuclear ZQC Spectroscopy	531
B. Heteronuclear ZQC Spectroscopy	535
V. Nuclear Overhauser Enhancement	535
A. Design and Data Handling for NOESY Experiments	537
B. NOE in the Rotating Frame	539
C. Interpretation of NOE Enhancement Results	541
D. Some Other Forms of NOE Experiments	541
References	542

In this chapter, we will consider a number of recently developed extensions and improvements in pulsed NMR methods for the liquid state. Some of the experiments to be discussed are based on more than one of the categories into which we have somewhat arbitrarily classified the methods included in this chapter, but we have attempted to describe them in relation to what appears to be the outstanding feature of each experiment.

I. RELAY and Observation by Hydrogen Detection

The term RELAY has been applied to a variety of different experiments in which there are successive transfers of polarization. In some of these, the successive transfers are homonuclear, or entirely within the hydrogen nuclear spin system, and in others, transfer is partially heteronuclear, involving as well a nucleus other than hydrogen. Related to the RELAY experiments are others in which the larger magnetogyric ratio of hydrogen is utilized as a means of enhancing sensitivity by detection of hydrogen magnetization rather than that of a nucleus such as ^{13}C or ^{15}N.

A. HOMONUCLEAR RELAYED COHERENCE TRANSFER (H—H—H)

The homonuclear RELAY extension of COSY is less sensitive than a corresponding simple COSY experiment, and optimization of spectral setup conditions is therefore even more important. In addition, the lineshapes in this experiment are somewhat unusual and the line intensities are complicated functions of the properties of the system and of the spectral parameters. Accordingly, Bax and Drobny (1) have presented an anlysis of RELAY, based on the product-operator treatment.

The sequence and the phase cycling procedure on which the analysis is based have been given in Fig. 6-6. For an AMX spin system with A not coupled to X, the following conclusions were reached: (a) The intensity of the relayed magnetization depends on the length of the interval τ in such a way that, providing $J(AM)$ and $J(MX)$ are within a factor of two of one another and both are much larger than $1/T_2$, the optimum intensity corresponds to τ of approximately $[J(AM)+J(MX)]^{-1}$. If J is of the order of $1/T_2$, τ should be reduced in length by about 10–30%; it should be approximately $2T_2$ if the J values are much smaller than $1/T_2$. What is embodied in these prescriptions is the balance between the increased magnetization transfer at longer times weighed against the increased loss of magnetization by transverse relaxation. (b) Intensities of diagonal and COSY peaks, the latter arising from single-step transfer, are minimized for $\tau = [J(AM)+J(MX)]^{-1}$. (c) Values of t_1 (max) and t_2(max) of the order of 100 to 150 msec are suggested; longer times may provide better resolution, but this is only required for molecules with severe overlap of cross multiplets, and the longer times reduce sensitivity if T_2 values are short as in macromolecules.

If protons A or X are coupled to other hydrogens, the same considerations apply so long as proton M is coupled to only one proton A and one proton X, but for systems such as AMX_2 or AMQX, the optimum value of τ depends on the nature of the spin system as well as on the magnitudes of the coupling constants. Although considerable work would be involved, this dependence could, at least in principle, be utilized to identify a spin-system type.

To minimize artifacts and diagonal and COSY peaks, it is recommended that the eight-step phase cycle of Fig. 6-6 be repeated four times (1), incrementing all phases by 90° for each repetition, à la CYCLOPS, a procedure which permits the transmitter frequency to be placed in the middle of the spectrum. If the absolute-value mode is used, resolution is optimized by multiplication by an unshifted sine-bell function in each dimension before Fourier transformation. This filter corresponds closely to the expected lineshape for RELAY peaks and also aids in suppressing diagonal and COSY peaks.

By adding another $-\tau_2-180°-\tau_2-90°$ segment to the pulse sequence before acquisition and expanding the phase cycle to 64 steps, the experiment may be extended to a two-step relay of coherence. The transfer efficiency of multiple RELAY is simply the product, for the successive steps, of the efficiencies as calculated for the individual transfers.

The transfer functions appropriate to single and double RELAY COSY for the common amino acid residues in proteins have been compiled (2). The transfer function depends strongly upon the type of spin system and the combination of J values. For short mixing times, however, the dependence on J values is small, and thus excitation not strongly dependent on the value of J may be achieved when τ is short. Longer values, in contrast, present opportunities for spectral simplification by selective excitation of specific system types. The transfer functions were also plotted, assuming typical values for coupling constants in amino acid residues, together with a T_2 value of 100 msec (2). In proteins, transfer to the backbone amide protons is advantageous because their region of absorption is free of artifacts from water suppression, and correlations to two or more side-chain protons can be simultaneously observed. Values of $^3J(HN-\alpha)$ were taken as 10 Hz for beta-type structures, 7 Hz for random coils, and 4 Hz for helices. As a consequence of this variation, the efficiency of relayed transfer from α-H to backbone NH is in general lower for helical secondary structure than for beta structure. For the plots, values of $^3J(\alpha-\beta)$ and $^3J(\beta-\gamma)$ were taken as 7.5 Hz, typical of free rotation, and 10 and 5 Hz, typical of restricted rotation, of side chains. Since the coupling of nonequivalent protons to NH is often sufficiently different so that both cannot be observed in the same single RELAY, it is recommended that a double RELAY be used, with the first period tuned to the larger coupling and the second tuned to the smaller coupling (2).

An extensive application of 2D homonuclear RELAY to the resolution of the overlapping peaks in the deoxyribofuranoside spectrum of short DNA oligomers has been made (3). A delay of 46 msec was used, based on an average coupling constant of 5.5 Hz in the sugar rings. The 3′ protons of the several rings in the oligomer sequence turned out to be resolved in the materials being investigated, and sections through the 2D spectra at the frequencies of these resonances enabled assignments of the other protons in each sugar ring to be made relatively easily.

Otter and Kotovych (4) have reported the appearance of unexpected cross peaks between the α and δ protons of proline and of arginine in a protein fragment. The intensities of these peaks were found to reach a maximum at a mixing time of 80 msec, corresponding to 50% or more of the intensities of the expected RELAY peaks, but the anomalous peaks disappeared for a mixing time of 40 msec. The cause of their appearance is possibly the existence of a range of T_2 values within the sample or the presence of four-bond coupling effects.

The homonuclear RELAY experiment has been modified by using a DANTE sequence to excite selectively only the A nucleus, for example, in an AMX system (5). The sequence may be represented as

$$\text{DANTE}_{\phi_1} - 90°_{\phi_2} - \tau/2 - 180°_{\phi_3} - \tau/2 - 90°_{\phi_4}, \text{ acquisition}$$

A four-step phase cycle is shown in Table 8-1, although only the first two steps are required. This cycle may be extended to eight steps by inverting the phase of the 180° pulse in order to remove any "90° character" the pulse might have because of its imperfect nature. The two-, four-, or eight-step cycle is then repeated four times with CYCLOPS. The fixed delay τ may be set according to the prescriptions of Bax and Drobny.

TABLE 8-1

PHASE CYCLING FOR THE SELECTIVE-EXCITATION HOMONUCLEAR RELAY EXPERIMENT

ϕ_1	ϕ_2	ϕ_3	ϕ_4	ϕ_{rec}
x	y	x	y	y
x	$-y$	x	y	y
x	y	x	$-y$	y
x	$-y$	x	$-y$	y

This experiment is now a one-dimensional experiment, useful in structural analysis if only a few connectivities need be determined. In addition to the saving of time and data space possible under these circumstances, an additional advantage is that it is easy to obtain a phase-sensitive spectrum, and the differing phase characteristics of directly connected and remotely connected spins may be utilized to differentiate them unambigously.

B. RELAY BY THE PATH H—H—X

One of the commonly applied versions of heteronuclear RELAY has been that described on page 455 of Chapter 6. This involves polarization transfer from a hydrogen nucleus H_A to a second hydrogen nucleus, H_B, to which it is coupled, followed by transfer from H_B to a heteronucleus X_B to which H_B is directly attached. If the spin system is represented as

$$\begin{array}{cc} H_A - H_B \\ | \quad | \\ X_A \quad X_B \end{array}$$

where the lines correspond to coupling pathways, then another route of transfer might be from H_B to H_A to X_A. The magnetization of X is now observed and the relations of the two hydrogens to the two heteronuclei, typically ^{13}C, can be deduced, although neither H need be directly coupled to the X nucleus more remote from it. Thus the method serves as a higher-sensitivity alternative to INADEQUATE for tracing out the skeleton of a molecule.

The behavior of the pulse sequence, shown in Fig. 5-45b, for this version of RELAY has been analyzed by Sarkar and Bax (6). As in the homonuclear case, the decreased sensitivity of RELAY compared to a corresponding shift correlation experiment imposes on the user the need to adjust carefully pulse sequence parameters in order to realize adequate signal-to-noise. For the system

$$\begin{array}{c} A-M-P \\ | \\ X \end{array}$$

where again the lines represent coupling relations, not chemical bonds, the transfer function is

$$f(\tau) = \sin[\pi J(AM)\tau] \cos[\pi J(MP)\tau] \exp(-\tau/T_{2,M})$$

Here τ is the duration of the mixing period and $T_{2,M}$ is the transverse relaxation time of spin M. The sensitivity depends strongly on the value of τ and on the sampling time in the t_1 dimension, as well as on the structure of the spin system. It was shown, for instance, that the sensitivity is much less for relay from a methine proton to an adjacent methylene or methyl carbon than from methylene or methyl protons to the methine carbon. For saturated cyclic hydrocarbons with many pairs of nonequivalent geminal protons, sensitivity is particularly poor. In general, intensities of RELAY peaks are 3 to as much as 20 times lower than those of heteronuclear shift-correlation spectra.

In a modified version of this experiment, Bolton (7) used semiselective refocusing of neighbor protons at the middle of the evolution time, so that these protons are then decoupled from remote protons. In order to eliminate the refocused signals of neighbor protons, the refocusing pulse segment is phase-cycled. An added delay time permits the remote protons to become antiphase with respect to the neighbor protons, so that polarization can be transferred from remote to neighbor protons.

The resulting 2D spectrum shows only correlations of carbon-13 (in the F_2 dimension) with remote protons (in the F_1 dimension). For remote protons which are at a chain terminus, and thus are coupled to no other protons than the neighbor protons, the signals are singlets. For chain-interior remote protons, the coupling to other than the neighbor protons is evident. Thus the method gives structural information, discrimination between neighbor and remote protons, and reduced multiplicity for the remote protons.

As a particularly effective means of exchanging polarization among the hydrogens in a spin system, homonuclear Hartman-Hahn transfer, discussed in detail in Section III, has been suggested. Bax, Davis, and Sarkar (8) employed for this purpose a spin lock, alternated along the $+/-$ axes of the proton rotating frame, with increments of 3-10 msec to a total time of 24-40 msec, at a power level of 6 W. Polarization is then transferred to ^{13}C by an INEPT-type sequence, ^{13}C acquisition is begun immediately after the 90°(C) pulse, and broadband decoupling of hydrogen is begun later, after a suitable refocusing interval. The signal-to-noise ratio in this experiment was found to be lower by a factor of 2 to 5 than for chemical shift correlation, but still quite a bit higher than for conventional RELAY.

Reynolds et al. (9) have devised a related sequence in which the two 180° pulses to hydrogen, one during the evolution period and the other in the middle of the refocusing period, each coincident with a 180° carbon pulse, are replaced by BIRD pulses. In addition, the pulse during the evolution period is stepped incrementally through the period, which is of fixed total length. For each carbon shift, there is then obtained a cross peak for each hydrogen two bonds removed (^1H—C—^{13}C), which is split by coupling to the hydrogens directly attached to the ^{13}C atom, and a singlet cross peak for hydrogens three bonds removed (^1H—C—^{13}C). Thus in addition to connectivity information, values of three-bond H—H coupling constants are obtained.

C. RELAY BY THE PATH H—C—C

Kessler and co-workers (10) proposed a version of RELAY as an alternative to INEPT-enhanced INADEQUATE for measuring ^{13}C-^{13}C coupling between a protonated carbon and a nonprotonated carbon. INEPT transfers polarization only to the protonated carbon, and the double-quantum coherence precession interval distributes that polarization between the two carbons. The RELAY method begins with proton excitation, coherence is then transferred by INEPT to the carbon bearing the proton, and then from that carbon to its nonprotonated neighbor, as shown in Fig. 8-1a. An alternative and perhaps better procedure may be to substitute for the first decoupling period a proton 180° pulse in the middle of the Δ_2 delay. In the 2D version of the experiment, sensitivity is lost because of proton–proton couplings active during t_1.

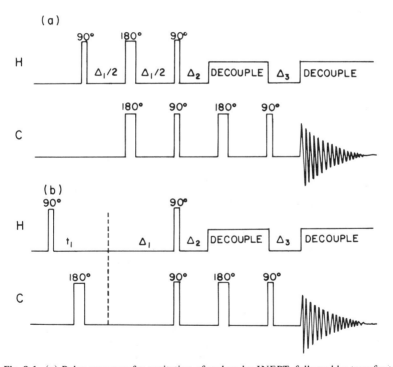

Fig. 8-1. (a) Pulse sequence for excitation of carbon by INEPT, followed by transfer to a nonprotonated neighboring carbon. The values of Δ_1 and Δ_2 are set in the usual way for refocused INEPT and Δ_3 is set to $1/2\,^1J(CH)$ to produce antiphase magnetization of protonated carbons, which cancels on decoupling. (b) Pulse sequence for 2D H—C—C RELAY, beginning with a COSY segment. Axes of the 2D plot represent shifts of ^{13}C and of ^{1}H. [After Kessler et al. (10).]

A further difficulty with this procedure is that direct transfer can also occur from a proton to a nonprotonated ^{13}C atom which neighbors the ^{12}C to which the proton is attached, and this interferes with suppression of "parent peaks." To overcome this problem, Lee and Morris (11) carried out the same experiment twice, with values of the delay Δ_1 of $1/2\,^1J(\text{CH})$ and $3/2\,^1J(\text{CH})$. The second spectrum is subtracted from three times the first one, based on the following reasoning: The INEPT-based signal derived from one-bond transfer is opposite in sign in the two spectra and is reinforced on subtraction. For the very much smaller longer-range coupling, the increase in the transfer function, which follows a sinusoidal variation, is nearly linear over this period, and the subtraction with 3/1 weighting leads to practically complete cancellation. This experiment can be extended to two dimensions (10) by substituting a COSY segment for the INEPT segment, as in Fig. 8-1b.

D. RELAY ORIGINATING ON A HETERONUCLEUS

If the system contains a heteronuclear label, either an isolated ^{31}P atom or a position enriched in ^{13}C, or possibly if a heteronucleus is selectively excited, then the transfer path X—H—H' might be used to pick out those protons in the vicintiy of the label. Field and Messerle (12), working with the ^{31}P case, started with an inverse INEPT experiment, which begins with a 90° pulse to the X nucleus. The time interval in the inverse INEPT part can be optimized for a particular spin system and value of J, usually that of a one-bond coupling. Phase cycling of one of the 90° pulses to the X nucleus alternates the phase of the proton mutliplet resulting from polarization transfer, so that other signals such as those from heteronuclear multiple-quantum coherence or those not arising from polarization transfer can be removed by subtracting FIDs. Between cycles, the protons are saturated to eliminate magnetization not arising from the proton transfer, as well as to provide an Overhauser effect. The X nuclei are excited by a 90° pulse, and their magnetization evolves during τ_1. During a subsequent interval, of length τ_2, the hydrogen multiplet components dephase and a second 90°(H) pulse then transfers magnetization to other hydrogens coupled to the attached one; the optimum value of τ_2 depends on both $J(\text{XH})$ and $J(\text{HH})$ and was found to be about 60 msec for α-D-gluocopyranose-1-phosphate, the molecule investigated. A pair of 180° pulses at the midpoint of the interval refocuses chemical shift effects.

To extend the method to two dimensions with the X shift as the second dimension, an additional period of variable length is introduced into the inverse INEPT segment, along with an additional 180° pulse. Successive steps in the transfer chain can be added for more-remote correlations by appending to the sequence additional units of the form

$$-\tau_n/2—180°(X), 180°(H)—\tau_n/2—90°(H)$$

Alternatively, a total coherence transfer method, such as homonuclear Hartman-Hahn or TOCSY, could be used to transfer coherence throughout the spin system, obviating the difficulty of optimizing the various interpulse time intervals (13).

E. Observation of Hydrogen Signals from Heteronuclear RELAY and Other Sequences

In principle, if magnetization is to be relayed via a heteronucleus, sensitivity is optimum if hydrogen serves as both the nuclear species to be initially excited and the species to be finally observed. A limitation on this procedure is that it requires that the hydrogen signals from those nuclei not involved in the polarization transfer be eliminated. There have been a variety of experiments designed for this purpose: we will first describe some involving polarization transfer from hydrogen to the heteronucleus and then consider others which begin with the generation of multiple quantum coherence.

As described in Chapter 5, pages 405-407, an early application of this method was to molecules in which the heteronuclear "label" is ^{31}P. The basic pulse sequence, used by several groups (14, 15) and shown in Fig. 8-2, consists of a heteronuclear shift correlation followed by a heteronuclear coherence transfer from the X nucleus to hydrogen. Undesired resonances are eliminated by phase inversion of either one of the pairs of simultaneous 90° pulses and subtraction of alternate scans. The 180° pulse in the middle of the evolution period serves to eliminate the effect of chemical shift on polarization transfer; if the ^{31}P carrier is on resonance, these pulses can be omitted. In the 2D spectrum, both frequencies represent proton shifts, and cross peaks connecting two protons indicate that they are coupled to a common ^{31}P nucleus.

Suppression of the unwanted signals by alternation of phase and parallel subtraction of FIDs requires excellent spectrometer stability and pulse quality. Rather than canceling these peaks after acquisition, it is better to null them during the pulse sequence or to avoid entirely their excitation. One trick is to invert, by means of a BIRD sequence, the magnetization of all protons *not* attached to ^{13}C (16). Just at the moment when these protons have relaxed back to the *xy* plane, so that their signals are saturated, the pulse sequence proper is begun. Of course, not all protons have the same relaxation time, so that some superfluous magnetization remains, but the task of the phase-cycle subtraction procedure is greatly eased. Carbon-13

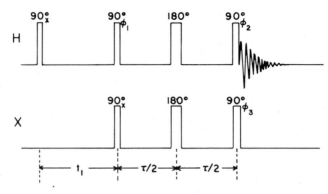

Fig. 8-2. Pulse sequence for ^1H—X—^1H heteronuclear RELAY. The value of τ is chosen to maximize the transfer coefficient

$$\sin \pi J(H_1 X)\tau \sin \pi J(XH_2)\tau \prod_{i \neq 1,2} \cos \pi J(iX)\tau \, e^{-\tau/T_2(X)}$$

One possible phase cycle is

ϕ_1	ϕ_2	Acquisition
x	x	x
x	−x	−x
−x	−x	x
−x	x	−x

decoupling, if begun at a time $1/{}^1J(CH)$ after the final $90°(^{13}C)$ pulse, suppresses one-bond connectivities, permitting remote connectivities to be selectively observed.

Another approach, suggested by Müller (17) and extended by Brühwiler and Wagner (18) is shown in Fig. 8-3. It begins with a segment in which the interval τ_0 is adjusted so that at point (1) all protons attached to ^{13}C are in antiphase along the x axis, while all other protons have precessed by a negligible amount away from the $-y$ axis in the rotating frame. For hydrogen-detected heteronuclear correlation (Fig. 8-3a), a $90°_{-x}(^1H)$ pulse turns this in-phase magnetization to the z axis, but does not affect the antiphase $I_x S_z$ magnetization, which is converted by the $90°_x(^{13}C)$ pulse into $I_x S_y$ multiple-quantum coherence, which evolves during the period of variable length t_1. A $180°(^1H)$ pulse in the middle of t_1 refocuses ^1H shifts and decouples ^1H from ^{13}C in F_1. A $90°(^{13}C)$ pulse just before acquisition regenerates single-quantum coherence. In the resulting 2D spectrum, heteronuclear multiplets appear in antiphase in the F_2 (or proton) dimension.

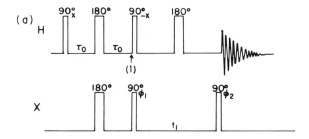

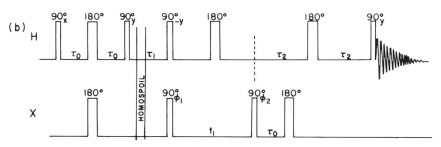

Fig. 8-3. (a) H—C—H RELAY, beginning with one-bond $J(CH)$ and yielding a coupled spectrum. (b) H—C—H—H RELAY, with elimination by a homospoil pulse of contributions to the initial step from spins not directly coupled. A suitable phase cycle for either sequence is that shown in Fig. 8-2, along with independent inversion of all 180° pulses. In (b), the phase of the last 90°(^{1}H) pulse was inverted every other four scans. [After Brühwiler and Wagner (18).]

For a ^{1}H—^{13}C—^{1}H—^{1}H RELAY pathway, a $90°_y$(^{1}H) pulse is applied after the initial segment, converting the antiphase magnetization $I_x S_z$ into longitudinal spin order, $-I_z S_z$. The undesired magnetization is left in the transverse plane where it can be destroyed by a homospoil pulse. After a short delay for the spectrometer to recover, a $90°_y$(^{1}H) pulse restores the antiphase situation for the protons coupled to ^{13}C, reversing the effect on them of the previous $90°_{+y}$ pulse. A 90°(^{13}C) pulse generates two-spin multiple-quantum coherence, which is converted to single-quantum coherence after the variable-length t_1. The last segment of the sequence, $-\tau_2-180°-\tau_2-90°$, is the usual proton–proton RELAY segment. This sequence is shown in Fig. 8-3b.

Hydrogen-detected chemical-shift correlation by multiple-quantum coherence has advantages of sensitivity but suffers from the usual problem of suppressing the intense signals from irrelevant hydrogens. One way of overcoming this difficulty (19) is by use of a BIRD pulse, in much the same way as described above for H—X—H RELAY (16).

Bolton (20) introduced sequences for heteronuclear RELAY with proton detection which start with a heteronuclear zero-quantum chemical-shift correlation, followed by a mixing period to allow proton magnetization to become antiphase with respect to both homo- and heteronuclear coupling. These methods are especially useful for samples selectively enriched in the heteronucleus. In the resulting 2D spectrum, the frequency of the heteronucleus appears as F_1 and the frequencies in F_2 are those of neighbor protons and remote protons, complete with heteronuclear couplings. An alternate version replaces the first 90° COSY pulse with a zero-quantum filter to generate selectively proton transverse magnetization. After a variable-length evolution time, a 90°(^{1}H) pulse transfers magnetization from neighbor to remote protons. In the 2D spectrum, the coupled resonances of neighbors appear on the diagonal, and the off-diagonal signals have F_2 frequencies corresponding to those of the remote protons and F_1 frequencies of the neighbor protons.

A method proposed by Bax and Summers (21) to select long-range ^{1}H—^{13}C connectivities is a modification of the hydrogen-detected multiple-quantum experiment, and the pulse sequence is shown in Fig. 8-4. The initial pulse to ^{13}C creates multiple-quantum coherence for protons that are coupled to ^{13}C. This is then removed from the spectrum by phase alternation of this pulse, keeping the receiver phase constant. The second ^{13}C pulse creates multiple-quantum coherence through the longer-range couplings. The 180°(^{1}H) pulse eliminates the effect of hydrogen chemical shift from F_1. The final 90°(^{13}C) pulse yields single-quantum coherence, and the 2D spectrum has correlation peaks which indicate the chemical shifts of protons which have long-range couplings to carbon atoms for which the shifts appear in the first dimension. Homonuclear H-H couplings remain in the peaks. Phase cycling of the second 90°(^{13}C) pulse removes the proton signals not coupled to ^{13}C.

For studies involving ^{15}N, where the dynamic range problem is even more severe than for ^{13}C, Mueller, Schicksnis, and Opella (22) utilized the more rapid transverse relaxation of protons attached to ^{14}N. A fixed interval of evolution between generation of multiple-quantum coherence involving ^{1}H and ^{15}N and recovery of single-quantum coherence is adjusted to take advantage of this selective relaxation. The constant-time feature also affords decoupling in F_1. A 180° "echo" pulse is "walked" through the multiple-quantum evolution period, and the time from the beginning of the period to the pulse becomes the variable for F_1. The attainment of essentially complete suppression of the peaks of protons attached to ^{14}N permits decoupling to be applied during acquisition; if this suppression were not achieved, the decoupled peaks for ^{15}N-attached protons could not be distinguished from the residual ^{14}N-attached proton peaks.

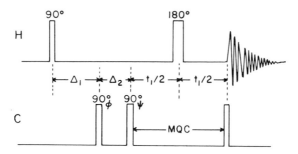

Fig. 8-4. Pulse sequence for detecting long-range ^{1}H-^{13}C connectivity using MQC as a low-pass J filter and hydrogen detection. The interval Δ_1 is set to $1/2\,^1J(CH)$ and Δ_2 is about 60 msec, much longer than Δ_1. Data is presented in the absolute value mode. A suggested phase cycle is

ϕ	ψ and acquisition
x	x
x	$-x$
$-x$	x
$-x$	$-x$
x	y
x	$-y$
$-x$	y
$-x$	$-y$

[After Bax and Summers (21).]

A rather interesting application of detection of mutliple-quantum coherence by higher-γ nuclei as a means of studying less-sensitive nuclei is the determination of the NMR parameters of ^{57}Fe by ^{31}P—^{57}Fe—^{31}P and ^{1}H—^{57}Fe—^{1}H experiments (23).

Addition of hydrogen observation to the ^{13}C-^{13}C INADEQUATE experiment (24) substantially improves the sensitivity. There are, of course, some limitations to the advantage to be gained by this procedure in which, following the multiple-quantum sequence, the carbon magnetization is transferred to protons by a modified inverse INEPT. The coupling to ^{13}C appears in the proton spectrum and doubles the numbers of signals. The cycle time is determined by the T_1 of the protonated carbons, and the proton population difference after the inverse INEPT is determined by the relatively small population difference attained by the ^{13}C nuclei. Furthermore, proton-proton couplings spread out the multiplet during the FID.

Finally, proton detection in a sequence in which pulses are applied to another nucleus may present hardware problems for spectrometers configured to observe signals of the same species irradiated by the broadband transmitter and not those of the protons which might be pulsed by the decoupler. For at least one instrument type, appropriate hardware modifications have been described (25). A phase-shifting and modulation unit has been designed for use when an independent heteronuclear channel is to be used in experiments of this sort (26).

II. Homonuclear Correlation

We first note that some workers have applied the acronym "COSY" to heteronuclear correlation experiments, but we prefer here to limit its compass to those correlations in which both nuclei are of the same species. A number of modifications and extensions have been made to the basic COSY experiment to extend its applicability and improve its performance, and its behavior has been analyzed in several ways. Some of these contributions were described in Chapter 5, and several others will be outlined here.

A. Delayed COSY

A delay before acquisition, originally suggested by Bax and Freeman (27) as a means of picking out long-range coupling constants, has been utilized by Nakashima and Rabenstein (28) to discriminate against those resonances of intact erythrocytes which have short T_2 values, and thus to uncover the spectra of small molecules. The sequence in this "delayed" COSY experiment is

$$90°_\theta - T_1 - 90°_\beta - D_2 - 180°_\beta - D_2 - \text{Acq}_\psi$$

The cycling of the phases was designed to enhance N-type peaks and to provide quadrature detection in both dimensions.

Delays may be inserted without the 180° refocusing pulse, and one of these delays is then between the second 90° pulse and the beginning of acquisition. It makes no difference, in the absence of 180° pulse, whether the other delay is said to follow the first 90° pulse, $90° - \Delta_1 - \vdots - t_1 - 90° - \Delta_2 - \text{Acq}$ (29), or to precede the second 90° pulse, $90° - t_1 - \vdots - \Delta_1 - 90° - \Delta_2 - \text{Acq}$. Mueller, Jeffs, and co-workers (30, 31) have reported analyses of rather large molecules, utilizing 2D methods which include delayed COSY with Δ_1 of 300 msec and Δ_2 of 50-60 msec. The sum of acquisition time and Δ_2 was chosen to be longer than the maximum value of t_1 and Δ_1, providing recording of the full coherence transfer echo, and thus including resonances covering a wide range of linewidths. Other spectral parameters used in this work were a width of 4812 Hz at 360 MHz, 4K points with quadrature detection, and 512 FIDs with zero filling to 1K in the t_1 dimension. Sine bell apodization was applied to each dimension before Fourier transformation.

The sequence SUPERCOSY (32-34) described in Chapter 5 contains two delays, with refocusing pulses in the middle of each, and is designed to produce in-phase cross-peak multiplets and antiphase diagonal multiplets so that poor resolution reduces the diagonal rather than the cross peaks. It has been pointed out (35) that effective use of this sequence requires preknowledge of the value of J and that a single value of J may not be appropriate in complex systems, although dependence on J may be utilized to achieve selectivity (33). The added delays may cause loss of signal by relaxation if T_2 is short or J small and lead to errors in phase which make the correction to absorption mode difficult. It has also been suggested that, since the experiment including 180° pulses gives the same results as that without refocusing pulses, provided the magnetic field is homogeneous, the simpler modification is preferred because it is less susceptible to artifacts (36).

B. Z-Filtered COSY

The procedure of z filtering (37) is the following. The desired component of transverse magnetization is turned into the z direction by a 90° pulse. The undesired components are allowed to precess in the xy plane for a period, the length of which is stochastically varied from experiment to experiment. As a consequence, the x and y components of this magnetization vary randomly at the end of the period of precession, and cancel upon time averaging the results. After the precession interval the desired mangetization is returned from the z direction to the transverse plane by a second 90° pulse.

A two-dimensional homonuclear correlation experiment incorporating this type of filtering has been labeled "z-COSY" (38). The sequence consists simply of three pulses, with flip angles of β, β', and β', in that order. The first two pulses are separated by the regularly varying interval t_1, and the second two are separated by the randomly varying filter interval τ_z, which must be short enough so that exchange processes do not contribute, but must be long enough to suppress zero-quantum effects. The tip angle of the first pulse can be any value up to 90°, as in the simple one-pulse FT experiment, using the Ernst angle appropriate to the relaxation and cycle times, typically 60°. The angle β' is small, usually 20°, so that coherence transfer is limited to directly connected transitions.

To suppress axial peaks, the phase of the first pulse is alternated as signals are added and subtracted. Coherences of all orders except 0, 4, 8, ... are eliminated by cycling the phase of the last pulse plus that of the receiver in 90° steps. The only significant transverse coherence remaining is that of zero order, which is eliminated in the z-filtering process by the variation of τ_z. If necessary, t_1 noise can be eliminated by symmetrization of the 2D array, as for other versions of COSY. The z-COSY method is somewhat less sensitive than conventional COSY but, because of the possible use of the Ernst angle, may require less time for equivalent results. An advantage of z-COSY is that both cross and diagonal peaks appear with pure absorption lineshapes in both dimensions, with simple patterns for the multiplets. In addition to the dominant peaks for directly connected transitions, which are positive if the connectivity is progressive and negative if it is regressive when the diagonal peaks are negative, there also appear much weaker peaks from coherence transfer between remotely connected transitions.

Because the flip angles in z-COSY are small, the effects of frequency offset are small and the response of the spin system is expected to be linear. Pfändler and Bodenhausen (39) analyzed the behavior of an ABX system and showed that, even for strong coupling, in-phase multiplets are obtained with the intensities following the simple rule that they are proportional to the product of the amplitudes of the corresponding lines in the one-dimensional spectrum. This is in contrast to AB systems with simple COSY and to larger strongly coupled spin systems with double-quantum filtered COSY, which yield phase-twisted lineshapes. The z-COSY analysis, however, showed that extraneous peaks corresponding to "virtual coupling" may appear for strongly coupled systems.

To obtain a spectrum complementary to z-COSY, an "anti-z-COSY" spectrum, the tip angles of the first two pulses are set to $180° \pm \beta$ and $\beta' + 180°$, respectively. Cross peaks now occur between the "remotest" transitions, which differ in the spin states of each of the passive spins. The "diagonal" peaks are now negative and are arranged as antidiagonal multiplets centered on the diagonal, and the cross-peak multiplets are reversed in position and amplitude, compared to the z-COSY spectrum. Further, the anti-z-COSY cross peaks are not necessarily symmetrically arranged about the diagonal. In addition to providing further information about connectivities in the spin system, the "complementary" spectrum is of importance in designing automated computer analysis of 2D spectra.

C. Heteronuclear-Filtered COSY

A technique, labeled the X filter, has been devised to pick out COSY resonances coupled to a particular "label" heteronucleus (40). The "X" does not refer to a spatial coordinate but to the presence and effect of the particular heteronucleus. The method consists simply of applying a 180°(X) pulse on every second cycle and storing the results of alternate scans separately. Subtraction eliminates all resonances but those from nuclei coupled to the label, while addition yields a spectrum for nuclei not attached to the label. The X filter simplifies complex spectra and, by use of labeled compounds, permits assignments of resonances in complex systems.

A half-filter method was applied to simplify the spectra of ^{15}N-enriched proteins in either of the two dimensions separately (41). A pulse segment $—\tau—180°(H)$, $180°(^{15}N)—\tau—180°(^{15}N)—$ is inserted in the COSY sequence. If it is placed before the evolution period, it is active in the F_1 dimension; if it is inserted before the detection period, it is effective in the F_2 dimension.

Gated decoupling has been employed to reduce complex overlap of signals in homonuclear COSY as well as in NOESY (42). The method is somewhat analogous to the use of gated decoupling in heteronuclear J spectroscopy, and is specially useful for samples enriched in a spin-$\frac{1}{2}$ nucleus such as ^{13}C or ^{15}N, in which the protons directly attached to the label nucleus can be selected. The X-nucleus decoupling can be active either during the COSY evolution period or during the acquisition period, and can be selective or broadband. The doublets are collapsed in one dimension or the other, depending on which decoupling period is chosen. Recognition of the multiplets is facilitated by the fact that the centers are shifted away from the diagonal for otherwise on-diagonal resonances.

D. Multiple-Quantum-Filtered COSY

With the aim of reducing overlap of cross-peak multiplets in complex spin systems, obtaining the cross peaks in absorption phase, and permitting better resolution of small couplings, an experiment called E. COSY—for exclusive correlation spectroscopy—has been designed (43, 44). It consists of taking appropriately weighted linear combinations of p-quantum filtered COSY spectra obtained with different values of the order p. The procedure utilizes the fact that pairs of connected transitions and pairs of transitions which are not connected have different dependences on the tip angle of the mixing pulse, or on an equivalent phase shift of the pulse, in order to restrict coherence transfer to directly connected transitions. The multiplets obtained have structures similar to those from the version of the COSY experiment using very small tip angles, with pure absorption lineshape in both dimensions, but with the advantage of higher intensities.

The sequence used is either $90°_x—t_1—\beta_y—t_2$ or $90°_\beta—t_1—90°_\beta 90°_{-x}—t_2$, where the value of β determines the coherence order. The highest order needed depends on the coupling network; in practice, combining spectra for $p = 2, 3,$ and 4 in the ratio 1:2:4 is often satisfactory. The experiment has been illustrated for a cyclic hexapeptide using a linear combination of +4 scans at $\beta = 0°$, -3 at $60°$, $+1$ at $120°$, $+1$ at $240°$, and -3 at $300°$. The diagonal peaks contain both absorption and dispersion contributions, and the weights of the zero- and one-quantum filtered spectra can be adjusted to reduce the spread of the diagonal multiplets or to reduce their intensity by allowing antiphase overlap.

Each cross-peak multiplet in E. COSY contains a four-component square split by coupling between the spins active in the coherence. The four-peak pattern is further split by coupling to passive spins into two patterns displaced from one another by distances parallel to the axes equal to the two passive-spin couplings. Comparison between line profiles from each of the displaced patterns permits very small, otherwise unresolved, coupling constants to be evaluated, and the relative directions of displacements depend on the relative signs of the coupling constants.

As in z-filtered COSY, complementary E.COSY spectra may be generated. This may be done by inserting a $180°_y$ pulse immediately after the β_y rotation or by inverting the phase of the third pulse in the second version of the sequence given above. Complementary spectra correlate "anticonnected" or remote transitions, as compared to the normal spectra, which correlate connected transitions. Complementary spectra contain the same information as normal spectra, but in mirror-image form; sometimes peaks which overlap in one spectral form are better resolved in the other form.

Several other 2D experiments involving the properties of multiple-quantum coherences have been suggested. One is "bilinear COSY," in which a bilinear pulse cluster $90°_x - \tau/2 - 180°_x - \tau/2 - 90°_x$ is used as the mixing sequence, with a mixing period in the demonstration example of 50 msec (45). Symmetry-imposed selection rules prevent some peaks from appearing in the resulting spectrum, simplifying it and reducing overlap. Further, the form of the spectrum provides information about relative signs of coupling constants.

UNCOSY uses the sequence shown in Fig. 8-5 (46), with a constant time between excitation and mixing, with resultant decoupling in the F_1 dimension. Addition of FIDs acquired with different values of the delay Δ leads to uniform excitation of cross peaks. This sequence also has the effect of suppressing diagonal peaks and acts as a double-quantum filter.

E. Difference Methods

Some of the COSY adaptations described above involve taking combinations of spectra obtained under varying conditions, but here we mention several experiments in which taking a difference between FIDs seems to be the key feature.

Cavanagh and Keeler (35) have devised a difference method for completely suppressing the diagonal peaks in COSY spectra. A normal COSY

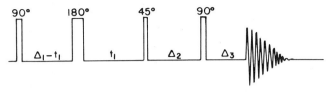

Fig. 8-5. The UNCOSY pulse sequence for uniform excitation. The intervals Δ_1, Δ_2, and Δ_3 are equal and are varied together. The phase is cycled according to the scheme

θ	ψ	Acquisition
x	x	x
x	$-y$	y
x	$-x$	$-x$
x	y	$-y$
y	x	y
y	$-y$	$-x$
y	$-x$	$-y$
y	y	x

[After Farmer and Brown (46).]

spectrum is obtained with two 90° pulses separated by t_1. A diagonal-only COSY spectrum is then obtained by application of a third 90° pulse at an interval Δ after the second 90° pulse. This causes the z magnetization to become observable, and phase cycling is used to select only that part of the signal which was z magnetization between the second and third pulses. However, there may also be zero-quantum coherence which cannot be eliminated by phase cycling, and so a z filter, effected by varying the interval between the second and third pulses over a range of values, is introduced. If care is taken so that both spectra are acquired under the same conditions, subtraction eliminates the diagonal peaks. In practice, a scaling factor may be required before subtraction, and differences in relaxation times may prevent complete elimination.

Mueller (47) has described a modification of the small-flip-angle COSY experiment in which a reference signal $S_{\text{ref}}(t_1, t_2)$ acquired without a mixing pulse, that is, by simply waiting a time t_1 after an initial 90° excitation, is subtracted from the small-angle COSY signal $S_x(t_1, t_2)$. This eliminates all components of the signal in phase with the mixing pulse. In a third experiment, the mixing pulse is shifted by 90°, yielding the signal $S_y(t_1, t_2)$ from which $S_{\text{ref}}(t_1, t_2)$ is also subtracted. After t_2 Fourier transformation, the absorptive component of $S_x(t_1, F_2) - S_{\text{ref}}(t_1, F_2)$ is stored as the real part and the absorptive component of $S_y(t_1, F_2) - S_{\text{ref}}(t_1, F_2)$ is stored as the imaginary part for the final t_1 Fourier transformation to get a phase-sensitive spectrum.

In the resulting 2D "P.E. COSY" spectrum, cross peaks from nonconnected transitions are suppressed to less than one-tenth the intensity of connected peaks if a 35° mixing pulse is used, and all cross peaks are in absorption mode. For coupled spin systems, dispersion components appear in the diagonal multiplets but they tend to cancel one another at a distance from the diagonal. The patterns obtained by this method are adapted to simple determination of sign and magnitude of J and to computer-assisted analysis of spectra.

F. ANALYSES AND APPLICATIONS

An extensive analysis of the cross-peak fine structure in multiple-quantum filtered COSY has been made by Boyd and Redfield (48), with special emphasis on the spin systems encountered in amino acids and on applications to proteins. The distribution of antiphase components in the cross-peak multiplets was shown to depend on the order of the multiple-quantum coherence selected between the two 90° pulses in the mixing sequence and on the number and relative magnitude of the spin–spin couplings involved in the cross peak. Good results were obtained for the spin systems in tryptophan and in aspartic acid, using either a 2Q or a 3Q filter, but the 4Q-filtered spectrum of tryptophan suffered from the decreased sensitivity associated with filters of higher order.

Müller, Ernst, and Wüthrich (49) have discussed multiple-quantum-filtered COSY spectra of proteins in detail, and presented a catalog of patterns associated with the common amino acid moieties, using 2Q, 3Q, and 4Q filters. In spectra with extensive overlap, spin systems may sometimes be identified from the pattern of peaks eliminated by MQ filtering.

Higher-order MQ filters give information to supplement that obtained by 2Q filtering, but are more sensitive to pulse imperfections, so that use of 90°-90° composite pulses is recommended. Odd- and even-quantum filtered spectra have different symmetry properties; for example, 2Q spectra have zero intensity in the center of an odd multiplet, while 3Q cross peaks have positive or negative peaks near the center, with the sign depending on the relative signs of the two coupling constants to passive nuclei. The 3Q filtered spectra have reduced diagonal intensities and permit resolution of cross peaks closer to the diagonal than possible for 2Q.

To eliminate t_1 and t_2 ridges, the FID for $t_1 = 0$ and the first point of each FID are set equal to zero, since a sinusoidal function is expected. Care must be taken with respect to the effects of strong coupling which leads to additional peaks, and of nonexponential relaxation in methyl groups which results in violations of the normal intensity rules, as well as of selection rules for cross-peak multiplets (50).

Further discussion of the applications of double- and triple-quantum filtering to the analysis of COSY spectra of proteins has been given by Rance and Wright (51).

Bain (52) has provided a general analysis of the rationale of phase cycling, with specific application to the simple COSY experiment. Computer simulations were employed to evaluate the effect of digital resolution on various COSY versions (36). In the simple experiment, cross peaks are predicted to appear even when the coupling constant is somewhat less than the digital resolution. However, peak intensities are unreliable if the coupling constant is less than three times the digital resolution. For qualitative detection of cross peaks resulting from small couplings, insertion of fixed delays may be helpful, but if J is less than the digital resolution, the intensity varies cyclically with the length of the delay.

Among the recent applications of homonuclear correlation has been the separation of polarization and multiplet effects in CIDNP experiments (53). Proton-proton 2D COSY of peptide lactones related to actinomycin demonstrated the existence of equilibria between two conformers and permitted the determination of relative concentrations (54). An inversion recovery experiment has been placed before a proton COSY sequence in order to measure T_1 values in overlapped spectra of molecules such as gramicidin A. Of course, these values may be complex functions of individual proton relaxation times and cross-relaxation rates, so interpretation of the results must be done with caution (55).

Extensions to other nuclei include ^{113}Cd-^{113}Cd correlations for the seven cadmium ions bound to the protein metallothionein, which confirmed the organization into clusters containing three and four members, with those ions in each cluster coupled to the other ions in the cluster (56). Confirmation of the presence of a cubic octamer containing one Ge atom was provided by ^{29}Si-^{29}Si correlation in a germanosilicate solution enriched to 93.5% in ^{29}Si (57). Structures of hexafluorotetracyclic compounds were determined by ^{19}F-^{19}F correlation (58).

A start has been made in development of methods for computer-assisted analysis of COSY spectra (59-62). For this purpose, it is best to have spectra of pure phase, with clean positive and negative peaks, such as provided by E. COSY or z-COSY. For the analysis, complementary pairs of spectra are employed, and use is made of the square or rectangular structure of cross peaks. In one approach, a positive and negative Boolean matrix is generated from each of the two complementary spectra, each matrix is symmetrized, and then the four arrays are searched for cross-peak structures with a mask set for successively increasing values of J (61). Each of the square or rectangular units found is then represented by its center of gravity, and the complementary spectra are similarly scanned in parallel to find repeated values of passive couplings. For these results, three-spin systems are picked out, and they are finally assembled into larger systems; these steps may require human intervention, especially if spectral overlap causes misidentification of clusters at an earlier stage. To this time, pattern-recognition methods have been limited to small molecules with weakly coupled spins.

III. Isotropic Mixing Experiments

There have been proposed several experiments based on elimination of the Zeeman terms from the Hamiltonian of an isotropically coupled spin system, or making the terms for the two partners equal, so that the system evolves solely under the effect of the scalar coupling. Recent versions of this approach have been labeled TOCSY, for total coherence transfer spectroscopy, or HOHAHA for homonuclear Hartman–Hahn experiment.

These methods are related to the original Hartman–Hahn experiment (63), much used in NMR of solids, in which the proton magnetization is spin-locked in the y direction of the rotating frame, and the rf field $B_{1,x}$ at the resonance frequency of another nucleus X is adjusted so that $\gamma_H B_{1,H} = \gamma_X B_{1,X}$. The two nuclear species then have the same precession frequency around B_1 and facile polarization transfer can occur, mediated by the direct dipolar interaction.

An adiabatic demagnetization version of this experiment, in which the rotating-frame energy levels were caused to cross slowly by an rf frequency-amplitude sweep, was applied to cross polarization via isotropic coupling in liquid-phase spectroscopy by Garroway, Chingas, and co-workers (64, 65). Because of the special hardware requirements, this method was never widely used.

More recently, several pulse methods for achieving homonuclear cross polarization or "isotropic mixing" have been developed. In Braunschweiler and Ernst's TOCSY (66), mixing is achieved by a series of equally spaced $180°_x$ pulses, following an evolution period of variable length t_1. These workers suggested that rf field inhomogeneity might be compensated by phase inverting the last two of the four pulses, and also suggested possible 10- and 16-pulse cycles that would have the same effect. Davis and Bax (67) used a spin lock for mixing. This has limited bandwidth unless very high power is employed, but the bandwidth may be extended by inverting the phase between $+x$ and $-x$ an even number of times. They also described a difference version of the experiment in which it is reduced to one dimension in order to conserve time and data space (68). A single multiplet is inverted by a selective 180° pulse in alternate scans, followed by a 90° excitation pulse and a spin-lock sequence of the form $(SL_y 60^\circ_{-y} 300^\circ_y SL_{-y} 60^\circ_y 300^\circ_{-y})n$. The 60°–300° pulse pairs, following the successive spin locks along the $+y$ and $-y$ axes, are composites to rotate the magnetization so that it is aligned with minimal loss for the next spin-lock phase.

Another basis for isotropic mixing is the broadband decoupling cycle MLEV-16 (69). An "MLEV-17" sequence consists of the 16 composite 180° pulses about the $+x$ and $-x$ axes with an extra 180°_x pulse added at the end to invert the phase error. MLEV-17 is repeated an even number of times, sandwiched between two spin-lock type trim pulses of a few milliseconds in length, which defocus any magnetization not parallel to the x axis. Use of MLEV-17 has the advantages of covering a wider frequency range, lengthening the effective T_1, and reducing NOE type magnetization transfer, which does occur with this sequence if the mixing time is prolonged. The overall experiment consists of a 90° excitation pulse which is cycled through all four phases, variable evolution period of length t_1, the mixing sequence just described, and the acquisition period. A selective version of this experiment is also possible (70).

Analysis of the behavior of a spin system under a Hamiltonian which includes only the isotropic coupling term—a situation corresponding in some ways to the limit of infinitely strong coupling rather than the weak coupling usually assumed for simplicity—shows that the individual single-spin operators take part in collective spin modes, leading to a transfer of coherence back and forth throughout the spin system, much as energy is transferred through a system of coupled mechanical oscillators. If the system is "examined" at the proper time, a *net transfer* of magnetization will be found, in contrast to sequences such as COSY, which in themselves produce no net magnetization transfer.

Braunschweiler and Ernst (66) showed that, for a two-spin system, the collective modes correspond to sums and differences of product-spin operators:

$$\Sigma_\alpha = \tfrac{1}{2}(I_{1\alpha} + I_{2\alpha}) \qquad \Delta_\alpha = \tfrac{1}{2}(I_{1\alpha} - I_{2\alpha})$$
$$\Sigma_{\alpha\beta} = (I_{1\alpha} I_{2\beta} + I_{1\beta} I_{2\alpha}) \qquad \Delta_{\alpha\beta} = (I_{1\alpha} I_{2\beta} - I_{1\beta} I_{2\alpha})$$

where α and β stand for the coordinates x, y, and z. The sum terms remain constant with time while the difference terms follow sinusoidal variations with time, so that the magnetization undergoes an oscillation with period $1/J(AX)$. Chandrakumar (71, 72) has analyzed the behavior of more complex systems and found differences in the response to Hartman–Hahn type mixing, in which only the spin-locked component of the transverse magnetization is conserved, and the response to pulsed isotropic mixing, where both components are conserved (72). For example, for an AX_N system with

$N > 2$, evolution under cross polarization is not periodic, but under isotropic mixing, evolution frequencies are rational multiples of one another and evolution is therefore periodic. Indeed, isotropic-mixing-produced frequencies are periodic for any $A_M X_N$ system, although not periodic in an AMX system (73).

Under isotropic mixing, magnetization is transferred first to directly coupled nuclei. For a mixing period of less than 20 msec, there is little relay, but increasing length of mixing time to 50 to 100 msec causes magnetization to be transferred to indirectly connected transitions and finally to all nuclei within the spin system, whence the acronym TOCSY. By carrying out a series of experiments with successively longer mixing times, it is possible to trace out connectivity pathways through the spin system.

In setting up a spin-lock experiment, 4–5 W of transmitter power is generally sufficient, but the length of the mixing period is greater than the high-power capabilities of most spectrometer amplifiers, so that it is generally better to take from the transmitter circuitry a lower-level signal and boost the level—perhaps through the decoupler amplifier—than to attenuate the high-power pulse output. One must also be somewhat careful about overheating the probe and the sample. Helpful details of experimental technique have been given by Bax and co-workers (74, 75).

The isotropic mixing experiment gives a 1D or 2D spectrum in which the peaks have a combination of in-phase absorption mode and antiphase dispersion mode, with the result that measurements of coupling constants are inaccurate because of phase distortions of the resonances. Braunschweiler and Ernst (66) suggested canceling the dispersion-mode component by adding the results of a series of experiments with different mixing times. Subramanian and Bax (70) have stated that this is effective only if the spin system is simple enough so that the magnetization transfer is truly oscillatory, and suggested using a z filter to provide pure absorption spectra. A $90°$—τ_z—$90°$ sequence is applied after mixing, and all antiphase components are converted by this filter to zero-quantum coherences, which are canceled by combining results for differing values of τ_z. The selective sequence with z filter is shown in Fig. 8-6.

Finally, Waugh (76) has pointed out that the *homonuclear* coherence transfer occurring in these experiments does not take place in accordance with the simple Hamiltonian $J\mathbf{I}_1 \cdot \mathbf{I}_2$, but that the magnetization exchange frequency has an offset dependence much more complex than has usually been assumed. This dependence arises from the presence of other bilinear terms which are a consequence of the "history of the scalar coupling operator" but which are not present in the heteronuclear case.

8. RECENT DEVELOPMENTS IN PULSED NMR METHODS

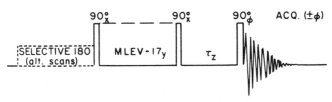

Fig. 8-6. Pulse sequence for the z-filtered one-dimensional HOHAHA experiment. The selective 180° pulse is omitted and data are subtracted in every other scan. The phase ϕ is stepped through all four 90° increments. The z filtering is accomplished by varying τ_z and adding the resulting spectra. The MLEV17$_y$ sequence is an integer number of repeats of (ABBA AABB BAAB BBAA 180$°_y$) where A is $90°_x 180°_y 90°_x$ and B is $90°_{-x} 180°_{-y} 90°_{-x}$. [After Subramanian and Bax (70).]

IV. Applications of Zero-Quantum Coherence

The generation and detection of multiple-quantum coherence, including zero-quantum coherence, which corresponds to a transition in which two coupled spins flip in opposite directions, was described in detail in Chapter 4, with applications in Sections VI and VII of Chapter 5 and Section III,B of Chapter 6. Zero- and double-quantum coherences arise as incidental effects in many pulse experiments and may be merely nuisances, but here we shall discuss some experiments which have been designed to take advantage of the desirable characteristics of zero-quantum coherence (ZQC).

The special features of ZQC include insensitivity of linewidth to magnetic-field inhomogeneity, spectral width which is relatively small because frequencies appear as differences rather than sums of the precession frequencies of the coupled nuclei (although negative magnetogyric ratios may complicate this situation), reduced overlap in complex spin systems, and the capability of providing information about spin systems which may complement results on connectivity and coupling obtained by other techniques.

A. HOMONUCLEAR ZQC SPECTROSCOPY

Some recent applications of ZQC include selection of the desired signals by phase cycling, which utilizes the difference in response of coherences of different orders to phase shifts, but others use an inhomogeneous magnetic field from field gradients in an imaging spectrometer to dephase all higher orders of coherence. The dephasing approach derives from an early one-dimensional experiment (76) in which two 90° proton pulses, separated by a suitable interval, generated multiple-quantum coherence, after which all orders except zero were destroyed by a field gradient in the z direction.

The magnetization was detected point by point in a complex fashion that is not required with present methods. For the vinyl group in triethoxyvinylsilane, there were obtained three pairs of zero-quantum lines, with the splitting between members of each pair equal to a difference between two of the three coupling constants in the system. From known magnitudes of J, the zero-quantum lines were assigned and the relative signs of the J values were determined.

Extensive expressions for zero-quantum frequencies in various spin systems were derived by Pouzard, Sukumar, and Hall (78). Using two-dimensional experiments based on phase cycling of the excitation pulses, they showed the utility of ZQC spectra in analysis of the hydrogen spectra of a tetra-acetylated pyranoside. A trial-and-error procedure was used to find the value of the interval between the two 90° excitation pulses optimum for ZQC generation, a value which depends in a complicated way on both chemical shifts and coupling constants and differs for various protons in the molecule. Insertion of a 180° pulse midway between the 90° pulses eliminates the effect of chemical shift on the excitation efficiency, so that the interval need only be optimized for magnitude of $J(H-H)$; unfortunately, this pulse also eliminates the excitation of ZQC for a *symmetrical* homonuclear spin system, that is, a system $A_n X_m$ where $n = m$ (79, 80).

Müller (79) devised a method, described in Chapter 5, which uses a 90°—$\tau/2$—180°—$\tau/2$—45° excitation sequence, and does generate ZQC with efficiency independent of chemical shift. This method has been modified by employing a 135° read pulse with delayed acquisition (81), as in Fig. 8-7a, which halves the data size in the F_1 dimension. For two coupled nuclei, k and l, signals are obtained at $[\omega_k, \omega_k/2]$, $[\omega_k, \omega_l/2]$, $[\omega_l, \omega_l/2]$, and $[\omega_l, \omega_k/2]$. Sum frequencies arising from two-spin-order terms are suppressed, and the result is a spectrum which appears like a COSY spectrum with p-type peak selection, but with a reduced F_1 dimension; the cross peaks are arranged about an "antidiagonal," $F_1 = -\frac{1}{2}F_2$. A phase cycle suitable for suppressing the undesired peaks is given in Table 8-2. To compensate for pulse imperfections, the cycle may be extended by changing the sign of ϕ_1 after every 8 pulses and the sign of ϕ_2 after every 16 pulses.

In one experimental variation of the 90°—$\tau/2$—180°—$\tau/2$—90° excitation experiment, Hall and Norwood (82) decoupled the nuclei in the coherence from all other nuclei by using a constant evolution time t_d between the excitation sequence and the read pulse, together with a 180° refocusing pulse which is stepped through the evolution period to provide the desired chemical shift modulation. Particular types of coherence may be emphasized or selected by the choice of the length of the evolution period or of the read-pulse tip angle. More-direct coupling information is obtained by varying the time of the final refocusing period in parallel with the effective

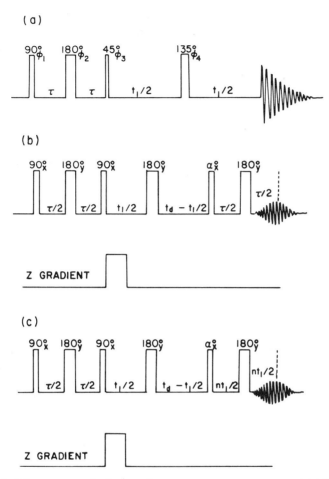

Fig. 8-7. Pulse sequences for homonuclear zero-quantum spectroscopy. (a) "SUCZESS" sequence of Santoro, Bermejo, and Rico (81). A phase cycle is given in Table 8-2. (b, c) Sequences of Hall and Norwood (82), using a pulsed magnetic field gradient. Sequence (c) yields a J-resolved spectrum.

evolution time, as shown in Fig. 8-7b. The single-quantum spectrum for a given spin system can then be reconstructed by looking for identical splittings in different multiplets which have the same structure and then shifting the frequency scale so as to superimpose the identical multiplets (82).

In yet another ZQC version (83), termed the ZECSY sequence, insertion of an additional pulse near the middle of the evolution period results in the appearance of correlation peaks between ZQCs which share a common spin, analogous to the cross peaks in SECSY. Thus if two ZQCs are excited,

TABLE 8-2
PHASE CYCLING FOR THE ZQC EXPERIMENT
SHOWN IN FIG. 8-6a

ϕ_1	ϕ_2	ϕ_3	ϕ_4	$\phi_{receiver}$
x	y	y	y	x
x	y	−y	−y	x
x	y	y	−y	−x
x	y	−y	y	−x
x	y	y	−x	y
x	y	−y	x	y
x	y	y	x	−y
x	y	−y	−x	−y

peaks result not only at the evolution precession frequency of each, but also at one-half the sum of, and one-half the difference of, the frequencies, unaffected by the phases of the pulses. If the magnetic field is inhomogeneous, presence of an extra delay in one part of the evolution period results in dephasing of other coherences so that phase cycling is not required. This method was used to trace out the AMX_3 spin system in L-threonine. Hall's experiments were designed in part with a view to applications in imaging where samples are large and inhomogeneous, a situation where ZQC has its greatest advantages. Therefore they tend to rely on inhomogeneity of the sample or an externally applied field gradient to eliminate higher coherence orders.

Müller and Pardi (84) combined MQC generation with a relay procedure for spin–spin coupling network analysis through remote connectivities. Excitation is by the sequence $90°_x$—τ_1—$180°_y$—τ_1—α_y, with τ_1 of 15 msec. Following the evolution period, a $90°_{\phi 4}$—τ_2—$180°_{\phi 5}$—τ_2—$90°_{\phi 6}$ sequence converts the multiple-quantum coherence back to transverse magnetization of the spins involved in the coherence and relays part of this magnetization to neighboring spins. As a result, signals correlating A and X in spin systems A—M—X, where A is coupled to M but not to X, are obtained.

Turner (85) generated homonuclear ZQC between pairs of protons, allowed the coherence to evolve for a variable period of time, and regenerated single-quantum coherence in such a way that it was antiphase for a particular value of $J(H-H)$. The magnetization was then transferred to ^{13}C, yielding after transformation a 2D correlation of $\nu(C)$ with the difference in frequencies of the protons attached to that carbon and to its neighbor, protons that are capable of entering into mutual multiple-quantum coherence. In this experiment, although the ZQC was homonuclear, the signals of the heteronucleus were utilized to provide information about the proton spin system as well as correlation between the two nuclear species.

B. Heteronuclear ZQC Spectroscopy

Müller (17) showed how zero-quantum and double-quantum coherences could be obtained in a 2D ^{13}C-^{1}H shift correlation experiment and detected by an echo technique, observing either ^{13}C or ^{1}H. Maximum sensitivity is, of course, provided by detection of the proton magnetization. By the heteronuclear ZQC experiment, which was analyzed for a number of different spin systems by Bolton (86), whose experimental results were described in Chapter 5, Section VII,A, a proton spectrum decoupled from the heteronucleus can be obtained and dynamic range problems from a solvent may be reduced (87). Labeling of a molecules, as by ^{15}N (86), makes this type of experiment particularly useful.

The selection of ZQC resonances for protons coupled to cadmium in the molecule metallothionein enriched in ^{113}Cd, with cancellation of other proton peaks in the complex overlapped spectrum of this molecule, has been described in two recent reports (88, 89). Three possible pulse sequences for generating heteronuclear ZQC are shown in Fig. 8-8. In version (a), the correlation obtained is that of (^{1}H) against [(^{113}Cd)+(^{1}H)]. In versions (b) and (c), the refocusing pulses lead to a direct chemical shift correlation, but may cause signal loss if there are multiple couplings of hydrogens to a given ^{113}Cd nucleus. If there is a range of J values of if the heteronucleus is coupled equally to several protons, the length of the interval denoted $1/2J$ or $1/4J$ in the diagrams may need to be set to a suitable compromise value; 10 msec and 30 msec were values used for $1/2J$ by the two groups in the H-Cd work. Metallothionein has as many as seven cadmium nuclei, but if there are only one or two heteronuclei in a molecule, the 2D version of the experiment is needlessly time-consuming, and a 1D version is usually quite satisfactory for generating zero-quantum coherence.

V. Nuclear Overhauser Enhancement

The enhancement produced by the nuclear Overhauser effect may be measured in several ways. In the 1D category of experiments, observations may be made in the *steady state*, irradiating one nucleus continuously at a constant level. Alternatively, in a *transient* experiment, one resonance is inverted or saturated and then the selective radiation is turned off and the spectrum taken after some time has elapsed. As Williamson and Neuhaus have pointed out (90), the results in the transient experiment are "symmetrical" or fully reciprocal between any pair of nuclei: irradiation of A affects B in the same way that irradiation of B affects A. The situation is that the intensity decrease depends on the rate at which magnetization is lost by T_1 processes to the surroundings, and this loss occurs for both nuclei in the transient experiment. so that the net effect is the same, whether A is

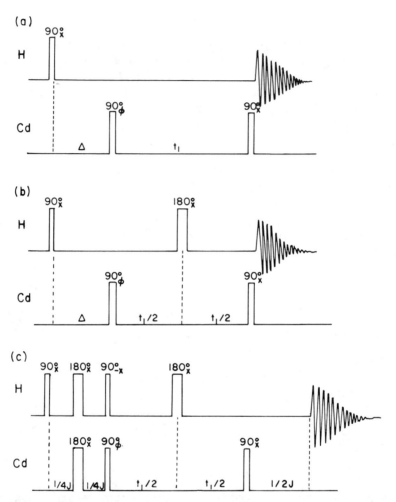

Fig. 8-8. Three versions of the heteronuclear pulse sequence for ^{1}H-^{113}Cd zero-quantum spectroscopy. Version (a) produces a correlation of the shift of ^{1}H versus the sum of the shifts of ^{1}H and ^{113}Cd. Versions (b) and (c) correlate the shift of ^{1}H with the shift of ^{113}Cd. The value of Δ is adjusted to bring the proton multiplet into antiphase, as $1/2J$(H-Cd) for a doublet, here about 30 msec. The phase ϕ is varied in four steps in parallel with the receiver phase. For each value of ϕ, the phases marked x are stepped through the four 90° intervals together, giving a 16-step cycle. [After Live et al. (88) and Otvos et al. (89).]

the "source" and B the "sink" or B is the source and A the sink. In contrast, in the steady-state experiment, the magnetization of the source is maintained by continuous irradiation, so that the particular T_1 of the sink determines the rate of loss of magnetization. Consequently, a difference in T_1 between

A and B means that the NOE observed between the two nuclei is not symmetrical (90).

A 2D NOESY experiment shows reciprocal effects between two nuclei and thus resembles the *transient* 1D experiment (90).

A. Design and Data Handling for NOESY Experiments

As described in Chapter 6, NOESY experiments for quantitative use in the determination of interatomic distances require that resonance "volumes" be determined, implying a three-dimensional instead of a two-dimensional integration (91). One approach is to carry out a two-dimensional integration over a number of parallel cross sections and sum the results (92). If variation in conditions between members of a series of experiments designed to measure the rate of change of enhancement is likely, accuracy may be improved by determining the ratios of areas of sections through cross peaks to sections through corresponding diagonal peaks (93). Another integration approach is to sum the intensities of all points within a given radius around each peak (94); errors may arise here if peaks overlap. Denk, Baumann, and Wagner (95) developed a procedure in which the observed spectrum is "deconvolved" with a set of reference lineshapes, yielding resolution of overlapping peaks, and a series of matrix manipulations is then performed to produce an array of cross-peak intensity values. The Gaussian apodization function has been shown to be superior to the exponential for one-dimensional integrations, in an analysis intended as a step toward accurate measurement of cross-peak volumes (96).

Experiments to obtain a series of spectra showing the rate of change of NOE enhancement have typically been performed at short mixing times in order to obtain the initial buildup rates, which have minimum complications from successive transfer steps or slower competing relaxation processes. However, sensitivity is low because of the short mixing time, and particular care must be taken to cancel J cross peaks as well as to remove the more serious interference of artifacts referred to as "t_1 noise" and "t_1 ridges," which make quantitative measurements difficult.

The normal phase cycling cancels cross peaks from higher orders of multiple-quantum coherence but does not eliminate those from zero-order coherence. These can be suppressed by random variation of the mixing time, or better by using a fixed mixing time with a 180° refocusing pulse inserted at a variable point in the mixing period. The latter procedure has been analyzed by Rance *et al.* (97), who suggest use of a discrete set of values of τ_1, the length of the *unrefocused* portion of the mixing period, Rance *et al.* (97), who suggest use of a discrete set of values of τ_i, the length of the *unrefocused* portion of the mixing period, incremented in parallel with t_1, the interval between the first two 90° pulses, according to the relation $\tau_i = \tau_i^0 + \chi t_1$. The results of a series of ten or more experiments in which τ_i^0 is systematically varied according to $\tau_i^0 = i\Delta\tau^0$, with $i = 0, 1, 2, \ldots, n-1$, are coadded, with that for $i = 0$ given half weight. If the spectrum is to be filtered in the frequency range between $\Delta\Omega_{min}$ and $\Delta\Omega_{max}$, then

$$\Delta\tau^0 = 2\pi(\Delta\Omega_{max} + \Delta\Omega_{min})^{-1}$$

Instrumental sources contribute to t_1 noise, and these have been analyzed for 2D spectroscopy in general (98). The t_1 ridges, which are bands extending parallel to the F_1 axis from strong, sharp resonances on the diagonal, may also have contributions from distortions arising from the data obtained at the first few t_1 values (99). Since instrumental instabilities are usually not under the control of the practicing spectroscopist, much attention has been given to the elimination of these artifacts, whether by experimental design or by data manipulation.

One possibility of reducing artifacts is suppression of the diagonal peaks. Denk and co-workers (99) have described a difference procedure for this purpose, which reduces t_1 noise as well, and is based on the work of Bodenhausen and Ernst (100). The following pulse sequence is used, with VD standing for "variable delay":

$$90°—t_1—90°—t_m—VD—90°, \text{acquisition} \qquad (a)$$

$$90°—t_1—90°—VD—90°, \text{acquisition} \qquad (b)$$

$$180°—t_m—90°—t_1—90°—VD—90°, \text{acquisition} \qquad (c)$$

One scan each, in succession, is acquired with (a), (b), (a), and (c). The (a) and (c) scans are added and the (b) scan subtracted. The combination (b) − (c) produces only diagonal peaks, and subtraction of this pair leaves the NOESY spectrum reasonably free of strong diagonal peaks and the accompanying ridges. Following each set of four scans, the phases of the 90° pulses and of the receiver are shifted in accordance with the usual NOESY sequence to suppress transfer from J coupling, and the variable delay period serves to eliminate zero-quantum coherence of J-coupled pairs.

Otting and co-workers (101) have attributed the presence of t_1 and t_2 ridges to improper scaling of the first data point before each step of the discrete Fourier transformation. This point corresponds to only one-half a dwell time period from time zero; consequently, if a cosine transform is used on a cosine modulated FID, a series of Lorentzian lines is obtained, corresponding to one of the ridges. This can, in principle, be corrected by dividing the first data point of each FID by two before the t_2 Fourier transform and dividing the spectrum corresponding to the first value of t_1

by two before the second FT. In practice, the initial point in t_2 may already be attenuated by the action of the hardware filter, and thus the proper correction factor for that point may be greater or less than unity, and must be found by trial. Since the sine function is small near zero, the recommended procedure is acquiring data with sine modulation, accomplished by a 90° phase shift of the receiver, and use of a sine FT.

After-the-fact data processing can reduce the magnitude of artifacts to some extent. Since NOESY spectra should be symmetrical, the symmetrization algorithm available on any commerical spectrometer may be applied (102), but this may yield false cross peaks if several strong ridges are present, and may produce some peak distortion as well. Methods of removing ridges include subtracting from each row in the matrix an "average row" obtained from the last 100 rows in the array (103), or subtracting with a suitable weighting factor a "mean t_1 ridge" obtained by projecting the 2D spectrum with appropriate scaling (104).

B. NOE IN THE ROTATING FRAME

An NOE experiment using a period of spin-locking in order to obtain enhancements in the rotating frame has been introduced by Bothner-By and co-workers (105) under the acronym CAMELSPIN, and has been termed ROESY by others (106). In the conventional NOE experiment, the enhancement is positive for small molecules with short correlation times, approaches zero as $\omega\tau_c$ approaches unity, and then becomes negative; for molecules of intermediate size it is substantially unobservable. For small molecules, the sense of the peaks is opposite to that for peaks from chemical exchange, but for macromolecules, the phases of the two types of peaks are the same. For the rotating-frame experiment, the NOE enhancement is always positive and increases monotonically with $\omega\tau_c$ toward a limit of 67.5%; as a consequence the phase of the peaks distinguishes them from those for chemical exchange, and, in addition, multipsin, or spin diffusion, effects are less significant than in the conventional experiment.

The 1D CAMELSPIN experiment is based on the difference between the results of two sequences. In the first, a $90°_x$ pulse is followed by a y-phase spin lock and then data acquisition; in the second, the $90°_x$ pulse is preceded by a selective 180° pulse which inverts the spins of one multiplet, which is then the source of the cross relaxation which produces the enhancement, and the remainder of the sequence is the same as the first. In a 2D version, a variable delay is inserted after the 90° pulse and before the spin lock.

Several types of artifacts may appear in CAMELSPIN or ROESY spectra, representing peaks in addition to those desired or peaks with a phase which may be misleading. COSY-type peaks, from transfer originating in pairs of scalar-coupled spins, are in antiphase and thus have zero integral, contributing in practice a negligible amount to the peak volume. More troublesome are peaks arising from homonuclear Hartmann-Hahn or HOHAHA transfer, described in Section III. These can be minimized by using limited rf power, 0.5 W or less, and by arranging the irradiation frequency so that it is not midway between two peaks, since equal offsets increase the transfer levels (106). HOHAHA peaks can be distinguished from NOE transfer peaks because the former have the same phase as the diagonal peaks whereas the NOE peaks have the opposite phase. Of course, to make this distinction, as with any phase distinction, the spectra must be in phase-sensitive rather than in absolute-value mode (107).

Accidental superposition of peaks of the two kinds leads to partial or complete cancellation. Furthermore, the phase rules cited apply only to single transfers; two-step processes of the type

$$P \xrightarrow[(J\text{-coupled})]{\text{HOHAHA}} Q \xrightarrow{\text{NOE}} R \quad \text{or} \quad R \xrightarrow{\text{NOE}} Q \xrightarrow[(J\text{-coupled})]{\text{HOHAHA}} P$$

where the atom R is physically close to Q but distant from P, may be misleading because the phase shows only one change, that of the NOE step, from that of the diagonal, and therefore appears to represent a single NOE transfer from R to P (108, 109). Such two-step peaks can be identified, because of the strong dependence of the HOHAHA step on the transmitter frequency, by carrying out a series of experiments with differing transmitter offsets.

Another type of two-step process, which appears in all NOE experiments and is often referred to as spin diffusion, leads to the same phase as that of the diagonal peaks:

$$P \xrightarrow{\text{NOE}} Q \xrightarrow{\text{NOE}} R$$

The effect of this process increases with longer mixing times and, like other two-step processes, interferes with quantitative analysis of interatomic distances by "stealing" intensity from the peak produced by the first step (110).

A possibility to be considered in the rotating-frame experiment is the potential effect of resonance offset on the cross-relaxation rate in that frame. Theoretical and experimental results of Farmer and Brown (111) indicate that the fall-off with frequency offset is much less than for transfer by HOHAHA, but for quantitative results, the value of θ for each nucleus should be 60° or more, where $\theta = \tan^{-1}[\gamma_i B_{sl}/(\omega_i - \omega_0)]$, implying that there is a minimum effective value of the spin-lock field B_{sl}.

C. Interpretation of NOE Enhancement Results

An approximation customarily made in the interpretation of NOE results in terms of internuclear distances is that the rotational motions of the pairs of nuclei involved are not correlated with one another. This neglect of cross-correlation effects appears justified if the initial linear buildup rates are used. The situation for times beyond this region has been analyzed by Bull (112) for a number of spin systems, with the conclusion that the errors are appreciable, although relatively small. Measurements of cross-correlation terms permit corrections to be made to the NOE values, and Bull has suggested experiments using multiple-quantum filters to pick out these terms.

The effects of strong coupling on NOE intensities have been analyzed by Keeler, Neuhaus, and Williamson (113). They found, both theoretically and experimentally, that for two spins which might be expected to have quite different enhancements from a third spin were they weakly coupled, the enhancement becomes more nearly equal as the mixing of the wave functions of the two spins is increased by strong coupling. The results apply both for equilibrium and for transient measurements. Indeed, the strong-coupling effect tends to overcome the usual negative "three-spin" effect.

To resolve ambiguities in NOESY results, van de Ven, Haasnoot, and Hilbers (114) have described an experiment in which double-quantum coherences are first created, then the magnetization is restored to single-quantum, and the resulting magnetization is transferred to other nuclei via the NOE. In the 2D spectrum resulting, only the double-quantum spectrum is evident in the diagonal peaks.

Methods of accounting for spin diffusion in order to interpret weak enhancements and thus provide additional parameters for geometrical analysis have been described (94). Prestegard and co-workers (115) have developed a program to simulate homonuclear NOESY spectra for a set of spin-$\frac{1}{2}$ nuclei of arbitrary size. The effects of strong coupling, multiple relaxation pathways, and T_1 relaxation are included, and the simulation may be carried out to show the effects of various pulse sequences and phase cycles on the intensity and phases of the resulting peaks.

D. Some Other Forms of NOE Experiments

Dimicoli, Vok, and Mispelter (116) have described a homonuclear proton–proton NOE sequence followed by an INEPT transfer to carbon-13. Although the experiment is handicapped by low sensitivity when carried out with ^{13}C at natural abundance, it has the advantage that the INEPT portion filters out resonances from homo-COSY transfer as well as the signal from the aqueous solvent.

In a series of papers, Kövér and Batta have discussed various aspects of NOE enhancement of ^{13}C by ^{1}H as a means of quantitative measurement of C–H distances. The choice of mixing time was considered and the optimum was found to depend strongly on the ^{13}C relaxation time (117). Three-spin effects transmitted from neighboring proton through the proton directly bound to the carbon observed were found to be important (118). The analysis was extended to the case of multiple-proton spin systems, and good agreement was found in a test in which calculated results of intensities based on distances measured on a model of a rigid molecule were compared with the experimental results for the molecule (119). A detailed analysis has also been made of the consequences of three-spin effects in heteronuclear AMX systems (120).

An improved method of suppressing the strong solvent resonance in proton–proton NOE experiments was suggested by Zuiderweg, Hallenga, and Olejniczak (121). The use of a decoupler rf field coherent with the carrier field eliminated t_1-dependent fluctuations of the baseline. Piveteau and co-workers (122) described suppression of the water signal in the CAMELSPIN experiment by a method analogous to the "jump and return" method. The carrier is placed at the water resonance frequency, all the signals are spin-locked, and, following the spin lock, the signals other than that of water are allowed to dephase during a fixed interval, after which the water magnetization is returned to the z direction by a 90° pulse.

Macura, Farmer, and Brown (123) showed how the low signal-to-noise limitations of measurements limited to short mixing times can be overcome by using a set of several mixing times surrounding the value for optimum intensity, along with measurements of the ratio of cross-peak to diagonal-peak intensity.

Finally, an innovative procedure used by Kerwood and Bolton (124) was said to overcome the problems produced by spin diffusion and by the variation of the correlation time for motion among various pairs of nuclei. The sample is equilibrated at high field, allowed to cross relax in a low field, corresponding to a frequency of less than 10 MHz, and then the signal is detected at high field.

References

1. A. Bax and G. Drobny, *J. Magn. Reson.* **61**, 306 (1985).
2. W. J. Chazin and K. Wüthrich, *J. Magn. Reson.* **72**, 358 (1987).
3. D. W. Hughes, R. A. Bell, T. Neilson, and A. D. Bain, *Can. J. Chem.* **63**, 3133 (1985).
4. A. Otter and G. Kotovych, *J. Magn. Reson.* **69**, 187 (1986).
5. J. Santoro, M. Rico, and F. J. Bermejo, *J. Magn. Reson.* **67**, 1 (1986).
6. S. K. Sarkar and A. Bax, *J. Magn. Reson.* **63**, 512 (1985).
7. P. H. Bolton, *J. Magn. Reson.* **63**, 225 (1985).

8. A. Bax, D. G. Davis, and S. K. Sarkar, *J. Magn. Reson.* **63**, 230 (1985).
9. W. F. Reynolds, D. W. Hughes, M. Perpick-Dumont, and R. G. Enriquez, *J. Magn. Reson.* **63**, 413 (1985).
10. H. Kessler, W. Bermel, and C. Griesinger, *J. Magn. Reson.* **62**, 573 (1985).
11. K. S. Lee and G. A. Morris, *J. Magn. Reson.* **70**, 332 (1986).
12. L. D. Field and B. A. Messerle, *J. Magn. Reson.* **62**, 453 (1985).
13. L. D. Field and B. A. Messerle, *J. Magn. Reson.* **66**, 483 (1986).
14. M. A. Delsuc, E. Guittet, N. Trotin, and J. Y. Lallemand, *J. Magn. Reson.* **56**, 163 (1984).
15. D. Neuhaus, G. Wider, G. Wagner, and K. Wüthrich, *J. Magn. Reson.* **57**, 164 (1984).
16. L. Lerner and A. Bax, *J. Magn. Reson.* **69**, 375 (1986).
17. L. Müller, *J. Am. Chem. Soc.* **101**, 4481 (1979).
18. D. Brühwiler and G. Wagner, *J. Magn. Reson.* **69**, 546 (1986).
19. A. Bax and S. Subramanian, *J. Magn. Reson.* **67**, 565 (1986).
20. J. A. Wilde, P. H. Bolton, N. J. Stolowich, and J. A. Gerlt, *J. Magn. Reson.* **68**, 168 (1986).
21. A. Bax and M. F. Summers, *J. Am. Chem. Soc.* **108**, 2093 (1986).
22. L. Mueller, R. A. Schicksnis, and S. J. Opella, *J. Magn. Reson.* **66**, 379 (1986).
23. R. Benn and C. Brevard, *J. Am. Chem. Soc.* **108**, 5622 (1986).
24. P. J. Keller and K. E. Vogele, *J. Magn. Reson.* **68**, 389 (1986).
25. L. D. Field and A. K. McPhail, *J. Magn. Reson.* **67**, 364 (1986).
26. R. W. Dykstra and G. D. Markham, *J. Magn. Reson.* **71**, 544 (1987).
27. A. Bax and R. Freeman, *J. Magn. Reson.* **44**, 542 (1981).
28. T. T. Nakashima and D. L. Rabenstein, *J. Magn. Reson.* **66**, 157 (1986).
29. R. W. Lynden-Bell, J. M. Bulsing, and D. M. Doddrell, *J. Magn. Reson.* **55**, 128 (1983).
30. P. W. Jeffs, L. Mueller, C. DeBrosse, S. L. Heald, and R. Fisher, *J. Am. Chem. Soc.* **108**, 3063 (1986).
31. S. L. Heald, L. Mueller, and P. W. Jeffs, *J. Magn. Reson.* **72**, 120 (1987).
32. A. Kumar, R. V. Hosur, and K. Chandrasekhar, *J. Magn. Reson.* **60**, 143 (1984).
33. R. V. Hosur, K. V. R. Chary, A. Kumar, and G. Govil, *J. Magn. Reson.* **62**, 123 (1985).
34. A. Kumar, R. V. Hosur, K. Chandrasekhar, and N. Murali, *J. Magn. Reson.* **63**, 107 (1985).
35. J. Cavanagh and J. Keeler, *J. Magn. Reson.* **71**, 561 (1987).
36. T. Allman and A. D. Bain, *J. Magn. Reson.* **68**, 533 (1986).
37. O. W. Sørensen, M. Rance, and R. R. Ernst, *J. Magn. Reson.* **56**, 527 (1984).
38. H. Oschkinat, A. Pastore, P. Pfändler, and G. Bodenhausen, *J. Magn. Reson.* **69**, 559 (1986).
39. P. Pfändler and G. Bodenhausen, *J. Magn. Reson.* **72**, 475 (1987).
40. E. Wörgötter, G. Wagner, and K. Wüthrich, *J. Am. Chem. Soc.* **108**, 6162 (1986).
41. G. Otting, H. Senn, G. Wagner, and K. Wüthrich, *J. Magn. Reson.* **70**, 500 (1986).
42. R. H. Griffey and A. G. Redfield, *J. Magn. Reson.* **65**, 344 (1985).
43. C. Griesinger, O. W. Sørensen, and R. R. Ernst, *J. Am. Chem. Soc.* **107**, 6394 (1985).
44. C. Griesinger, O. W. Sørensen, and R. R. Ernst, *J. Chem. Phys.* **85**, 6837 (1986).
45. M. H. Levitt, C. Radloff, and R. R. Ernst, *Chem. Phys. Lett.* **114**, 435 (1985).
46. B. T. Farmer and L. R. Brown, *J. Magn. Reson.* **71**, 365 (1987).
47. L. Mueller, *J. Magn. Reson.* **72**, 191 (1987).
48. J. Boyd and C. Redfield, *J. Magn. Reson.* **68**, 67 (1986).
49. N. Müller, R. R. Ernst, and K. Wüthrich, *J. Am. Chem. Soc.* **108**, 6482 (1986).
50. N. Müller, G. Bodenhausen, K. Wüthrich, and R. R. Ernst, *J. Magn. Reson.* **65**, 531 (1985).
51. M. Rance and P. E. Wright, *J. Magn. Reson.* **66**, 372 (1986).
52. A. D. Bain, *J. Magn. Reson.* **56**, 418 (1984).
53. R. Boelens, A. Podoplelov, and R. Kaptein, *J. Magn. Reson.* **69**, 116 (1986).
54. A. B. Mauger, O. A. Stuart, J. A. Ferretti, and J. V. Silverton, *J. Am. Chem. Soc.* **107**, 7154 (1985).

55. A. S. Arseniev, A. G. Sobol, and V. F. Bystrov, *J. Magn. Reson.* **70**, 427 (1986).
56. J. D. Otvos, H. R. Engeseth, and S. Wehrli, *Biochemistry* **24**, 6735 (1985).
57. C. T. G. Knight, R. J. Kirkpatrick, and E. Oldfield, *J. Am. Chem. Soc.* **108**, 30 (1986).
58. A. W. H. Jans, E. M. Osselton, C. P. Eyken, B. Griewe, and J. Cornelisse, *J. Magn. Reson.* **70**, 169 (1986).
59. B. U. Meier, G. Bodenhausen, and R. R. Ernst, *J. Magn. Reson.* **60**, 161 (1984).
60. P. Pfändler, G. Bodenhausen, B. U. Meier, and R. R. Ernst, *Anal. Chem.* **57**, 2510 (1985).
61. P. Pfändler and G. Bodenhausen, *J. Magn. Reson.* **70**, 71 (1986).
62. P. H. Bolton, *J. Magn. Reson.* **70**, 344 (1986).
63. S. B. Hartmann and E. L. Hahn, *Phys. Rev.* **128**, 2042 (1962).
64. R. D. Bertrand, W. B. Moniz, A. N. Garroway, and G. C. Chingas, *J. Magn. Reson.* **32**, 465 (1979).
65. A. N. Garroway and G. C. Chingas, *J. Magn. Reson.* **38**, 179 (1980).
66. L. Braunschweiler and R. R. Ernst, *J. Magn. Reson.* **53**, 521 (1983).
67. D. G. Davis and A. Bax, *J. Am. Chem. Soc.* **107**, 2820 (1985).
68. D. G. Davis and A. Bax, *J. Am. Chem. Soc.* **107**, 7197 (1985).
69. A. Bax and D. G. Davis, *J. Magn. Reson.* **65**, 355 (1985).
70. S. Subramanian and A. Bax, *J. Magn. Reson.* **71**, 325 (1987).
71. N. Chandrakumar and S. Subramanian, *J. Magn. Reson.* **62**, 346 (1985).
72. N. Chandrakumar, G. V. Visalakshi, D. Ramaswamy, and S. Subramanian, *J. Magn. Reson.* **67**, 307 (1986).
73. N. Chandrakumar, *J. Magn. Reson.* **71**, 322 (1987).
74. M. W. Edwards and A. Bax, *J. Am. Chem. Soc.* **108**, 918 (1986).
75. M. F. Summers, L. G. Marzilli, and A. Bax, *J. Am. Chem. Soc.* **108**, 4285 (1986).
76. J. S. Waugh, *J. Magn. Reson.* **68**, 189 (1986).
77. A. Bax, T. Mehlkopf, J. Smidt, and R. Freeman, *J. Magn. Reson.* **41**, 502 (1980).
78. G. Pouzard, S. Sukumar, and L. D. Hall, *J. Am. Chem. Soc.* **103**, 4209 (1981).
79. L. Müller, *J. Magn. Reson.* **59**, 326 (1984).
80. L. D. Hall and T. L. Norwood, *J. Magn. Reson.* **69**, 391 (1986).
81. J. Santoro, F. J. Bermejo, and M. Rico, *J. Magn. Reson.* **64**, 151 (1985).
82. L. D. Hall and T. J. Norwood, *J. Magn. Reson.* **69**, 397 (1986).
83. L. D. Hall and T. J. Norwood, *J. Magn. Reson.* **69**, 585 (1986).
84. L. Müller and A. Pardi, *J. Am. Chem. Soc.* **107**, 3484 (1985).
85. D. L. Turner, *J. Magn. Reson.* **65**, 169 (1985).
86. P. H. Bolton, *J. Magn. Reson.* **57**, 427 (1984).
87. A. Bax, R. H. Griffey, and B. L. Hawkins, *J. Magn. Reson.* **55**, 301 (1983).
88. D. Live, I. M. Armitage, D. C. Dalgarno, and D. Cowburn, *J. Am. Chem. Soc.* **107**, 1775 (1985).
89. J. D. Otvos, H. R. Engeseth, and S. Wehrli, *J. Magn. Reson.* **61**, 579 (1985).
90. M. P. Williamson and D. Neuhaus, *J. Magn. Reson.* **72**, 369 (1987).
91. M. S. Broido, T. L. James, G. Zon, and J. W. Keepers, *Eur. J. Biochem.* **150**, 117 (1985).
92. P. A. Mirau and F. A. Bovey, *J. Am. Chem. Soc.* **108**, 5130 (1986).
93. G. T. Montelione, P. Hughes, J. Clardy, and H. A. Scheraga, *J. Am. Chem. Soc.* **108**, 6765 (1986).
94. E. T. Olejniczak, R. T. Gampe, and S. W. Fesik, *J. Magn. Reson.* **67**, 28 (1986).
95. W. Denk, R. Baumann, and G. Wagner, *J. Magn. Reson.* **67**, 386 (1986).
96. G. H. Weiss, J. A. Ferretti, and R. A. Byrd, *J. Magn. Reson.* **71**, 97 (1987).
97. M. Rance, G. Bodenhausen, G. Wagner, K. Wüthrich, and R. R. Ernst, *J. Magn. Reson.* **62**, 497 (1985).
98. A. F. Mehlkopf, P. Korbee, T. A. Tiggelman, and R. Freeman, *J. Magn. Reson.* **58**, 315 (1984).

99. W. Denk, G. Wagner, M. Rance, and K. Wüthrich, *J. Magn. Reson.* **62**, 350 (1985).
100. G. Bodenhausen and R. R. Ernst, *Mol. Phys.* **47**, 319 (1982).
101. G. Otting, H. Widmer, G. Wagner, and K. Wüthrich, *J. Magn. Reson.* **66**, 187 (1986).
102. R. Baumann, G. Wider, R. R. Ernst, and K. Wüthrich, *J. Magn. Reson.* **44**, 402 (1981).
103. R. E. Klevit, *J. Magn. Reson.* **62**, 551 (1985).
104. S. Glaser and R. H. Kalbitzer, *J. Magn. Reson.* **68**, 350 (1986).
105. A. A. Bothner-By, R. L. Stephens, and J. Lee, *J. Am. Chem. Soc.* **106**, 811 (1984).
106. A. Bax and D. G. Davis, *J. Magn. Reson.* **63**, 207 (1985).
107. D. G. Davis and A. Bax, *J. Magn. Reson.* **64**, 533 (1985).
108. D. Neuhaus and J. Keeler, *J. Magn. Reson.* **68**, 568 (1986).
109. B. T. Farmer, S. Macura, and L. R. Brown, *J. Magn. Reson.* **72**, 347 (1987).
110. A. Bax, V. Sklenar, and M. F. Summers, *J. Magn. Reson.* **70**, 327 (1986).
111. B. T. Farmer and L. R. Brown, *J. Magn. Reson.* **72**, 197 (1987).
112. T. E. Bull, *J. Magn. Reson.* **72**, 397 (1987).
113. J. Keeler, D. Neuhaus, and M. P. Williamson, *J. Magn. Reson.* **73**, 45 (1987).
114. F. J. M. van de Ven, C. A. G. Haasnoot, and C. W. Hilbers, *J. Magn. Reson.* **61**, 181 (1985).
115. L. E. Kay, J. N. Scarsdale, D. R. Hare, and J. H. Prestegard, *J. Magn. Reson.* **68**, 515 (1986).
116. J. L. Dimicoli, A. Volk, and J. Mispelter, *J. Magn. Reson.* **63**, 605 (1985).
117. K. E. Kövér and G. Batta, *J. Magn. Reson.* **69**, 344 (1986).
118. K. E. Kövér and G. Batta, *J. Magn. Reson.* **69**, 519 (1986).
119. K. E. Kövér, G. Batta, and Z. Mádi, *J. Magn. Reson.* **69**, 538 (1986).
120. G. Batta, K. E. Kövér, and Z. Mádi, *J. Magn. Reson.* **73**, 477 (1987).
121. E. R. P. Zuiderweg, K. Hallenga, and E. T. Olejniczak, *J. Magn. Reson.* **70**, 336 (1986).
122. D. Piveteau, M. A. Delsuc, E. Guittet, and J. Y. Lallemand, *J. Magn. Reson.* **71**, 347 (1987).
123. S. Macura, B. T. Farmer, and L. R. Brown, *J. Magn. Reson.* **70**, 493 (1986).
124. D. J. Kerwood and P. H. Bolton, *J. Magn. Reson.* **68**, 588 (1986).

Appendix: Reading List

For the reader who may wish background material to assist in the understanding of the contents of this volume, the following books are among those which may be recommended as providing a variety of excellent introductory treatments.

R. J. Abraham and P. Loftus, "Proton and Carbon-13 NMR Spectroscopy." Heyden, London, 1978.

J. W. Akitt, "NMR and Chemistry," 2nd ed. Chapman and Hall, London, 1983.

E. D. Becker, "High-Resolution NMR," 2nd ed. Academic Press, New York, 1980.

A. Carrington and A. D. McLachlan, "Introduction to Magnetic Resonance." Harper and Row, New York, 1967; Halsted-Wiley (paperback, 1979).

T. C. Farrar and E. D. Becker, "Pulse and Fourier Transform NMR." Academic Press, New York, 1971.

E. Fukushima and S. B. W. Roeder, "Experimental Pulse NMR—A Nuts and Bolts Approach." Addison-Wesley, Reading, Massachusetts, 1981.

D. G. Gadian, "Nuclear Magnetic Resonance and its Application to Living Systems." Oxford, London and New York, 1982.

H. Günther, "NMR Spectroscopy" (translated by R. W. Gleason). Wiley, New York, 1980.

R. K. Harris, "Nuclear Magnetic Resonance Spectroscopy." Pitman, London, 1983.

A. G. Marshall (ed.), "Fourier, Hadamard, and Hilbert Transforms in Chemistry." Plenum, New York, 1982.

M. L. Martin, G. L. Martin, and J. J. Delpuech, "Practical NMR Spectroscopy." Heyden, London, 1980.

D. Shaw, "Fourier Transform NMR Spectroscopy," 2nd ed. Elsevier, Amsterdam and New York, 1984 (paperback, 1987).

Index

A

Accordion experiment, for exchange rate determination, 415–417
Acenaphthalene, SEMINA carbon-13 spectrum, 239–240
Acetic acid, SLAP spectrum of carbon-13 enriched, 200
2-Acetonaphthalene, heteronuclear relayed coherence transfer, 406, 410
Acrylonitrile, double-quantum filtered 1D spectrum, 335
Active spin, 266
Acyl carrrier protein, structure determination, 475–478
Alcohols
  separation of proton spectrum of, 177
  z-filtered proton spectrum of mixture, 176
Amine-boranes and aminoboranes, INEPT spectra, 61
Amino acids, 2D J spectra, 362
3-Aminopropanol, 2D double-quantum spectrum, 332
Ammonium perfluorooctanoate, fluorine-19 COSY spectrum of, 496–499
Amplitude modulation, 16
5-α-Androstane, SEMINA carbon-13 spectrum, 238

Angular momentum
  basis for product operators, 125
  quantum number, 260
Antiecho
  from coherence transfer, 327
  in multiple-quantum evolution, 284
  response in carbon-13–carbon-13 connectivity mapping, 291
APT
  editing of carbon-13 spectrum by, 207–208
  phase cycling in, 25, 98
  sequence for multiplicity determination, 22
Artifacts
  in COSY, 267
  in 2D exchange experiments, 414

B

Baseline correction, by phase cycling, 96
Basic pancreatic trypsin inhibitor
  double-quantum filtered COSY spectrum, 301–302
  relay correlations in proton spectrum, 404
  relayed NOESY spectrum, 417–418
  2D spectrum, 436

Benzene, deuterium-carbon polarization transfer in d_6, 78
Bilinear pulse cluster
　for decoupling in 2D heteronuclear correlation, 398
　in heteronuclear RELAY, 515–517
　in RELAY experiment, 512
　selective inversion of nuclei by, 100–101
　in 2D J spectroscopy, 360
　use as mixing sequence in COSY, 523
Bilinear rotation, effect on angular momentum operator, 321–323
Bilirubin, nitrogen-15 spectrum, 191
BIRD pulse, see Bilinear pulse cluster
Bloch equations, 113–114, 260–261
Boron-11
　homonuclear 2D spectrum of polyhedral boranes, 381
　polarization transfer from hydrogen, 76–77
　2D correlation with protons, 391, 395
BPTI, see Basic pancreatic trypsin inhibitor
2-Bromothiazole
　carbon-13 spectra edited by SEMUT GL, 219
　SLAP spectrum and coupling constants, 204–205
3-Bromothiophene-2-aldehyde, J spectrum, 13
Brucine, carbon-13 spectra edited by SEMUT GL, 219
Bullvalene, INEPT in exchange studies on, 64

C

Cadmium-113
　COSY in metallothionein, 527
　zero-quantum coherence with protons, 535
Calibration of decoupler pulses, 102
CAMELSPIN, see Nuclear Overhauser effect, in rotating frame
Carbohydrates, see also Saccharides
　COSY spectrum of, 380
　heteronuclear 2D shift correlation, 383
　1-O-methyl-β-D-galactose, 437
　2D J spectra, 362
Carbon-13
　correlation spectra, 383–395, 400–402, 405–411, 461–464
　DEPT spectra, 70–73, 182–185
　INEPT in relaxation measurements, 63
　INEPT spectra, 53–64
　intensities of multiplets resulting from polarization transfer, 40
　inverse polarization transfer, 75
　J spectra, 353, 361–363
　magnetization transfer spectra, 419–420
　multiple-quantum spectra, 289–292, 422–426
　NOE spectra, 542
　polarization transfer spectra, 31–37, 78–81
　relayed spectra, 509–518
　SEMUT spectra, 209–247
　sign determination of homonuclear coupling constants, 197
Carr–Purcell–Meiboom–Gill experiment, measurement of T_2 by, 9–10
Cartesian product operator, see Operator, Cartesian
Cedrol, carbon–carbon connectivity in, 423
Chemical shift, evolution of density operator under, 130, 160–161
2-Chloronaphthalene, heteronuclear 2D J spectrum, 362
Cholesterol
　edited proton-coupled spectrum, 229
　edited SEMUT carbon-13 spectra, 221
　proton-coupled carbon-13 spectrum by GL⁺ sequences, 230
CIDNP, COSY applied to, 527
Coherence, 116
　definition, 265
　multiple-quantum, 155–156
　multiple-quantum, operator representation, 164–166
　order, 165
　order, in E.COSY, 523
　order, excitation of, 267
　order, filtering of, 292–310
　order, and phase cycling, 136–140
　partial-spin multiple-quantum, 265
　total-spin multiple-quantum, 265
　zero-quantum, 155–156
Coherence transfer, 266
　coefficients of, for pulse rotations, 328–330
　selection rules, 329
Coherence-transfer echo, 305–311
Coherence-transfer pathway
　analysis of, 99
　description by propagators, 297–298

INDEX 551

in echo formation, 308–310
map for double-quantum coherence
  detection, 295–296
in phase-cycled experiments, 99
selection of order in, 314
simplification by composite rotations,
  319–324
Collective modes, in isotropic mixing,
  529–530
COLOC, sequence for long-range
  heteronuclear correlation, 389, 394
Commutators, algebra of, 248
Composite pulses
  application of, 88–90
  based on arbitrary phase shifts, 88
  based on 90° steps, 84–86
  with dual compensation, 86–87
  in proton-flip heteronuclear 2D J
    spectrum, 359
Composite refocusing sequence, 334
Composite rotations, for coherence pathway
  simplification, 319–324
Computer-assisted analysis, COSY spectra, 527
Convolution, function for FID in 2D
  spectra, 457–460
Correlation
  heteronuclear, from 1D polarization
    transfer, 185–192
  homonuclear, see COSY
  INEPT pulse sequences for heteronuclear,
    65
  spectra, chemical shift, 364–411
  spectra, heteronuclear, 382–402
COSLAP, see Sign-labeled polarization
  transfer
COSY
  COSY-45, 197
  delayed acquisition in, 520
  E.COSY, 523
  F_1 decoupling, 379
  for fluorine-19 spectroscopy of
    fluorocarbons, 496–503
  multiple-quantum filtered, 301–305, 425,
    459, 523–524, 526
  P.E. modification, 525–526
  spectra in biological systems, 442–446
  in structure determination of
    oligosaccharides, 482–485
  theory and applications, 364–381
  X-filtered, 523
  z-filtered, 521
Coumarin, heteronuclear 2D J spectrum, 353

Coupled spectra, editing of, 227–228
Coupling constants
  long-range nitrogen-15–proton by
    modified INEPT, 64
  relative signs, by polarization transfer,
    192–202
  sign determination of, 45–49
  sign determination of carbon-13–
    carbon-13, 197–202
Cross relaxation, see also Nuclear
    Overhauser effect
  matrix for, 472
Cross talk
  in edited spin-echo spectra from
    distribution of J values, 27–28
  in SEMUT GL or DEPT GL subspectra,
    211–218
CYCLOPS
  in COSY, 366
  scheme for quadrature detection, 138
Cytidine 3′-phosphate, heteronuclear
  correlation, 421
Cytochrome c, RELAY spectrum, 468

D

2D spectroscopy, see Two-dimensional
    spectroscopy
DANTE
  selective excitation of carbon-13 in
    polarization transfer to protons, 75
  in selective excitation in homonuclear
    RELAY, 510
  selective excitation of multiplet
    component, 170–171
  sequence for selective excitation, 91–92
  in sign-labeled polarization transfer, 196–197
Data processing, by sine-bell, in COSY, 369
Decalin, INEPT in exchange studies on, 64
Decoupler, calibration of pulses from, 102
Decoupling
  in F_1, 377–379
  in heteronuclear J spectra, 19–24
  homonuclear, in homonuclear 2D J
    spectra, 358
  in INEPT experiments, 52
  of protons from fluorine-19, 491–493
  in SPT experiments, 51
  in 2D heteronuclear correlation, 398–400
DEFT, see Driven-equilibrium Fourier
    transform

Degenerate system, SPT in, 37–41
Delay
  in acquisition, 25, 520
  in COSY for long-range coupling, 371–373
  mixing, in NOESY, 448
Density matrix, 115
  for multiple-quantum experiment, 450
Density operator, see Operator
DEPT
  comparison with INEPT, 184
  composite pulses in, 89
  deuterium–carbon polarization transfer by, 80
  editing of carbon-13 spectrum by, 209–214
  family of sequences, 66–73
  in heteronuclear correlation, 247
  inverse version, 75
  for multiplicity selection in 2D heteronuclear correlation, 400–401
  operator analysis of, 175–180
  operator analysis, for I_2S system, 252–253
  for polarization transfer to spin-1 nuclei, 77
  pulse sequence and density operators, 179–183
  selection of experimental parameters for, 82–83
  use in relayed coherence spectroscopy, 411
DEPT$^+$, 179–183
DEPT^{++}, 179–183
DEPT GL$^+$, 227
  operator analysis of editing by, 253–256
Detection of multiple-quantum coherence, 280–287
Deuterated molecules, modulated carbon-13 spin-echo spectra, 24
Deuterium
  double-quantum spectra, 424
  interaction in liquid crystals by SPT, 48
  modulated carbon-13 spin echoes in molecules containing, 24
  polarization of carbon-13 by transfer from, 80–81
  polarization transfer involving, 76, 78–82
  2D correlation with protons, 391, 395
1,3-Dibromobutane
  carbon-13–carbon-13 coupling constants from SEMINA subspectra, 234
  INEPT and DEPT spectra, 182–183
  SEMINA carbon-13 spectrum, 235, 236

tip-angle effects in carbon-13 spectrum, 171–172
2,3-Dibromopropionic acid, 2D multiple-quantum spectra, 264, 304–306
2,3-Dibromothiophene
  determination of signs of coupling constants in, by SPT, 45–48
  double-quantum coherence in proton spectrum, 273–276
  double-quantum filtered 1D spectrum, 335
  evolution of double-quantum coherence in inhomogeneous field, 309
  SLAP spectrum and coupling constants, 205–206
  tip-angle dependence of coherence transfer in spectrum, 285
Difference method, see also Subspectra, generation of
  in COSY spectroscopy, 524–526
  for diagonal peak cancellation in exchange experiments, 415
  in H—C—C RELAY, 514
  heteronuclear spin-echo experiments involving, 28–30
  in heteronuclear SPT spectroscopy, 43
  in proton-decoupled fluorine-19 spectra, 493
  z-filtered COSY, 522–523
  in zero-quantum-coherence spectra, 421
Diffusion, measurement of, by spin echoes, 8
N,N-Dimethylformamide, INEPT in exchange studies on, 64
Double-quantum coherence, see also INADEQUATE
  carbon–carbon connectivity by, 422–424
  detection, 280–288
  evolution, 276–280
  excitation, 268–276
  operator representation, 119, 123
Double-quantum spectroscopy, see also INADEQUATE
  scheme for two-spin-$\frac{1}{2}$ system, 137
Driven-equilibrium Fourier transform, 11

E

E. coli lac repressor headpiece, triple-quantum filtered proton spectrum, 299–300

Echo, *see also* Spin echo
  coherence-transfer, 305–310
E.COSY, sequence for simplifying
  correlation spectra, 523
Editing, spectral, *see* Subspectra, generation of
Enaminones, INEPT spectra, 61
EPT
  inverse version, 73
  pulse sequence for single spin-$\frac{1}{2}$ to
    single spin-$\frac{1}{2}$ polarization transfer, 70
ESCORT, for cross-talk reduction in edited
  spin-echo spectra, 28
Estrone 3-methyl ether, decoupler
  calibration by carbon-13 spectrum of,
  103–104
Evolution
  of magnetization, 4–6
  of multiple-quantum coherence, 166,
    276–280, 324–326
  of product operators, 130–135, 159–164
Exchange, study by 2D Jeener experiment,
  412–419
Excitation
  of coherence orders, 267
  of double-quantum coherence, 268–276
  of multiple-quantum coherence in AMX
    system, 310–319
  of zero-quantum coherence, 268–276
Exorcycle
  phase cycle for eliminating artifacts, 98
  in 2D J spectra, 358
Exponential operators, 248

F

Ferredoxin, 2D proton–carbon-13
  correlation spectrum, 463
Field-gradient pulse
  in order-selective detection, 292–293
  for phase cycling, 137
  for zero-quantum selection, 99
Filtering
  double-quantum, connectivity mapping
    by, 290
  of heteronuclear coherence transfer by J
    filtering, 407
  multiple-quantum, 289, 292–293
  multiple-quantum, for COSY, 300–305
  multiple-quantum, higher order, 425
Flip-angle effect
  basis of, 49
  selection of multiplets by, 50
Fluorine-19
  COSY in hexafluorotetracyclic
    compounds, 527
  signs of coupling by DEPT with 2D
    correlation, 401
  SPT to carbon-13 from, 44–45
  2D J spectrum of trichlorotrifluoro-
    ethane, 19
  2D shift correlation with protons,
    395–399
  2D spectra of compounds, 489–505
Formamide, INEPT spectrum, 61
Free induction decay, weighting function
  for, 457–460

G

Galactose, 1-*O*-methyl-β-D-
  COSY spectrum, 443
  homonuclear relayed coherence for,
    455–456
  J spectrum, 439–440
  multiple-quantum spectrum, 453
  structure, 441
Gated-decoupler method
  in COSY or NOESY, 523
  for heteronuclear modulation of spin
    echoes, 352
  for modulation of heteronuclear spin
    echoes, 16
  in multiplicity determination, 18–24
Gaussian pulses, 92
GL⁺ editing sequences, 227–228
Glucose, RELAY spectrum, 412
Glycylglycine, INEPT spectrum, 61
Gramicidin-S
  INEPT spectra, 61
  nitrogen-15–proton multiple-quantum
    spectrum, 427

H

Hahn experiment, spin-echo, 6
Hartman–Hahn transfer
  homonuclear, followed by INEPT, 512
  homonuclear, for isotropic mixing, 528
  for J cross polarization in liquids, 151
  peaks from, in NOE rotating-frame
    experiments, 540

Hartman–Hahn transfer *(continued)*
  z-filtered 1D experiment, 537
Hemoglobin, carbonmonoxyleg-, RELAY spectrum, 405
Hermite polynomial pulse shaping, 92
HOHAHA, *see* Hartman–Hahn transfer
Homonuclear correlation spectroscopy, *see* COSY; SECSY
Homonuclear spin–spin interactions
  effects in heteronuclear spin-echo difference spectra, 30
  scaling of, 167
Hydrocarbon mixtures, editing of spin-echo carbon-13 spectra, 26
Hydrogen bond, proton exchange rates in polynucleotides, 479
Hydrogen detection, *see also* Indirect observation
  in carbon-13–carbon-13 INADEQUATE, 519
  in heteronuclear RELAY, 515
  of relayed multiple-quantum coherence, 517
9-α-Hydroxyestrone methyl ether, COSY spectrum, 379
Hyperbolic secant for pulse shaping, 93

I

INADEQUATE, *see also* Multiple-quantum coherence
  carbon–carbon connectivities by, 422–424
  composite pulses in, 89
  DEPT for sensitivity enhancement with, 71
  editing of, by SEMUT–SEMINA sequences, 229–242
  INEPT for sensitivity enhancement in, 64
  operator analysis of, 140–145
  phase cycle for, 98
Indirect observation
  of coupling constants through heteronucleus, 360
  of nitrogen-15 by multiple-quantum 2D, 426–428
INEPT
  comparison with DEPT, 184
  composite pulses in, 89
  editing of carbon-13 spectrum by, 207–209
  extension to correlation experiment, 64–65

  inverse, in RELAY from a heteronucleus, 514
  inverse version of, 73
  nature of polarization transfer in, 37
  polarization transfer in RELAY, 513
  polarization transfer to higher-spin nuclei by, 77–78
  product-operator analysis of, 175–183
  pulse sequence for, 54–56
  refocused, 59–61, 70, 181
  selection of experimental parameters for, 82–83
  selective versions of, 62–63
  in 2D exchange spectra, 419
  use with NOE sequence, 541
  vector model of, 57
INEPT$^+$
  density operator for, 181
  proton-coupled carbon-13 spectrum of 1,3-dibromobutane, 182
  proton-coupled nitrogen-15 spectrum of cyclic hydrazide, 185
  pulse sequence for, 179
Inverse polarization transfer, 73–75
Isochromat, spin, 6
Isotope effects, of carbon-13 on fluorine-19, 44–45
Isotropic mixing, 528–531

J

J cross polarization, 151
J filtering, 407, 411
J spectroscopy
  application to biological systems, 438–441
  composite pulses in, 90
  heteronuclear, 14–17
  homonuclear 1D, 12–14
  2D, based on INEPT, 64
  two-dimensional versions, 351–364
Jeener–Broekaert experiment, 122
Jeener experiment, *see* COSY

L

Lexicon, multiple-quantum spectroscopy, 265–266
Limited data sets, for INEPT carbon-proton correlation, 65
Lineshape, in 2D spectra, 457

Linewidth, narrowing of, in J spectra, 13
Liquid crystals
  determination of coupling constants in, by SPT methods, 48-49
  2D J spectra of AB and AB_2 systems, 358
Long-range coupling
  assignment by Bauer, Freeman, Wimperis sequence, 390, 395
  carbon-13–carbon-13, 290
  carbon–hydrogen shift correlation by, 387-391
  from COSY experiment, 368-371
  determination by multiple-quantum coherence and relay, 534
  from extended DEPT GL editing, 242-245
  from heteronuclear 2D J spectra, 359, 360
  selection in hydrogen-detected multiple-quantum experiment, 517-518
Longitudinal spin order, 155-156
Lysozyme, coupling constants, by J spectroscopy, 14

M

Macromolecules
  homonuclear correlation spectra, 377-379
  J spectra, 358, 362
Magnitude spectrum, in 2D plotting, 459
Menthol
  edited phase-sensitive 2D carbon–hydrogen correlation, 247
  edited proton-coupled carbon-13 spectra, 231-232
  INEPT carbon-13 spectrum of, 58
  multiplet-separated heteronuclear correlation, 401
  SEMINA carbon-13 spectrum, 237, 241
Mercury-199, correlation experiment for indirect shift determination, 402
Metal ions, polarization transfer from phosphorus, 62
Metallothionein, heteronuclear proton–cadmium zero-quantum coherence, 535-536
Methyl group, proton–carbon-13 polarization transfer in, 37-39
Methyl isocyanide, polarization of nitrogen-14 in, 42
3-Methylpyridine, edited proton-coupled carbon-13 spectra, 231-232

Mixing time, in 2D NOESY, 448-449
MLEV, use for isotropic mixing, 529
Modulation
  amplitude and phase, of spin echoes, 15-16
  of echoes by spin-spin coupling, 262
  of pulses, for selective excitation, 91-94
  of spin echoes, 351-364
  by spin–spin coupling, in heteronuclear correlation, 383
Multiple-quantum coherence
  for biological systems, 449-455
  density matrix interpretation, 450
  editing of INADEQUATE spectra, 229-242
  error compensation in INADEQUATE, 89
  evolution, 276-280, 324-326
  hydrogen-detected INADEQUATE, 519-520
  nature of, 66-68
  operator representation, 119-124, 132, 156, 164-166
  in SEMUT GL and DEPT GL, 227
  techniques applied to relay spectroscopy, 407
Multiplets
  intensities from polarization transfer between spin-$\frac{1}{2}$ nuclei, 40-41
  intensities from polarization transfer from spin-1 to spin-$\frac{1}{2}$, 79
  spin-spin, rate of dephasing, 20-22
Multiplicity determination, *see also* Subspectra, generation of
  in decoupled carbon-13 spectra, 151
  by INEPT, 59
  by modulation of spin echoes, 17-25
  by 2D J spectroscopy, 361
Muons, spin-polarized, 114

N

Nitrogen-15
  bilirubin IX-α spectrum, by INEPT and SINEPT, 191
  coupling constants from modified INEPT, 64
  DEPT GL spectra of dipeptide esters, 245
  indirect detection by proton magnetization, 462-464
  indirect observation by multiple-quantum 2D, 426-428

Nitrogen-15 (*continued*)
  INEPT in measurements of relaxation of, in peptides, 63
  intensities of multiplets resulting from polarization transfer, 40
  proton-coupled INEPT spectrum of cyclic hydrazide, 185
  proton-decoupled multiple-quantum spectra, 518
  pulse sequence for extended DEPT GL editing of decoupled spectra, 243
  SPT spectrum in 1-phenylpyrazole, 50
  transfer of polarization to, 31
  2D shift correlation with protons, 384–387
NOESY, *see* Nuclear Overhauser effect
Noise
  in NOESY, suppression of, 538–539
  t_1, reduction in COSY, 373
  t_1, reduction in 2D spectra, 460–461
  t_1 and t_2, reduction, 526
Nonselective polarization transfer, coupling constant signs from, 195–207
Nuclear Overhauser effect
  data processing in, 539
  design of experiments, 537–538
  in polynucleotide structure determinations, 479–482
  relation to protein structure, 470–475
  in rotating frame, 539–540
  for silicon-29, 41–42
  study by 2D magnetization transfer, 413–419
  2D spectra in determination of oligosaccharide structure, 482–485
  2D spectroscopy for biological systems, 446–449
  z filter in, 174
Nucleic acid
  nitrogen-15–proton multiple-quantum shift correlation, 428
  RELAY experiment on DNA, 510
Nucleotide
  carbon–hydrogen correlation, 384
  2D J spectrum, 363
  2D phosphorus–proton correlation in, 396
  poly-, structure determination of, 479–482

O

Off-resonance decoupling, selection of nonprotonated carbons by, 11
Operator
  analysis of DEPT for I_2S system, 252–253
  analysis of INEPT and DEPT, 175–180
  analysis of SEMUT–DEPT GL⁺ editing, 253–256
  base, 127, 129–130, 153–154
  Cartesian, 113, 116–135, 272, 276–280
  Cartesian, in INADEQUATE sequence, 141–142
  density, 115, 126, 129, 152–169, 270–280
  density, shortcuts in calculation of, 166–169
  evolution under chemical shift, 130
  evolution under free precession, 130
  evolution under selective pulse, 134–135
  evolution under spin-spin coupling, 131
  evolution under strong pulse, 133–134
  product, 153–169
  product, diagrams for, 155–157
  product, formulation of density operator, 266
  single-element, 113, 118–120, 127, 130, 136, 277–278
  single-element, description of phase shifts, 295
  single-transition, 112, 327–329
  spherical tensor, 112
Order, transition, *see* Coherence, order
Oxytocin
  nitrogen-15–proton double-quantum-correlated spectrum, 464–465
  structure of, 466

P

Panamine, carbon-13–carbon-13 connectivity by double-quantum spectroscopy, 290–291
Parameter selection
  for J spectroscopy, 439–441
  for 2D NOESY, 447–449
Partial-spin coherences, evolution of, 325–326
Passive spin, 266
P.E. COSY, 525
Peptides
  aromatic residues in, by COSY, 369, 373
  carbon–hydrogen correlation, 384, 409
  COSY spectrum, 380
  cross-relaxation interactions, 476
  extended DEPT editing of carbon-13 and nitrogen-15 spectra, 243–245
  heteronuclear 2D shift correlation, 384–387, 397–399

INDEX

INEPT in measurements of nitrogen-15
    relaxation in, 63
INEPT spectra, 61
    long-range correlation in, 389–391,
        393–394
    NOE spectra, 418–419
Perfluoromethylcyclohexane, COSY
    spectrum, 500
Phase cycling
    in APT experiment, 25
    coherence-order selection by, 136–140
    in COSY, 369, 445, 527
    in INADEQUATE, 98
    methods of, 96–99
    for multiple-quantum filtered COSY, 459
    for multiple-quantum filtering, 303, 519
    in multiple-quantum solvent suppression,
        471
    in multiple-quantum spectrum, 453
    of proton purging pulses in editing
        sequences, 226
    in quadrature detection, 97
    in relayed COSY, 456
    in selective homonuclear RELAY, 510
    in SLAP, 203
    in 2D NOE experiments, 413
    in z-filtered COSY, 521
    in zero-quantum coherence, 531–534
Phase modulation, 15
Phase-sensitive 2D spectra
    in COSY, 374–375
    States, Haberkorn, Ruben method, 374,
        460
    time-proportional phase incrementation
        for, 375
Phase shift, 4
    in Carr–Purcell–Meiboom–Gill sequence,
        9
    hardware for, 94–95
    in selection of coherence pathway, 294–295
    by z pulse, 95–96
Phase-twist lineshape
    in COSY, 366
    in homonuclear 2D J spectrum, 355
    in quadrature 2D experiments, 145
Phenylethanol, deuterium–carbon
    polarization transfer in deuterated, 78
1-Phenyl[^{15}N$_2$]pyrazole, flip-angle effect in
    nitrogen-15 spectrum of, 50
Phosphorus-31
    elimination of distortions in phospholipid
        bilayer spectra, 11

heteronuclear proton coherence transfer
    through, in phosphite and and
    phosphate, 407
homonuclear 2D J spectrum, 363
homonuclear 2D spectra of phosphates,
    381
polarization transfer to metals from, 62
2D exchange spectra, 419
2D shift correlation with protons, 396
zero-quantum heteronuclear correlation,
    420–421
Pinene, spin-echo J-modulated carbon-13
    spectra, 23
Piperidine, multiple-quantum spectrum and
    carbon-13–carbon-13 coupling, 289
Plastocyanin, copper, homonuclear relayed
    spectrum, 404
Polarization transfer
    benefits of experiments utilizing, 31–32
    in degenerate systems, 37–43
Polymers, containing fluorine-19, spectra of,
    489–505
Poly(vinyl fluoride), proton-decoupled
    fluorine-19 COSY spectrum, 501–504
Poly (vinylidene fluoride)
    proton-decoupled fluorine-19 COSY
        spectrum, 499–501
    proton-decoupled fluorine-19 spectra,
        491–497
Population transfer, see Polarization
    transfer
Product operator formulation, see
    Operator
Propagator, see also Operator, evolution
    definition of, 512
    description of coherence-transfer steps,
        297
Prostaglandin, 2D J spectra, 362
Proteins
    COSY spectrum, 380
    heteronuclear 2D shift correlation,
        384–387, 461–465
    homonuclear double-quantum spectra,
        424
    homonuclear relayed correlation, 403–405
    multiple-quantum filtered spectra, 526
    multiplet decoupling in COSY, 377
    multiplicity selection in 2D multiple-
        quantum spectrum, 454–455
    relayed coherence transfer in, 509
    structural analysis by COSY and NOE
        experiments, 475–479

Proteins (*continued*)
  2D exchange spectra, 417–419
  2D *J* spectra, 362, 441
  2D NOE measurement in, 413–419
Proton detection, *see also* Indirect observation
  in heteronuclear multiple-quantum coherence, 428
Proton-flip method
  for heteronuclear modulation of spin echoes, 14, 352
  for heteronuclear 2D *J* spectra, 359
Pseudo-echo, *see also* Sine bell processing of FID
  weighting of 2D *J* spectra, 357
Pulse
  with arbitrary phase, effect of, 250–251
  cascade, 273
  decoupler, calibration of, 102–104
  evolution of density operator under, 162–164
  hard, 3
  nature and properties of rf, 83
  nonselective rf, evolution under, 162
  radio frequency, 3
  selective, 3
  selective, evolution under, 134–135, 163
  semiselective, 3
  with tilted axis, effect of, 251–252
  vector diagram of effect on nuclei, 5
Purging
  in DEPT, 182–183
  in heteronuclear experiments, 169–171
  in homonuclear experiments, 172–175
  in multiple-quantum filter, 337
  z filter for, 173–175, 241, 337
Pyridines, INEPT spectra, 61
Pyrimidines, INEPT spectra, 61

Q

Quadrature detection
  in COSY experiment, 365, 444
  in INADEQUATE, 423
  in multiple-quantum coherence experiment, 451
  phase cycling for, 97
Quadrupolar nuclei, polarization transfer involving, 76–82

R

Radio frequency, nonuniformity of field, 3
RECAPT, sequence for multiplicity determination, 23
Refocusing
  delay, values to give maximum signal, 51
  in INEPT, 59
  periods, significance of types, 30
  pulse, in multiple-quantum excitation, 271, 334
  to produce echoes, 6
Relaxation
  cross-, matrix for, 472
  elimination of effect of, in modulated spin-echo experiments, 28
  INEPT in measurement of, 63
  measurement for protons by 2D heteronuclear correlation, 397
RELAY
  correlation by, heteronuclear, 405–411
  correlation by, homonuclear, 402–405
  correlation spectrum of cytochrome *c*, 467–468
  F_1 decoupling in, 379
  heteronuclear, H—C—C, 513
  heteronuclear, H—H—X, 511
  homonuclear coherence transfer by, 508–511
  homonuclear 2D experiments, 455–456
  multiple-step, 509
Rephasing, of spins in INEPT, 58
Retinal, heteronuclear 2D *J* spectrum, 363
RNA, nitrogen-15 multiple-quantum correlation, 428
ROESY, *see* Nuclear Overhauser effect, in rotating frame
Rotary echo, order-selective detection by, 292–293
Rotating frame, 2, 4
  nuclear Overhauser effect in, 539–540

S

Saccharides, *see also* Carbohydrates
  oligo-, small couplings in, 369
  oligo-, structure determination of, 482–485
  proton-carbon-proton coherence transfer in triglucoses, 407

2D J spectra, 362
Second-order effects, in 2D J spectra, 361, 439
Secondary structure of proteins, 478
SECSY
 advantages, 364
 F_1 decoupling in, 379
 flip-angle effect in, 364
 spectra of macromolecules, 444–445
Selection rules, for coherence transfer, 329–330
Selective excitation
 DANTE sequence for, 91–92
 tailored pulses for, 92–94
Selective polarization transfer
 applications of, 43–44
 in AX spin system, 32–37
 combination with INEPT, 63
 by DEPT sequence, 73
 relative signs of coupling constants from, 193–195
 sensitivity enhancement by, 150–151
Selective population transfer, see Selective polarization transfer
Selective pulse
 in heteronuclear 2D J spectra, 359
 operator evolution under, 134–135, 163–164
SEMINA sequences, for editing of INADEQUATE spectra, 229–242
SEMUT
 for decoupler calibration, 102–103
 editing of carbon-13 spectrum by, 209–214
 editing of INADEQUATE spectra by, 229–242
SEMUT GL⁺, 227
 operator analysis of editing by, 253–256
Sensitive-point method, use of spin-echo signal in, 11
Sensitivity
 advantages of polarization transfer and NOE, 82
 in carbon–carbon connectivity experiments, 424
 of 2D spectra, 348–349
Sign-labeled polarization transfer, coupling constant signs from, 195–207
Signal-to-noise, in edited carbon-13 spectra, 222–225
Silicon-29

COSY in germanosilicate, 527
 homonuclear 2D J spectrum, 363
 intensities of multiplets resulting from polarization transfer, 40
 relaxation measurements by INEPT, 63
 signal-to-noise in spectra with SPT or with NOE, 42
 transfer of polarization to, 31, 61
Silver-109, transfer of polarization to, 31, 61
Sine-bell processing of FID, 355, 357, 369, 457
SINEPT
 for decoupler calibration, 102–103
 pulse sequences for 1D heteronuclear chemical shift correlation, 186–192
SKEWSY, sequence for rate of magnetization transfer, 417
SLAP, see Sign-labeled polarization transfer
Solvent suppression
 in COSY and NOESY, 466–470
 by homonuclear refocused INEPT, 62
 multiple-quantum, phase cycle for, 471
 in proton–proton NOE Experiments, 542
 in 2D exchange spectra, 417
 in 2D NOE experiments, 414
 in 2D spectra, 380
Spectrometer requirements for 2D spectroscopy, 345–350
Spin diffusion, effects in NOE spectra, 541
Spin echo
 effect of T_2 on, 8–10
 Hahn experiment, vector representation, 6–8
 in J spectroscopy, 355
 modulation of, by spin–spin coupling, 12–30, 262
 sequence, operator transformation in, 166–167
Spin-echo methods, uses of, 10–11
Spin-flip method, heteronuclear modulation of spin echoes by, 15
Spin lock, for isotropic mixing, 528–530
Spin–spin coupling
 carbon-13–carbon-13, measurement by double-quantum coherence, 287–292
 effect of homonuclear on spin echoes, 12
 evolution of density operator under, 161–162
 SLAP for determination, 195–207
Spin–spin relaxation, effect on spin echoes, 8

Spin system
  AA'XX', 364
  AB, 358
  ABC, 358
  AMQX, 509
  AMX, 48, 266, 302–305, 310–319, 331, 335–337, 358, 364, 508, 542
  AMX_2, 509
  AX, 19–20, 32, 51, 60, 67, 267–280, 287–291, 351–352, 358, 364
  A_2X, 19–20, 60, 331, 358, 364
  A_2X_2, 364
  A_2X_3, 365
  A_3X, 19–20, 60
  types in proteins, 475
  zero-quantum frequencies in various, 532
Spinor behavior, 135
SPT, see Selective polarization transfer
States–Haberkorn–Ruben method, 374, 460
Stern–Gerlach experiment, 114
Steroid, 379
  homonuclear 2D J spectrum of, 356
Strong coupling
  effect in NOE spectrum, 474
  effects on NOE intensities, 541
  in homonuclear 2D J spectra, 361
Subspectra, generation of
  by APT, INEPT, DEPT, or SEMUT, 207–214
  by DEPT, 70–71
  by DEPT with 2D J spectrum, 71, 72
  for large molecules, 438–439
  for long-range coupling, 242–245
  principles of 1D, 151–152
  from refocused INEPT, 59–61
  by SEMUT, DEPT, and GL versions, aspects of, 220–226
  by SEMUT GL and DEPT GL, 214–220
  by spin echo modulation, 26–28
  in 2D spectra, 245–247
SUCZESS sequence, 523
Sugars, editing of 2D J-resolved carbon-13 spectra of, 71–72
SUPERCOSY, 521
  pulse sequence, 376
Symmetrization, of homonuclear 2D correlation spectra, 367, 461

T

TANGO, bilinear pulse sequence for selection by J value, 101

Terpene, heteronuclear 2D shift correlation, 383
Tertiary structure of proteins, 478
Tetramethyldisilazane, INEPT in measurements of relaxation of nuclei in, 63
Tilting, in homonuclear 2D J spectrum, 354–355
Time-proportional phase incrementation
  in COSY experiment, 444
  in INADEQUATE, 143
  for separation of double-quantum coherence orders, 138–139
Tin, polarization by INEPT, 61
Tip-angle dependence
  of coherence transfer with three-spin operator, 316–319
  in excitation of multiple-quantum coherence, 273, 275–276, 314
  of transfer between orders of coherence, 284–286
TOCSY
  extended versions, 528–530
  pulse sequence for, 375
TPPI, see Time-proportional phase incrementation
Transfer coefficients, for mixing pulse in multiple-quantum coherence, 284
Transfer function, in relayed coherence transfer, 509
Transitory selective irradiation, 150
Tree procedure
  for evolution of Cartesian-product operators, 132
  for spectral representation of product operators, 157–159
Trichloroethane, modulated spin-echo spectra of protons, 18
Trichlorotrifluoroethane, 2D fluorine J spectrum of, 19
1,1,2-Trichloro-3,3,3-trifluoropropene, SPT carbon-13 spectrum of, 44–45
Trimethylsilyl groups
  polarization of silicon-29 nuclei by INEPT, 61–62
  silicon-29 shifts in metal complexes, 44
Tungsten-183, homonuclear 2D spectrum of heteropolytungstates, 381
Two-dimensional spectroscopy
  general scheme for, 262–264, 344–345
  instrumental requirements, 345–350
  scheme for J spectra, 17
Two-spin order, 279

U

Ulithiacyclamide, carbon–hydrogen correlation, 393
UPT
   deuterium–carbon polarization transfer by, 80
   pulse sequence for multiplet selection involving spins greater than $\frac{1}{2}$, 68, 78

V

Vinyltrimethylsilane, silcon-29 SPT spectrum, 39, 53

W

Water, signal elimination, *see* Solvent suppression

Z

Z-COSY, *see* z-Filtered COSY
z Filter
   purging by, 241
   for purging homonuclear experiments, 173–176
   for purging multiple-quantum-filtered spectra, 337
z-Filtered COSY, 519–520
z Rotation, by composite pulse, 140, 95–96
Zero-quantum coherence
   in 2,3-dibromothiophene, 288
   detection, 280–288
   evolution, 276–280
   excitation of, 268–276
   operator representation, 119, 123
   in phosphorus–hydrogen correlation, 420–421
   recent applications of, 531–535
zz Signals, 414

Lanchester Library